Risk Analysis for Prevention of Hazardous Situations in Petroleum and Natural Gas Engineering

Davorin Matanovic
University of Zagreb, Croatia

Nediljka Gaurina-Medjimurec
University of Zagreb, Croatia

Katarina Simon
University of Zagreb, Croatia

A volume in the Advances in Environmental Engineering and Green Technologies (AEEGT) Book Series

ENGINEERING SCIENCE REFERENCE
An Imprint of IGI Global

Managing Director: Lindsay Johnston
Production Manager: Jennifer Yoder
Development Editor: Allyson Gard
Acquisitions Editor: Kayla Wolfe
Typesetter: Christina Henning
Cover Design: Jason Mull

Published in the United States of America by
Engineering Science Reference (an imprint of IGI Global)
701 E. Chocolate Avenue
Hershey PA 17033
Tel: 717-533-8845
Fax: 717-533-8661
E-mail: cust@igi-global.com
Web site: http://www.igi-global.com

Library of Congress Cataloging-in-Publication Data

Risk analysis for prevention of hazardous situations in petroleum and natural gas engineering / Davorin Matanovic, Nediljka Gaurina-Medjimurec, and Katarina Simon, editors.
pages cm
Includes bibliographical references and index.
Summary: "This book explains the potential hazards of petroleum engineering activities, emphasizing risk assessments in drilling, completion, and production, and the gathering, transportation, and storage of hydrocarbons"-- Provided by publisher.
ISBN 978-1-4666-4777-0 (hardcover) -- ISBN 978-1-4666-4778-7 (ebook) -- ISBN 978-1-4666-4779-4 (print & perpetual access) 1. Petroleum industry and trade--Safety measures. 2. Petroleum industry and trade--Accidents--Risk assessment. 3. Gas industry--Accidents--Risk assessment. 4. Oil well drilling--Safety measures. 5. Gas well drilling--Safety measures. I. Matanovic, Davorin, 1949- II. Gaurina-Medjimurec, Nediljka, 1957- III. Simon, Katarina, 1962-
TN871.R55 2014
363.11'96655--dc23
2013027527

This book is published in the IGI Global book series Advances in Environmental Engineering and Green Technologies (AEEGT) (ISSN: 2326-9162; eISSN: 2326-9170)

British Cataloguing in Publication Data
A Cataloguing in Publication record for this book is available from the British Library.

Advances in Environmental Engineering and Green Technologies (AEEGT) Book Series

ISSN: 2326-9162
EISSN: 2326-9170

Mission

Growing awareness and focus on environmental issues such as climate change, energy use, and loss of non-renewable resources have brought about a greater need for research that provides potential solutions to these problems. The field of environmental engineering has been brought increasingly to the forefront of scholarly research and, alongside it, environmentally-friendly, or "green," technologies as well.

Advances in Environmental Engineering & Green Technologies (AEEGT) Book Series is a mouthpiece for this research, publishing books that discuss topics within environmental engineering or that deal with the interdisciplinary field of green technologies.

Coverage

- Alternative Power Sources
- Biofilters & Biofiltration
- Contaminated Site Remediation
- Green Transportation
- Industrial Waste Management & Minimization
- Policies Involving Green Technologies & Environmental Engineering
- Pollution Management
- Renewable Energy
- Sustainable Communities
- Waste Management

IGI Global is currently accepting manuscripts for publication within this series. To submit a proposal for a volume in this series, please contact our Acquisition Editors at Acquisitions@igi-global.com or visit: http://www.igi-global.com/publish/.

The Advances in Environmental Engineering and Green Technologies (AEEGT) Book Series (ISSN 2326-9162) is published by IGI Global, 701 E. Chocolate Avenue, Hershey, PA 17033-1240, USA, www.igi-global.com. This series is composed of titles available for purchase individually; each title is edited to be contextually exclusive from any other title within the series. For pricing and ordering information please visit http://www.igi-global.com/book-series/advances-environmental-engineering-green-technologies/73679. Postmaster: Send all address changes to above address.

Titles in this Series

For a list of additional titles in this series, please visit: www.igi-global.com

Risk Analysis for Prevention of Hazardous Situations in Petroleum and Natural Gas Engineering
Davorin Matanovic (University of Zagreb, Croatia) Nediljka Gaurina- Medjimurec (University of Zagreb, Croatia) and Katarina Simon (University of Zagreb, Croatia)
Engineering Science Reference • copyright 2014 • 415pp • H/C (ISBN: 9781466647770) • US $185.00 (our price)

Marine Technology and Sustainable Development Green Innovations
Oladokun Sulaiman Olanrewaju (University Malaysia Terengganu, Malaysia) Abdul Hamid Saharuddin (University Malaysia Terengganu, Malaysia) Ab Saman Ab Kader (Universiti Teknologi Malaysia, Malaysia) and Wan Mohd Norsani Wan Nik (University Malaysia Terengganu, Malaysia)
Information Science Reference • copyright 2014 • 338pp • H/C (ISBN: 9781466643178) • US $195.00 (our price)

Sustainable Technologies, Policies, and Constraints in the Green Economy
Andrei Jean-Vasile (Petroleum and Gas University of Ploiesti, Romania) Turek Rahoveanu Adrian (Institute of Research for Agricultural Economics and Rural Development, Romania) Jonel Subic (Institute of Agricultural Economics, Belgrade, Serbia) and Dorel Dusmanescu (Petroleum and Gas University of Ploiesti, Romania)
Information Science Reference • copyright 2013 • 390pp • H/C (ISBN: 9781466640986) • US $180.00 (our price)

Energy-Aware Systems and Networking for Sustainable Initiatives
Naima Kaabouch (University of North Dakota, USA) and Wen-Chen Hu (University of North Dakota, USA)
Information Science Reference • copyright 2012 • 469pp • H/C (ISBN: 9781466618428) • US $180.00 (our price)

Green and Ecological Technologies for Urban Planning Creating Smart Cities
Ozge Yalciner Ercoskun (Gazi University, Turkey)
Information Science Reference • copyright 2012 • 404pp • H/C (ISBN: 9781613504536) • US $180.00 (our price)

Environmental Modeling for Sustainable Regional Development System Approaches and Advanced Methods
Vladimír Olej (University of Pardubice, Czech Republic) Ilona Obršálová (University of Pardubice, Czech Republic) and Jirí Krupka (University of Pardubice, Czech Republic)
Information Science Reference • copyright 2011 • 492pp • H/C (ISBN: 9781609601560) • US $180.00 (our price)

Rethinking Sustainable Development Urban Management, Engineering, and Design
Tan Yigitcanlar (Queensland University of Technology, Australia)
Information Science Reference • copyright 2010 • 416pp • H/C (ISBN: 9781616920227) • US $180.00 (our price)

IGI GLOBAL
DISSEMINATOR OF KNOWLEDGE
www.igi-global.com

701 E. Chocolate Ave., Hershey, PA 17033
Order online at www.igi-global.com or call 717-533-8845 x100
To place a standing order for titles released in this series, contact: cust@igi-global.com
Mon-Fri 8:00 am - 5:00 pm (est) or fax 24 hours a day 717-533-8661

Table of Contents

Preface xv

Chapter 1
General Approach to Risk Analysis 1
Davorin Matanovic, University of Zagreb, Croatia

Chapter 2
Risk Due to Wellbore Instability 23
Nediljka Gaurina-Medjimurec, University of Zagreb, Croatia
Borivoje Pasic, University of Zagreb, Croatia

Chapter 3
Risk Due to Pipe Sticking 47
Nediljka Gaurina-Medjimurec, University of Zagreb, Croatia
Borivoje Pasic, University of Zagreb, Croatia

Chapter 4
Lost Circulation 73
Nediljka Gaurina-Medjimurec, University of Zagreb, Croatia
Borivoje Pasic, University of Zagreb, Croatia

Chapter 5
Simultaneous Operations 96
Zdenko Kristafor, University of Zagreb, Croatia

Chapter 6
The Macondo 252 Disaster: Causes and Consequences..115
Davorin Matanovic, University of Zagreb, Croatia

Chapter 7
Risk Analysis of Completion and Production Systems 132
Davorin Matanovic, University of Zagreb, Croatia

Chapter 8
Risk and Remediation of Irreducible Casing Pressure at Petroleum Wells 155
Andrew K. Wojtanowicz, Louisiana State University, USA

Chapter 9
Risk Analysis in the Process of Hydraulic Fracturing 181
Sonja Koščak Kolin, University of Zagreb, Croatia
Marin Čikeš, University of Zagreb, Croatia

Chapter 10
Workover Impact on Accidental Risk 199
Bojan Moslavac, University of Zagreb, Croatia

Chapter 11
Gathering Systems and Processing Facilities Risk Analysis 218
Svijetlana Dubovski, INA Plc, Croatia

Chapter 12
Activities in Oil and Gas Processing for Avoiding or Minimizing Environmental Impacts 247
Svijetlana Dubovski, INA Plc, Croatia

Chapter 13
Transportation Risk Analysis 264
Dragan Crnčević, INA Plc, Croatia

Chapter 14
Oil and Gas Storage Tank Risk Analysis 303
Katarina Simon, University of Zagreb, Croatia

Chapter 15
CO_2 Underground Storage and Wellbore Integrity 322
Nediljka Gaurina-Medjimurec, University of Zagreb, Croatia
Borivoje Pasic, University of Zagreb, Croatia

Chapter 16
Petroleum Industry Environmental Performance and Risk 358
Lidia Hrncevic, University of Zagreb, Croatia

Compilation of References 388

About the Contributors 410

Index..413

Detailed Table of Contents

Preface xv

Chapter 1
General Approach to Risk Analysis 1
Davorin Matanovic, University of Zagreb, Croatia

Broadly accepted methodology that is implemented in the oil industry when dealing with risks includes as the first step the identification of possible hazards. That is done by gathering information about degree of risk according to working procedures, processes, and individuals involved in the operation of the process. That is the first step in risk management, an iterative process that must lead to the use of proper measurements in the way of protecting people, facilities and environment. The analysis is done based on the combination of probability and severity of undesirable events, and the final consequences. Explanation of basic terms, their interdependence, dilemmas, and methods of risk analysis are introduced. Each method is shortly described with main anteriority and shortcomings. Differences between quantitative methods, qualitative methods, and hybrid methods (the combination of qualitative-quantitative or semi-quantitative methods) are elaborated. The impact, occurrence, and the consequences are at the end compared to risk acceptance criteria concept. The ALARP (As Low as Reasonably Practicable) framework is explained with some observation on the quality and acceptance in petroleum industry. Finally, the human impact on the risk and consequences is analyzed.

Chapter 2
Risk Due to Wellbore Instability 23
Nediljka Gaurina-Medjimurec, University of Zagreb, Croatia
Borivoje Pasic, University of Zagreb, Croatia

Exploration and production as one of the most important parts of the petroleum industry encounters different problems, usually resulting in nonproductive time and additional expenses. The most common and most expensive of them are related to wellbore instability and associated problems. Wellbore instability problems are usually related to drilling operation, but they can also appear during completion, workover, or the production stage of a certain well. The traditional solution for wellbore instability problems is composed from the early recognition of specific wellbore instability problems, the main cause identification and swift response. For more effective solution it is necessary to incorporate wellbore stability and risk assessment in the early phase of well design. This chapter gives one general overview of wellbore instability problems and their causes as well as an overview of actual approaches and methods in wellbore stability and risk assessment.

Chapter 3
Risk Due to Pipe Sticking 47
Nediljka Gaurina-Medjimurec, University of Zagreb, Croatia
Borivoje Pasic, University of Zagreb, Croatia

A stuck pipe is a common worldwide drilling problem in terms of time and financial cost. It causes significant increases in non-productive time and losses of millions of dollars each year in the petroleum industry. There are many factors affecting stuck pipe occurrence such as improper mud design, poor hole cleaning, differential pressure, key seating, balling up of bit, accumulation of cuttings, poor bottom hole assembly configuration, etc. The causes of a stuck pipe can be divided into two categories: (a) differential sticking and (b) mechanical sticking. Differential-pressure pipe sticking occurs when a portion of the drill string becomes embedded in a filter cake that forms on the wall of a permeable formation during drilling. Mechanical sticking is connected with key seating, formation-related wellbore instability, wellbore geometry (deviation and ledges), inadequate hole cleaning, junk in hole, collapsed casing, and cement related problems. Stuck pipe risk could be minimized by using available methodologies for stuck pipe prediction and avoiding based on available drilling parameters.

Chapter 4
Lost Circulation 73
Nediljka Gaurina-Medjimurec, University of Zagreb, Croatia
Borivoje Pasic, University of Zagreb, Croatia

Lost circulation is defined as the uncontrolled flow of mud into a thief zone and presents one of the major risks associated with drilling. The complete prevention of lost circulation is impossible, but limiting circulation loss is possible if certain precautions are taken. Failure to minimize lost circulation can greatly increase the cost of drilling, as well as the risk of well loss. The risk of drilling a well in areas known to contain potential zones of lost circulation such as fractured, cavernous, or high permeability formations is a key factor in making a decision to approve or cancel a drilling project. The successful management of lost circulation should include identification of potential loss zones, optimization of drilling hydraulics, and remedial measures when lost circulation occures.

Chapter 5
Simultaneous Operations 96
Zdenko Kristafor, University of Zagreb, Croatia

Simultaneous operations can be defined as conducting independent operations in which the course of any operation may impact the safety of personnel, equipment, and environment of another operation. Operations are meant in terms of performing drilling, completion, workover, and construction or production activities simultaneously, mainly at offshore facilities. In general, they have an impact on the installation safety procedures and contingency planning program. Thus, for the purpose of performing simultaneous operations, they must comply with organizational safety and technological requirements. They are to be coordinated through joint planning efforts by supervisors who plan and direct activities. Typical chain-of-command as well as simultaneous operations decision making process flow diagrams are presented in this chapter. Basic steps to be regarded are performing risk assessment, assess and control risks, monitor the simultaneous tasks and communicate the control measures. Communication is the key to performing these tasks safely.

Chapter 6
The Macondo 252 Disaster: Causes and Consequences 115
Davorin Matanovic, University of Zagreb, Croatia

The disaster of the Deepwater Horizon platform, while drilling the Macondo 252 well in the Gulf of Mexico in 2010 is for now the last of numerous tragedies, blowouts, and oil spills resulting from petroleum engineering activities. After the accident, several commissions, investigation groups, advisory committees, and company reports were prepared. They investigate causes and consequences of the disaster from different standpoints, but mainly come to the same conclusions. The "nth" approach is presented in this chapter.

Chapter 7
Risk Analysis of Completion and Production Systems 132
Davorin Matanovic, University of Zagreb, Croatia

A variety of definitions are available through the literature, but the universal one is in defining the well completion as the optimal path for the reservoir fluids to be produced. That means to achieve a desired production with minimal costs. Wells represent the greatest part of expenditure when developing the reservoir. For a long period of time it was defined to be simple, reliable, and safe with enough flexibility to allow future operations. Nowadays, so called "intelligent completions" appear to give more financial benefits, flexibility, and control. The reliability of system components is essential for long-lasting production. In addition, the differences according to natural flowing well risk and artificial lift are given.

Chapter 8
Risk and Remediation of Irreducible Casing Pressure at Petroleum Wells 155
Andrew K. Wojtanowicz, Louisiana State University, USA

Oil well cement problems such as small cracks or channels may result in gas migration and lead to irreducible pressure at the casing head. Irreducible casing pressure also termed, Sustained Casing Pressure (SCP) is hazardous for a safe operation and the affected wells cannot be terminated without remedial operations. It is believed that even very small leaks might lead to continuous emissions of gas to the atmosphere. In the chapter, the author describes physical mechanisms of irreducible casing pressure and qualifies the associated risk by showing statistical data from the Gulf of Mexico and discussing the regulatory approach. This chapter also introduces a new approach to evaluate risk of casing pressure by computing a probable rate of atmospheric emissions from wells with failed casing heads resulting from excessive pressure. Also presented is a new method for assessing potential for self-plugging of such wells flowing wet gas as the gas migration channels could be plugged off by the condensate.

Chapter 9
Risk Analysis in the Process of Hydraulic Fracturing 181
Sonja Koščak Kolin, University of Zagreb, Croatia
Marin Čikeš, University of Zagreb, Croatia

This chapter focuses on risk to the environment from hydraulic fracturing operations, starting with transport of materials and ending when the well is routed to the production facilities. The initial assumption for the fracturing risk analysis is that the well is new and was constructed correctly so that all producible formations are securely isolated behind the barriers of casing and competent cement. The justification for this assumption is that the vast majority of fracturing is the first major stimulation in a well and occurs immediately after completing a new well. Although many well development problems are blamed on fracturing, there are only excluded problems that are real and worthy of the discussion to help define boundaries of the fracturing risk (King, 2012).

Chapter 10
Workover Impact on Accidental Risk 199
Bojan Moslavac, University of Zagreb, Croatia

Certain risks are very common to any application of a downhole control system during workover while others will be field-specific or at least increased or decreased by the given well conditions. Workover anomalies may be caused by erosion, corrosion, mechanical errors, temperature effects on electronics, wear and tear on the dynamic seals, or seizure of moving components. Obviously, the simpler the system and the fewer moving parts, the fewer components are available to fail. Procedures and supporting control software must be developed to ensure optimum system use. Blowouts are not very common today as most blowout preventing systems serve the purpose. However, occasionally, these systems may fail due to either technical problems on the wellbore equipment or human error. Both reasons can cause enormous damage on wellbore equipment, reservoir, surface facilities, or even human loss. The right accidental risk assessment approach can reduce the chances of repeating bad scenarios that happened in the past.

Chapter 11
Gathering Systems and Processing Facilities Risk Analysis 218
Svijetlana Dubovski, INA Plc, Croatia

Gathering system is defined as one or more segments of pipeline, usually interconnected to form a network that transports oil and natural gas from the production wells to one or more production facilities, gas processing plant, storage facility, or a shipping point. There are two types of pipeline networks: radial and trunk system. Produced well fluids are often complex mixtures of the liquid hydrocarbons, gas, and some impurities that can have detrimental effects on the integrity of the gathering pipelines. It is necessary to eliminate most of the impurities before oil and natural gas can be stored and sold. Complexity of the processing facility depends on the treated fluid composition. Environmental impacts during the oil and gas transportation and processing phase will cause long-term habitat changes. To minimize that, it is very important to implement appropriate activities across the designing, construction, operational, and decommissioning phases.

Chapter 12
Activities in Oil and Gas Processing for Avoiding or Minimizing Environmental Impacts 247
Svijetlana Dubovski, INA Plc, Croatia

Formal risk assessments are necessary at various phases of the asset life cycle as they help personnel identify, evaluate, and control hazards that could result in loss of life, injury, pollution, property damage, or business disruption. Hazard evaluations of production development concepts or facility design are well-defined processes, for which much literature is available as guidance. Such evaluations are mandated in some jurisdictions for project regulatory approval. This chapter provides guidance on activities implementation from designing phase, construction phase, operational phase, and decommissioning phase of gathering and processing systems. For each phase, activities that can be implemented to avoid or minimize environmental impact are described, emphasizing the implementation purpose.

Chapter 13
Transportation Risk Analysis 264
Dragan Crnčević, INA Plc, Croatia

Petroleum is transported across the water in barges and tankers, and on land, using pipelines, trucks, and trains. Natural gas is moved, mainly, by pipelines. The most common causes of tanker accidents are: fire/explosions, loading/offloading, structural damage, collision, and grounding. Pipeline accidents are due to: corrosion, third parties activities, mechanical damage, natural events, and operational error. Some of the most commonly applied preventive activities that reduce spills in waterborne transportation are: double-hulled tanker, navigation safety and radio communications equipment, tanker exclusion zone, etc. The pipeline condition can be recorded by using various nondestructive measurement techniques or by chemical analysis of fluid flows. Different types of sensors can be used to locate and determine the size of an anomaly in the pipeline geometry. Mayor methods for detecting leaks are measuring the hydrodynamic parameters or registering abnormal conditions in the fluid flow and detecting phenomena in the immediate vicinity of the pipeline.

Chapter 14
Oil and Gas Storage Tank Risk Analysis 303
Katarina Simon, University of Zagreb, Croatia

Storage tanks are widely used in the oil refinery and petrochemical industry in storing a multitude of different products ranging from gases, liquids, solids, and mixtures. Design and safety concerns have become a priority due to tank failures causing environment pollution as well as fires and explosions, which can result in injuries and fatalities. The chapter illustrates different types of crude oil and oil product storage tanks as well as the risks regarding the storage itself. Considering that the natural gas, in its gaseous state, is stored in underground storages like oil and gas depleted reservoirs, aquifers or salt caverns, and there are numerous publications and books covering the subject in detail, this chapter only illustrates the storage of liquefied natural gas and the risks posed by its storage.

Chapter 15
CO_2 Underground Storage and Wellbore Integrity 322
Nediljka Gaurina-Medjimurec, University of Zagreb, Croatia
Borivoje Pasic, University of Zagreb, Croatia

Geologic storage is the component of Carbon Capture and Storage (CCS) in which the carbon dioxide (CO2) is disposed in the appropriate underground formation. To successfully inject CO2 into the subsurface to mitigate greenhouse gases in the atmosphere, the CO2 must to be trapped in the subsurface and must not be allowed to leak to the surface or to potable water sources above the injection zone. For the purposes of risk assessment, a priority is to evaluate what would happen if CO2 migrated unexpectedly through the confining unit(s), potentially resulting in undesirable impacts on a variety of potential receptors. One of the main risks identified in geological CO2 storage is the potential for CO2 leakage through or along wells. To avoid leakage from the injection wells, the integrity of the wells must be maintained during the injection period and for as long as free CO2 exists in the injection zone.

Chapter 16
Petroleum Industry Environmental Performance and Risk 358
Lidia Hrncevic, University of Zagreb, Croatia

The petroleum industry holds long- and short-term environmental risks. Besides production fluids, all petroleum industry activities involve either use of fluids, which contain abundant substances, or waste generation, both associated with potential risk to the environment. The principal environmental risk associated with the petroleum industry is the risk of fluid spill/emission to the environment. Although in recent decades the risk analysis methodologies have matured, to date there is still no universally accepted methodology for environmental risk assessment in petroleum industry. In this chapter, the petroleum industry's environmental incident history and statistics are presented. The environmental impact of the petroleum industry's activities, its extent, and trends are analyzed. The overview of pollution sources with associated environmental risk is given along with the analysis of the causes and consequences of incidents in the petroleum industry.

Compilation of References 388

About the Contributors 410

Index...413

Preface

The accelerated growth of the world population and their living standard also includes the growth of energy needs. New developments in knowledge and technology change the prediction of Hubbert's oil peak curve philosophy in the direction of much higher hydrocarbon reserves that can be exploited. These reserves are spread all over the world, but the largest reserves are concentrated in several areas that correspond with the known spacing of the source rocks (Middle East, Northern and Central Africa, Siberia-Russia, North Sea, Alaska, Central USA, Gulf of Mexico, South America, Indonesia, China, etc.). The estimated quantity of oil which engineering and economic data demonstrate that is recoverable with reasonable certainty, under existing economic and operating conditions, has been estimated about $1.27 \cdot 10^{12}$ barrels ($202 \cdot 10^9$ m^3) in 2003, and $1.39 \cdot 10^{12}$ barrels ($216 \cdot 10^9$ m^3) in 2011, but the different estimations are going up to $3.9 \cdot 10^{12}$ barrels ($620 \cdot 10^9$ m^3). At the same time, the gas reserves are estimated to be $6.4 \cdot 10^{15}$ ft^3 ($1.8 \cdot 10^{14}$ m^3).

Such growing demands demand new paths for supply from the layer/deposit to the users. The supply path starts with the well (that exists or must be drilled) and continues with completion, production, gathering, and transportation to the storage systems. All of that can be a potential source of hazard for people and the environment.

The Macondo disaster, the last known worldwide oil spill, is only the fourth of the world's largest oil spills. The first, the Gulf War oil spill in Kuwait ($11 \cdot 10^6$ barrels; $1.75 \cdot 10^6$ m^3), was the result of the destruction of thousands of wellheads that prevented blowout and controlled production from the well when functioning correctly. Several others on the list were the result of blowouts, starting with the first one, Shaw Gusher (Canada, 1862), and the Lucas Gusher (USA, 1901). Qom Wildcat Gusher (Iran, 1956) was the largest one with a spill of about $10.8 \cdot 10^6$ barrels ($1.71 \cdot 10^6$ m^3) of oil.

The second largest oil spills are the result of tanker disasters. The biggest was the Atlantic Empress (West Indies, 1979), with $2.1 \cdot 10^6$ bbl ($0.33 \cdot 10^6$ m^3) of oil spilled. Several other large tanker disasters are known, but the most publicity was given to Exxon Valdez (Alaska, 1989) with "only" 260,000 barrels (41,340 m^3) of oil spilled.

When talking about petroleum transport, it is mainly through the pipelines. The leakages of such systems are usually not visible because they are buried in the ground. In some areas (Africa), human greediness or poverty can result in tragedies, when the stealing of oil from the pipeline finishes with explosion or fire.

As the last point of the pathway is the storage system, which can differ in volume. They have to store all the produced hydrocarbons before the final use. Statistically, it is possible that some or many of them could catch fire, but much more of them can leak.

All of that shows that the path of the oil or gas from the layer/deposit to the user can seriously endanger people and the environment.

The essential part of the book will be the engineering analysis of potential hazards and risk assessment in three areas: (1) drilling, (2) completion, production, workover, and formation treatments, and (3) gathering, transportation, and storage of hydrocarbons. In addition, the sources and triggers of the hazards are determined, and remedial or controlling actions elaborated.

The aim of the book is to point out the potential risk of any of those three segments of petroleum engineering activities. The risk assessment and the designing and working approach in direction of avoiding accidents are elaborated.

The book gives a short introduction to the problem with the approach to risk analysis in chapter 1. Explanation of basic terms, their interdependence, dilemmas, and methods of risk analysis are introduced. Each method is shortly described with main anteriority and shortcomings. The impact, occurrence, and the consequences are at the end compared to the risk acceptance criteria concept. The ALARP (As Low as Reasonably Practicable) framework is explained with some observation on the quality and acceptance in petroleum industry. Finally, the human impact on the risk and consequences is analyzed.

Wellbore instability problems are usually related to drilling operation, but they can also appear during completion, workover, or the production stage of certain wells. Chapter 2 gives one general overview of wellbore instability problems and their causes as well as an overview of actual approaches and methods in wellbore stability and risk assessment.

A stuck pipe is a common worldwide drilling problem in terms of time and financial cost. It causes significant increases in non-productive time and losses of millions of dollars each year in the petroleum industry. Stuck pipe risk could be minimized by using available methodologies for stuck pipe prediction and avoiding based on available drilling parameters as is stated in chapter 3.

In chapter 4, lost circulation is defined as the uncontrolled flow of mud into a thief zone and presents one of the major risks associated with drilling. Successful management of lost circulation should include identification of potential loss zones, optimization of drilling hydraulics, and remedial measures when lost circulation occurs.

Simultaneous operations as given in chapter 5 are to be coordinated through joint planning efforts by production, workover/completion, drilling and construction supervisors, and/or engineers, who plan and direct activities. Typical chain-of-command as well as simultaneous operations decision making process flow diagrams are presented in this chapter. In general, they have an impact on the installation safety procedures and contingency planning program. Once the simultaneous operations have been identified, there are basic steps to be regarded: performing risk assessment, assess and control risks, monitor the simultaneous tasks, and communicate the control measures.

After the accident on the Deepwater Horizon platform, while drilling the Macondo 252 well in the Gulf of Mexico in 2010, several commissions, investigation groups, advisory committees, and company reports have been prepared. The author's approach is presented in chapter 6.

Well completion is defined as the optimal path for the reservoir fluids to be produced. The reliability of system components is essential for long lasting production. In addition, the differences according to natural flowing well risk and artificial lift are given. Nowadays, so called "intelligent completions" appear to give more financial benefits, flexibility, and control, but also a new range of risks, as explained in chapter 7.

Irreducible casing pressure, also termed Sustained Casing Pressure (SCP), is hazardous for a safe operation, and the affected wells cannot be terminated without remedial operations. In chapter 8, physical mechanisms of irreducible casing pressure and qualification of the associated risk by showing statistical data from the Gulf of Mexico and discussing the regulatory approach are introduced, with new approach to evaluate the risk of casing pressure by computing a probable rate of atmospheric emissions from wells with failed casing heads resulting from excessive pressure.

Chapter 9 is focused on the risk to the environment from hydraulic fracturing operations. Although many well development problems are blamed on fracturing, there are only excluded problems that are real and worthy of the discussion to help define boundaries of the fracturing risk. The initial assumption for the fracturing risk analysis is that the well is new and was constructed correctly so that all producible formations are securely isolated behind the barriers of casing and competent cement.

Workover risk and anomalies may be caused by erosion, corrosion, mechanical errors, and temperature effects on electronics, wear and tear on the dynamic seals, or seizure of moving components. Obviously, the simpler the system and the fewer moving parts, the fewer components are available to fail. The right approach and operating system selection is essential, as shown in chapter 10.

Gathering system as defined in chapter 11, include one or more segments of pipeline, usually interconnected to form a network that transports oil and natural gas from the production wells to one or more production facilities as well as from production facility to the inlet of a gas processing plant, storage facility, or a shipping point. Complexity of the processing facility depends on the treated fluid composition. Environmental impact during the oil and gas transportation and processing phase will cause long-term habitat changes. Such impact would also occur when surface facilities are removed after their useful life in a process of decommissioning. To avoid or minimize the environmental impact of gathering systems and surface facilities, it is very important to implement appropriate activities across the various phases: designing phase, construction, operational, and decommissioning phase.

Formal risk assessments are necessary at various phases of the asset life cycle as they help personnel identify, evaluate, and control hazards that could result in loss of life, injury, pollution, property damage, or business disruption. Hazard evaluations of production development concepts or facility design are well-defined processes, for which much literature is available as guidance. Such evaluations are mandated in some jurisdictions for project regulatory approval. Chapter 12 provides guidance on activity implementation from the designing phase, construction phase, operational phase, and decommissioning phase of gathering and processing systems.

Petroleum and natural gas must be moved from the production site to refineries or to users. These movements are made by using a number of different modes of transportation. Petroleum is transported across the water in barges and tankers. On land, petroleum is moved using pipelines, trucks, and trains. Natural gas is moved, mainly, by pipelines. Most of the time petroleum and natural gas are transported quietly and safely. However, accidents do occur. Chapter 13 describes the causes of incidents during oil and gas transportation both on land and across water.

Chapter 14 illustrates different types of crude oil and oil product storage tanks as well as the risks regarding the storage itself. Considering that the natural gas, in its gaseous state, is stored in underground storages like oil and gas depleted reservoirs, aquifers or salt caverns, and there are numerous publications and books covering the subject in detail, this chapter will only illustrate the storage of liquefied natural gas and the risks posed by its storage.

Geologic storage as the component of Carbon Capture and Storage (CCS) is elaborated in chapter 15. For the purposes of risk assessment, a priority is to evaluate what would happen if CO_2 migrated unexpectedly through the confining unit(s), potentially resulting in undesirable impacts on a variety of potential receptors. One of the main risks identified in geological CO_2 storage is the potential for CO_2 leakage through or along wells. To avoid leakage from the injection wells, the integrity of the wells must be maintained during the injection period and for as long as free CO_2 exists in the injection zone.

In chapter 16, the petroleum industry's environmental incident history and statistics are presented. In addition, the environmental impact of the petroleum industry's activities, its extent, and trends is analyzed. The overview of pollution sources with associated environmental risk is given along with the analysis of the causes and consequences of incidents in the petroleum industry. The impact on live organisms, soil, water, and air are discussed in general.

Davorin Matanovic
University of Zagreb, Croatia

Chapter 1
General Approach to Risk Analysis

Davorin Matanovic
University of Zagreb, Croatia

ABSTRACT

Broadly accepted methodology that is implemented in the oil industry when dealing with risks includes as the first step the identification of possible hazards. That is done by gathering information about degree of risk according to working procedures, processes, and individuals involved in the operation of the process. That is the first step in risk management, an iterative process that must lead to the use of proper measurements in the way of protecting people, facilities and environment. The analysis is done based on the combination of probability and severity of undesirable events, and the final consequences. Explanation of basic terms, their interdependence, dilemmas, and methods of risk analysis are introduced. Each method is shortly described with main anteriority and shortcomings. Differences between quantitative methods, qualitative methods, and hybrid methods (the combination of qualitative-quantitative or semi-quantitative methods) are elaborated. The impact, occurrence, and the consequences are at the end compared to risk acceptance criteria concept. The ALARP (As Low as Reasonably Practicable) framework is explained with some observation on the quality and acceptance in petroleum industry. Finally, the human impact on the risk and consequences is analyzed.

DOI: 10.4018/978-1-4666-4777-0.ch001

INTRODUCTION

The best approach in defining the risk is the implementation of risk management. It includes at the same time measures to avoid the occurrence of the hazard and other to reduce possible harms. Reason for that is the cognition that risk cannot be eliminated; so it has to be managed.

In decision-making process or system design; the hazard analysis, the risk analyses, and at last the risk assessment should be conducted as the part of the risk management process (Kavianian, 2003). They are used to identify possible hazards and treats, analyze the causes and consequences and give the description of possible risk. To understand the meaning of risk analysis, some differences and meanings attached to the term should be precisely defined. The term "risk analysis" has been adopted by petroleum industry rather than "hazard analysis" adopted by some authors involved in chemical industry problems (Kletz, 1999). Nevertheless the term would be (risk analysis or hazard analysis) it is used to describe methods that are used to identify hazards and help to estimate the probability and possible consequences of possible accident. The relation of probability and severity of the undesirable event, can determine if the risk is acceptable or not. When it is not acceptable, must the system be modified to lower one or another or both causes of the accident. To repeat in other words; the term "risk analysis" in this book will be addressed to description and identification of the undesirable events and characterization the cause and effects of "hazards". Two important items in the analysis terminology are "risk" and "hazard".

The risk assessment will indicate the appropriate process hazard analysis methodology that should be applied to the process. This will add in identifying of the different types of hazards that influence the system components acting. Also that will help to select possible solutions to eliminate the hazard. A hazard can be defined as a substance, event or situation, with the potential to cause direct harm or initiate a sequence of events leading to harm. Harm is defined as a physical injury or damage to the health of people or damage to property or the environment (ISO/IEC, 2012). In petroleum engineering it can include release of flammable or explosive substances or vapors, chemical spills, leaking pipes or valves, falling objects, etc. The effects or the consequences of the hazard can be immediate or long-term. They should impact people, the environment or the economic resources. The problems themselves are many and varied, and different methods are required in order to deal with them. Most of the techniques developed to date are applied during the development of a specific project. Therefore they should be discussed according to the normal sequence of project development phases: conceptual design and planning; detailed analysis; construction; commissioning and operation. The identification of a hazard is referred as the most important step in a risk assessment process. It relies on knowledge retention, which means, to be able to store and retrieve the information and knowledge generated earlier, learning through experience and accidents involved. It also includes the ability to predict hazards and combinations of hazards that have not yet been encountered. Figure 1 is an illustration of the hazard identification process.

Probabilistic risk assessment (PRA) methodology, that was developed by nuclear industry, and can be implied on oil and gas industry, in fact answers three questions: (1) What can go wrong in the process?; (2) How likely is the accident scenario to occur as the probability or a frequency?; and (3) What should be the consequences? In quantitative risk assessment (QRA), hazard identification can be the most important step. The explana-

Figure 1. Illustration of the hazard identification process (Brandsæter, 2002)

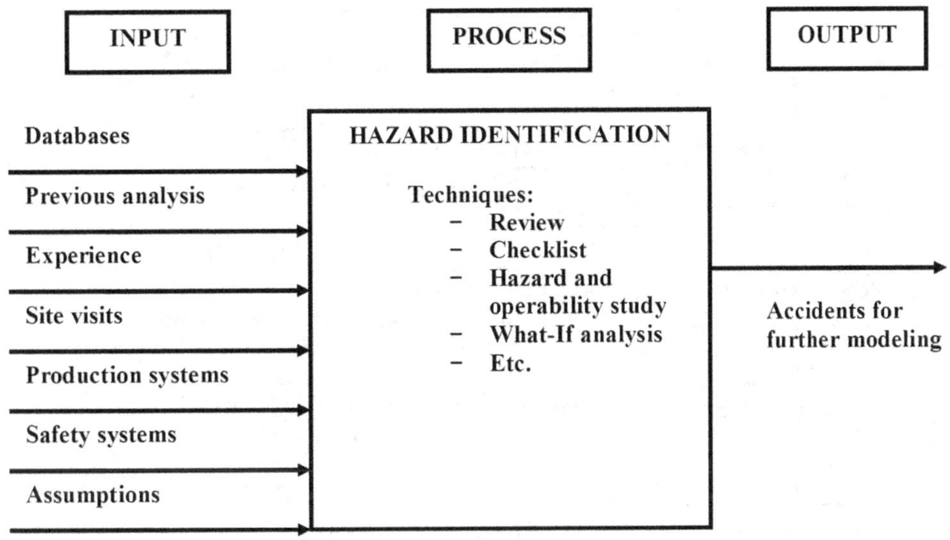

tion can be that something not identified, can not be evaluated nor mitigated. The process reclines on stored and retrieved knowledge through information gathered in previous work or project designs. Also experiences when coping with accidents, blow-outs or other are beneficial. Site visits to drilling rigs, production or other systems are important too. They enable the control of system functioning, and the most important are the safety systems. Hazard identification can be accomplished using different techniques; reviews, checklists, using hazard and operability studies (HAZOP), what-if analysis and other. The outputs are the possible accidents that should be further analyzed.

The confusion in terminology can lead to misunderstanding. That is because according to the definition of the risk it implies that for quantitative expression of risk, the probability or expected frequency is used. The probability (P) scale ranges from 0.0 to 1.0 where the end points of the scale; $P = 1.0$ means 100% certainty and $P = 0.0$ means 0% certainty. The opposite of probability is risk. Much more, the fraction in interval between 0 and 1 is used. In fact the probability of a given occurrence or event is equal to 1 minus the risk for this event not occurring. When there are simultaneous occurrences of several independent events, the probability is equal to the product of their individual probabilities. If there are occurrences of several mutually exclusive events, the probability of occurrence of at least one event is equal to the sum of the probabilities of each individual event. Defining of probability of either one or both of two independent events is done using combination rule. That means that the overall probability is the sum of independent probabilities minus their multiplication.

The best results can be achieved when used through the development of the project. The approach is illustrated in Figure 2.

The idea is to analyze the project through design and development stages. Doing so, it makes possible to make necessary changes that arise from prior analysis.

Risk refers to the likelihood of occurrence of events that would have adverse consequences upon the safety of people, the environment, or economic resources. To be able to specify

Figure 2. Illustration of hazard analyses during the development of a project (Yokel & Simiu, 1985)

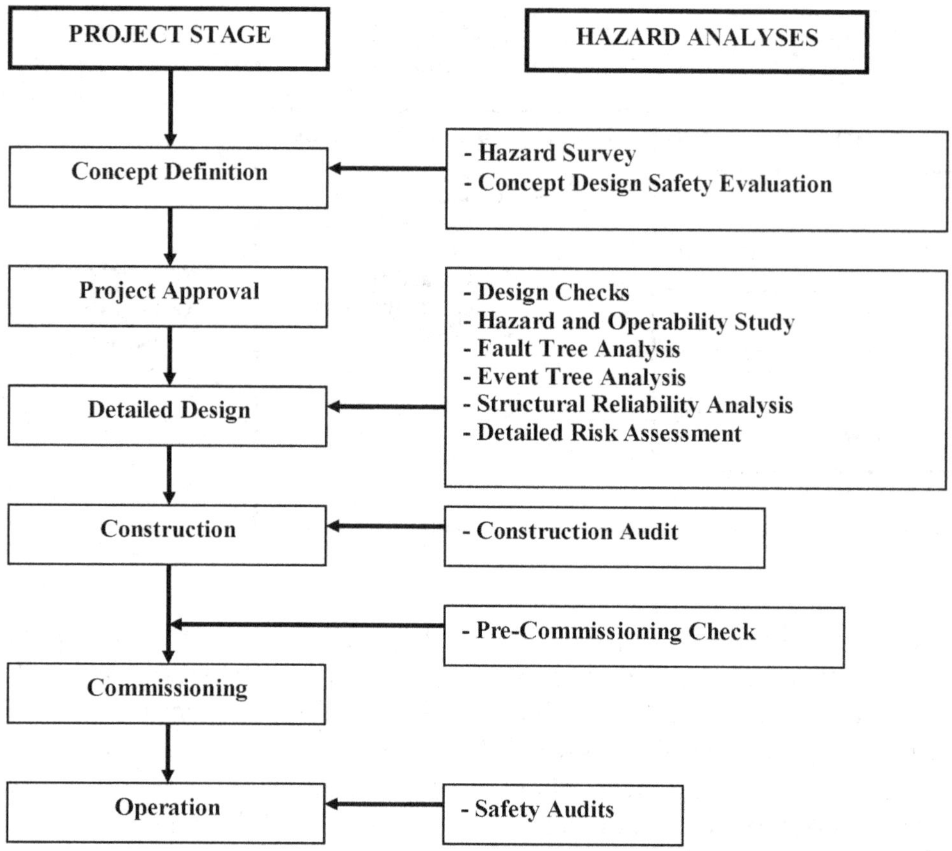

the risk, it is necessary to specify the event that must be considered, the possibility of the event to occur at the specific site over a given period of time, and the possible consequences of the event according to value or degree of losses which my be incurred. The probability of such undesirable events to occur can be expressed in two ways; qualitatively (e.g. rare, occasional, frequent) or quantitatively, as a normalized frequency or probability.

In the petroleum industry the favorable definition of risk (R) involves the frequency of an undesired event (F) and the consequences of the event (C); (It is an combination of the probability of an event and its consequences:

$$R=f(F,C) \tag{1}$$

Also the term exposure (E) should be a part of the expression of a risk, because the setting in which the event frequency and consequences were considered, to be able to directly relate risk estimates to specific activities considered for safety action, standard or regulations.

$$R=f(F, C, E) \tag{2}$$

When exposure is included, risk is expressed in terms of probable consequences per unit of exposure through the period of time that was analyzed for specific activity. The consequences of a hazard can be immediate or act on long-term. They can also be expressed quantitatively through economic loss or fatality rate, or qualitatively as severe or minor.

Risk when defined as the consequence per unit of time is in fact the multiplication of the frequency expressed as event per unit of time and the magnitude of the consequence per event.

$$Risk = \sum_{i=1}^{n} C_i F_i \quad (3)$$

where (C_i) is the consequence of undesirable event that can be expressed in currency, lives or volume of hydrocarbons; (F_i) is the frequency of occurrence expressed per unit of time, per installation or per operation, and (i) is the ith sequence.

When dealing with people and the rest of the environment as the subjects in a process, the term "vulnerability" can play significant role in understanding hazard impact. In such context risk is the product of hazard and vulnerability. The impact may be lethal, what includes loss of lives or non-lethal with just the exposure of biological organisms to the damage. The idea is too find the mostly endangered species and analyze the way it can sustain the hazard. If it shows that the reproduction potential and natural mortality rate is related, the recovery time should be determined.

Risk analysis is a relatively promising approach which might be used to identify, analyze, and manage the hazards associated with complex technological projects such as onshore and offshore oil and gas operations. A full risk analysis of a design or an operating procedure requires a number of steps, as shown in Figure 3.

For a specified system, with definite design and operating procedures, the first step in risk analysis is the identification of the hazards, risks and consequences. Knowing the acceptance criteria and other requirements if any, risks are evaluated as acceptable or unacceptable. When risks can not be accepted, the whole procedure or the design should be revised or acceptance criteria and requirements

Figure 3. Flow diagram of risk analysis

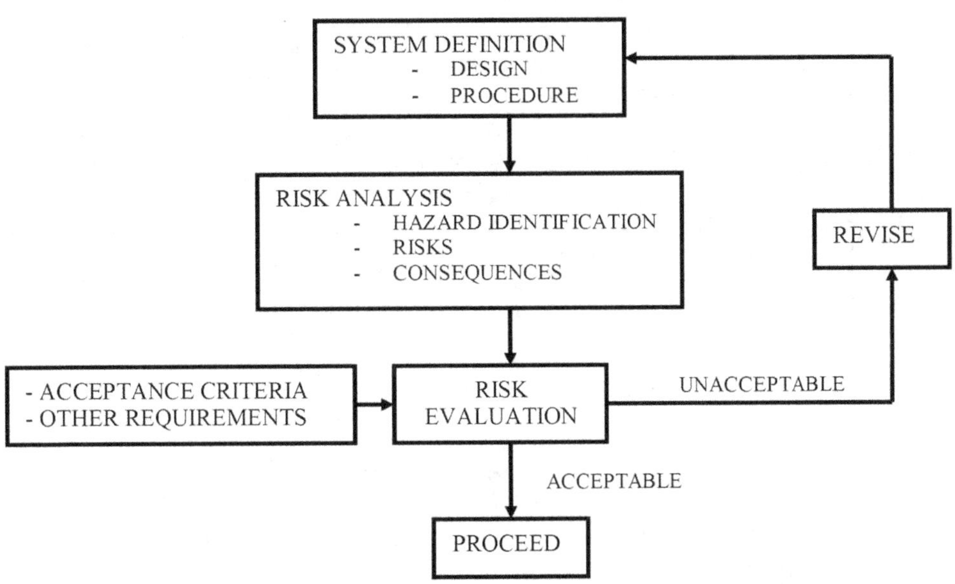

modified, until one can proceed with the project.

When hazards and possible threats are identified, cause and consequence analysis done and risk is described, the next step is the evaluation of obtained results. The evaluation process is defined as risk assessment. Next step connote the treatment of risk in the way of reduction or other way of treatment. Such process is called risk management and means all activities with regard to risk. According to Aven &Vinnenm (2007) risk management means to achieve "an appropriate balance between realizing opportunities for gains while minimizing losses". Figure 4 is the flow diagram of the risk assessment and reduction process as a combination of two concepts based on ISO (International Organization for Standardization) and IEC (International Electro technical Commission) recommendations.

The first step is to analyze risk through three steps. First is the hazard identification through the identification of sources of potentially hazardous situations. The aim is to define likelihood of such occurrences. Next step is the identification of sequence of events that can lead to hazards. There is always some kind of initiator or trigger that starts the sequence of events and the result can be hazards and exposure to harmful impacts. Simply, any outcome having potential to cause harm or damage relevant to the risk acceptance criteria must be considered. Such incidents can lead to losses and damages. They can be human lives, damage of the environment or material losses of different kind. Such possible consequences should be identified and

Figure 4. Flow diagram of risk management process

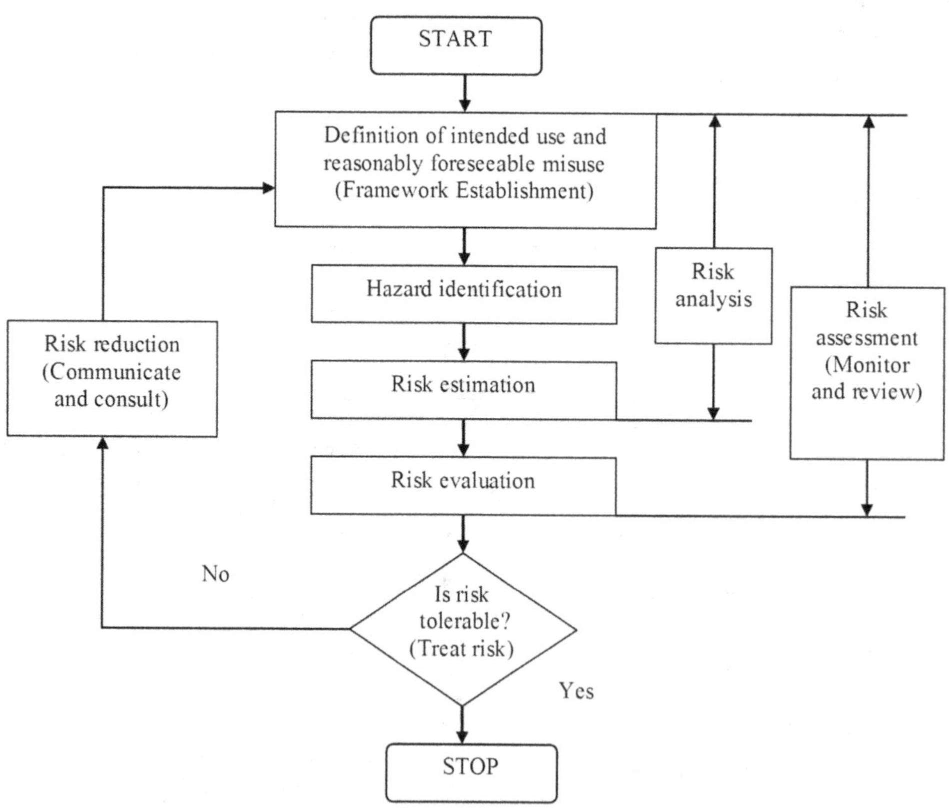

evaluated. Through that the risk is evaluated as well. The derived risks are than compared to the criteria or guidelines or legislative to determine the tolerability of the achieved risk. That is the step of making the decision in the way of risk reduction measures if risk is not tolerable. In petroleum engineering there is a wide range of possible failure cases that should lead to hazards. The main can be defined as: (1) blowouts through the drilling, completion, workover and well abandonment, (2) all kind of well control incidents that are not so severe as blowouts, (3) any leak of gas, oil or both in production, gathering, storage and transportation processes, (4) fires regardless to cause and place, (5) spills of other dangerous materials – chemicals, fuels, bottled gas leaks, radioactive materials etc., (6) accidental explosive detonations – perforator charges, (7) dropped objects on rigs and platforms, through construction, derrick rigging-up, crane transfer etc., (8) transport accidents, (9) marine events and collisions including diving accidents, and (10) structural events due to material fatigue, design error, foundation failure, different collapses and natural disasters.

RISK ANALYSIS AND ASSESSMENT TECHNIQUES

Lot of different techniques can be applied in the risk analysis or assessment process. According to Mahravilas et al. (2011), they can be classified into three groups, as shown in Figure 5. The original representation is slightly modified to include analysis that are also possible but are not mentioned originally (gray shaded). That are: qualitative, quantitative and hybrid techniques. Qualitative techniques include studies based on the personnel experience. No mathematical expressions and estimations are included. In fact they represent the checklist reviews with questions and process parameters that are used to discuss the process design, operation or malfunction. Quantitative techniques are represented by mathematical evaluation, based on historical evidence through data bases, about failures. Using such data it is than possible to predict the occurrence of undesired event or hazard. Hybrid techniques represent the combination of those mentioned before. They are more complex due to their ad hoc character that changes the combination of used techniques according to the problem that must be analyzed.

The first group – *Quantitative Techniques* that are widely and mostly in use (relative frequency is over 65%) are based on numerical values of the probability, vulnerability and consequences, that as a result gives a numerical value of the risk.

The Proportional Risk-Assessment (PRAT) Technique: (Ayyub, 2003) Calculates the risk using a proportional formula with included consideration of potential consequences of the accident, the probability of the accident and the possible exposure. The risk is the product of the probability factor, severity factor and the frequency or exposure factor. It strongly relies on the ability to validly estimate input parameters.

The Decision Matrix Risk-Assessment (DMRA) Technique: (Haimes, 2009) Is a combination (product) of severity and likelihood, that categorizes risk on an informed judgment basis. The interpretation can be quantitative, because the risk is measured and graphical. The severity and probability ratings in petroleum engineering (PE) are addressed to the identified hazard by using the risk matrix and environmental consequence description (Table 1). The risk description can be based on the product of the severity and likelihood with adequate designations.

Risk Measures of Societal Risk: Associated with petroleum engineering technical

Figure 5. The classification of main risk analysis and assessment (RAA) methodologies

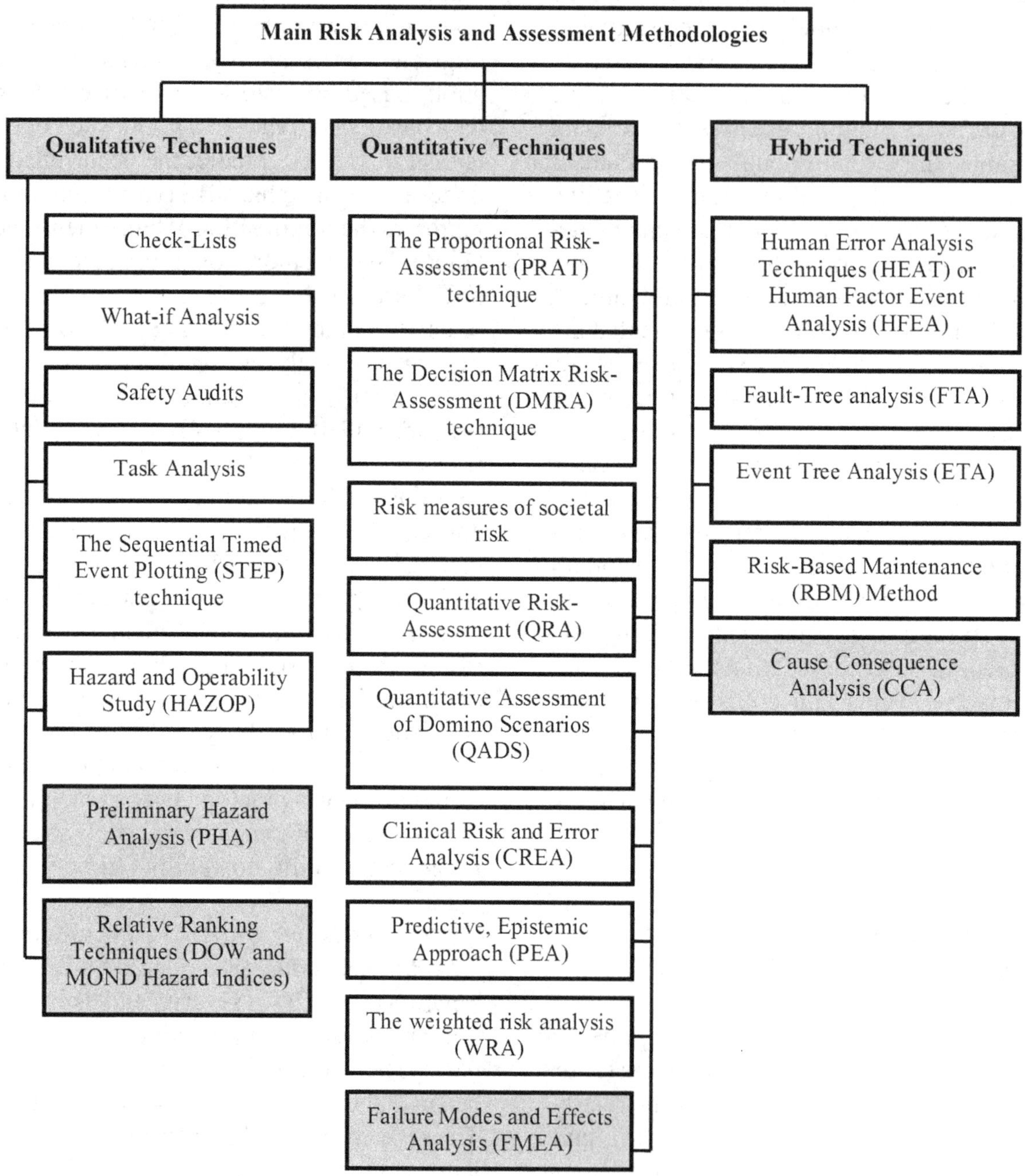

systems can also be evaluated in the way presented on the basis of accident scenario (the category of the accident), frequency of the scenario (probability per time unit) and the consequence of the scenario (number of injuries or fatalities or financial losses). The representation of the analysis is drawn as the F-N curve (Figure 6). It represents the relation of the fatalities cumulative frequency (*F*) with respect to number of fatalities (*N*) in log-log relationship. Also the tolerable and intolerable criteria lines are drawn. The position of

Table 1. Risk matrix

CONSEQUENCE				LIKELIHOOD					
				0	A	B	C	D	E
SEVERITY	HEALTH	SAFETY	ENVIRONMENT	Non-credible to occur Could happen in PE	Rare to occur Reported for PE	Unlikely to occur Once in Company	Credible to occur Several times in Company	Probable to occur Several times/year in company	Frequent to occur Several times/ year in one location
1	Slight impact	Slight impact	Slight impact			LOW			
2	Minor impact	Minor impact	Minor impact				MEDIUM		
3	Big impact	Big impact	Local impact					HIGH	
4	Mayor impact	Mayor impact	Mayor impact						
5	Extensive impact	Extensive impact	Extensive impact						

DESCRIPTOR	DESCRIPTION
Slight impact	- small discharges, limited and occasional impact, not visible, small greenhouse gasses emissions (GHG), good materials selection
Minor impact	- discharges that impact the area but have transitory effect, less than one week for clean up and 1 to 2 years for natural recovery, slight impact of GHG emissions, adequate materials selection
Big/local impact	- discharges of limited amount that affect nearby environment with longer effects, short term wider impact on water, soil and biodiversity, les than 1 month for clean up and 2 to 5 years for natural recovery, limited GHG emission, inadequate materials selected
Mayor impact	- large discharges that impact the environment with severe and long lasting damage, widespread impact even on some threatened species or functions, 1 to 5 months needed for clean up and 5 to 10 years for natural recovery, significant GHG emission, financially significant input to restore the damage, poor materials selected
Extensive impact	- large discharges with persistent damage of the environment, long-term and broader impact with possible permanent loss of species or ecosystem functions, more than 5 months for clean up and more than 10 years for natural recovery, extremely poor material selection

F-N between these two criteria lines means that the technical system is acceptable according to social preferences (Kosmovski, 2006). The ways to reduce risk if necessary, can be determined by the use of ALARP (as low as reasonably practicable) principle (Holmberg & Knochenhauer, 2007).

Quantitative Risk-Assessment (QRA): Have been developed to enable risk assessment from large blowouts with fires (dust explosion hazards). It is based on calculations that are performed to select appropriate scenario. Also methodology and tools have to bee used to estimate consequences and give fair representation of expected accident sce-

Figure 6. F-N curve with criteria for societal risk determination

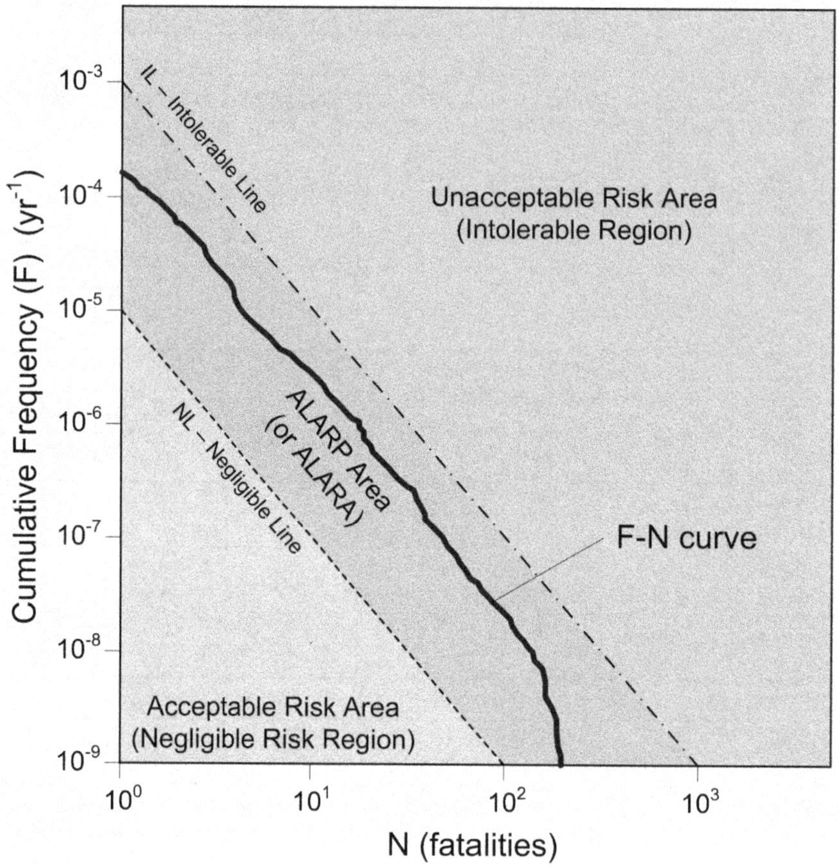

nario. The tools are recalibrated and validated against experience and test results and real measurements (van der Voort et al., 2007).

Quantitative Assessment of Domino Scenarios (QADS): Assumes that an accident with a primary triggering event is propagating and causing several secondary events resulting with the more severe consequences that those from primary event alone. That results from the growth and acceleration of the overall accident due to so called "domino effect". In the analysis it is necessary to investigate and determine the primary accidental scenario, than realize and determine the propagation effects with respect to the vectors of propagation and escalation. Also other associated scenarios should be analyzed with regard to primary event.

Clinical Risk and Error Analysis (CREA): Is related to medical domain and enables the determination of risk index (or the sum of risk indexes) by combining the probability of occurrence and the severity with included the occurrence of error mode. All of that is based on available data and expert's judgment.

Predictive, Epistemic Approach (PEA): Is based on state-of-knowledge and gives the opportunity to combine real data and subjective information to predict accidental situations. The damages from abnormal actions can be predicted by deterministic or probabilistic analytical approach.

The weighted risk analysis (WRA): Enables the comparation of different risks (loss of lives, economic losses) with respect to the same reference that can be money or some other. Not only technical aspects are analyzed but also the environmental, economical, political and societal as well. The weighing factors are attributed to related scenarios according to consequences giving the opportunity to compare them and determine the measures to be taken for risk reduction (Suddle, 2009).

Failure Modes and Effects Analysis (FMEA): Gives a systematic tabular presentation with evaluation results of causes and effects of recorded or known types of equipment or component failures represented with annual estimation.

Second group – *Qualitative Techniques* include an analytical estimation process and the engineers' ability, to interpret the risk using a word form, rating scales of descriptive or numeric kind to describe the magnitude and likelihood of potential consequence.

Checklist Analysis: Is the first technique in a group and is a list of questions or a worksheet about operation, maintenance, installation safety concerns, and is a simplest method for hazard identification. It can be described as a systematic approach built on the historical knowledge. Application is possible in all systems or activities regardless to equipment or human impact. The most important in the implementation is that it is carried out by the trained and experienced person or a small group. Quality of the analysis strongly depends on the experience and quality of the checklist creators and the users.

What-if-Analysis: Is simply compiled of a number of investigative questions that should be asked by an experienced team about the system components under examination (here we are talking about hydrocarbon systems, generally petroleum engineering systems). The idea is to determine what can go wrong and what can be the consequences. The technique is systematic, but depends on expert team quality through the questions generation and comprehensiveness of the review. Also they have to control and ensure that the adequate safeguards are implemented and in place. When conducted by more than one team of experts that work together it seams to be a high-level and detailed risk-assessment technique. The final report is a review of potential problems and a list of recommendations for their prevention.

Safety Audits: Include different procedures prepared for inspection of installations, processes or plants according to the safety. The auditors review and verify the implementation of appropriate approach in system design, in controlling operational conditions and in safety measures. The result should be a list of recommendations about safety procedures, possible improvements and awareness of the operating personnel (Reiniers et al., 2005).

Task Analysis: Its primary use is in identification of the interaction between the system and personnel involved. The principle is in process analysis and the personnel job performance and mutual interaction (Doytchev & Szwillus, 12009). The result of the task analysis is so called "Task Model". The analysis process comprises collection of data about human interventions and system demands, graphical representation of results and comparison between system demands and operator capabilities in order to ensure their compatibility (Landau et al., 1998).

The Sequentially Timed Event Plotting Technique: Combines the timing and sequence of events or actions that can be a trigger or can lead to the accident (Hendrick & Benner, 1987). To do so, the sequences of events that can contribute to the accident are plotted, which provides overview over the initiation of the accident or changes that can disrupt the system or the process. The event building blocks those are used to identify the event contain the information about the time at which the event has started, his duration, the trigger (agent) that caused the event, the description of the event and the source of such information. At the end all events are connected with incoming and outgoing results showing relationships between events in term of proceeding or following sequence (Kontogiannis et al., 2000).

Hazard and Operability Study: Includes systematical approach for identification and documentation of hazards. That is in fact critical safety study on deviations intended through the design of each system component. Such deviations are formulated and analyzed from a standardized list. The expression of risk is in qualitative series (e.g., numerical from 1 to 5) relative one to another. Also some kind of sets of guide words can be used (e.g., NONE – complete negation to design intention, MORE THAN – quantitative increase, LESS THAN – quantitative decrease, PART OF – only part of intention is fulfilled, AS WELL AS – something else out of the design intention, REVERSE – the occurrence of logical opposition of design intention, OTHER THAN – complete substitution). In that way the corresponding process variable deviations are obtained (Khan & Abbasi, 1997). The approach gives the insight in potential problems in the process or a system, because the basic principle of the HAZOP study is that hazard arises in the system due the deviations that are not expected in process normal behavior. The analysis of the system should be done as a typical sequence starting with the selection of a study node (Yang & Yang, 2005). By applying guidewords the possible hazard and operability problems are identified. If there are none next node is used. If problems appear the causes and consequences are recorded and solutions suggested.

Preliminary Hazard Analysis (PHA): Identifies each hazard according to potential causes and effects. The list of recommendations or protective measures (known and possible) to be used are listed there. The "preliminary" term means that it is usually upgraded by performing additional studies. The energy sources and hazard materials are in focus and special attention is dedicated to identify those hazardous features of a system or plant that can result in unacceptable risk (Kavianian, 2003).

Relative Ranking Techniques (DOW and MOND Hazard Indices): Method is based on penalties or award points that are assigned for hazard and protective measures. All of them are listed in a checklist accounting form and at the end combined into an index. Such index is an indication of the relative ranking of the system risk.

Third group - *hybrid techniques* are the combination of qualitative and quantitative techniques or are of semi-quantitative type.

Human Error Analysis Techniques (HEAT) or Human Factor Event Analysis (HFEA): Have been introduced because it has been recognized that people and their errors have major contribution in accidents in petroleum industry. It has been stated that about 85-90% of accidents are due to human factor. The reasons for errors can be the lack of human reliability and consistency. The fluctuation and deficiency of qualified working personal that has been recognized resulted with workers that are not familiar with normative treatments, some times unpredictable according to misdiagnoses and wrong actions. Error concealment is in human nature in attempt to avoid blame or material consequences. Also to be prominent many engineers make wrong solutions trying to be dominant, which leads to various kinds of failures or losses. One problem has also been realized through the efficacy of different jobs. That is the performance of routine tasks that after a while leads to unsafe behavior and poor efficiency. The example can be the performance of different jobs with coiled tubing that have been analyzed (Engel & Mackey, 2001). When coiled tubing was first introduced as a new technology, major advances were made with clean/wash out operations, to assist with the sand clean out from the wells. They have covered about 32% of coiled tubing jobs at that time. The analysis covered about 1200 runs over the 23 month period and showed an overall success of 82% by addressing these issues. The database has suggested that there are three fundamental causes of failure. Clean out (sand wash out), along with tools and drilling are among them. Detailed planning and training has been identified as the primary solution of these problems.

The usual errors in human action are slips (the action is not performed in time or is performed incorrectly), lapses (actions that are commissioned or not in time), mistakes (doing something in wrong way or not doing necessary at all), and omission of detection (because of the lack of detection nothing or irrelevant acts are performed (Kumamoto, 2007). The reasons for human errors can result from (a) changes in teamwork according to actions performance, (b) changes in skill level needed for the action that have not been followed by proper training and adequate set of information, (c) change in communication demands due the complexity of the equipment or the procedure, and (d) change in environmental conditions that can influence negatively the working capabilities of the personnel. Because of that there is a necessity to prepare personnel by adequate training to increase human reliability. The environment is such as it is, but the working place can be adjusted to be friendlier (heating in the cold areas and the opposite, air-conditioning in tropical areas). Working interfaces should be self-understanding and visible, procedures according to the actions prepared in written shape and distributed to the personnel. That should be followed by adequate training and at last supervision.

In the petroleum industry, not only the personnel but also the technique and technology are changed all the time. That means that the people should be prepared to understand and accept new knowledge and technologies. The influence of technical and technological progress on economical efficiency can be closely examined through the scientific and technical and technological progress.

Technical progress is defined as an economical appearance that values all changes in means for production, technological progress and organization of production and distribution that leads to increase of production and increase of product value. It also leads to increase of producer's knowledge and improvement of their work (Matanovic et al., 2001).

Stress factors that influence the human action can be psychological, physical or organizational. Regardless the reason the stress level drastically changes the ability of personnel to perform tasks. The dependence of stress level on human performance shows that extremes are not optimal, but that there is an optimal peak in between when the performance is the best.

The stress level can be lowered and so the human-error-probability by fulfilling several requirements (ASME, 2003). Written procedures (for performing tasks) should been prepared and the quality of human-machine interface balanced with the personnel training and abilities. Lowering of stress level can also be accomplished by increasing of operators training and experience (better quality and frequency of practice), higher quality of written procedures and controls, proper installation and functionality control of securing and signaling instrumentation, and by raising the available time to response according the required complexity. Also acting with explosives (perforating guns), radioactive materials (nuclear logging devices), environment with lethal gases (hydrogen sulfide), high pressures and in the area of great potential for explosion or fire generation, rises the stress level from the beginning.

Fault-Tree Analysis (FTA): Can be presented as the combination of graphical and mathematical model that enables to combine points of failures of greatest importance and give expression of their probability. Through visualization of interrelations among different system elements or constituents, equipment failure is related to human error or environmental factors that can lead to the accident. The term "Fault Three" in fact shows the way of analyzing systems or interrelations. The top event (something unwanted) is a starting point and is represented by a rectangle. Through different branches according to the system intricacy, possible causes of the hazard are identified and ranked. Elements of the system, process and conditions are described through logic gates with only two possibilities ("AND or "OR"). The main point is called the top event, and must be fully described. Than the system is analyzed by personnel that understand the relations and interdependences in the system that can cause the top event. For the construction of a fault three diagram different symbols are used with specified meaning. The cause of a top event (some kind of leakage usually in petroleum industry) can be a miss function or bad material selection of one or more system components. "AND" gates are used when all indicated components or conditions can contribute for top event to exist. The "OR" gates show that only one of the system parts can lead to the top event if not functioning well. Undeveloped events are considered to be out of scope of interest, there is not enough information, or can not affect the system. Basic faults or events are those that lead to the top event because of fatigue or malfunction and are represented with circle (Harms-Ringdahl, 2001). The other symbols that are used in fault tree construction can be a "diamond" representing an undeveloped event, triangle representing the point of transfer in or out depending on connecting line. Some other different symbols can be used to represent inhibit gate, external event or conditional event (Ostebo, 1991).

Event Tree Analysis (ETA): Are also mathematical and graphical presentation of combination of events and circumstances according to possible accident sequence according to annual estimation. It uses decision tree to logically develop and visualize models that will lead from an initial event to the outcomes. The name of the system arises from the presentation shape. It starts with initiating event and grows up like a tree as the number of sequence events increases. The initial event and subsequent events can diverse to probable events finishing with final results. There is no relation between probable subsequent events. All events are also time related because the sequence is of importance like in domino effect. The results can be as qualitative description of problems coming from combination of events resulting from initial event. Also the quantitative estimation of event frequency and various failure sequences is possible. Much more the list of recommendations how to reduce risk can be prepared in design, construction or operating stage of the system (Hong et al., 2009).

Risk-Based Maintenance (RBM) Method: Is also the combination of qualitative and quantitative techniques. Quantitative description arises from the quality of prepared consequence study. Based on those estimations of the probability of failure are determined. Method consists of three modules. First one is risk determination by the event identification and estimation. Second one analyzes consequences and acceptance criteria. Finally the maintenance is planed with consideration of risk factors included (Khan & Haddara, 2003).

Cause Consequence Analysis (CCA): Follows the diagram presentation of events that have been started by the critical event. The sequence of events that develop from the critical event and the performance of the controlling and safety systems should be taken into account. Analysis can be expressed qualitatively, quantitatively or in both ways according the defined objectives of the analysis. Being an illustrative logic diagram it gives a good visualization of the development from the initiation to different consequences of critical event, especially in systems where such sequences of occurrence are vital in development of critical event to the hazard.

The similar overview of the risk analysis and risk management methods (Andersen & Mostue, 2012) that are especially applicable in petroleum industry (e.g. integrated operations – IO). In the overview some other resources have been used (Everdij & Bloom, 2006; Sklet, 2002) grouping and characterizing the methods in some different way.

The overview starts with *technical reliability analyses* as a quantification of the technical reliability as the input to quantitative risk analyses represented by the FMEA, Failure mode, effects and criticality analysis and Reliability and availability of computer based safety system. The methods are mainly focused on technology and applied in the design phase of the project.

Consequence Analyses: Identify affects of the accidents by the use of physical phenomena models. They are represented by FTA, ETA and Consequence and escalation models, and focused on humans and technology.

Quantitative Risk Analyses: Give the hazard evaluation and identification through quantitative data. The methods are QRA and Total risk analysis ad main viewpoint is technology.

Task analyses (Hierarchical task analysis HTA) are used to analyze human actions and decisions and are also used as the base for human reliability analyses (Kontogiannis, 2003).

Organizational Risk Analyses: (e.g. Barrier and operational risk analysis – BORA, System-action-management - SAM) Are used in quantitative approach when analyzing effect of organizational and management factors. All aspects (human, technology and organizational conditions) are in the focus of the analysis but the main application phase is the operational phase.

Qualitative Risk Analyses: Verify unwanted events qualitatively (e.g. Preliminary hazard analysis – PHA, Structured what if technique – SWIFT, Hazard identification – HAZID) with all aspects in focus but mostly applied in design phase).

Systemic Model Based Analyses: Are focused on flexibility of socio-technical systems (Function resonance analysis method – FRAM, System theoretic accident model and process – STAMP) with same aspect according to human, organization and technology. The application is usually in operational phase.

Verification Analyses: Validate human, technical and organizational conditions (Crisis intervention in offshore production – CRIOP) used in design and operational phase according to all aspects.

RISK ACCEPTANCE CRITERIA

The petroleum activities almost always involve risk. Aspects of acceptance criteria for risk differ according the countries and their regulations. Because of that there is a need for generally accepted criteria. In fact the universal goal is to eliminate risk, but this can never be completely achieved. So the criteria of risk acceptability must be adopted. Risk is only acceptable if there is a kind of benefit when deciding of the acceptance (Fischhoff et. al., 1981). Through the acceptance of risk criteria the numerical risk estimates determined using quantitative risk assessment are translated in terms of being low enough to proceed with the project or give some economic benefit. The problems appear when pre-determined quantitative risk acceptance criteria are used because it can lead to focusing in wrong direction. Trying to satisfy criteria the possibility to obtain overall applicable and cost-effective solution or measure can be overlooked. Also the tools to check the risk to meet criteria are not always sufficiently accurate to be used without additional review. That is especially true when talking about environmental risk, because it always includes some extent of political influence on the acceptability. The broadly used framework used for risk criteria states that risk should be reduced to the level that is as low as reasonably practicable (ALARP). The framework (Figure 7) is divided in three bands, defining two levels of risk criteria. First band represents the unacceptable region with almost no tolerance to the risk except in specially determined circumstances. In this region the risk reduction measures are essential. So called ALARP or tolerable region enables to determine to use or not to use risk reduction measures according to the proportion between costs and benefits. The broadly acceptable region at the bottom states that no risk reduction measures are needed. So the two levels of criteria are the maximum tolerable criterion above which risk can not be tolerated and the other below whish the risk has no significance.

To obtain optimal results through the ALARP determination process it is necessary to follow several steps. The first one is the identification and assess of the hazard. Then the confirmation that minimum acceptance criteria are met follows. Third step is mentioned to be critical. That is the identification of complete range of possible risk reduction measures. Optimal approach is the identifica-

Figure 7. Framework for risk tolerability (Dumitran & Onutu, 2010)

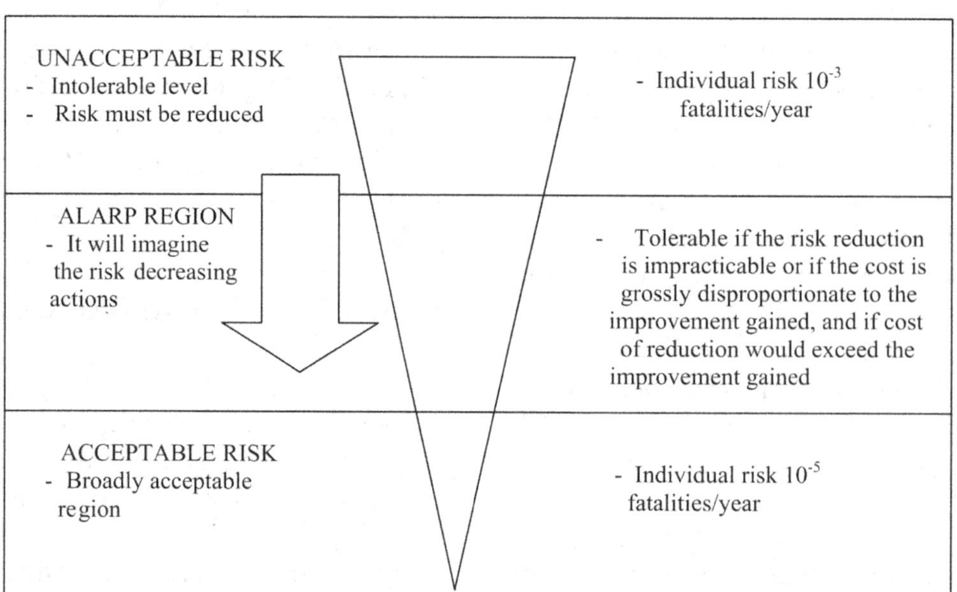

tion of feasible improvements that can be applied to eliminate hazards, reduction to the exposure of the personnel, reduction of the frequency of occurrence, diminishing of the consequences and improvement of the necessary evacuation (Lewis, 2007). Different tools can be used to demonstrate ALARP in regard with the complexity and risks. The lowest level is satisfied through codes and standards. Than at the next level the good practice and engineering judgment will do. The broadest region is covered by risk assessment and cost benefit analysis. The highest two levels are searching for peer reviews, benchmarking and at the end consultations with stakeholders.

The acceptance criteria are used for evaluation of results obtained through risk analyses and must be given for the personnel on the installation as a whole, and those groups or individuals mostly endangered or exposed to risk. The loss of main safety functions and the pollution from the installation must also be evaluated. If cost-benefit is a guiding principle instead of mechanistic approach that can lead to encouragement of the operating companies to consider if further risk reduction is possible and reasonable. Two aspects of criteria are widely accepted. One is the individual risk that is calculated through the identification of all sources of fatality risk to an individual. The contribution from the source is than defined and at the end all of them are summarized to obtain the overall risk. In petroleum engineering primary risk sources are occupational accidents, transportation related accidents and petroleum related leading to fires or explosions. Individual risk criteria (IRPA) expressed annually differs in various countries (AIChE, 2009). For workers the maximum tolerable criterion is 10^{-3} per year, and broadly acceptable criterion is 10^{-6} per year. For the members of public the range is from 10^{-4} to 10^{-6} per year. It is possible that companies use more strict criteria for new facilities designs. Talking about societal risk it is necessary to point out that criterion are defined to limit the risk of major or huge accidents to avoid restriction of activities or land use. The statistical value of the life regarding to the societal investments in risk reducing measures for offshore petroleum industry is about 6,000,000 GBP (HSE, 2006). The analysis of costs and benefits of risk reduction measures appear to be about 7.5 times more per expected life saved.

REFERENCES

AIChE. (2009). *Guidelines for developing quantitative safety risk criteria: Appendix B survey of worldwide risk criteria application*. New York: American Institute of Chemical Engineers, Inc. Centre for Chemical Process Safety.

Andersen, S., & Mostue, B. D. (2012). Risk analysis and risk management approaches applied to the petroleum industry and their application to IO concepts. *Safety Science*, *50*, 2011–2019. doi:10.1016/j.ssci.2011.07.016

ASME. (2003). *Standard for probabilistic risk assessment for nuclear power plant applications*. ASME RA-Sa-2003. Addenda to ASME RA-S-2002.

Aven, T., & Vinnem, J. E. (2007). *Risk management with applications from the offshore petroleum industry*. London, UK: Springer-Verlag.

Ayyub, B. M. (2003). *Risk analysis in engineering and economics*. London: Chapman & Hall. doi:10.1201/9780203497692

Brandsæter, A. (2002). Risk assessment in the offshore industry. *Safety Science*, *40*, 231–269. doi:10.1016/S0925-7535(01)00048-0

Doytchev, D. E., & Szwillus, G. (2009). Combining task analysis and fault tree analysis for accident and incident analysis: A case study from Bulgaria. *Accident; Analysis and Prevention*, *41*(6), 1172–1179. doi:10.1016/j.aap.2008.07.014 PMID:19819365

Dumitran, C., & Onutu, I. (2010). Environmental risk analysis for crude oil soil polution. *Carpathian Journal of Earth and Environmental Sciences*, *5*(1), 83–92.

Engel, S. P., & Mackey, P. (2001, March). *Opportunities to improve the success rate of coiled-tubing operations*. Paper presented at the SPE/ICoTA Coiled Tubing Roundtable. Houston, TX.

Everdij, M. H. C., & Bloom, H. A. P. (2006). *Safety methods database*. Paper presented at the 8th International Conference on Probabilistic Safety Assessment and Management (PSAM8). New Orleans, LA.

Fischhoff, B., Lichtenstein, S., Slovic, P., Derby, S. L., & Keeney, R. L. (1981). *Acceptable risk*. New York: Cambridge University Press.

Haimes, Y. Y. (2009). Risk modeling, assessment and management (3rd ed.). Chichester, UK: A John Wiley & Sons Inc. Publication.

Harms-Ringdahl, L. (2001). *Safety analysis, principles and practice in occupational safety* (2nd ed.). CRC Press. doi:10.1201/9780203302736

Hendrick, K., & Benner, L. Jr. (1987). *Investigating accidents with STEP*. New York: Marcel Dekker Inc.

Holmberg, J.-E., & Knochenhauer, M. (2007). Probabilistic safety goals, phase 1 – Status and experiences in Sweden and Finland- Nordic nuclear safety research. ISBN 978-87-7893-216-7

Hong, E.-S., Lee, I.-M., Shin, H.-S., Nam, S.-W., & Kong, J.-S. (2009). Quantitative risk evaluation based on event tree analysis technique: Application to the design of shield TBM. *Tunnelling and Underground Space Technology*, *24*(3), 269–277. doi:10.1016/j.tust.2008.09.004

HSE. (2006). *Offshore installations*. Retrieved from http://www.hse.gov.uk/offshore/is2-2006.pdf

ISO/IEC. (2012). *Guide 51.2, safety aspects – Guidelines for the inclusion in standards* (3rd ed.). Author.

Kavianian, H. R. (2003). *Process safety management: A legal and technical overview.* Paper presented at the ASSE Professional Development Conference and Exposition. Denver, CO.

Khan, F. I., & Abbasi, S. A. (1997). Mathematical model for HAZOP study time estimation. *Journal of Loss Prevention in the Process Industries*, *10*(4), 249–251. doi:10.1016/S0950-4230(97)00010-7

Khan, F. I., & Haddara, M. R. (2003). Risk-based maintenance (RBM), a quantitative approach for maintenance/inspection scheduling and planning. *Journal of Loss Prevention in the Process Industries*, *16*(6), 561–573. doi:10.1016/j.jlp.2003.08.011

Kletz, T. A. (1999). HAZOP and HAZAN, identifying and assessing process industry hazards. Warwickshire, UK: Institution of Chemical Engineers.

Kontogiannis, T. (2003). A petri net-based approach for ergonomic task analysis and modelling with emphasis on adaptation to system changes. *Safety Science*, *41*(10), 803–835. doi:10.1016/S0925-7535(02)00035-8

Kontogiannis, T., Leopoulos, V., & Marmaras, N. (2000). A comparison of accident analysis techniques for safety-critical man-machine systems. *International Journal of Industrial Ergonomics*, *25*, 327–347. doi:10.1016/S0169-8141(99)00022-0

Kosmovski, K. T. (2006). Functional safety concept for hazardous systems and new challenges. *Journal of Loss Prevention in the Process Industries*, *19*, 298–305. doi:10.1016/j.jlp.2005.06.003

Kumamoto, H. (2007). *Satisfying safety goals by probabilistic risk assessment.* London, UK: Springer-Verlag.

Landau, K., Rohmert, M., & Brauchler, R. (1998). Task analysis, part 1 – Guidelines for the practitioner. *International Journal of Industrial Ergonomics*, *22*(1-2), 3–11. doi:10.1016/S0169-8141(97)00064-4

Lewis, S. (2007). *Risk criteria – When is low enough good enough?* Risktec Solutions Limited. Retrieved from http://www.risktec.co.uk/media/43520/risk%20criteria%20-%20when%20is%20low%20enough%20good%20enough%20-%20saudi.pdf

Marhavilas, P. K., Koulouriotis, D., & Gemeni, V. (2011). Risk analysis and assessment methodologies in the work sites: On a review, classification and comparative study on the scientific literature of the period 2000-2009. *Journal of Loss Prevention in the Process Industries*, *24*, 477–523. doi:10.1016/j.jlp.2011.03.004

Matanovic, D., Gaurina-Medjimurec, N., Rajkovic, D., & Kristafor, Z. (2001). *Economy and technology balance in well life cycle management.* Paper presented at the SPE 3rd International Symposium on Oilfield Scale. Aberdeen, UK.

Ostebo, R., Tronstad, L., & Fikse, T. (1991). *Risk analysis of drilling and well operations.* Paper presented at the SPE/IADC Drilling Conference. Amsterdam, The Netherlands.

Reniers, G. L. L., Dullaert, W., Ale, B. J. M., & Soudan, K. (2005). Developing an external domino prevention framework. Hazwim. *Journal of Loss Prevention in the Process Industries*, *18*, 127–138. doi:10.1016/j.jlp.2005.03.002

Sklet, S. (2002). *Methods for accident investigation*. Trondhaim, Norway: Norwegian University of Science and Technology.

Suddle, S. (2009). The weighted risk analysis. *Safety Science*, *47*(5), 668–679. doi:10.1016/j.ssci.2008.09.005

van der Voort, M. M., Klein, A. J. J., de Maaijer, M., van den Berg, A. C., van Deursen, J. R., & Versloot, N. H. A. (2007). A quantitative risk assessment tool for the external safety of industrial plants with a dust explosion hazard. *Journal of Loss Prevention in the Process Industries*, *20*(4-6), 375–386. doi:10.1016/j.jlp.2007.04.024

Yang, S.-H., & Yang, L. (2005). Automatic safety analysis of control systems. *Journal of Loss Prevention in the Process Industries*, *18*, 178–185. doi:10.1016/j.jlp.2005.04.003

Yokel, F. Y., & Simiu, E. (Eds.). (1985). *Application of risk analysis to offshore oil and gas operations – Proceedings of an international workshop*. Gaithersburg, MD: National Bureau of Standards.

ADDITIONAL READING

Ahammed, M., & Melchers, R. E. (2006). Gradient and parameter sensitivity estimation for systems evaluated using Monte Carlo analysis. *Reliability Engineering & System Safety*, *91*(5), 594–601. doi:10.1016/j.ress.2005.04.005

Al-Ghamdi, A. S. (2002). Using logistic regression to estimate the influence of accident factors on accident severity. *Accident; Analysis and Prevention*, *34*(6), 729–741. doi:10.1016/S0001-4575(01)00073-2 PMID:12371778

Ale, B. J. M., Baksteen, H., Bellamy, L. J., Bloemhof, A., Goossens, L., & Hale, A. et al. (2008). Quantifying occupational risk: the development of an occupational risk model. *Safety Science*, *46*(2), 176–185. doi:10.1016/j.ssci.2007.02.001

Aneziris, O. N., Papazoglou, I. A., Mud, M. L., Damen, M., Kuiper, J., & Baksteen, H. et al. (2008). Towards risk assessment for crane activities. *Safety Science*, *46*(6), 872–884. doi:10.1016/j.ssci.2007.11.012

Apeland, S., Aven, T., & Nilsen, T. (2002). Quantifying uncertainty under a predictive, epistemic approach to risk analysis. *Reliability Engineering &. System Safety*, *75*(1), 93–102. doi:10.1016/S0951-8320(01)00122-3

Bartolozzi, V., Castiglione, L., Picciotto, A., & Galluzzo, M. (2000). Qualitative models of equipment units and their use in automatic HAZOP analysis. *Reliability Engineering & System Safety*, *70*(1), 49–57. doi:10.1016/S0951-8320(00)00042-9

Beim, G. K., & Hobbs, B. F. (1997). Event tree analysis of lock closure risks. *Journal of Water Resources Planning and Management*, *123*, 137–198. doi:10.1061/(ASCE)0733-9496(1997)123:3(169)

Cagno, E., Di Giulio, A., & Trucco, P. (2001). An algorithm for the implementation of safety improvement programs. *Safety Science*, *37*(1), 59–75. doi:10.1016/S0925-7535(00)00050-3

Chen, G., & Zhang, X. (2009). Fuzzy-based methodology for performance assessment of emergency planning and its application. *Journal of Loss Prevention in the Process Industries*, *22*(2), 125–132. doi:10.1016/j.jlp.2008.10.003

Cilingir, C., & Mackhieh, A. (1998). Effects of performance shaping factors on human error. *International Journal of Industrial Ergonomics*, *22*, 285–292. doi:10.1016/S0169-8141(97)00079-6

Fine, W. T., & Kinney, W. D. (1971). Mathematical evaluation for controlling hazards. *Journal of Safety Research*, *3*(4), 157–166.

Guikema, S. D., & Paté-Cornell, M. E. N. (2002). Component choice for managing risk in engineered systems with generalized risk/cost functions. *Reliability Engineering & System Safety*, *78*(3), 227–238. doi:10.1016/S0951-8320(02)00155-2

Jo, Y. D., & Park, K. S. (2003). Dynamic management of human error to reduce total risk. *Journal of Loss Prevention in the Process Industries*, *16*(4), 313–321. doi:10.1016/S0950-4230(03)00019-6

Khan, F. I., Amyotte, P. R., & DiMattia, D. G. (2006). HEPI: a new tool for human error probability calculation for offshore operation. *Safety Science*, *44*(4), 313–334. doi:10.1016/j.ssci.2005.10.008

Kirchsteiger, C. (2005). A new approach to quantitative assessment of reliability of passive systems. *Safety Science*, *43*(10), 771–777. doi:10.1016/j.ssci.2005.08.016

Kosmowski, K. T. (2006). Functional safety concept for hazardous systems and new challenges. *Journal of Loss Prevention in the Process Industries*, *19*, 298–305. doi:10.1016/j.jlp.2005.06.003

Kouabenan, D. R. (2009). Role of beliefs in accident and risk analysis and prevention. *Safety Science*, *47*(6), 767–776. doi:10.1016/j.ssci.2008.01.010

Labeau, P. E., Smidts, C., & Swaminathan, S. (2000). Dynamic reliability: towards an integrated platform for probabilistic risk assessment. *Reliability Engineering &. System Safety*, *68*(3), 219–254. doi:10.1016/S0951-8320(00)00017-X

Landau, K., Rohmert, W., & Brauchler, R. (1998). Task analysis. Part I - Guidelines for the practitioner. *International Journal of Industrial Ergonomics*, *22*(1-2), 3–11. doi:10.1016/S0169-8141(97)00064-4

Lind, N. (2002). Social and economic criteria of acceptable risk. *Reliability Engineering &. System Safety*, *78*(1), 21–25. doi:10.1016/S0951-8320(02)00051-0

Maes, M. A., & Faber, M. H. (2006). Bayesian framework for managing preferences in decision-making. *Reliability Engineering & System Safety*, *91*(5), 556–569. doi:10.1016/j.ress.2005.04.003

Majdara, A., & Nematollahi, M. R. (2008). Development and application of a risk assessment tool. *Reliability Engineering &. System Safety*, *93*(8), 1130–1137. doi:10.1016/j.ress.2007.09.007

Marhavilas, P. K., Koulouriotis, D. E., & Voulgaridou, K. (2009). Development of a quantitative risk assessment technique and application on an industry's worksite using real accidents' data. *Scientific Journal of Hellenic Association of Mechanical & Electrical Engineers*, *416*, 14–20.

Melchers, R. E. (2001). On the ALARP approach to risk management. *Reliability Engineering & System Safety*, *71*(2), 201–208. doi:10.1016/S0951-8320(00)00096-X

Rao, K. D., Gopika, V., Sanyasi Rao, V. V. S., Kushwaha, H. S., Verma, A. K., & Srividya, A. (2009). Dynamic fault tree analysis using Monte Carlo simulation in probabilistic safety assessment. *Reliability Engineering &. System Safety*, *94*(4), 872–883. doi:10.1016/j.ress.2008.09.007

Røed, W., Mosleh, A., Vinnem, J. E., & Aven, T. (2009). On the use of the hybrid causal logic method in offshore risk analysis. *Reliability Engineering &. System Safety*, *94*(2), 445–455. doi:10.1016/j.ress.2008.04.003

Rouvroye, J. L., & van den Bliek, E. G. (2002). Comparing safety analysis techniques. *Reliability Engineering & System Safety*, *75*(3), 289–294. doi:10.1016/S0951-8320(01)00116-8

van Duijne, F. H., Aken, D., & Schouten, E. G. (2008). Considerations in developing complete and quantified methods for risk assessment. *Safety Science*, *46*(2), 245–254. doi:10.1016/j.ssci.2007.05.003

Vaurio, J. K. (2001). Modelling and quantification of dependent repeatable human errors in system analysis and risk assessment. *Reliability Engineering & System Safety*, *71*(2), 179–188. doi:10.1016/S0951-8320(00)00098-3

KEY TERMS AND DEFINITIONS

ALARP (As Low as Reasonably Practicable): The principle that states that the risk tolerability depends on the level of risk (when lower than the limit of tolerability) compared to practicability or costs to gained improvement.

Consequence: An impact or the outcome that may result from a hazard, that may be expressed quantitatively as the amount of money, as a category of high medium or low level, or in descriptive manner.

Hazard: Any potentially damaging physical event, human activity or phenomenon characterized by its probability according to timing, location and intensity that may cause particular danger to occur within a given period of time. The danger can be in a way of loss of lives, injuries, environmental degradation or property damage.

Probability: It can be of statistical or subjective interpretation. Nevertheless it represents a measure of the degree of certainty that a future event may occur, with value of zero as impossibility and 1.0 as certainty.

Risk: Shows the interdependence between the probability of a hazard and the consequences to life, health, property or the environment.

Risk Analysis: A process or a methodology of objective determination of risk or the likelihood of an event by combining probabilities and the consequences or hazards and vulnerability to help in making risk management decisions. It can be qualitative (that uses word form or rating scales to describe the consequence magnitude or likelihood of the consequence to occur), or quantitative (based on numerical values and resulted in numerical evaluation of risk).

Risk Assessment: Represents the extent of risk analysis and evaluation process to estimate risk posed by hazard. It includes making the decision recommendation on whether probable risks are tolerable and risk control measures adequate.

Risk Management: The systematic approach in identifying, analyzing, assessing, mitigation and monitoring of the risk.

Vulnerability: Shows the degree of fragility of a natural or social community towards a hazard, based on resulted risk and the potential to react or withstand it (based on adaptability or coping capacity).

Chapter 2
Risk Due to Wellbore Instability

Nediljka Gaurina-Medjimurec
University of Zagreb, Croatia

Borivoje Pasic
University of Zagreb, Croatia

ABSTRACT

Exploration and production as one of the most important parts of the petroleum industry encounters different problems, usually resulting in nonproductive time and additional expenses. The most common and most expensive of them are related to wellbore instability and associated problems. Wellbore instability problems are usually related to drilling operation, but they can also appear during completion, workover, or the production stage of a certain well. The traditional solution for wellbore instability problems is composed from the early recognition of specific wellbore instability problems, the main cause identification and swift response. For more effective solution it is necessary to incorporate wellbore stability and risk assessment in the early phase of well design. This chapter gives one general overview of wellbore instability problems and their causes as well as an overview of actual approaches and methods in wellbore stability and risk assessment.

INTRODUCTION

The petroleum industry as well as other industries, encounters different problems because of the specific nature of the job itself and always presents a possible human fault. Unlike other industries, the petroleum industry employees, especially engineers in the exploration and production department (petroleum engineers, geologists, geomechanical engineers, geophysics etc.) work in very specific conditions and environment, and with limited data availability. The main objectives of this part of the petroleum industry are discovering hydrocarbons reservoirs and creating facilities for their economical, safe and ecological acceptable production. One of the most important parts in the exploration and production activities is

DOI: 10.4018/978-1-4666-4777-0.ch002

drilling a borehole, a subsurface object that connects a reservoir with the surface facility and provides a pathway for hydrocarbon production. In general, drilling is a specific technical and technological process and because of that, it requires specific engineer's skills and the usage of specific equipment and instruments. On the other hand, drilling is very expensive. Aldred et al. in their work published in 1999 estimate that oil and gas companies spend about $20x10^9$ dollars annually on drilling worldwide. Today, drilling implies drilling in harsh surface environment (artic areas, deep seas, etc.) as well as drilling though complex lithology (tectonically active areas, fracture and faulted zones, areas with salt or magma intrusions etc.). Additionally, drilling an extended reach or slim hole wells especially in such environment make drilling process more complex and expensive.

From the beginning of the drilling up to nowadays, oil and gas companies have probably continuously encountered a certain wellbore instability problem or problems. Wellbore instability can simply be explained as any change in wellbore diameter in comparison to the bit diameter used for the hole drilling. The appearance of the wellbore instability in a certain well can result in different problems like hole cleaning problems (cuttings volume increase, caving appearance, hole pack-off etc.), in the inability to run a casing string into the hole, or the inability of performing the cementing or logging operation and etc. In some cases, wellbore instability problems can cause the abandonment of one portion of the hole or in the extreme case of a whole wellbore. The most important consequence of wellbore instability problems are Non Productive Time (NPT) and additional (unscheduled) expenses. The petroleum industry spends more than $1x10^9$ of dollars annually for solving wellbore instability problems (Tare et al., 2002; Zhang et al., 2004; Khodja et al., 2010) what is about one tenth of the entire drilling expenses today. From the comparison of this data with the available date from the last decade of twentieth century it can be concluded that unscheduled expenses are continually increasing and that they have increased for about 100% in the last 20 years. Additionally, based on ten year researching and data collecting, Muniz et al. (2005) concluded that 30% of all additional expenses during drilling have been spent for solving wellbore instability problems.

In the past, the wellbore instability problems were solved most frequently on trial/failure bases (van Oort et al., 1996; Tare et al., 2002). This method can provide a possible solution for the wellbore instability problems on the certain well, but it doesn't provide the ultimate solution for a particular wellbore instability problem or wellbore instability problems as a whole. There isn't a unique solution for all wellbore instability problems because each well represents an individual case and demands an individual approach. The well individuality is the result of difference in lithology (sandstone, shale, limestone etc.), rock properties (compressive/tensile strength, in-situ stress conditions, fractures, anisotropy etc.), local tectonics (faulted zones, local magmatic and salt intrusion etc.) and wellbore properties (depth, trajectory, azimuth, inclination etc.). Because of this, wells on the same field sometimes experience quite different wellbore instability problems.

As mentioned earlier, wellbore instability problems can occur during drilling or during a different operation on a certain well (e.g. completion, workover, production, stimulation etc.). Since the wellbore instability problems become more complicated with elapsed time (time between problem spotting and the solution), the key factor for efficient solving of a distinct wellbore instability problem is in time detection. Possible indicators (symptoms) of wellbore instability which are primarily caused by wellbore collapse or convergence are shown in Table 1. McLellan (1994) has classified the indicators of wellbore instability into two groups: direct and indirect indicators respectively. A direct indicator includes observations such as an oversized or undergauged hole (readily observed from caliper logs), caving at the surface and etc., as the unmistakable evidence of existing wellbore instability problems. The caving from the wellbore wall, circulated to the surface, and the hole fill after tripping confirm that spalling processes are occurring in the wellbore. Large volumes of cuttings and/or cavings, in an excess of the volume of rock which would have been excavated in a gauge hole, similarly attest to hole enlargement. If the fracture gradient was not exceeded and vuggy or naturally fractured formations were not encountered, a requirement for a cement volume in excess of the calculated drilled hole volume is also a direct indication that the enlargement has occurred. On the other hand indicators such excessive drillstring vibration, high torque and drag friction, deviation control problems etc. may be a result of wellbore instability problems, but also may be a result of improper drilling regime, equipment failure or a technical mistake. For example, drillstring failure may be a result of wellbore instability problem, fatigue of materials or unsatisfactory drillstring design.

THE CAUSES OF WELLBORE INSTABILITY

Solving of certain wellbore instability problems requires two basic elements, recognition and understanding of given wellbore instability mechanisms as well as defining their major causes. According to Darly (1969), the majority of main wellbore instability causes were defined in 1938. by Halbouty & Kaldenbach. Today, the classification of the wellbore instability causes can be done in two ways. The first classification of the wellbore instability causes is based on the possibility to control them. This classification given by McLellan classifies them into two groups of

Table 1. Indicators of wellbore instability (McLellan, 1996)

Direct indicators	Indirect indicators
Oversize hole	High torque and drag friction
Undergauge hole	Hanging up of drillstring, casing, or coiled tubing
Excessive volume of cuttings	Increased circulating pressures
Excessive volume of cavings	Stuck pipe
Cavings at surface	Excessive drillstring vibrations
Hole fill after tripping	Drillstring failure
Excess cement volume required	Deviation control problems
	Inability to run logs
	Poor logging response
	Annular gas leakage due to poor cement job
	Keyhole seating
	Excessive doglegs

factors: uncontrollable and controllable factors respectively (Table 2). Uncontrollable factors are mostly related to lithology, local tectonic activities (faulting and folding) and rock properties. Unlike uncontrollable factors the group of controllable factors is mostly related to drilling process itself and may be successfully controlled through proper drilling process design and good engineer practice. Similar classification can be found in the work of Mohiuddina et al. published in 2005.

Except for the above-mentioned classification of wellbore instability causes given by McLellan, there is also the second classification which is broadly accepted by different authors. Based on this classification, the causes of wellbore instability were divided into mechanical and physico-chemical causes respectively (Chenevert & Pernot, 1998; Osuji et al., 2008). In spite of this very clear classification, it is fairly difficult to identify the exact cause of wellbore instability. The majority of wellbore instabilities are the result of the joint acting of the either groups. This is because one group may cause changing of the rock properties or in-situ conditions around wellbore, generating, in this way, a prerequisite for the other wellbore instability mechanisms which are caused by totally different groups of causes. For example, a shale/drilling fluid interaction (physico-chemical cause) may cause clay minerals hydration and pore pressure increasing in a near wellbore region. Moreover, these processes and related process such as hydration stress, rock softening and e.g. may cause a change of the rock properties which are important for mechanical integrity and stability of rock. In the next sections, physico-chemical and mechanical causes and related mechanisms of wellbore instability will be briefly described (Zhang et al., 2004; Rojas et al., 2006; Ewy & Morton, 2009).

The Mechanical Causes of Wellbore Instability

Throughout history, in the geological sense Earth was being changed as consequences of different geological processes. These processes such sedimentation, volcano activities, different tectonic activities (earthquake, folding, faulting) etc., had an influence on the present underground lithology and rock properties. Before the borehole drilling, rock at the certain depth is in an equilibrium state as a result of above mentioned processes. This equilibrium in-situ stress state may be presented with three main stresses: 1) maximum main stress (σ_v), 2) maximum horizontal stress (σ_H) and 3) minimum horizontal stress (σ_h).

The mentioned classification of in-situ stresses implied normal conditions, where

Table 2. The causes of wellbore instability (McLellan, 1994)

Uncontrollable (Natural) Factors	Controllable Factors
Naturally Fractured or Faulted Formations	Bottom Hole Pressure (Mud Density)
Tectonically Stressed Formations	Well Inclination and Azimuth
High In-situ Stresses	Transient Pore Pressures
Mobile Formations	Physico-chemical rock-Fluid Interaction
Unconsolidated Formations	Drill String Vibrations
Naturally Over-Pressured Shale Collapse	Erosion
Induced Over-Pressured Shale Collapse	Temperature

overburden stress is the maximum main stress at the same time. Likewise, there are areas with specific conditions (tectonically active area) where horizontal stress exceeds overburden stress and becomes main in-situ stress (Lake, 2006). During borehole drilling, certain rock volume is substituted with equal mud volume of a certain density (less then rock) causing the development of a new stress regime (McLean & Addis, 1990). This new stress regime developed on the wellbore wall is consisted of three each other perpendicular stresses: 1) axial stress (σ_a), 2) radial stress (σ_r) and 3) tangential stress (σ_t) (Figure 1).

From a rock mechanics´ point of view, wellbore stability depends on the relation between rock in-situ stresses and stresses developed on the wellbore wall as well as the mechanical properties of rock (rock strength, rock properties anisotropy etc.). Depending on these relations, one of the two main wellbore instability failure mechanisms can be

Figure 1. Stress redistribution around the wellbore wall during drilling

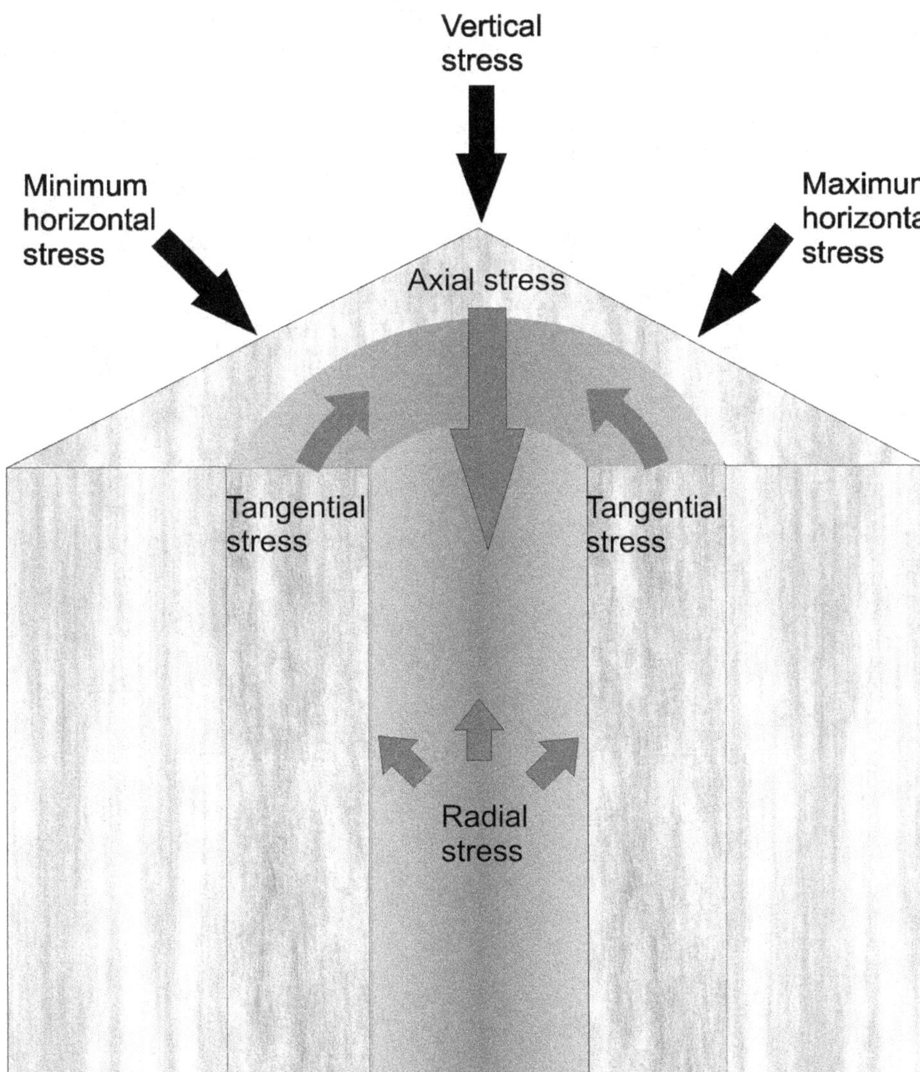

developed in a certain well: shear (collapse) or tensile failure. The appearance of the certain rock failure also depends on the rock drilling fluid interaction and its influence on the original (in-situ) mechanical and petrophysical rock properties (Lal, 1999). According to Zhang et al. (2006a), stresses magnitude and distribution on the wellbore wall are being depended on mechanical, chemical, thermal and hydraulic effect. It is important to emphasize the possibility of the concurrently appearances of different wellbore failure mechanisms in the same borehole, as result of different lithology. Mechanical wellbore stability also extremely depends on well characteristics, on well trajectory in particular, then azimuth and inclination (Labenski et al., 2003). The relation between in-situ stresses, well inclination and azimuth displayed in the coordinate system was shown in Figure 2.

The placement of the wellbore with certain azimuth and inclination and existing in-situ stresses in the coordinate system is important for the calculation of stresses developed on the wellbore wall. At a vertical wellbore, there

Figure 2. Relation between in-situ stresses and coordinate systems for vertical and deviated well

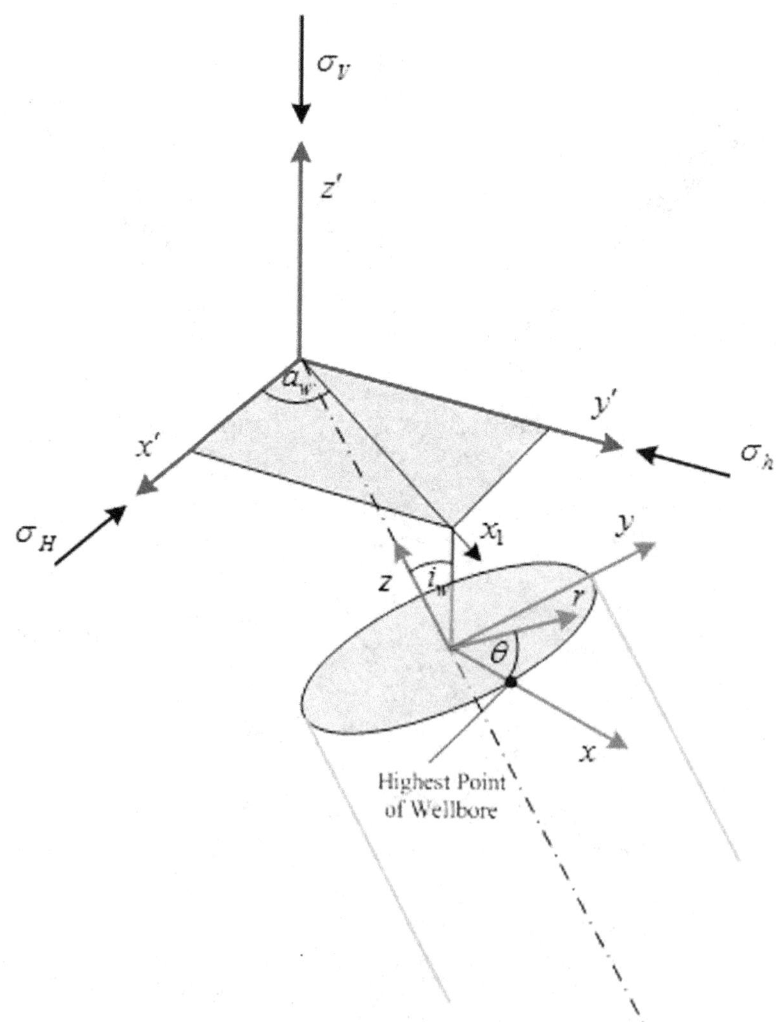

is no difference between the coordinate system (x´, y´, z´) and the coordinate system (x, y, z) which make the calculation of stresses on the wellbore wall simplest. Unlike the vertical well, in the case of deviated or horizontal well drilling calculation of stresses developed on wellbore wall are more perplexed as a result of difference between the two mentioned coordinate systems. Local stresses induced by in-situ stress and hydraulic effects at the wellbore wall ($r = r_w$), for the vertical well can be described as follows (Pasic et al., 2007):

$$\sigma_r = p_w \quad (1)$$

$$\sigma_t = (\sigma_H + \sigma_h) - 2\cdot(\sigma_H - \sigma_h)\cdot\cos2\theta - p_w \quad (2)$$

$$\sigma_a = \sigma_v - 2\cdot(\sigma_H - \sigma_h)\cdot\nu\cdot\cos2\theta \quad (3)$$

where:

σ_r - Radial normal stress at wellbore [Pa],
p_w - Wellbore pressure [Pa],
σ_t - Hoop stress at wellbore [Pa],
σ_H - Maximum in-situ horizontal stress [Pa],
σ_h - Minimum in-situ horizontal stress [Pa],
θ - Point location angle [°],
σ_a - Axial stress at wellbore [Pa],
σ_v - Vertical (overburden) stress at wellbore [Pa],
ν - POISSON ration [-].

According to the previous equations it can be concluded that the radial stress (σ_r) depends on the wellbore pressure or mud weight. The tangential (σ_t) and axial (σ_a) stress depends on σ_H, σ_h, p_w, ν and θ.

With the increasing of distance from the wellbore wall radial, axial and tangential stress are changed. Stress calculation for a certain point on large distance from wellbore wall ($r \rightarrow \infty$) can be calculated with the fallowing equations (Pasic et al., 2007):

$$\sigma_r = ½\cdot(\sigma_H + \sigma_h) + ½\cdot(\sigma_H - \sigma_h)\cdot\cos2\theta \quad (4)$$

$$\sigma_t = ½\cdot(\sigma_H + \sigma_h) - ½\cdot(\sigma_H - \sigma_h)\cdot\cos2\theta \quad (5)$$

$$\sigma_a = \sigma_v \quad (6)$$

Unlike the equations for stresses calculation on the wellbore wall, for the point on a certain distance from the wellbore wall radial and tangential stresses depends on minimum and maximum in-situ stresses and a point location angle.

According to the above mentioned equations it can be concluded that radial, axial and tangential stress tend to equalize with in-situ stresses with increasing distance from wellbore wall. The total equalization between radial, axial and tangential stresses on one side and maximum and minimum horizontal and vertical stresses on the other side, the examining point on certain distance from the wellbore in the maximum horizontal stress direction ($\theta = 0$) can be observed. In this case, radial stress well is equal along with maximum horizontal stress, tangential with minimum horizontal stress and axial along with vertical stress respectively.

Unlike the vertical stress, in the case of deviated and horizontal well drilling, it is much heavier to calculate stresses at the wellbore wall. Local stresses induced by in-situ stress and hydraulic effects at the wellbore wall ($r = r_w$), can be calculated with the following equations (Awal et al., 2001, Simangunsong et al., 2006):

$$\sigma_r = p_w \quad (7)$$

$$\sigma_t = (\sigma_x + \sigma_y) - 2\cdot(\sigma_x - \sigma_y)\cos2\theta - 4\cdot\tau_{xy}\cdot\sin2\theta - p_w \quad (8)$$

$$\sigma_a = \sigma_z - \nu\,[2\cdot(\sigma_x - \sigma_y)\cdot\cos2\theta + 4\cdot\tau_{xy}\cdot\sin2\theta] \quad (9)$$

$$\tau_{\theta z} = 2\cdot(\tau_{yx}\cdot\cos\theta - \tau_{xz}\cdot\sin\theta) \quad (10)$$

$$\tau_{r\theta} = \tau_{rz} = 0 \qquad (11)$$

where:

σ_x - normal (in-situ) stress in x-direction [Pa],
σ_y - normal (in-situ) stress in y-direction [Pa],
τ_{xy} - in-situ shear stress in (x,y,z) coordinated system [Pa],
σ_z - normal (in-situ) stress in z-direction [Pa],
$\tau_{\theta z}$ - shear stress at wellbore [Pa],
τ_{yx} - in-situ shear stress in (x,y,z) coordinated system [Pa],
τ_{xz} - in-situ shear stress in (x,y,z) coordinated system [Pa],
$\tau_{r\theta}$ - shear stress at wellbore [Pa],
τ_{rz} - shear stress at wellbore [Pa].

Besides the mentioned influence of in-situ stress and hydraulic effect, stresses at the wellbore wall also strongly depend on temperature difference between rock and drilling fluid as well as their physico-chemical interaction. The stresses alterations at the wellbore wall ($r = r_b$) due to thermal and physic-chemical effect can be expressed by the following equations (Chen et al., 2001; Zhang et al., 2006a):

$$\sigma_r = 0 \qquad (12)$$

$$\sigma_t = ((\alpha_p \cdot (1-2 \cdot \nu))/(1-\nu)) \cdot (p_w - p_p) + ((E \cdot \alpha_p)/(3 \cdot (1-\nu))) \cdot (T_w - T_f) \qquad (13)$$

$$\sigma_t = ((\alpha_p \cdot (1-2 \cdot \nu))/(1-\nu)) \cdot (p_w - p_p) + ((E \cdot \alpha_p)/(3 \cdot (1-\nu))) \cdot (T_w - T_f) \qquad (14)$$

where:

α_p - Biot´s constant [-],
p_p - initial pore pressure [Pa],
E - Young´s modulus [Pa],
α_t - volumetric-thermal-expansion-coefficient [1/°C],
T_w - wellbore wall temperature [°C],
T_f - initial formation temperature [°C].

During drilling, drilling mud is circulated through a borehole while cooling and heating different parts of the borehole at the same time. This means that wellbore has a thermal neutral point, below which drilling mud cools the formations and above which the same fluid heats formations. A result is a more stable hole below the thermal neutral point and less stable above. Cooling or heating of the certain rock results in the change in collapse and fracture pressure whether drilling a vertical, deviated or horizontal well (Chen et al., 2001). Although different laboratory investigations and field observations were conducted, numerous discussions upon the influence the thermal effects have on wellbore stability still exist. According to Chen et al. (2001), when determining minimum mud weight required, the influence of the thermal effect in the case of horizontal well drilling is smaller as compared to the effect on a vertical well. In order to investigate the effect of cooling and heating of rock during drilling on its stability, Zhang et al (2006a) conducted a comprehensive laboratory study. They demonstrated that both the lower and upper critical mud weight decrease with cooling and increase with mud heating. In other words mud cooling is beneficial to preventing compressive failure and mud heating is beneficial to avoiding lost circulation. They also found that the magnitude of thermal effects on wellbore stability depends on volumetric-thermal-expansion coefficient. Both the lower and upper critical mud weights decrease with increasing matrix volumetric-thermal-expansion coefficient. Since the bottom hole mud temperature for this particular case is less than the formation temperature, a lower thermal compressive stress is generated when the matrix volumetric thermal expansion coefficient is high.

Besides the hydraulic and thermal effect, wellbore stability of vertical, deviated and horizontal wells strongly depends on regional tectonic activity and prevailing in situ stress (normal, overthrust/reverse or strike-slip). According to Awal et al. (2001), during drilling in normal stressed formation ($\sigma_v > \sigma_H > \sigma_h$) a vertical well is more stable than a horizontal or deviated one, until, while drilling through formation under a strike slip stress regime ($\sigma_H > \sigma_v > \sigma_h$) the vertical well is less stable than the deviated and horizontal ones. There is also recommendation that the wellbore should not be drilled parallel to maximum horizontal stress (σ_H) in order to minimize wellbore instability problems (Zhang et al., 2006a).

Because shale formations represent the majority of all drilled formation, a physico-chemical interaction between shale and drilling mud also plays an important role in the assurance of wellbore mechanical stability. Moreover, it is well-known that controlling of water and cations movement into or out the shale formation during drilling is one of the key factors in controlling wellbore stability. A water flux into shale formation causes stress redistribution around wellbore, shale strengthens and Young´s modulus decreases, and pore pressure increases. These changes may have detrimental effect on wellbore stability. (Zhang et al., 2004). Unlike hydration, shale dehydration to a certain level can be beneficial to wellbore stability. Shale over-dehydration results in fractures in the near wellbore formation and may also disturb wellbore stability, especially during drilling through natural fractured shale (See Table 4). Water desorption in fractured shales may widen the fractures and destabilize the wellbore. Some authors point out the beneficial effect of small degree water adsorption on wellbore stability during drilling through naturally fractured shale. In a fractured stressed shale formation it is particularly important to control water movement because the in-situ stresses are in a critical state (Rojas et al. 2006).

There are two main wellbore failure types, shear failure and tensile failure type respectively (Table 3). The development of a specific failure type depends upon the magnitude of induced stresses developed on a wellbore wall as well as rock compressive and tensile strength.

There are several different peak-strength criteria that are commonly used to predict mechanical wellbore instability. According to McLean & Addis (1990) these criteria can be divided in four groups (Table 4), based on principal stresses used for wellbore stability criterion definition and assumed rock behavior (linear or non-linear).

The peek-strength criterion can be determined directly from laboratory testing if core is available or from a back analysis of hole conditions recorded from calipers and drilling logs. More sophisticated numerical analyses which include non-linear anisotropic materiel behavior may be conducted only when the data from core laboratory testing are available.

The Physico-Chemical Causes of Wellbore Instability

Drilling and stress redistribution always results at the same time as interaction between used mud and rock on the wellbore wall and in near wellbore zone. Most frequently, this undesirable interaction is the result of inappropriate drilling fluid design and mineralogical composition and petro physical properties of rock (porosity, permeability, pore throat size etc.). In most cases, wellbore instability problems caused by physico-chemical interaction between rock and drilling fluid are associated with shale.

Shale is fine-grained, clastic sedimentary rock with a certain portion of different clay minerals (smectites, kaolinite, illite, chlorite and mixed layer clay minerals). Shale represents 75% of all drilled formation worldwide

Table 3. Shear and tensile failure types (Pasic et al., 2007)

Failure type	Geometry and Orientation
Shear failure types	
Shear Failure Shallow Knockout $\sigma_a > \sigma_t > \sigma_r$	The failure will occur in the radial/axial plane because the maximum (σ_a) and minimum (σ_r) stresses are oriented in this plane (a vertical plane).
Shear Failure Wide Breakout $\sigma_t > \sigma_a > \sigma_r$	The failure will occur in the radial/tangential plane because the maximum (σ_t) and minimum (σ_r) stresses are oriented in this plane (the horizontal plane).
Shear Failure High-Angle Echelon $\sigma_a > \sigma_r > \sigma_t$	The failure will occur in the axial/tangential arc because the maximum (σ_a) and minimum (σ_t) stresses are oriented in this arc (the arc of the borehole wall).
Shear Failure Narrow Breakout $\sigma_r > \sigma_a > \sigma_t$	The failure will occur in the radial/tangential plane because the maximum (σ_r) and minimum (σ_t) stresses are oriented in this plane (the horizontal plane).
Shear Failure Deep Knockout $\sigma_r > \sigma_t > \sigma_a$	The failure will occur in the radial/axial plane because the maximum (σ_r) and minimum (σ_a) stresses are oriented in this plane (a vertical plane).
Shear Failure Low-Angle Echelon $\sigma_t > \sigma_r > \sigma_a$	The failure will occur in the axial/tangential arc because the maximum (σ_t) and minimum (σ_a) stresses are oriented in this arc (the arc of the borehole wall).
Tensile failure types	
Tensile Failure Cylindrical $\sigma_r \leq -To$	This failure is concentric with the borehole. A low mud weight would favor the failure due to the magnitude of σ_r being lower.
Tensile Failure Horizontal $\sigma_a \leq -To$	This failure creates horizontal fractures.
Tensile Failure Vertical $\sigma_t \leq -To$	This failure creates a vertical fracture parallel with the maximum horizontal stress direction. This is because is this orientation the tangential stress has to overcome the smallest formation tensile strength.

Table 4. The categorization of peak-strength criterion (McLean & Addis, 1990)

Criterion	Principal stresses used for defining of wellbore stability criterion	
	Maximum (σ_1), intermediate (σ_2) and minimum (σ_3) principal stresses	Maximum (σ_1) and minimum (σ_3) principal stresses
Linear Criterion	**Category A**	**Category B**
	e.g. Drucker-Prager	e.g. Mohr-Coulomb
Non-Linaer Criterion	**Category C**	**Category D**
	e.g. Pariseau	e.g. Hoek-Brown

and causes about 90% of all wellbore instability problems (Tan et al., 2002). Besides mineralogical composition, shale behavior during interaction with water base mud strongly depends on shale permeability. Shale has very low permeability comparing to sandstone, ranging from a few to few tens nanoDarcies (Labenski et al., 2003). Because of the shale low permeability and low pore throat size (the range from 0.01 to 0.001 μm) mud filter cake is not generated at the wellbore wall. So there are no physical barriers between mud and shale.

Shales are primary composed from a certain portion of quartz (SiO_2) and different types of clay minerals, but there is also a small content of calcite, dolomite, pyrite and other. From wellbore stability point of view the most important is clay mineral content, especially content of water sensitive clay minerals. Depending on the total clay content and the content of a specific type of clay mineral (montmorilonite, illite, kaolinite, chlorite and mixed layer minerals) shale formations demonstrate different behavior during contact with water from drilling muds (dispersion, swelling, caving tendencies etc.). Based on these characteristics, shales are divided into five classes, ranging from soft (gumbo) shale to very hard shale (O´Brein & Chenevert, 1973) (Table 5).

Crystalline and Osmotic Hydration (Swelling) of the Clay Minerals

All above mentioned suggest that mud and shale can develop one or more different interaction mechanisms. These mechanisms can be developed simultaneously, sometimes acting in opposite direction and causing water and ions movement in/out from shale. This movement of water molecules in or out from shale results in hydration or dehydration process.

There are two swelling mechanism of clay minerals. The first mechanism, also known as crystalline swelling, is related to surface hydration (water molecules adsorption) and can occur in all types of clay minerals. This type of clay minerals hydration causes slight swelling as a result of small increase in clay minerals interlayer space (in the range from 9 to 20 Å). For wellbore stability assurance the most important process is osmotic swelling or water absorption into clay mineral structure. Osmotic swelling is addressed only to specific clay minerals and causes a tremendous increase of clay interlayer space in the range of 20 to 130 Å. The Intensity of osmotic swelling strongly depends on an exchangeable cation appearance in clay minerals and shale membrane efficiency (shale acts as an imperfect semi-permeable membrane). In the 1983, Fritz and Marine first researched membrane reflection coefficient, and osmotic potential of compacted clay minerals (Ewy & Morton, 2009). Today, laboratory equipment and methods are developed based on that knowledge.

Besides the shale volume increase, clay minerals hydration and swelling also cause a change in shale pore pressure and shale mechanical properties (e.g. rock strength). These changes, which are directly the outcome of mud/shale interaction, affect the mechanical stability of the rock. For these reasons, wellbore stability, especially in a water sensitive

Table 5. Shale clasifications (O´Brein & Chenevert, 1973)

Class	Shale	Clay content	Characteristics
1	Soft (Gumbo)	High in montmorillonite, some illite	High dispersion
2	Soft	Fairly high in montmorillonite, High in illite	fairly high dispersion
3	Medium-hard	High in interlayered clays, High in illite, chlorit	Moderate dispersion, sloughing tendencies
4	Hard	Moderateillite, Moderate chlorite	little dispersion, sloughing tendencies
5	Very hard	High in illite, moderate chlorite	brittle, no significant dispersion, caving tendencies

Figure 3. Wellbore instability during drilling through shale formation (Adapted from Chenevert & Osisanya, 1989)

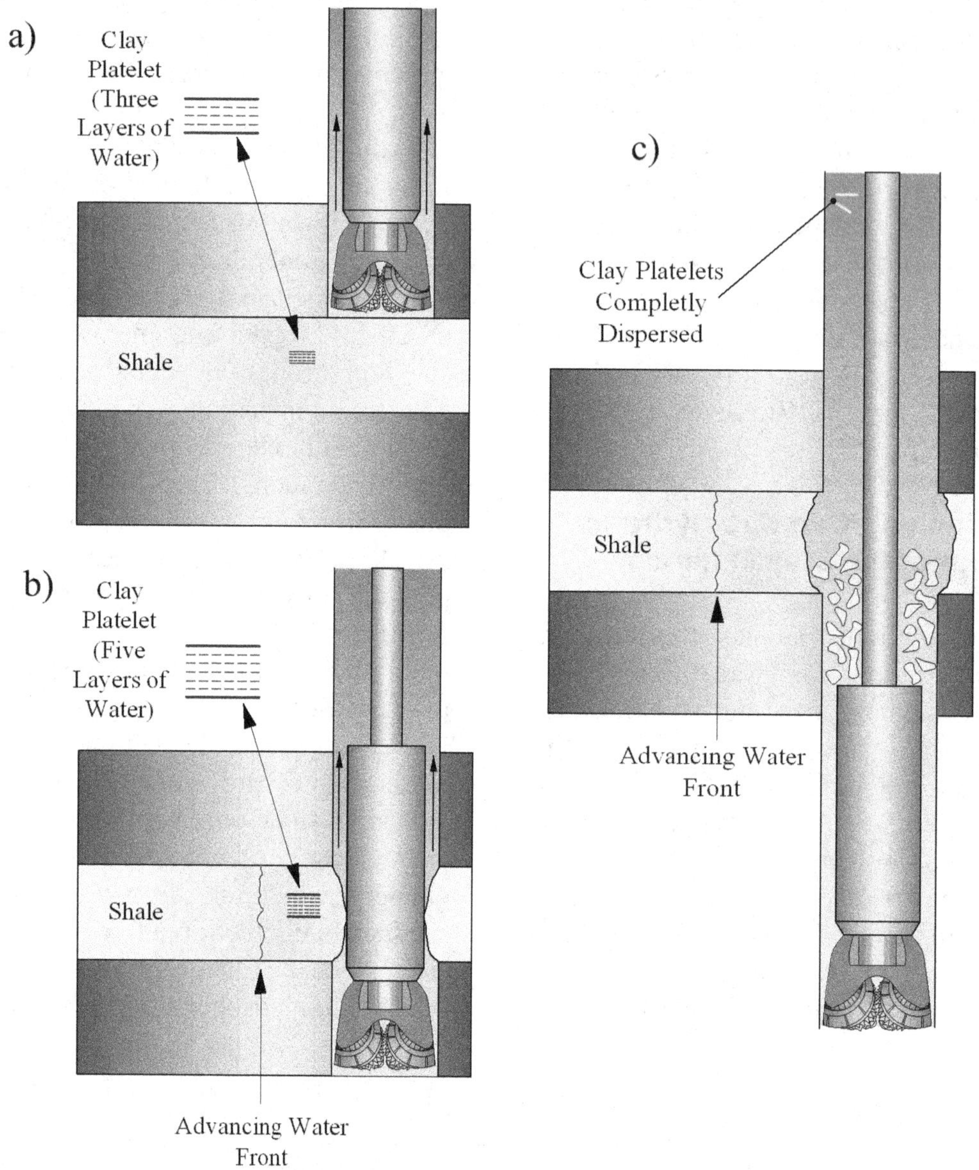

formation like shale, generally results in the concurrent action of mechanical and physico-chemical cause.

There are four main mechanisms related to water and ion movement during shale/drilling fluid interaction: chemical osmosis, hydraulic flow, capillary pressure and diffusion flow.

Wellbore instability during drilling through shale is shown in Figure 3. Before drilling, clay platelets are only partially hydrated within shale (Figure 3a). During drilling, shale draws water from drilling fluids (Figure 3b) and causes its swelling and dispersion (Figure 3c).

Chemical Osmosis

By definition chemical osmosis can be described as water molecules movement (passage) between two media with different solutes concentration (water activities) through a semi-permeable membrane. This process presents the main water driving force during shale/drilling fluid interaction. In shale/drilling fluid system shale is acting as a semi-permeable membrane and the whole process is determined by the water activity difference between drilling mud and shale pore fluid at in-situ conditions. The flow direction is always from high water activity (low salt concentration) to low water activity (high salt concentration) (Abass et al., 2006). This phenomenon can occur in water-based drilling or emulsion drilling fluids as the latter includes a water phase. Water flow from shale formation into drilling fluid (shale dehydration) can be induced only by increasing of salt content in drilling fluid (drilling fluid activity decreasing) with respect to shale water activity.

All solutes of a solution to which a membrane is exposed will be fully or partially "reflected" by the membrane. An ideal semi-permeable membrane (i.e., one that allows passage of the solvent only) has a reflection coefficient of 100% (or 1). Nonideal membranes, which allow a partial passage of solute, have reflection coefficients of less than 1 and are, therefore, referred to as "leaky." (Schlemmer et al., 2003). According to Al-Bazali et al. (2006) shale in contact with water based muds develop a low membrane with a very low reflection coefficient in the range from 0.18% to 4.23%. Unlike the results of researching which was conducted by Al-Bazali et al. (2006), Tan et al. (2002) stated that new generation water based muds are capable of generating a membrane with efficiencies in the range from 55% to 85%. Shale membrane reflection coefficient depends on drilling mud properties, shale properties (e.g. pore throat radius, cation exchange capacity) and other factors like shale compaction and diagnosis degree (van Oort et al., 1995). Schlemmer et al. (2003) recognized and described three distinct types of shale membranes. *Type 1* membranes are generally characterized by coupled flows of water and solutes between fluid and shale. *Type 2* membranes are generally characterized by greatly reduced near-wellbore shale permeability which result in water and solutes flow decreasing. Invert-emulsion fluids tend to form efficient *Type 3* membranes. Type 3 membranes transport water more selectively, but shale permeability and fluid chemistry may alter performance measurements. However, under certain conditions, these fluids can yield lower capillary pressures than described previously and invade the interstitial fabric of high permeability shales. Theoretical osmotic pressure and shale membrane reflection coefficient can be described by the following equations (van Oort et al., 1995; Ewy & Morton, 2009):

$$\Delta\pi = -((R\cdot T)/V)\cdot\ln(a_{sh}/a_{df}) \quad (15)$$

where:

$\Delta\pi$ – theoretical osmotic pressure [Pa]
V - partial molar volume of water [0.018 m^3/mol],
R - gas constant [8.314·J/(mol·°K)]
T - absolute temperature [°K]
a_{sh} - water activity of shale [-],
a_{df} - water activity of drilling fluid [-].

$$\alpha_m = \Delta p_{os}/\Delta\pi \quad (16)$$

where:

α_m - membrane reflection coefficient [-],
Δp_{os} - osmotic pressure difference accounting for membrane efficiency [Pa],

Hydraulic Flow

Hydraulic flow (convective, Darcian flow) is drilling mud flow due to difference between drilling mud pressure and shale pore pressure. During the overbalance drilling operation (drilling fluid pressure is higher than pore fluid pressure), the drilling fluid flow is usually from wellbore to shale formation. In comparison with water based fluids, oil based drilling fluids have an advantage in this type of flow because of the interfacial tension capillary pressure. When oil based mud is used, the radial stress on a wellbore wall is higher in comparison with water based muds because of the pressure dissipation of the later as it is miscible with the shale pore water (Abass et al., 2006). The hydraulic flow is controlled by Darcy's law and can be expressed with the following equation (Zhang et al., 2006b):

$$v_o = - (k/\mu)\cdot(p - \rho\cdot g) \quad (17)$$

where:

v_o - bulk flow velocity [m/s],
k - permeability [m^2],
μ - viscosity of drilling fluid [Pa·s],
p - pressure gradient between the mud hydraulic pressure and pore pressure [Pa/m],
ρ - drilling fluid density [kg/m^3],
g - gravitational acceleration [m/s^2].

Capillary Pressure

One of the possible solutions for solving wellbore instability problems in shale is the use of oil based muds. Because of a very small pore throat radius and surface tension between two immiscible fluids (oil based mud and pore water), capillary pressure is developed on wellbore walls. This capillary pressure prevents drilling fluid flow into shale pore space and at the same time it neutralizes drilling fluid flow due to overbalance pressure. Capillary pressure developed on wellbore wall can by expressed by the following equation (van Oort et al., 1996):

$$p_c = (2\cdot\sigma\cdot\cos\gamma)/r_p \quad (18)$$

where:

p_c – Capillary pressure [Pa],
σ – Interfacial tension [N/m],
γ – Contact angle [°],
r_p – pore throat radius [m].

Diffusion Flow

As mentioned above, shale acts as a non-ideal or leaky membrane with a very low membrane reflection coefficient, which means that it's leaky for certain ions presents in both, pore water and drilling fluid. In this situation there is an additional flow in the shale/drilling fluid interaction, called a diffusion flow. This flow is governed by the difference in ions concentration between the drilling fluid and the pore water. The direction of the diffusion flow is always from a system with higher to a system with lower ion concentration, which means the opposite direction of the osmotic flow. The movement of ions through shale can be restricted due to the narrow pore throats, cation exchange capacity, dimension of hydrate ions and negatively charged clay surfaces (Zhang et al., 2006b). The diffusion of ions can be expressed by using Fick's law (Equation 19).

$$J_i = -D_{si}\cdot((C_{i,sh} - C_{i,m})/\Delta x) \quad (19)$$

where:

J_i - mass flux of ith ion [$mol/(m^2\cdot s)$],
D_{si} - diffusion coefficient of the ith ion in the shale [m^2/s],

$C_{i,sh}$ - concentration of ith ion in pore fluid [mol/m^3],
$C_{i,m}$ - concentration of ith ion in mud [mol/m^3],
Δx - length of shale [m].

WELLBORE INSTABILITY RISK ASSESSMENT

In oil and gas exploration and production operation as well as in other different industries where uncertainties appear risk analysis can be a very useful tool (Cunha, 2004). The first article related to risk analysis and risk assessment in drilling operation was published in 1968 by Newendorp and Root. In this fundamental article, the authors considered the possibility of risk analysis in drilling investment (Cunha et al., 2005). After that, wellbore instability risk analysis and assessment were discussed by various authors. According to Cunha (2004) risk analysis has been widely implemented in all aspects of exploration and production activity, especially in decision making process related to specific drilling operations such as: optimum depth to set casing, directional drilling, wireline operation, special remedial operation, borehole stability, the prediction of pore pressure and fracture gradient etc. Besides, applying the risk model to new technology such for example casing while drilling can be very difficult because there is no experience base for quantifying either the probability of an event occurring or the economic consequences of the event. Risk assessment can also be a very useful tool when techniques and approaches different from well-known engineer practices are considered. For example, operators in Gulf of Mexico considered much more aggressive casing design in order to reach the target sand with the optimum borehole size and the minimum number of strings (Brehm et al., 2004). A similar application of quantitative risk assessment in casing design was conducted by Moos et al. and the results were published in 2003.

Risk is composed of two components: the probability of an event occurring and the economic consequences if it occurs. Personnel involved in different aspects of drilling process make decisions based on the risk assessment every day. Decisions making process in the past has been based more on personal experience and intuition than science. Today companies implement more explicit risk models that quantify both the risk and the potential for economic impact on drilling projects (Houtchens et al., 2007). There are a number of techniques which have been developed to predict optimal drilling operational parameters (mud weights or drilling trajectories). Traditionally, these techniques have been limited to deterministic analyses that are based on the assumption that in situ conditions and rock properties are precisely known (Moos et al., 2003, Tutuncu et al., 2005). Whereas, due to insufficient geomechanical data and problem related to the uncertainty (error) associated with measurement, data collecting and interpretation it was started with utilizing of probabilistic methods. In 1999 Ottesen et al. presented a new statistical approach based on Quantitative Risk Analysis (QRA). Quantitative risk analysis is a technique for establishing the uncertainty of a given outcome as a function of the uncertainties in the input parameter. This technique was applied to wellbore stability assessment by a number of authors (Moos et al., 2004).

Wellbore stability assessment represents one of the key parts of modern wellbore design approach, and it is incorporated at early stage of well planning. Fully wellbore stability risk assessment consisted from different part like, developing of geomechanical earth model, well planning, drilling regime optimization, collecting and interpretation of real-time drilling data and geomechanical earth model improvement (Alderd et al., 1999).

The first step in wellbore stability assessment is 1D or 3D mechanical or geomechanical earth model developing. The mechanical or geomechanical earth model is a numerical representation and computer (software) visualization of the geomechanical and other (geological, petro physical etc.) properties of the rock regarding to depth. The accuracy and complexity of developed geomechanical earth model strongly depends on available data (Brehm et al., 2004). Usually, the development of geomechanical earth model has started with tree main components: in-situ stress magnitude and orientation pore pressure and effective rock strength. According to Moos et al., 2004 quantifiable sources of uncertainty include

- Uncertainties in the velocities derived from the seismic data,
- Uncertainties in the functional form of the transforms between velocities and other parameters such as density that are intermediate results in the analysis,
- Uncertainties in the parameters used in the transform equations, and
- Depth uncertainties from time-to-depth calculations.

During the development of a new filed in an unexplored area, there are always much more uncertainties than in drilling a well on an existing field. Uncertainties are primarily related to the quality and quantity of available data. Data reliability can be increased through appropriate correlations between date obtained by seismic measurement and data obtained by well logging measurement and laboratory testing on core sample. Data correlation and calibration cannot be done prior to drilling the first (exploration) well. Another source of uncertainty in the computations results from a lack of information of different geological structural and stratigraphic anomalies, such as existing of magmatic intrusion (e.g. laccolite) or salt dome. The presence of salt formation (intrusions) causes changes in in-situ stresses (Dusseault, et al., 2004). Unlike typical sand/shale formations (in-situ stress in the vertical direction is equal to the overburden stress and the horizontal stress is somewhat less than the vertical stress) in a salt formation, the in-situ stress is generally assumed to be equal in all directions and is equal to the weight of the overburden stress (Baker et al., 1994).

Quantitative risk assessment is used to assess uncertainties in input data and results present probability to achieve a desired degree of stability as a function of mud weight. It is also a method for identification of the critical parameters that contribute the most to the uncertainties in the results. A mud weight window represents a range of allowable equivalent mud densities or pressures that provide wellbore stability on the specific depth. The minimum mud pressure (lower boundary) must be higher than wellbore collapse pressure and maximum mud pressure (upper boundary) lower than lost circulation pressure, at a certain depth. Wellbore collapse pressures is the mud pressure below which the entire wellbore wall becomes unstable, and lost circulation pressure is the pressure at which a significant amount of mud is lost into the formation as a result of the initiation and propagation of hydraulic fractures. Both pressures strongly depend on orientations and magnitudes of in-situ stresses, wellbore orientation, rock strength and pore pressure. Quantitative risk analysis can be done commonly by using Monte Carlo simulation as well as competitive neural network.

Collecting of real-time data during drilling is useful for geomechanical model improving, as well as the improvement of the decision making process of overall wellbore stability risk assessment. At the real-time data collecting, their quality mast be considered as well as the purpose of collecting certain data. In this regard, Millheim et al. (1998) concluded the following: first, company and drilling organization collect drilling data without questioning its quality and without a clear picture of what their data needs are, and second, drilling data are not collected with the thought that they can be analyzed later and for specific purpose (Lagreca et al., 2008). Sadilier et al. (2011) suggest the introduction of computer alarm system in the entire drilling process in order to filter and distribute the relevant information to personnel more effectively. The main objective of this approach is ensuring that the right person receives the right information at the right time to make the correct decision.

CONCLUSION

Wellbore instability has been one of the most intriguing problems in the petroleum industry, especially in the drilling and exploration department. Anomalies appearing within wellbore instability are mainly caused by the shale formations represented with 75% of all drilled formations. Problems involving wellbore instability include tight hole spots, wellbore diameter enlargement, caving appearance, the inability of carrying out wireline operations, poor hole cleaning and unsuccessful wellbore cementing operations and others. Sometimes, these problems cause even the abandonment of a certain sections or the whole well. According to recent studies, wellbore instability and associated problems costs are more than $1x10^9$ US dollars per year.

In the past, solutions for wellbore instability problems were composed from the early recognition of specific wellbore instability problems, main cause identification and swift response. A more effective solution is necessary to incorporate quantitative wellbore stability and risk analysis and assessment in the early phase of well design. Relatively new technologies like underbalanced drilling, slimhole completions, re-entry wells with open hole build sections, and multilateral wells have to take into consideration, during the well planning stage and wellbore instability risk assessment. The objective of a wellbore stability assessment is to quantify the influence of those parameters that affect the integrity of a given well such as a lack of sufficient wellbore pressure, pore pressure transmission, hole inclination and others. A wide variety of analytical and numerical models exist for prediction wellbore stresses and modes of instability for nearly all possible loading conditions, well geometries, rock properties and wellbore fluids. Dedicated laboratory tests and in-situ stress measurements are desirable to have more confidence in predictions achieved with analytical or numerical modeling tools. Every well should be evaluated individually based on the next criteria: the type of anticipated problems, their potential severity, the quantity and quality of data needed for a proper analysis, time and budget, and the success of previous analyses of this type. An advanced quantitative risk analysis of wellbore instability can be done commonly by using Monte Carlo simulation as well as competitive neural network.

REFERENCES

Abass, H., Shebatalhamd, A., Khan, M., Al-Shobaili, Y., Ansari, A., Ali, S., & Mehta, S. (2006). *Wellbore instability of shale formation, Zuluffield, Saudi Arabia*. Paper presented at Technical Symposium of Saudi Arabia Section. Dhahran, Saudi Arabia.

Al-Bazali, T. M., Zhang, J., Chenevert, M. E., & Sharma, M. M. (2006). *Factors controlling the membrane efficiency of shales when interacting with water-based and oil-based muds*. Paper presented at SPE International Oil & Gas Conference and Exhibition. Beijing, China.

Aldred, W., Plumb, D., Bradford, I., Cook, J., Gholkar, V., & Cousins, L. et al. (1999). Managing drilling risk. *Oilfield Review*, *11*(2), 2–19.

Awal, M. R., Khan, M. S., Mohiuddin, M. A., Abdulraheem, A., & Azeemuddin, M. (2001). *A new approach to borehole trajectory optimisation for increased hole stability*. Paper presented at the SPE Middle East Oil Show. Bahrain.

Barker, J. W., Feland, K. W., & Tsao, Y-H. (1994, September). Drilling long salt sections along the U. S. Gulf Coast. *SPE Drilling & Completion*, 185-188.

Brehm, A. D. K., Davis, J. B., Ward, C. D., & Bowman, G. R. (2004). *Incorporating geomechanics into the well design process - A Gulf of Mexico case study*. Paper presented at Gulf Rocks 2004, the 6th North America Rock Mechanics Symposium (NARMS), Rock Mechanics Across Borders and Disciplines. Houston, TX.

Chen, M. Y. G., Chenevert, M. E., & Sharma, M. M. (2001). *Chemical and thermal effects on wellbore stability of shale formations*. Paper presented at SPE Annual Technical Conference and Exhibition, New Orleans, LA.

Chenevert, M. E., & Osisanya, S. O. (1989, September). Shale/mud inhibition defined with rig-site methods. *SPE Drilling Engineering*, 261–268.

Chenevert, M. E., & Pernot, V. (1998). *Control of shale swelling pressures using inhibitive water-base muds*. Paper presented at SPE Annual Technical Conference and Exhibition. New Orleans, LA.

Cunha, J. C. (2004). *Risk analysis application for drilling operation*. Paper presented at the Petroleum Society's 5th Canadian International Petroleum Conference (55th Annual Technical Meeting). Calgary, Canada.

Cunha, J. C., Demirdal, B., & Gui, P. (2005). *Use of quantitative risk analysis for uncertainty quantification on drilling operations - Review and lessons learned*. Paper presented at the SPE Latin American and Caribbean Petroleum Engineering Conference. Rio de Janeiro, Brazil.

Darley, H. C. H. (1969, July). A laboratory investigation of borehole stability. *Journal of Petroleum Technology*, 883–892.

Dusseault, M. B., Maury, V., Sanfilippo, F., & Santarelli, F. J. (2004). *Drilling around salt: Risks, stresses and uncertainties*. Paper presented at Gulf Rocks 2004, the 6th North America Rock Mechanics Symposium (NARMS), Rock Mechanics across Borders and Disciplines. Houston, TX.

Ewy, R. T., & Morton, E. K. (2009, September). Wellbore-stability performance of water-based mud additives. *SPE Drilling & Completion*, 390-397.

Houtchens, B., Foster, J., & Tessari, R. (2007). *Applying risk analysis to casing while drilling*. Paper presented at the 2007 SPE/IADC Drilling Conference. Amsterdam, The. Netherlands.

Khodja, M., Canselier, J. P., Bergaya, F., Fourar, K., Khodja, M., Cohaut, N., & Benmounah, A. (2010). Shale problems and water-based drilling fluid optimisation in the Hassi Messaoud Algerian oil field. *Applied Clay Science*, *49*(4), 383–393. doi:10.1016/j.clay.2010.06.008

Labenski, F., Reid, P., & Santos, H. (2003). *Drilling fluids approaches for control of wellbore instability in fractured formations.* Paper presented at SPE/IADC Middle East Drilling Technology Conference & Exhibition. Abu Dhabi, Saudi Arabia.

Lagreca, A. J., Damski, C., Peden, J. M., & Nakagawa, E. Y. (2008, December). The role of knowledge, tools and systems for drilling analysis. *SPE Drilling & Completion*, 330-337.

Lake, L. W. (Ed.). (2006). *Petroleum engineering handbook – Drilling engineering* (Vol. 2). Society of Petroleum Engineers.

Lal, M. (1999). *Shale stability: Drilling fluid interaction and shale strength.* Paper presented at SPE Latin American and Caribbean Petroleum Engineering Conference. Caracas, Venezuela.

McLean, M. R., & Addis, M. A. (1990). *Wellbore stability analysis: A review of current methods of analysis and their field application.* Paper presented at IADC/SPE Drilling Conference. Houston, TX.

McLellan, P. (1994). *Assessing the risk of wellbore instability in horizontal and inclined wells.* Paper presented at SPE/CIM/CANMET International Conference on Recent Advances in Horizontal Well Applications. Calgary, Canada.

McLellan, P. J. (1996). Assessing the risk of wellbore instability in horizontal and inclined wells. *The Journal of Canadian Petroleum Technology*, *35*(5), 21–32.

Millheim, K., Maidla, E., & Kravis, S. (1998). *An example of the drilling analysis process for extended reach wells.* Paper presented at the 1998 SPE Annual Technical Conference and Exhibition. New Orleans, LA.

Mohiuddin, M. A., Khan, K., Abdulraheem, A., Al-Majed, A., & Arifullah, V. (2005). *Field based criteria for the design of safe mud weight window.* Paper presented at SPE Technical Symposium of Saudi Arabia Section. Dhahran, Saudi Arabia.

Moos, D., Pavel, P., Finkbeiner, T., & Zoback, M. (2003). Comprehensive wellbore stability analysis utilizing quantitative risk assessment. *Journal of Petroleum Science Engineering*, *38*, 97–109. doi:10.1016/S0920-4105(03)00024-X

Moos, D., Peska, P., Ward, C., & Brehm, A. (2004). *Quantitative risk assessment applied to pre-drill pore pressure, sealing potential, and mud window predictions from seismic data.* Paper presented at Gulf Rocks 2004, the 6th North America Rock Mechanics Symposium (NARMS). Houston, TX.

Muniz, E. S., da Fontoura, S. A. B., & Lomba, R. F. T. (2005). *Rock-drilling fluid interaction studies on the diffusion cell.* Paper presented at SPE Latin American and Caribbean Petroleum Engineering Conference. Rio de Janeiro, Brazil.

O'Brien, D. E., & Chenevert, M. E. (1973, September). Stabilizing sensitive shales with inhibited, potassium-based drilling fluids. *Journal of Petroleum Technology*, 1089–1100.

Osuji, C. E., Chenevert, M. E., & Sharma, M. M. (2008). *Effect of porosity and permeability on the membrane efficiency of shales*. Paper presented at SPE Annual Technical Conference and Exhibition. Denver, CO.

Ottesen, S., Zheng, R. H., & McCann, R. C. (1999). *Borehole stability assessment using quantitative risk analysis*. Paper presented at the 1999 SPE/IADC Drilling Conference. Amsterdam, The Netherlands.

Pasic, B., Gaurina-Medjimurec, N., & Matanovic, D. (2007). Wellbore instability: Causes and consequences. *The Mining-Geological-Petroleum Engineering Bulletin*, *19*, 87–98.

Rojas, J. C., Clark, D. E., & Zhang, J. (2006). *Stressed shale drilling strategy-water activity design improves drilling performance*. Paper presented at SPE Annual Technical Conference and Exhibition. San Antonio, TX.

Sadlier, A., Wolfe, C., Reese, M., & Says, I. (2011). *Automated alarms for managing drilling pressure and maintaining wellbore stability – New concepts in while –Drilling decision making*. Paper presented at the SPE Annual Technical Conference and Exhibition. Denver, CO.

Schlemmer, R., Friedheim, J. E., Growcock, F. B., Bloys, J. B., Headley, J. A., & Polnaszek, S.C. (2003, December). Chemical osmosis, shale, and drilling fluids. *SPE Drilling & Completion*, 318-331.

Simangunsong, R. A., Villatoro, J. J., & Davis, A. K. (2006). *Wellbore stability assessment for highly inclined wells using limited rock mechanics data*. Paper presented at the 2006 SPE Annual Technical Conference and Exhibition. San Antonio, TX.

Tan, C. P., Wu, B., Mody, F. K., & Tare, U. A. (2002). *Development and laboratory verification of high membrane efficiency water-based drilling fluids with oil-based drilling fluid-like performance in shale stabilization*. Paper presented at SPE/ISRM Rock Mechanics Conference. Irving, TX.

Tare, U. A., Mody, F. K., & Tan, C. P. (2002). *New generation of membrane efficient water-based drilling fluids: Pragmatic and cost-effective solutions to borehole stability problems*. Paper presented at the Petroleum Society's Canadian International Petroleum Conference. Calgary, Canada.

Torres, M. E., Frydman, M., Casalis, D., Ramirez, A., León, M. F., & Villalba, E. (2005). *3D analysis for wellbore stability: reducing drilling risk in Oriente basin, Ecuador*. Paper presented at the SPE Latin American and Caribbean Petroleum Engineering Conference. Rio de Jeneiro, Brazil.

Tutuncu, A. N., Geilikman, M., Couzens, B., & van Duyvenboode, F. (2005). *Integrated wellbore quality and risk assessment study guides successful drilling in Amazon jungle*. Paper presented at SEG/Houston 2005 Annual Meeting. Houston, TX.

van Oort, E., Hale, A. H., & Mody, F. K. (1995). *Manipulation of coupled osmotic flows for stabilisation of shales exposed to water-based drilling fluids*. Paper presented at SPE Annual Technical Conference & Exhibition. Dallas, TX.

van Oort, E., Hale, A. H., Mody, F. K., & Roy, S. (1996, September). Transport in shales and the design of improved water-based shale drilling fluids. *SPE Drilling & Completion*, 137-146.

Zhang, J., Chenevert, M. E., Al-Bazali, T., & Sharma, M. M. (2004). *A new gravimetric-swelling test for evaluating water and ion uptake in shales*. Paper presented at SPE Annual Technical Conference and Exhibition. Houston, TX.

Zhang, J., Clark, D. E., Al-Bazali, T. M., Chenevert, M. E., Rojas, J. C., & Ong, S. (2006b). *Ion movement and laboratory technique to control wellbore stability*. Paper presented at the AADE 2006 Fluids Conference. Houston, TX.

Zhang, J., Yu, M., Al-Bazali, T. M., Ong, S., Chenevert, M. E., Sharma, M. M., & Clark, D. E. (2006a). *Maintaining the stability of deviated and horizontal wells: Effects of mechanical, chemical, and thermal phenomena on well designs*. Paper presented at SPE International Oil & Gas Conference and Exhibition. Beijing, China.

ADDITIONAL READING

Ahmed, K., Khan, K., & Mohamad-Hussein, M. A. (2009, March). *Prediction of wellbore stability using 3D finite element model in a shallow unconsolidated heavy-oil sand in a Kuwait field*. Paper SPE 120219 presented at the 2009 SPE Middle East Oil & Gas Show and Conference, Bahrain International Exhibition Centre, Kingdom of Bahrain.

Barker, J. W., Feland, K. W., & Tsao, Y-H. (1994). Drilling long salt sections along the U. S. Gulf Coast. *SPE Drilling & Completion, September*, 185-188.

Chen, X., Tan, C. P., & Haberfield, C. M. (1996, October). *Wellbore satability analysis guidelines for practical well design*. Paper SPE 36972 presented at the 1996 SPE Asia Pacific Oil and Gas Conference, Adelaide, South Australia, Australia.

Chen, X., Tan, C. P., & Haberfield, C. M. (1997). Guidelines for efficient wellbore stability analysis. *International Journal of Rock Mechanics & Mining Sciences, 34(3-4)*, 50.e1–50.e14.

Clyne, N., & Bonsall, D. (1998, June). *Workplace hazard management in the drilling industry*. Paper SPE 46650 presented at the 1998 SPE International Conference on Health, Safety and Environment in Oil and Gas Exploration and Production, Caracas, Venezuela.

Coelho, D. K., Roisenberg, M., de Freitas Filho, P. J., & Jacinto, C. M. C. (2005). Risk assessment of drilling and completion operations in petroleum wells using A Monte Carlo and Neutral Network approach. In M. E. Kuhl, N. M. Steiger, F. B. Armstrong and J. A. Joines (Eds.), *Proceedings of the 2005 Winter Simulation Conference* (pp. 1892-1897), Winter Simulation Conference ©2005.

Dowson, S. L., Willson, S. M., Wolfson, L., Ramos, G. G., & Tare, U. A. (1999, October). *An integrated solution of extended-reach drilling problems in the Niakuk Field, Alaska: Part I - Wellbore stability assessment.* Paper SPE 56563 presented at the 1999 SPE Annual Technical Conference and Exhibition, Houston, Texas, U.S.A.

Engevik, M. O. (2007). *Risk assessment of underbalanced and managed pressure drilling operations*. Master thesis, Faculty of Engineering Science and Technology, Department of Production and Quality Engineering, Trondheim, Norway.

Gallant, C., Zhang, J., Wolfe, C. A., Freeman, J., Al-Bazali, T., & Reese, M. (2007, November). *Wellbore stability considerations for drilling high-angle wells through finely laminated shale. A case study from Terra Nova.* Paper SPE 110742 presented at the 2007 SPE Annual Technical Conference and Exhibition, Anaheim, California, U.S.A.

Hassan, S., Klimentos, T., Badri, M., Sengul, M., & Zeid, A. (1999, November). *Optimizing drilling performance by wellbore stability evaluation and directional drilling practices.* Paper SPE/IADC 57575 presented at the 1999 SPE/IADC Middle East Drilling Technology Conference, Abu Dhabi, UAE.

Johnstone, J. A., & Gruenhagen, H. (2001, September). *Using rotary closed-loop drilling to increase operational efficiency and reduce operational risk.* Paper SPE 71840 presented at the Offshore Europe Conference, Aberdeen, Scotland. (2004). *Well integrity in drilling and well operations.* Norsok Standard, Standards Norway, Norway, from http://www.standard.no/PageFiles/1315/D-010r3.pdf.

Kang, Y., Yu, M., Miska, S., & Tajach, N. E. (2009, October). *Wellbore stability: A critical review and introduction to DEM.* Paper SPE 124669 presented at the 2009 SPE Annual Technical Conference and Exhibition, New Orleans, Louisiana, U.S.A.

Khatlan, A., Jafari, A., & Milani, A. A. (2007, March/April). *An integrated system for risk assessment of drilling operations in oil and gas wells.* Paper SPE 106714 presented at the 2007 SPE Production and Operations Symposium, Oklahoma City, Oklahoma, U.S.A.

Koupriantchik, D., Meyers, A. G., & Hunt, S. P. (2004, June). *3D geomechanical modelling towards understanding stress anomalies causing wellbore instability.* Paper ARMA/NARMS 04-493 presented at the Gulf Rocks 2004, the 6th North America Rock Mechanics Symposium (NARMS), Rock Mechanics across Borders and Disciplines, Houston, Texas. U.S.A.

Kristiansen, T. G. (2004, March). *Drilling wellbore stability in the compacting and subsiding Valhall field.* Paper IADC/SPE 87221 presented at the IADC/SPE Drilling Conference, Dallas, Texas, U.S.A.

Kristiansen, T. G., Mandziuch, K., Heavey, P., & Kol, H. (1999, March). *Minimizing drilling risk in extended-reach wells at Valhall using geomechanics, geoscience and 3D visualization technology.* Paper SPE/IADC 52863 presented at the 1999 SPE/IADC Drilling Conference, Amsterdam, Holland.

Liu, J., Li, Q., & Wang, Y. (2013). Risk analysis in ultra deep scientific drilling project-a fuzzy synthetic evaluation approach. *International Journal of Project Management, 31*(3), 449–458. doi:10.1016/j.ijproman.2012.09.015

Malloy, K. P. (2011). *Risk profile of dual gradient drilling. Final report.* Stress Engineering Services Inc. - Mohr Engineering Division.

Mclntyre, B., Hibbert, T., Keir, D., & Dixon, R. ORourke, T., Mohammed, F., Donald, A., Chang, L., Syed, A., & Biran, V. (2009, September). *Maganing drilling risk in a mature North sea field.* Paper SPE 124666 presented at the 2009 SPE Offshore Europe Oil & Gas Conference & Exhibition, Aberdeen, UK.

Mohiuddin, M. A., Khan, K., Abdulraheem, A., Al-Majed, A., & Awal, M. R. (2007). Analysis of wellbore instability in vertical, directional, and horizontal wells using field data. *Journal of Petroleum Science Engineering*, *55*, 83–92. doi:10.1016/j.petrol.2006.04.021

Mostafavi, V., Aadnoy, B. S., & Hareland, G. (2011, June). *Model-based uncertainty assessment of wellbore stability analyses and downhole pressure estimations*. Paper ARMA 11-127 presented at the 45th US Rock Mechanics/Geomechanics Symposium, San Francisco, CA, U.S.A.

Ong, S., & Baim, A. S. Ibrahim, M. Z., & Zheng, Z. (2000, September). *Geomechanical analysis for Resak's extended-reach drilling - A case example*. Paper IADC/SPE 62727 presented at the 2000 IADC/SPE Asia Pacific Drilling Technology, Kuala Lumpur, Malaysia.

Ottesen, S., & Kwakwa, K. A. (1991, March). *A multidisciplinary approach to in-situ stress determination and its application to wellbore stability analysis*. Paper SPE/IADC 21915 presented at the 1991 SPE/IADC Drilling Conference, Amsterdam, The Netherlands.

Parra, J. G., Celis, E., & De Gennaro, S. (2003). Wellbore stability simulations for underbalanced drilling operations in highly depleted reservoirs. *SPE Drilling & Completion*, June, 146-151.

Plumb, R. A., Hooyman, P., Veeningen, D., Dutta, N., Ritchie, G., & Bennaceur, K. (2004, June). *A new geomechanics process reduces operational risk from exploration to production*. Paper ARMA/NARMS 04-616 presented at Gulf Rocks 2004, the 6th North America Rock Mechanics Symposium (NARMS), Rock Mechanics Across Borders and Disciplines, Houston, Texas. U.S.A.

Prassl, W. F., Peden, J. M., & Wong, K. W. (2005). A process-knowledge management approach for assessment and mitigation of drilling risk. *Journal of Petroleum Science Engineering*, *49*, 142–161. doi:10.1016/j.petrol.2005.05.012

Rabbani, E., Sharif, F., Koolivand Salooki, M., & Moradzadeh, A. (2012). Application of neural network technique for prediction of uniaxial compressive ctrength using reservoir formation properties. *International Journal of Rock Mechanics and Mining Sciences*, *56*, 100–111. doi:10.1016/j.ijrmms.2012.07.033

Salehi, S., Hareland, G., Dehkordi, K. K., Ganji, M., & Abdollahi, M. (2009). Casing collapse risk assessment and depth prediction with a Neural network system approach. *Journal of Petroleum Science Engineering*, *69*, 156–162. doi:10.1016/j.petrol.2009.08.011

Salehi, S., Hareland, G., & Nygaard, R. (2010). Numerical simulations of wellbore stability in under-balanced-drilling wells. *Journal of Petroleum Science Engineering*, *72*, 229–235. doi:10.1016/j.petrol.2010.03.022

Santarelli, F. J. (1994, August). *Rock mechanics characterization of deep Formations: A technico-economical overview*. Paper SPE 28021 presented at the SPE/ISRM Rock Mechanics in Petroleum Engineering Conference, Delft, The Netherlands.

Seymour, K. P., Rae, G., Peden, J. M., & Ormston, K. (1993, September). *Drilling close to salt diapirs in the North sea*. Paper SPE 26693 presented at the Offshore European Conference, Aberdeen, UK.

Singh, S. K., Subekti, H., Al-Asmakh, M., & Al-Samarraie, L. (2012, May). *An integrated approach to well integrity evaluation via reliability assessment of well integrity tools and methods: results from Dukhan field, Qatar*. Paper SPE 156052 presented at the SPE International Production and Operations Conference and Exhibition, Doha, Qatar.

Skogdalen, J. E., & Vinnem, J. E. (2012). Quantitative risk analysis of oil and gas drilling, using deepwater horizon as case study. *Reliability Engineering & System Safety*, *100*, 58–66. doi:10.1016/j.ress.2011.12.002

Tan, C. P., Rahman, S. S., Chen, X., Willoughby, D. R., Choi, S. K., & Wu, B. (1998, September). *Wellbore stability analysis and guidelines for efficient shale instability management*. Paper IADC/SPE 47795 presented at the IADC/SPE Asia Pacific Drilling Technology ´98 Conference and Exhibition, Jakarta, Indonesia.

Tan, C. P., & Willoughby, D. R. (1993). *Critical mud weight and risk contour plots for designing inclined wells*. Paper SPE 26325 presented at the 68th Annual Technical Conference and Exhibition of Society of Petroleum Engineers, Houston, Texas, U.S.A.

Willson, S. M., Edwards, S., Heppard, P. D., Li, X., Coltrin, G., Chester, D. K., et al. (2003, October). *Wellbore stability challenges in the deep water, Gulf of Mexico: case history examples from the Pompano field*. Paper SPE 84266 presented at the SPE Annual Technical Conference and Exhibition, Denver, Colorado, U.S.A.

Zeynali, M. E. (2012). Mechanical and physico-chemical aspects of wellbore stability during drilling operations. *Journal of Petroleum Science Engineering*, *82-83*, 120–124. doi:10.1016/j.petrol.2012.01.006

KEY TERMS AND DEFINITIONS

Clay Minerals Hydration: Represents a process of water molecules adsorption or absorption on inner or outer surface of the clay minerals. Intensity and magnitude of clay minerals hydration depend on clay minerals type and in-situ conditions.

Crystalline Swelling: Short-range swelling and it is characteristic of all clay minerals. Crystalline swelling of a certain clay mineral causes slightly increasing of interlayer space which lies between 9 to 20 Å.

Mechanical Earth Model (MEM): Numerical representation and computer (software) visualization of the geomechanical and other (geological, petro physical etc.) properties of the rock regarding to depth.

Mud Weight Window: Represents a range of allowable equivalent mud densities or pressures that provide wellbore stability on the specific depth. The minimum mud pressure (lower boundary) must be higher than wellbore collapse pressure and maximum mud pressure (upper boundary) lower then lost circulation pressure, at the certain depth.

Osmotic Swelling: Clay swelling type limited to certain clay minerals which contain exchangeable cations in interlayer space. This type of swelling causes tremendously increases in interlayer space and clay volume (much larger than at crystalline swelling).

Rock in-Situ Stress: Represents native stress of the certain rock in undisturbed condition as a result of different geological processes (sedimentation, falding, foulting etc.) and local tectonic activities.

Shale Swelling: Represents volume increase of the shale rock as a result of clay hydration.

Wellbore Instability: Can be defined as any change in borehole diameter by comparison with diameter of drilling bit used for its drilling.

Chapter 3
Risk Due to Pipe Sticking

Nediljka Gaurina-Medjimurec
University of Zagreb, Croatia

Borivoje Pasic
University of Zagreb, Croatia

ABSTRACT

A stuck pipe is a common worldwide drilling problem in terms of time and financial cost. It causes significant increases in non-productive time and losses of millions of dollars each year in the petroleum industry. There are many factors affecting stuck pipe occurrence such as improper mud design, poor hole cleaning, differential pressure, key seating, balling up of bit, accumulation of cuttings, poor bottom hole assembly configuration, etc. The causes of a stuck pipe can be divided into two categories: (a) differential sticking and (b) mechanical sticking. Differential-pressure pipe sticking occurs when a portion of the drill string becomes embedded in a filter cake that forms on the wall of a permeable formation during drilling. Mechanical sticking is connected with key seating, formation-related wellbore instability, wellbore geometry (deviation and ledges), inadequate hole cleaning, junk in hole, collapsed casing, and cement related problems. Stuck pipe risk could be minimized by using available methodologies for stuck pipe prediction and avoiding based on available drilling parameters.

DOI: 10.4018/978-1-4666-4777-0.ch003

INTRODUCTION

Pipe sticking is, for most drilling organizations, the greatest drilling problem worldwide. It results in a significant amount of non-productive time and ends up as one of the major causes of increased well costs (Yarim et al., 2008; Reid et al., 2000; Pal et al., 2000). It may result in abandonment of the current hole and force a sidetrack. It is estimated that the cost of stuck pipe in deep oil and gas wells can be approximately 25% of the overall budget. In some areas, events related to differentially stuck pipe can be responsible for as much as 40% of the total well cost (Reid et al., 2000). The causes of stuck pipe can be divided into two categories (Isambourg, et al., 1999): (a) mechanical (key seating, formation-related wellbore instability, wellbore geometry (deviation and ledges), inadequate hole cleaning, junk in a hole or collapsed casing, cement related) and (b) differential pressure (wall sticking). Differential pressure sticking is usually indicated when the drill string cannot be rotated, raised or lowered, but full circulation at normal pressure can be established (Bushnell-Watson & Panesar, 1991). The force required to pull the pipe free can exceed the strength of the pipe. Usually, even if the stuck condition starts with the possibility of limited pipe rotation or vertical movement, it will degrade to the inability to move the pipe at all. Many oil and gas reservoirs are mature and are becoming increasingly depleted of hydrocarbons, which increases the risk involved with the stuck pipe. This is due to the fact that decreasing pore pressure increases the chance of stuck pipe. Therefore, the risk of differentially stuck pipe increases when drilling depleted reservoirs and avoids when drilling underbalanced. The increased use of deep, highly-deviated and tortuous wells has increased the risk of drill pipe and wireline logging tool strings getting stuck downhole, too. If this risk is not appropriately managed and effectively mitigated, significant financial exposure can result from the cost of the multi-day fishing operations (Prasad et al., 2012). The physical mechanisms of sticking wireline tools are similar to the mechanisms involved in drill-pipe sticking. The complexity of the wells has increased significantly in later years. Reach has been more than doubled, and high inclination and fully 3-D well paths are common. However, statistics shows that sidetracking the boreholes due to stuck pipe has also shown a significant increase, and is presently a high cost factor. The margins between success and failure are now much smaller (Aadnøy et al., 1999). Traditionally, stuck pipe problems are solved by using some standard methods and techniques after they occur, but the real key to savings and success is in the avoidance of the risks associated with the stuck pipe. Minimizing the risks of stuck pipe while drilling has been the goal of many operators recently.

Many researchers attempt to identify the parameters and their corresponding effects to minimize the risk of stuck pipe (Chamkalani et al., 2013; Jahanbakhshi et al., 2012; Al-Baiyat & Heinze, 2012; Meqeem et al., 2012; Shoraka et al., 2011; Meschi et al., 2010; Murillo et al., 2009; Miri et al., 2007; Aadnøy et al., 1999; Hopkins & Leicksenring, 1995; Howard & Glover, 1994; Hempkins et al., 1987; Courteille & Zurdo, 1985; Kingsborough et al., 1985). They proposed models or techniques to identify and diagnose the stuck pipe early and prevent its occurrence. The accuracy of the predictive model depends on the size of database and the variables selected for analysis. Even if different techniques and guidelines have been developed to reduce the probability of occurrence of stuck pipe and these have saved drilling industry millions of dollars, they suffer from exclusive prediction of this event. In some cases where the pipe is not freed, fishing and even sidetracking should be employed which imparts a huge financial load on the final cost of drilling the well.

DIFFERENTIAL STICKING

Differential sticking by definition is a situation in which the drilling assembly (pipe, drill collars and bottom hole assembly) is stuck in a filter cake that was previously deposited on a permeable zone (Figure 1a). A significant overbalance must also exist and the pipe must be stationary, or almost stationary, to allow the bond between the cake and the pipe to develop. The pipe is held in the cake by a difference in pressures (ΔP) between the hydrostatic pressure of the mud (P_m) and the formation (pore) pressure (P_f) in the permeable zone (Santos, 2000). This pressure (ΔP) acts upon the area of the pipe in contact with the filter cake and isolated from the pressure of the mud in the hole by a new filter cake (Figure 1b). There must be a minimum penetration of the drill pipe before a significant change of pressure at pipe/cake interface can occur. Therefore if filter cake can be made thin enough, sticking can be avoided. Three differential forces have been used to free the pipe: axial (equivalent to working the pipe up or down), radial (equivalent to pulling the pipe across the wellbore) and torque (rotation of the pipe) (Figure 1c). In practice it is likely to be a combination of an axial force and torque which is used to free the pipe. The force required to free differentially stuck pipe must overcome adhesion and the differential pressure exerted by the mud. It has a tendency to increase with time until all water is expelled from the filter cake. Once sticking is established, a significant force is required to free the pipe, even if the mud overbalance is removed. At best, several hours of rig time can be spent in various freeing operations. In more serious cases, the pipe cannot be freed and the well has to be sidetracked or abandoned. Pullout force needed to free a stuck pipe is equal to:

$$F = \Delta P.A.\mu \tag{1}$$

Figure 1. Pipe before stuck (a), stuck pipe (b) and forces to free the pipe (c)

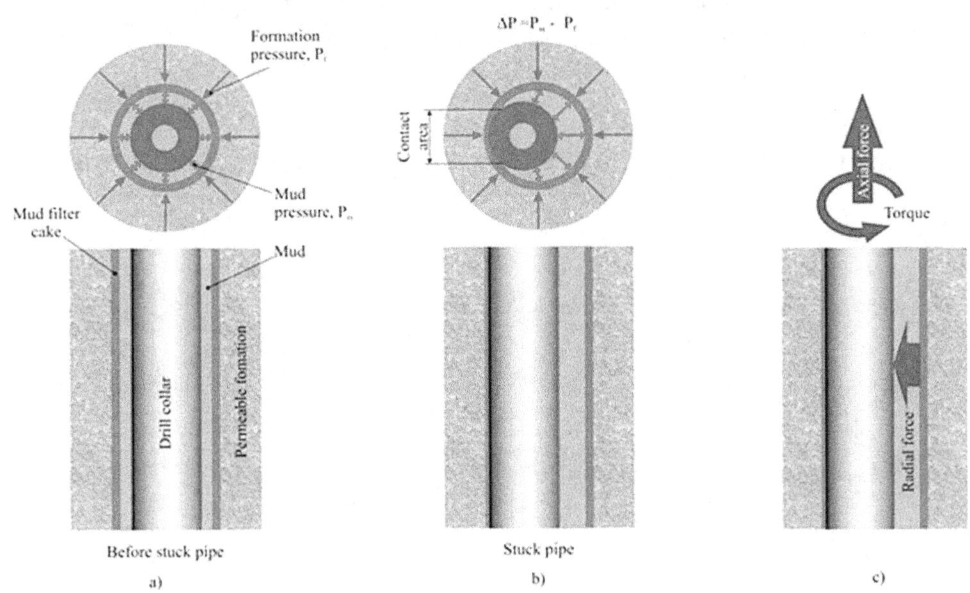

where:

F - pullout force [N]
ΔP - differential pressure [Pa]
A - contact area [m^2]
μ– coefficient of friction [-]

The value of pullout force (F) is also increased with compressibility and thickness of the filter cake, hole deviation, and diameter of the drill collars. It is decreased with an increase in diameter of the hole. The pullout force is time-dependent since contact area (A) and coefficient of friction between the collars and the cake (μ) increases with time.

Differential Sticking Causes

Differential sticking causes could be: (a) relatively high differential pressure and (b) mud cake characteristics (thickness, permeability, and lubricity). In the situations when it is not possible to reduce the differential pressure by reducing the mud weight the option is to act on the mud cake (Outmans, 1974). In a situation when a pipe is rotating, a dynamic filter cake is formed and a drill collar penetrates only a short distance into a mud filter cake (Figure 2a) but when the pipe is stationary, a static mud filter cake is formed and the drill collar is pushed into the mud filter cake by differential pressure (Figure 2b). In highly deviated wellbores, with pipe being stationary, pressure between the mud filter cake and the drill collar varies from zero to the value of differential pressure (ΔP) (Figure 2c).

Methods of Reducing the Risk of Differential Sticking

Differential sticking tendencies of mud depend on mud filter cake properties: thickness, shear strenght, and lubricity. These filter cake properties are influenced by a combination of variables such as: mud overbalance, solids content of the mud (both high-gravity and low-gravity solids), mud type (e.g., oil-based, polymer water-based, gel water-based), specific mud composition, and fluid loss. Early detection of differential pressure sticking risks could be made through the observation of torque and drag levels while drilling to detect any sign of deviation from a normal trend for the well. The objective of the stuck pipe avoidance practices is to ensure conditions are maintained at all times that allow pulling force to exceed sticking force (shear strength) (Dupriest et al., 2010). To mitigate differential pressure sticking events, operators often (Montgomery et al., 2007; Simon et al., 2005):

Figure 2. Situation in wellbore with pipe rotating (a) and pipe stationary (b and c) (Caenn et al., 2011)

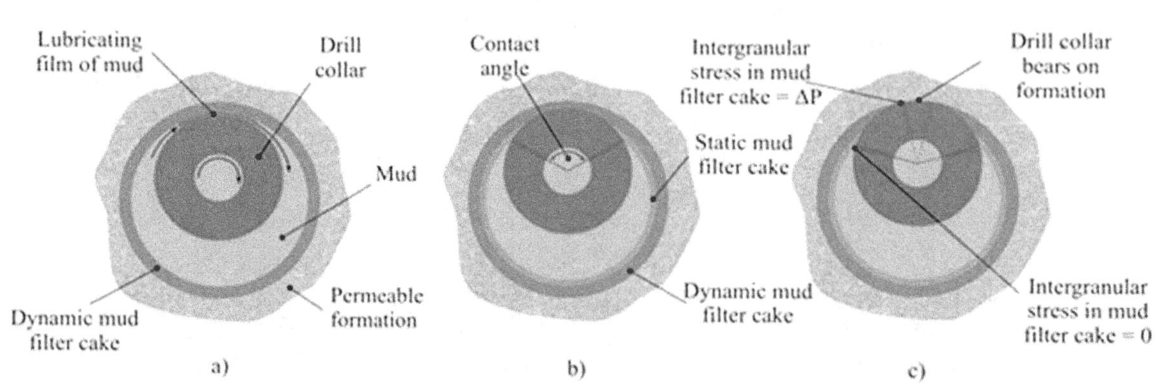

- Minimize the overbalance (by decreasing mud weight),
- Maintain a low filtration rate,
- Keep low-gravity solids to a minimum,
- Minimize stationary time (drill string rotates at all times),
- Minimize drilled length through low pressure formations,
- Increase drill collar and drill string stabilization,
- Optimize fluid properties in attempts to minimize the risk of sticking, and
- Select a drilling fluid that will yield a smooth filter cake with low coefficient of friction.

The addition of certain lubricants to water- and oil- based muds will reduce the risk of differential sticking and, should sticking still occur, reduce the force needed to free the stuck pipe or tool. Lubricants are designed to reduce the coefficient of friction of drilling fluid which decreases torque and drag. Depending on their chemical composition and the state of dispersion or solubility in the base mud, lubricants can: (a) coat metal surfaces, reducing the adhesion of steel to the mud cake, (b) be incorporated into the filter cake and provide better fluid-loss control (resulting in thinner cakes), and (c) be incorporated into the filter cake to reduce the yield stress of the cake.

However, despite the best efforts of operators a differential pressure sticking event may still occur (Ayers et al., 1989). The best cure for differential sticking is to prevent it by use of drill-collar stabilizers and, more important, conscientiously shortening the intervals of rest when the pipe is opposite permeable formations (Dupriest et al., 2010).

Methods of Freeing the Stuck Pipe

Methods used to get the pipe free, in addition to pulling and torqueing the pipe, include: (a) lowering hydrostatic pressure in the wellbore (*by reducing the mud weight; this will reduce the differential pressure; should not be used if well control is a problem*), (b) placing a spotting fluid next to the stuck zone and (c) applying shock force just above the stuck point by mechanical jarring, or (d) all the above. The most common approach, however, to getting pipe free is to pump a chemical spotting fluid in a wellbore annulus. Before the placement of the spotting fluid, the depth (free point) to where the drill string is free and where sticking starts must be determined. This free point can be calculated using measurements of the drillstring stretch taken on the rig floor. Knowing the stretch (ΔL) and the forces applied (F_1) and (F_2), Hooke's law, the length of the drill string from the surface to the free point (L_f) is equal to:

$$L_f = \frac{E \cdot A \cdot \Delta L}{F_2 - F_1} \tag{2}$$

where:

E - Elastic Modulus (Young's Modulus) of steel [i.e., 200 GPa],
A - cross-sectional area of the pipe body [m^2],
ΔL - stretch distance (elastical stretch of the free portion of the drill string) (m),
F_1 - force to place the entire drill string in tension (N),
F_2 - force greater than F_1 but less that the force limited by the yield stress of the pipe grade [N].

Spotting Fluid

A spotting fluid (spot) is a small volume or pill of any substance, oil or water base that is positioned in the wellbore to achieve a specific purpose. Most service companies provide multiple spotting fluid options. The purpose of the spotting fluid is to dissolve or break down the filter cake so the pipe can be freed. Possible mechanisms of spotting fluid action to help free differentially stuck pipe are (Montgomery et al., 2007):

- Breaks the capillary forces that hold the drillstring against the wellbore wall;
- Penetrates, dehydrates and cracks (breaks up) the mud filter cake;
- Reduces the contact (stuck) area between pipe and wall;
- Reduces the forces needed to work the pipe free;
- Increases drillstring lubricity throughout stuck zone;
- Allows pipe to be pulled free.

Spotting fluids need to be in place as quickly as possible (within six hours after a pipe becoming stuck). Figure 3 shows a free drill collar in the center of the hole (a), initial (b) and final (c) position of the drill collar in contact with the wall, before spotting the fluid. When pipe sticking occurs, pumping of spotting fluid has to be as quickly as possible to minimize the continued filter cake buildup which leads to a higher contact angle and pull free forces (Figure 3c). The mud filter cake shrunk after spotting fluid (Figure 3d) and the drill collar is freed (Figure 3e).

Spotting fluids could be divided into four main categories: (a) water-based spotting fluids, (b) diesel-based spotting fluids, (c) synthetic-based spotting fluids, and (d) acid based spotting fluid. They could be unweighted or weighted (spotting fluid, viscosifier and weighting material) and consist of detergents, soaps, oils, surfactants and other chemicals (wall cake cracking material). Oil-based mud is the traditional spotting fluid (Krol, 1981; Ayers et al., 1989). Because of concern about

Figure 3. The position of a drill collar in the hole before and after spotting fluid

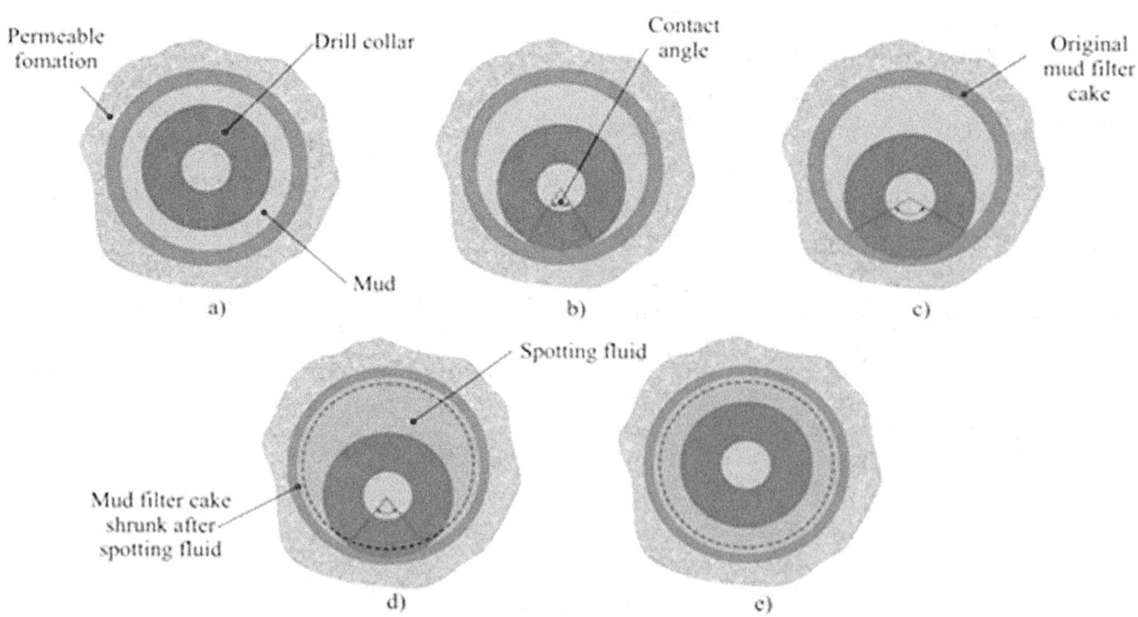

mud disposal, spotting fluids used offshore are either synthetic-based fluids or benign water-based fluids (such as drill-in fluids and salt solutions) (Kercheville et al., 1986). Drill-in fluids are low and ultra-low solids fluids; the sealing mechanism is generated inside the rock, so they are leaving just a thin film on the borehole walls. Salt solutions with a low activity coefficient combined with environmentally-safe lubricants (two-phase spot) produce low torque levels.

MECHANICAL STICKING

The term mechanically sticking is broadly used to describe all sorts of sticking problems except differential sticking. Mechanical sticking occurs as a result of one or more of the following conditions: key seating, formation instability, inadequate hole cleaning, running into under a gauge hole, accumulation of cuttings, drilling plastic formations, large boulders falling into the hole, and running tools with too large a gauge for the hole diameter (Bowes & Procter, 1997; Chamkalani et al., 2013).

Key Seating

Key seating is a situation frequently encountered in deviated holes when the drill pipe wears into the wall. It is caused when the drill pipe, which is of smaller diameter than the drill collars, rubs against the side of the hole and wears a slot (Figure 4a). The body and tool joints of the drill pipe wear a groove in the rock about the same diameter as the tool joints. Key seating is diagnosed when the drill pipe can be reciprocated within the range of tool joint distances or until the collar reaches the key seat, while pipe rotation and circulation remain normal. It is still possible to drill ahead when key seating has occurred. However, when pulling out of the hole, the larger diameter drill collars will hang up on the key seat and the pipe will become stuck.

Figure 4. Key seating (a) and reaming (b)

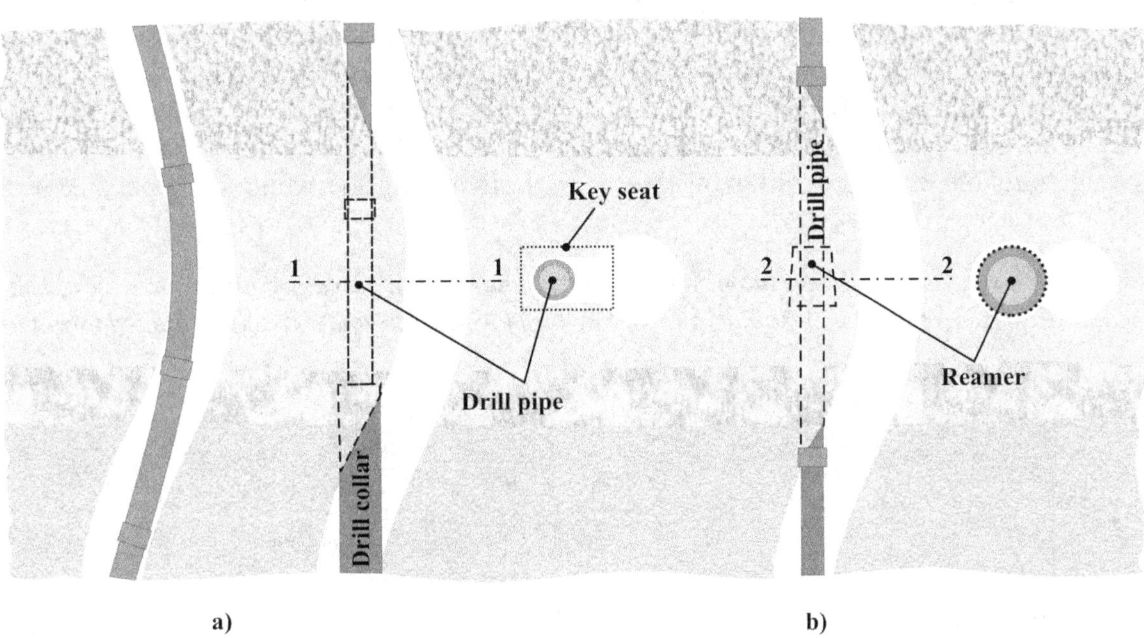

The greater the dogleg severity, the greater the side load and the faster the key seat can be developed. This type of sticking is likely to happen in a soft formation while dropping angle. Key seats can also be developed at ledges and casing shoes where the groove is worn into the metal instead of the formation. The development of key seats is directly affected by the number of rotating hours. A preventive measure is to carefully control upper hole deviation and dogleg severity throughout the well path. This action will eliminate the force that leads to key seat creation. Once a key seat is formed, the best solution is to ream out the small-diameter portion of the hole with reaming tools (Figure 4b). This action will solve the immediate stuck-pipe problem, but the key seat can be formed again unless preventive steps are taken.

Formation-Related Wellbore Instability

Formation related sticking is caused by the constriction of the borehole by mechanically or chemically unstable formations. This includes naturally fractured or faulted formations, tectonically stressed formations, mobile salt formations, unconsolidated sands, swelling shales, etc (Pasic et al., 2007; Prasad et al., 2012). The best practices for maintaining a stable borehole during drilling involves geomechanical analysis, stress modeling, and ensuring optimal mud weight and good borehole clean up. A natural fracture system in the rock can often be found near faults. Rock near faults can be broken into large or small pieces. If they are loose they can fall into the wellbore and jam the string in the hole (Nguyen et al., 2007). Even if the pieces are bonded together, impacts from the BHA due to drill string vibrations can cause the formation to fall into the wellbore (Figure 5a) and stuck pipe can occur while drilling. Drill string vibrations have to be minimized to help stabilize these formations (Bowes & Procter, 1997).

When a hole is drilled in an area of high tectonic stresses the rock around the wellbore will collapse into the wellbore and produce splintery cavings similar to those produced by over-pressured shale (Figure 5b). In the tectonic stress case the hydrostatic pressure required to stabilize the wellbore may be much higher than the fracture pressure of the other exposed formations (Bowes & Procter, 1997). This mechanism usually occurs in or near mountainous regions. Planning to case off these formations as quickly as possible and maintaining adequate drilling fluid weight can help to stabilize these formations. In tectonically stressed areas with fractured shales, drilling practices can contribute to destabilizing the wellbore. Mechanical dis-

Figure 5. Pipe stuck caused by unstable formations: a) naturally fractured or faulted formations b) tectonically stressed formations, c) mobile (plastic flowing) formations, and d) unconsolidated formations

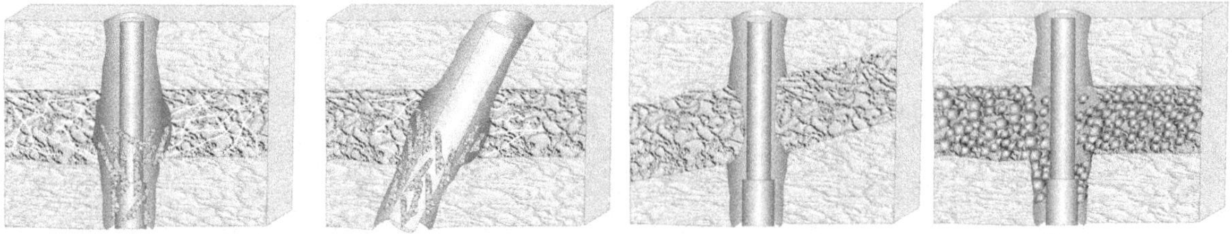

a) b) c) d)

turbance of yielded rock caused by backreaming, excessive drill string vibrations and excessive pressure fluctuations mobilized large volumes of cavings that resulted in stuck pipe if they were not transported out of the well (Yarim et al., 2008). This material and also the solids removed from wash out zones can result in a sudden pack off situation. The mechanical action of the drill string while backreaming will cause agitation to the already weakened rock. It will also cause an enlarged annulus and more solids to remove from the wellbore. This can be a challenging situation as it may not be possible to provide adequate hole cleaning without pipe rotation which then causes more problems. The solution in this case may be to pump at the maximum rate without rotation and very slowly pull out of the hole. An additional problem occurs when the BHA moves from a failing rock to a competent rock (over gauge to gauge hole). In this case, a significant concentration of cavings ahead of the BHA can cause stuck pipe at the interface between the over gauge and gauge hole. Prevention actions during drilling through overpressured formations include monitoring cuttings and the Rate of Penetration (ROP) and the use of adequate hole cleaning procedures.

The mobile (plastic flowing) formation squeezes into the wellbore because it is being compressed by the overburden forces (Figure 5c). It behaves in a plastic manner, deforming under pressure. This mechanism normally occurs while drilling salt because the mud weight is not sufficient to prevent the formation squeezing into the wellbore (Bowes & Procter, 1997). The deformation results in an undergauge hole, causing problems of running BHA´s, logging tools, and casing. An appropriate drilling fluid and maintaining sufficient drilling fluid weight are required to help stabilize these formations.

Some formations are particularly weak and unconsolidated and are prone to swelling (e.g., shale) and/or sloughing (eg., coal or shale). In general, shale is the formation that causes the most problems regarding stuck pipe. Shales are weaker along the formation bedding plane than across it. For this reason, hole stability may vary greatly in holes through the same formation but at different inclinations and directions. During drilling operations, swelling and sloughing can grip the pipe and cause sticking. Brittle/sloughing shales slough into the hole after breaking into pieces. Sloughing shale indications are: large amounts of shale at the shaker screens after circulating bottoms up, excess drag during trips, and high levels of fill. Some shales chemically react with water and swell (Pasic et al., 2007). All swelling shales are potential causes of stuck pipe whether they are highly reactive gumbos, or shales which hydrate very quickly. Even though the hole may be cleaned by sufficient flow rates during drilling, it may need to be re-drilled after each stand as the swelling continues. The indications of swelling are: torque increases during drilling, difficulty in sliding the Bottom Hole Assembly (BHA) when orienting, drag in connections, excessive drag when Pull Out of Hole (POOH), requirement for repeated reaming of sections, and an increase in Methylene Blue Test (MBT) of water based muds. To avoid stuck pipe in shales mud properties must be maintained. Time spent conditioning mud is not time wasted even if it means tripping back to the shoe because it may prevent a stuck pipe incident. Tripping with caution through swelling formations is necessarily. Reaming of each single in tight hole is recommended. Also, when using a top drive, stopping rotation, picking up midway through each stand and reaming down is advisable. If hole conditions are severe, more frequent reaming may be required. A top drive allows tight sections to be tripped through using slow rotation and circulation. After pulling into a tight spot, running back into gauge hole and circulating before back reaming out is advantageous. Sections which proved to be tight during a trip out of the hole should always be reamed on the trip in. Drilling top hole in unconsolidated formations usually results in large pieces of rock (boulders) falling into the hole (Figure 5d). Boulders can be dislodged and fall into the hole or against the drill string causing erratic torque, erratic drag on connections, and tendency to stick when pulling out of the hole.

Inadequate Hole Cleaning

Cuttings which are not adequately removed out of a hole will settle on the lower side of the hole and may cause the hole to pack off, resulting in stuck pipe. Due to the tendency of the cuttings to fall to the lower side of the hole, high angle holes are more difficult to clean than vertical wells. When a downhole motor is being used in an inclined well, without rotating the drill string, it is probable that the cuttings beds are not being disturbed. In addition, for effective hole cleaning, the circulation rate must be higher than the slip velocity of the cuttings. This is more difficult to achieve in a deviated hole than in a vertical hole. If backreaming operations are conducted too fast, solids from washouts and cavings are introduced into the circulating system at a faster rate than the hole is being cleaned. This can then result in a pack off, too. It should not be assumed that any resistance is always at the bit; stabilizers and drill collars contact may be indicative of a buildup of loose material in the hole and a potential pack off situation. As the wellbore inclination increases, especially between 30° to 60°, hole cleaning becomes more critical, the well becomes more prone to cutting beds and avalanches (Yarim et al., 2008). Drilling-fluid systems with poor suspension characteristics exhibit strong packing off tendencies. The indications of inadequate hole cleaning are: mass balance incorrect while drilling, excessive fill after trips, lack of cuttings on shakers, increase in pump pressure, excessive overpull on connections and trips, erratic and increasing torque while drilling, sudden smooth increase in drag whilst pull out of hole, and stuck shortly after pumps off. Poor hole cleaning will cause hole conditions to steadily deteriorate. Consequently, there should be an opportunity to recognise and react to the problem. Prior to starting a trip, the hole should be circulated until it is as clean as is practically possible. A trip should not be started if there are still significant quantities of cuttings coming over the shakers at that time. It may be beneficial to rotate and reciprocate the string while circulating in inclined wells, as the movement assists hole cleaning by disturbing cuttings beds. A lower rate of penetration results in less cuttings so they can be removed efficiently.

Undergauge Hole

Undergauge hole occurs when the gauge protection on the bit and stabilizer has become ineffective through drilling long sections of abrasive formations. If care is not taken when tripping in the new bit or it can become jammed in the undergauge hole. If a full gauge BHA is tripped into an undergauge hole, a stuck pipe situation can occur, too. Greater care in keeping the hole full gauge will prevent this problem. Also, running large gauge tools will most likely result in stuck pipe, and are normally associated with a mistake in the selection of a downhole component. The undergauge hole is indicated by: an increase in down drag on run in hole (RIH), increasing or erratic torque, bit stalling, and a reduction in rate of penetration (ROP). Preventive actions include: the identification of abrasive/

hard formations, checking bit and stabilizers gauge before and after each trip, slow trips and reaming the hole. If the hole is thought to be under gauged, extreme caution must be applied when tripping into the hole. Additionally, the undergauge hole can result from any of the following causes: (1) plastic flowing formations, (2) wall-cake buildup in a permeable formation, and (3) swelling shales. The importance of maintaining a gauge hole and efficient removal of cuttings has long been recognized. An in-gauge, stable hole not only allows for trouble free drilling but also helps to ensure quality logs, better cement jobs and hence increased profit.

Remedial Procedures for Stuck Pipe

Freeing stuck pipe can be undertaken in a number of ways, depending on what caused the sticking (Table 1). The decision on how long to continue attempting to free stuck pipe vs. back off, plug back, and then sidetrack is an economic issue.

STUCK PIPE RISK ASSESMENT

Many authors have tried to determine how and why a drill pipe becomes stuck and to recommend changes in operating practices in order to reduce the frequency of stuck pipe occurrences and the associated costs. Kingsborough et al. (1985) and Hempkins et al. (1987) were the first in demonstrating the application of *multivariate statistical analysis* to examine a large body of drilling data to find prevailing differences in drilling variables between wells which were non-stuck and wells which were stuck, and to predict the occurrence of stuck drill pipe based upon patterns in drilling parameters. They studied Gulf of Mexico stuck pipe problems, and collected all information from stuck pipe occurrences that might be pertinent to the phenomenon. Data collected for the 131 cases of sticking (experienced in period between 1981 and 1984) included 96 drilling parameters recorded on a daily basis for at least 3 days before and including the day of sticking and 77 items recorded once for each case. The parameters collected included mud properties, depths, hole-geometry information, costs, hydraulics, bit data, bottomhole assembly (BHA) data, operational remarks, etc. Most data came from daily reports and consisted of normal drilling parameters collected and measured in accordance with the American Petroleum Institute (API) recommended practices. A multivariate statistical approach was taken to analyze these data. The stuck pipe occurrences were classified as either differentially stuck or mechanically

Table 1. Stuck pipe causes and recommended freeing methods

Causes		Steps to free stuck pipe
Differential sticking	High differential pressure. Mud cake characteristics (thickness, permeability and lubricity).	Slump string and rotate. Reduce mud weight. Use spotting fluid.
Mechanical sticking	Key seating	Work string down and rotate. Ream the key seat.
	Undergauge hole	Increase mud weight. Underream. Circulate fresh water (in case of salt formation).
	Inadequate hole cleaning (Packing off)	Rotate and reciprocate drillstring. Increase flow rate without exceeding the maximum allowed Equivalent Circulating Density (ECD). Back off and wash over.

stuck on the basis of the observed behavior of the well and operations at the time of sticking. In addition, a third set of data was gathered from wells that had not become stuck for use as a control in the analysis. The nonstuck data were collected from randomly selected wells in the same general areas of operation as the stuck wells.

Nineteen runs were made with the discriminant analysis program. A total of 28 variables (drilling parameters) were investigated including Measured Depth (MD), True Vertical Depth (TVD), average drill gas, maximum drill gas, connection gas, trip gas, survey angle, mud weight, plastic viscosity, yield point, 10-second gel, 10-minute gel, water loss, filtrate pH, filtrate chloride, filtrate calcium, oil percent, water percent, hole size, flow rate, drill-collar outside diameter, bit depth, BHA length, drill-collar length, time stuck, torque, drag, and percent solids (Hempkins et al., 1987). Each run used different combinations of these variables to evaluate the effects of each parameter on the discrimination of the three groups. The final result yielded an overall correct classification of 87%, which means that 87% of the time the discriminant analysis was able to identify correctly the wells belonging to the appropriate predetermined group (differentially stuck, mechanically stuck, and nonstuck). This final result used 20 commonly reported variables. These variables describe most drilling phenomena and relationships including measured depth, true vertical depth, open hole, survey angle, mud weight, plastic viscosity, yield point, 10-second gel, 10-minute gel, API water loss, filtrate pH, filtrate chlorides, water percent, hole size, flow rate, drill-collar outside diameter, drill-collar length, torque, drag, and percent solids. The multivariate discriminant analysis, with these 20 variables, can be used to differentiate between mechanically, differentially, and nonstuck wells with an 81 to 87% success rate. In addition, the stuck pipe discriminant plots can be used to make risk evaluations and decisions when a well is drilled in order to enhance drilling efficiency and reduce stuck pipe cost.

Howard & Glover (1994) used *Multivariate Discriminant Analysis* (MDA) to develop stuck pipe predictive models for wells drilled in the Gulf of Mexico and North Sea areas. Two stuck pipe data bases have been developed with data from more than 1000 wells drilled in the Gulf of Mexico and in the European North Sea. A total of 20 oil operating companies were involved in project, and each company collected about 50 data sets. For the Gulf of Mexico each data set contains three days of information, which was extended to five days for the North Sea data base. Each company submitted half stuck wells and half not stuck wells drilled in the previous five years. Statistical techniques were used to develop stuck pipe predictive models with two classification types. When classifying in three categories (differentially, mechanically or not stuck) - an accuracy of 75% is achieved, rising to 80% for two-class (stuck or not stuck) prediction. Using these models, graphical tracking of the probability of stuck pipe occurring may be undertaken, to monitor drilling operations for stuck pipe avoidance. Multivariate Discriminant Analysis was also used to develop models to predict the probability of freeing pipe once stuck. They also introduced the Freeing Index (FI) as a linear combination of 9 variables identified as being most important such as stuck time, time to position spotting fluid, hours soak before jarring, hours jar functioned, well profile (straight, build & hold, S-type), hole angle and over pull. If the value of the FI is less than -0.7353 (Gulf of Mexico) or 0.2745 (North Sea), the model predicts that the stuck drill string can be freed (Howard & Glover, 1994). The FI correctly predicts 83% freed wells, and 81% not freed for the Gulf of

Mexico, and 90% freed wells, and 86% not freed for the North Sea. Mechanically stuck wells have a higher probability of freeing than differentially stuck; 46% vs. 37% for the Gulf of Mexico, 74% vs. 60% for the North Sea.

Hopkins & Leicksenring (1995) described ways for reducing the risk of stuck pipe in the Netherlands. All the wells drilled in 1993 were thoroughly analyzed to identify the causes of the stuck pipe incidents. Well bore instability and the inefficient removal of solids (cuttings and cavings) accounted for nearly 60% of the events and occurred mainly in the 311.15 mm (12 1/4") and larger hole sizes. If stuck logging tools and casings are excluded, then the share rises to 70%. In more than 90% of these cases, the drill string was being pulled out of hole. Differential sticking was observed mainly in the smaller hole sizes, across reservoir sections, and was identified to be the main cause for stuck logging tools. The *Cost of Stuck Pipe events* and *Stuck Pipe Incident Frequency* (SPIF) were introduced as the primary performance indicators to measure and control Stuck Pipe Incidents. Stuck equipment awareness has been shown to be essential in ensuring faster, safer drilling. Drag charts are an excellent tool to monitor hole conditions, and to assist in interpreting the causes of deteriorating hole. The use of computer programs cart aids the choice of mud weights. Input data can be derived from offset wells and when available, cores and/or logs. The active approach to designing wells by providing mud weight predictions necessary for borehole stability based on offset data, providing advice on hole cleaning (flow rates and rheology), and introducing drag charts and other trend analysis methods has resulted in a significant improvement in well cost effectiveness.

Aadnøy et al. (1999) applied a mechanistic analysis of differentially stuck pipe in a deviated well. The following elements were analyzed: (1) the forces developed during differential sticking, (2) pipe strength under combined loads; tension, torque and pressure, (3) effects of buoyancy under various conditions like equal or different mud densities in drill pipe and annulus, (4) wellbore friction as related to torque and drag. They derived new equations to estimate the depth to the stuck point in deviated wellbores, based on pull and torsion (rotation) tests. In particular it is shown that bends in the wellbore leads to more friction, which with the new equations results in a deeper stuck point in a deviated well compared to a similar vertical well. Equations for pipe strength under combined loading are also defined. In particular, it is shown that the drill pipe can be loaded towards uniaxial pipe strength even if a full pump pressure is applied on the inside. The differential pressure across the stuck interval is the dominating factor. The authors presented three methods to free the pipe, which were developed from the analysis: (1) maximum mechanical force method, (2) minimum density method, and (3) maximum buoyancy method. The most important remedy to free the pipe is to reduce the bottom hole pressure. This can be done by displacing the well with a lighter mud. One can also increase buoyancy by displacing the inside of the drill string with seawater.

There are various predictive models for the assessment and reduction of risks associated with wireline conveyance (Underhill et al., 1998; Prasad et al., 2012.). The likelihood of tool and cable sticking during formation testing operations is a function of multiple parameters that include high overbalance, poor mud quality, long and heavy tool string in highly deviated well, a small difference between the borehole and the tool diameter, duration of stationary times, existence of depleted reservoirs in the open-hole, etc (Prasad et al., 2012.). Computer programs are available to model the forces acting on the tool string during the logging run and evaluate the sticking risk quantitatively at any depth. Such modeling can identify the nature and magnitude of tool deployment risks, and should, in particular, be used in situations where the well geometry, trajectory, mud properties and field experience suggest that the tool string is likely to become stuck. Another computer program that allows quantitative assessment of the risks is Cerberus for Wireline™ (Trademark of NOV CTES). This program analyzes the forces acting on the tool string as a function of depth, and enables risk assessment through management of key parameters like the cable compression, overpull capacity, and weak point strength that are essential for a safe and successful operation (Castillo et al., 2008; Prasad et al., 2012.).

Many stuck pipe incidents occur while tripping and backreaming out of a hole, and it is clear that this is a high risk activity and requires special attention (Yarim et al., 2007 and 2008). Backreaming is the practice of pumping and rotating the drillstring while simultaneously pulling out of the hole. When reliable top drive drilling systems (TDS) on conventional drilling rigs were introduced over 30 years ago, the practice of backreaming has become a popular technique in the driller's toolbox for tripping out of the hole initially in deviated wells. In general, backreaming operations have become a popular solution to poor hole conditions while pulling out of the hole, but they are also known for causing the very same problems they are supposed to prevent, such as stuck pipe (Yarim et al., 2008). Backreaming is generally performed: (1) when there is a problem to trip out of the hole normally without circulation and rotation, (2) to prepare the wellbore to run logs, (3) to prepare the wellbore to run casing, (4) to eliminate tight hole conditions, and (5) to clean the hole to eliminate cuttings beds and hole cavings. The impact of backreaming on stuck pipe incidents can be estimated based on solids induced packoffs which accounted for about 65% of stuck pipe incidents from 2004 to 2006 (Yarim et al., 2007). The majority of these solids induced pack off incidents occurred while tripping out of the hole and backreaming. In addition, most of the drill string components, especially BHA components, are designed to work in compression while drilling with weight on bit, and backreaming exerts some additional loads and forces that can lead to tool failures. The paper published by Yarim et al. (2008) focuses on backreaming operations, reviews the implications of backreaming by using real-time data and case histories, and suggests proven procedures to trouble-free backreaming. Backreaming should not be practiced as a first action if there are indications of a large amount of cuttings in the annular space, or if the indications of wellbore instability exist. The first action should be taken by going down with the drillstring and circulating to clean the hole. If backreaming is not done properly, it may complicate the operations, causing wellbore stability issues, higher equivalent circulating densities (ECDs), and it may also cause stuck pipe incidents due to packoffs. They recommend the conditions that require backreaming and also suggest wellbore conditions that are not recommended for backreaming. More importantly, they propose key drilling parameters that need to be monitored when backreaming. These include pump pressure, torque, hook load, cuttings rate and downhole measurements (if available) to detect and prevent backreaming induced operational problems. In addition, they provide guidelines for successful backreaming operations and define conditions that may be resolved without needing to backream. There are several wellbore conditions that may require backreaming but there are also conditions where backreaming should be avoided. The conditions that may require backreaming are: (1) when tripping out of the hole cannot be carried out on the elevators without excessive overpull and the risk of stuck pipe, (2) if there are concerns about swabbing, especially with balled up bit/BHA (although pumping out of hole could be an alternative), (3) where there is insufficient mud weight to hold back plastic formations e.g., salt, mobile shales in order to allow normal tripping procedures, and (4) if there are known wellbore mechanical issues, e.g., tight hole conditions that cannot be resolved by circulating. The conditions where backreaming should be avoided (if possible) are: (1) when tripping out of the hole is possible without circulation and rotation (i.e., backreaming should not be automatic especially in wells with inclination <30°), (2) when backreaming may destabilize the formation, e.g., where there is pre-existing fractured/failed rock, (3) where cuttings beds exist in high angle wells and the risk of packoff is high, and (4) where rig pumps have insufficient capacity for adequate hole cleaning.

The *fuzzy logic* and *neural network modelling* are successfully used to estimate and predict differential pipe sticking (Siruvuri et al., 2006; Murillo et al., 2009). These methods are widely used in other industries and in the petroleum field, especially in reservoir and core analyses, and make it possible to estimate the risk of stuck pipe occurrence in the well planning procedure and during drilling in real time. The no-convolutional analysis of the model is based on the constraints of different drilling variables. Discriminant function analysis and fuzzy logic were used to classify drilling and mud variables. Extensive simulations were performed. In a study conducted by Murillo et al. (2009) there were 185 datasets generated from the obtained drilling and mud reports. The input parameters used in the fuzzy and neural network model were: measured depth, true vertical depth, bit flow rate, 10 min gel, 10 sec gel, mud weight, plastic viscosity, yield point, calcium filtrate, chloride filtrate, torque, circulating pressure, weight on bit, drag, bit size, bottomhole assembly, the rate of penetration, revolutions per minute, discriminant function F1, discriminant function F2 (Murillo et al., 2009). The output parameters (variables) in the fuzzy and neural network model were: Fuzzy Function, Fuzzy Function 2, Neural Network Function 1, Neural Network Function 2, Stuck, Differential Stuck, Mechanical Stuck, Non-stuck.

The variables of the dataset for modeling development were chosen. These variables were common to both stuck and non-stuck wells. Three major groups of data were identified: mechanical, differential, and non-stuck. These groups form the basis of the entire analysis in this project. The wells were classified into 59 differentials, 68 mechanical stuck pipe, and 58 non-stuck as a control input in the entire analysis. A discriminant analysis is used to generate a predictive model of group membership, based on similar characteristics of each group. Discriminant functions can be built based on a linear combination of the predictor variables. Given a set of independent variables, the discriminant analysis attempts to find linear combinations of those variables that best separate the groups of cases. These combinations are called discriminant functions (F1 and F2) and have the form displayed in the equation (3) (Murillo et al., 2009):

$$F_{1,2} = b + b_1 x_1 + ... + b_n x_n \tag{3}$$

where:

$F_{1,2}$ - Discriminant score
b - Discriminant coefficient

x - Independent variable

The number of possible discriminant functions can be computed by finding the minimum of either number of groups minus 1 or the number of independent variables (Hempkin et al., 1987). In this case, the number of groups being 3 minus 1 is 2 and the number of independent variables (predictor variables) is 18. The final result yielded an overall correct classification of 98.4% cases. Differential stuck and non-stuck cases were 100% correctly classified and the mechanically stuck cases exhibited some misclassification error. Approximately 2.9% of the mechanically stuck cases were misclassified as non-stuck and 1.5% was misclassified as differentially stuck.

Fuzzy logic is an analytical statistical technique developed to handle the concept of "partial truth"; truth values that lie between „completely true"and „completely false". In fuzzy logic, any interpretation is possible, but some are more probable than others. The generated datasets will have error and uncertainty. Fuzzy logic uses this error and uncertainty as a powerful predictive tool to complement conventional techniques. Therefore, this tool is used to complement the discriminant function classification of stuck and non-stuck pipe occurrences with associated probabilities.

The fuzzy model was used to predict stuck pipe occurrences and to provide the optimal values of the variables necessary to move a well from the stuck region into the non-stuck region. The 18 variables were reduced into five dimensionless forms (depth ratio, gel ratio, RB (Rheology Factor * Bit Factor), salt ratio, BR (Bit Force Factor * ROP Factor)). The redefined variables were used as independent variables for the fuzzy input and F1, F2 discriminant functions from the discriminant analysis were used as control input.

A neural network is a powerful data modeling tool that is able to capture and represent complex input/output relationships. The true power and advantage of neural networks lies in their ability to represent both linear and non-linear relationships and in their ability to learn these relationships directly from the data being modeled. Traditional linear models are simply inadequate when it comes to modeling data that contains non-linear characteristics (Murillo et al., 2008). It can be used to extract patterns and detect trends that are too complex to be noticed by either human or other computer techniques (Siruvuri et al., 2006). The motivation for the development of neural network technology stemmed from the desire to develop an artificial system that could perform "intelligent" tasks similar to those performed by the human brain. Mathematically, neural networks can be viewed as a multi-variable non-linear regression. It initially assumes a random relationship between all the inputs and the desired outputs. By comparing its first attempt at an answer to the desired output, it self-modifies this initial random relationship into a relationship that best fits the outputs. The network is presented with many example of inputs and outputs, and relationships are learned after reviewing the examples over and over again, as many as ten million times (Boomer, 1995). The neural network model was also used to predict stuck pipe occurrences and to provide the optimal values of the variables necessary to move a well from the stuck region into the non-stuck region. The redefined variables were also used as the input curves F1, F2 discriminant functions from the discriminant analysis as the output curves. The results showed small error between the predicted and actual functions. There were misclassifications of the type of stuck for both fuzzy logic and neural network (Murillo et al., 2008). Data availability is a potential problem when carrying out an analysis to predict stuck pipe. Insufficient amount of data to train/test and cross validate a network can lead to a poor generalization (Siruvuri et al., 2006).

Meschi et al. (2010) introduced a new parameter called *Reducing Stuck Index* (RSI). They studied mud logging and daily reports of 75 wells in one of Iranian Southwest oil fields. The performance of wells that did not experience stuck pipe was compared with the performance of those that led to stuck pipe. Mud weight, yield point, plastic viscosity, initial gel strength, Marsh funnel viscosity, dial reading at 600 rpm, solid content, temperature, washout and the period that pumps do not need maintenance were employed to introduce RSI which contains all the mentioned parameters. It was found that RSI is proportional to the mud weight, initial gel strength; yield point, solid content and temperature and inversely proportional to plastic viscosity, Marsh funnel viscosity and dial reading at 600 rpm to the power of 0.3, in stuck and non-stuck conditions. The value of RSI for non-stuck well will be greater than stuck well at the particular depth. Monitoring RSI helps the driller to know whether the drilling process results in stuck pipe. If the situation is leading to stuck pipe, the drilling parameters can be managed in such a way that RSI lies in the safe range and stuck pipe is prevented.

Shoraka et al. (2011) developed model for stuck drill pipe in Iranian southern oil field drilling wells for Ilam and Sarvak formations. For this purpose, drilling data including: measured depth, true vertical depth, mud weight, percent oil and water, formation pressure, etc. from forty (40) wells drilled in one of the field were selected (Ilam and Sarvak formation in Ahwaz and Abb-Taymoor oil field). Using the *multivariate statistical regression analysis* an equation representing stuck pipe has been developed. The developed equation can be used for the prevention of risk due to the stuck drill string. Next, using the multivariate statistical discriminant analysis techniques two function were found and they detect differences in drilling variables between wells. The data were gathered for the one hundred fifteen (115) stuck cases and one hundred five (105) non–stuck cases in the study. From one hundred fifteen (115) stuck case, fifty (50) were mechanically stuck and sixty five (65) were differentially stuck, that induced twenty four (24) variables typically reported on morning drilling reports. The drilling data were gathered for the actual sticking day at the time of sticking. Regression analysis and Discriminant analysis are two types of multivariate techniques.

Regression analysis is a statistical tool for evaluating the relationship of one or more independent variables x_1, x_2, x_k to a single, continuous dependent variable Y. In most of statistical approach, one variable is usually taken to be a response or dependent variable that is a variable to be predicted from other variables. The other variables are called predictor or independent variables. One of the first steps of calculating an equation with several dependent variables is to calculate a correlation matrix for all the variables. This matrix shows the correlation between the dependent variable (DIFPR-Differential Pressure, psi) and any other independent variable (GPM-Flow rate, gal/min; AAN-Cross Section of Annulus, in^2; GL_1-10 Second Gel, lbs/100 ft^2; GL_2-10 Minute Gel, lbs/100 ft^2; MW-Mud Weight, ppg; TVD-True Vertical Depth, ft; Formation Pressure, Differential Pressure, psi; ANGLE- Angle of Well, degree; DRCODIA-Drill Collar Outside Diameter, in; DCLENGHT- Drill Collar Length, ft; YP-Yield Point, lbs/100 ft^2, and MD-Measured Depth, ft), and also the correlation among the independent variables. The high correlation among variables is shown by a greater number so that this number ranges from 0 to 1 and the closer the number is to 1, those two variables show greater correlation to each other. The minus sign indicates that those two variables have reverse relationship. The most appropriate model is obtained which can be seen in Equation (4) (Shoraka et al., 2011):

DIFPR = (0.515) TVD + (577.893) MW – (10.893) GL1 + (12.424) GL2 + (1.031) AAN + (0.107) ANGLE – (0.944) FRPR + (5.056) DRCODIA + (0.03577) DCLENGHT – (0.642) YP – (0.01056) GPM – (0.0019190) MD (4)

Using the model obtained above and applying the values related to its variables, the possibility of stuck pipe according to drilling conditions can be estimated, and by changing the values of the above variables, there could be prevention from this occurrence. In a regression analysis, if a number related to DIFPR is between 689.47 – 2757.9 kPa (100 - 400 psi), drilling conditions will be safe and the less predictive of stuck pipe may exist, but, if is outer of this range, risk of stuck pipe will exit (Shoraka et al., 2011).

In a discriminant analysis the statistical problem is to develop a rule, or a discriminant function. Two discriminant functions (Function 1 and Function 2) were proposed, and by using these two functions and substituting the values of the identified variables, stuck probability can be predictable according to drilling conditions (Shoraka et al., 2011):

Function 1:

(-0.001) TVD – (0.993) MW + (0.015) PV – (0.028) YP + (0.371) GL1 – (0.327) GL2 + (0.014) VIS + (0.001) FL + (0.005) DIFPR – (0.004) GPM – (0.062) AAN + (0.003) RPM – (0.001) MD + (0.011) ANGLE + (0.016) SOLID – (0.019) WATER – (0.028) OIL + (0.398) HOLE + (0.003) FRPR – (0.042) DRCODIA – (0.058) PH + (7.939) (5)

Function 2:

(-0.483) MW – (0.013) PV – (0.019) YP + (0.155) GL1 – (0.162) GL2 + (0.115) VIS + (0.105) FL – (0.001) DIFPR + (0.005) GPM – (0.102) AAN + (0.019) RPM + (0.001) CA + (0.004) SOLID + (0.082) WATER + (0.072) OIL – (0.026) HOLE – (0.001) FRPR – (0.218) DRCODIA + (0.002) DCLENGHT + (0.032) PH – (6.239) (6)

By putting the quantities of the variables in these functions for the next three days, two numbers of the above mentioned functions will result, by plotting which in figure 6, a unit point will be reached. The drilling conditions for the next 3 days can be predicted.

A discriminate analysis can be used to differentiate between three groups of wells, using plotting in the graph with a 80% to 86% success rates. For applying the calculated models in the actual field, the position of drilling can be located in the region of "non-stuck" in the graph of discriminant analysis, and by locating the differential pressure in a regression model in the range of 689.47 – 2757.9 kPa (100 - 400 psi), the predicted stuck pipe is prevented.

Chamkalani et al. (2013) describe a robust *support vector regression* (SVR) methodology that offers a superior performance for stuck pipe prediction either mechanically or differentially by using available drilling parameters. This diagnosis guides drilling managers to choose appropriate solutions and remedial operations. A new hybrid Least Square Support Vector Machine (LSSVM) model is developed by using drilling parameters such as measured depth, mud weight, plastic vis-

Figure 6. Position of three day drilling (Shoraka et al., 2011)

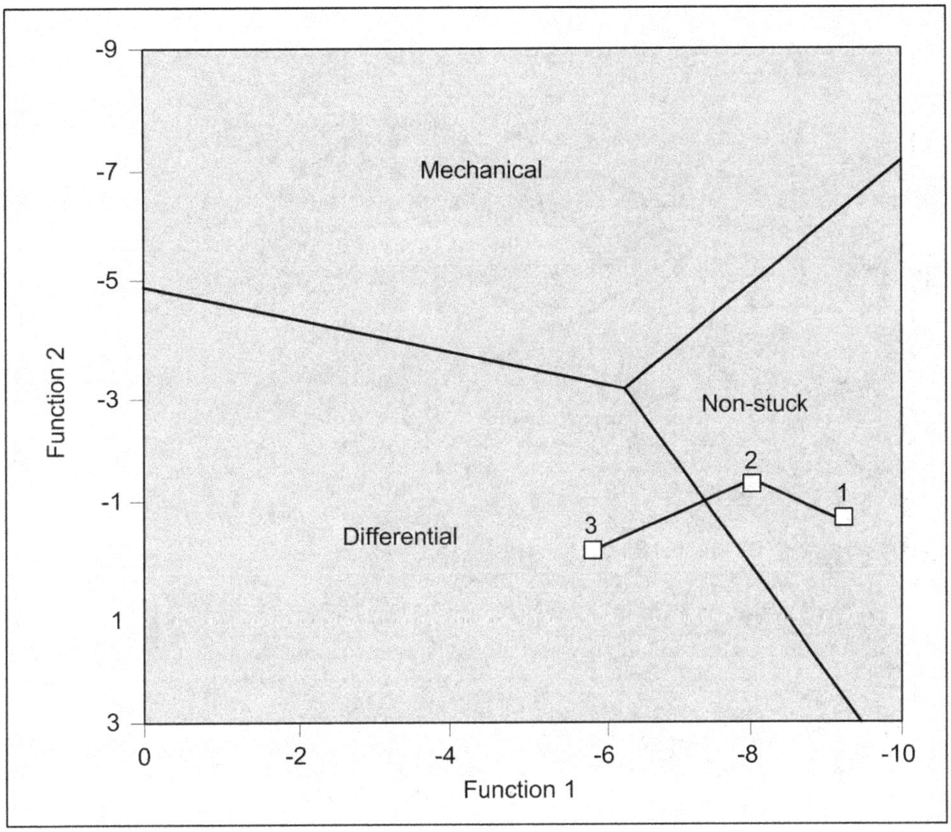

cosity, yield point, gel strengths, PH and solid percent from different wells to predict the pipe sticking. LSSVM model identifies nearly all the cases except some cases which it might be subjected to field observation or human faults and errors. A performance analysis shows that LSSVM classifier has high accuracy. The data provided for this study were gathered from Iranian oilfields in order to classify and predict the pipe sticking (Shoraka et al., 2011). The model is constructed on the basis of classifying the stuck pipe challenge to non-stuck and stuck conditions. Consecutively, the stuck condition was considered as differential and mechanical modes. In the study, 219 sets of data were gathered in which non-stuck had a portion of 109 data, mechanical acquired 51 dataset, and the remaining, 59 data, were allotted to differential sticking. Due to different parameters affecting stuck pipe prediction, it is not possible to visualize the LSSVM's output. Chamkalani et al. (2013) developed a two dimensional correlation which utilized the two discriminant functions proposed by Shoraka et al. (2011). All the inputs are 23 parameters which are transformed into two functions where the real inputs of model are these two functions. The Differential, Mechanical, and Non-Sticking titles were set up for the output. Figure 7 represents a universal

Figure 7. Predictor for determination of stuck pipe into mechanical, differential, and non-stuck

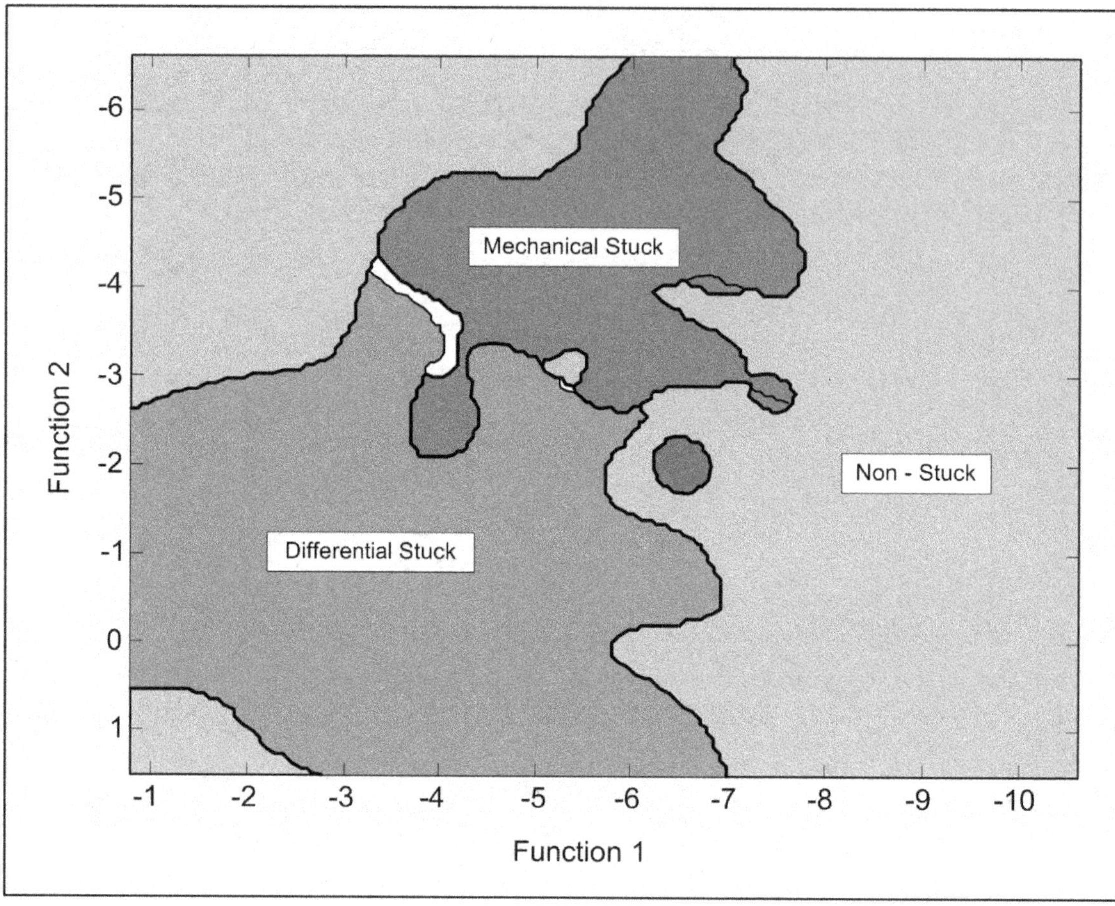

classifier which has validated its performance and preciseness. According to the results, the proposed methodology derives a very precise prediction for diagnosing pipe sticking with accuracy of 95% and higher (Chamkalani et al., 2013).

CONCLUSION

Pipe sticking is a big problem for the petroleum industry and will continue in the future. The recent increase in drilling activity, the shortage of experienced personnel and equipment, and drilling in higher-risk areas has increased the risk of stuck pipe events in all drilling operations. It can occur while drilling, making a connection, logging, testing, or during any kind of operation which involves leaving the equipment in the hole. This problem is known to occur in a form of several major mechanisms including: wellbore instability, differential sticking forces, improper hole cleaning and the forming of drill-cutting beds, especially in high angle wells. The consequences of a stuck pipe are very costly and include: lost drilling time when freeing the pipe, time and cost of fishing, trying to pull out of the hole the broken part of the BHA, and abandonment of the tool in the hole because it is very difficult or too expensive to remove it. The petroleum industry has widely recognized that mud design and training of well site personnel is essential to achieve a reduction of stuck pipe. Many incidents could be avoided by more careful planning or greater care at the rig site. While stuck pipe continues to be a problem much has been done to reduce the occurrence of such events and to minimize the time and cost impact when such events occur. If the risks associated with the stuck pipe are identified in advance, better solutions can be provided to reduce the cost for drilling operation (tripping, fishing) and for drilling mud, to minimize time for drilling a well, and to minimize damage of BHA and surface equipment. Traditionally, stuck pipe problems have been solved by using some standard methods and techniques after they occur, but the real key to savings and success is in the avoidance of the risks associated with the stuck pipe. Many researchers attempt to identify the parameters and their corresponding effects to minimize the risk of stuck pipe by using different models such as Multivariate Statistical Analysis, Discriminated Analysis, Artificial Neural Network Model, Support Vector Machine Model, and Adaptive Fuzzy Logic Model.

Improved planning and risk analysis through computer modeling can significantly reduce the chances of pipe sticking. Such a proactive approach also helps in identifying options and solutions in case of a sticking event. The accuracy of the predictive model depends on the size of database and the variables selected for analysis.

REFERENCES

Aadnøy, B. S., Larsen, K., & Berg, P. C. (1999). *Analysis of stuck pipe in deviated boreholes*. Paper presented at the 1999 SPE Annual Technical Conference and Exhibition. Houston, TX.

Al-Baiyat, I., & Heinze, L. (2012). *Implementing artificial neural networks and support vector machines in stuck pipe prediction*. Paper presented at the SPE Kuwait International Petroleum Conference and Exhibition. Kuwait City, Kuwait.

Ayers, R. C., O'Reilly, J. E., & Henry, L. R. (1989). *Offshore operators committee Gulf of Mexico spotting fluid survey*. Paper presented at the 1989 SPE/IADC Drilling Conference. New Orleans, LA.

Boomer, R. J. (1995). *Predicting production using a neural network*. Paper presented at the Petroleum Computer Conference. Houston, TX.

Bowes, C., & Procter, R. (1997). *Drillers stuck pipe handbook, 1997 guidelines & drillers handbook credits*. Ballater, Scotland: Schlumberger.

Bushnell-Watson, Y. M., & Panesar, S. S. (1991). *Differential sticking laboratory tasts can improve mud design*. Paper presented at the 66th Annual Technical Conference and Exhibition of the Society of Petroleum Engineers. Dallas, TX.

Caenn, R., Darley, H. C. H., & Gray, G. R. (2011). *Composition and properties of drilling and completion fluids* (6th ed.). London: Elsevier.

Castillo, H., Turner, D., & Smalley, E. (2008). *Extreme production logging in long horizontal wells with high flow rates using downhole tractors*. Paper presented at the SPE Annual Technical Conference and Exhibition. Denver, CO.

Chamkalani, A., Pordel Shahri, M., & Poordad, S. (2013). *Support vector machine model: A new methodology for stuck pipe prediction*. Paper presented at the SPE Middle East Unconventional Gas Conference and Exhibition. Muscat, Oman.

Courteille, J. M., & Zurdo, C. (1985). *A new approach to differential sticking*. Paper presented at the SPE Annual Technical Conference and Exhibition. Las Vegas, NV.

Dupriest, F. E., Elks, W. C., Jr., & Ottesen, S. (2010). *Design methodology and operational practices eliminate differential sticking*. Paper presented at the 2010 IADC/SPE Drilling Conference and Exhibition. New Orleans, LA.

Hempkins, W. B., Kingsborough, R. H., Lohec, W. E., & Nini, C. J. (1987, September). Multivariate statistical analysis of stuck drillpipe situations. *SPE Drilling Engineering*, 237–244.

Hopkins, C. J., & Leicksenring, R. A. (1995). *Reducing the risk of stuck pipe in The Netherlands*. Paper presented at the 1995 SPE/IADC Drilling Conference. Amsterdam, The Netherlands.

Howard, J. A., & Glover, S. B. (1994). *Tracking stuck pipe probability while drilling*. Paper presented at 1994 IADC/SPE Drilling Conference. Dallas, TX.

Isambourg, P., Ottesen, S., Benaissa, S., & Marti, J. (1999). *Down-hole simulation cell for measurement of lubricity and differential pressure sticking*. Paper presented at the 1999 SPE/IADC Drilling Conference. Amsterdam, The Netherlands.

Jahanbakhshi, R., Keshavarzi, R., Shoorehdeli, M. A., & Emamzadeh, A. (2012, December). Intelligent prediction of differential pipe sticking by support vector machine compared with conventional artificial neural networks: An example of iranian offshore oil fields. *SPE Drilling & Completion,* 586-595.

Kercheville, J. D., Hinds, A. A., & Clements, W. R. (1986). *Comparison of environmentally acceptable materials with diesel oil for drilling mud lubricity and spotting fluid formulations*. Paper presented at the 1986 IADC/SPE Drilling Conference. Dallas, TX.

Kingsborough, R. H., Lohec, W. E., Hempkins, W. B., & Nini, C. J. (1985). *Multivariate statistical analysis of stuck drill pipe situations*. Paper presented at the 60th Annual Technical Conference and Exhibition of the SPE. Las Vegas, NV.

Krol, A. D. (1981). *Laboratory evaluation of stuck pipe spotting fluid effectiveness*. Paper presented at the 56th Annual Fall Technical Conference and Exhibition of Society of Petroleum Engineers of AIME. San Antonio, TX.

Meschi, M. R., Shahbazi, K., & Pordel Shahri, M. (2010). *A new criteria to predict stuck pipe occurrence*. Paper presented at the SPE North Africa Technical Conference and Exhibition. Cairo, Egypt.

Miri, R., Sampaio, J., Afshar, M., & Lourenco, A. (2007). *Development of artificial neural networks to predict differential pipe sticking in Iranian offshore oil fields*. Paper presented at the International Oil Conference and Exhibition in Mexico. Veracruz, Mexico.

Montgomery, J. K., Keller, S. R., Krahel, N., & Smith, M. V. (2007). *Improved method for use of chelation to free stuck pipe and enhance treatment of lost returns*. Paper presented at the 2007 SPE/IADC Drilling Conference. Amsterdam, The Netherlands.

Muqeem, M. A., Weekse, A. E., & Al-Hajji, A. A. (2012). *Stuck-pipe best practices – A challenging approach to reducing stuck pipe cost*. Paper presented at the SPE Saudi Section Technical Symposium and Exhibition. Al-Khobar, Saudi Arabia.

Murillo, A., Neuman, J., & Samuel, R. (2009). *Pipe sticking prediction and avoidance using adaptive fuzzy logic and neural network modeling*. Paper presented at the 2009 SPE Production and Operations symposium. Oklahoma City, OK.

Nguyen, V., Abousleiman, Y., & Hoang, S. (2007). *Analyses of wellbore instability in drilling through chemically active fractured rock formations: Nahr Umr shale*. Paper presented at the 15th SPE Middle East Oil & Gas Show and Conference, Bahrain International Exhibition Centre. Kingdom of Bahrain.

Outmans, H. D. (1974, July 15). Spot fluid quickly to free differentially stuck pipe. *Oil & Gas Journal*, 65–68.

Pal, K., Mangla, V. K., Joshi, N. P., Tewari, P. P., & Bose, U. N. (2000). *A new approach solves differential sticking problems in Geleki and Ahmedabad fields*. Paper presented at the 2000 SPE Annual Technical Conference and Exhibition. Dallas, TX.

Pasic, B., Gaurina-Medjimurec, N., & Matanovic, D. (2007). Wellbore instability: Causes and consequences. *The Mining-Geological-Petroleum Engineering Bulletin*, *19*, 87–98.

Prasad, T., Castillo, H. C., & Elshahawi, H. (2012). *Effective mitigation of tool sticking risk in formation testing and fluid sampling operations.* Paper presented at the SPE Annual Technical Conference and Exhibition. San Antonio, TX.

Reid, P. I., Meeten, G. H., Way, P. W., Clark, P., Chambers, B. D., Gilmour, A., & Sanders, M. W. (2000). Differential-sticking mechanisms and a simple wellsite test for monitoring and optimizing drilling mud properties. *SPE Drilling & Completion*, *15*(2), 97–104. doi:10.2118/64114-PA

Santos, H. (2000). *Differentially stuck pipe: Early diagnostic and solution.* Paper presented at the IADC/SPE Drilling Conference. New Orleans, LA.

Shoraka, S. A. R., Shadizadeh, S. R., & Pordel Shahri, M. (2011). *Prediction of stuck pipe in Iranian south oil fields using multivariate statistical analysis.* Paper presented at the Nigeria Annual International Conference and Exhibition. Abuja, Nigeria.

Simon, K., Gaurina-Medjimurec, N., & Pasic, B. (2005). Drilling fluids differential sticking tendency determination. *The Mining-Geological-Petroleum Engineering Bulletin*, *17*, 31–35.

Siruvuri, C., Nagarakanti, S., & Samuel, R. (2006). *Stuck pipe prediction and avoidance: A convolutional neural network npproach.* Paper presented at the IADC/SPE Drilling Conference. Miami, FL.

Underhill, W. B., Moore, L., & Meeten, G. H. (1998). *Model-based sticking risk assessment for wireline formation tools in the US Gulf Coast*. Paper presented at the SPE Annual Technical Conference and Exhibition. New Orleans, LA.

Yarim, G., Ritchie, G. M., & May, R. B. (2008). *A guide to successful backreaming: Real-time case histories.* Paper presented at the 2008 SPE Annual Technical Conference and Exhibition. Denver, CO.

Yarim, G., Uchytil, R., May, R. B., Trejo, A., & Church, P. (2007). *Stuck pipe prevention – A proactive solution to an old problem.* Paper presented at the 2007 SPE Annual Technical Conference and Exhibition. Anaheim, CA.

ADDITIONAL READING

Aarrestad, T. V., & Blikra, H. (1994, June). *Torque and drag: key factors in extended reach drilling.* Paper IADC/SPE 27491 presented at the 1994 IADC/SPE Drilling Conferencce, Dalas, Texas, U.S.A

Adams, N. (1977, October). *A field case study of differential pressure sticking*. Paper SPE 6716 presented at the 1977 SPE Annual Technical Conference and Exhibition, Denver, U.S.A.

Addis, M. A., Barton, N. R., Bandis, S. C., & Henry, J. P. (1990, September). *Laboratory studies on the stability of vertical and deviated borehole*. Paper SPE 20406 presented at the 65th Annual Technical Conference and Exhibition of the Society of Petroleum Engineers, New Orleans, LA, U.S.A.

Al-Dossary, A. (2013, March). *New offshore sidetrack practices reduced significantly stuck pipes in KSA*. Paper SPE 16461 presented at the 6th International Petroleum Technology Conference, Beijing, China.

Ayers, R. C., & O'Reily, J. E. (1989, February-March). *Offshore operators committee Gulf of Mexico spotting fluid survey*. Paper SPE 18683 presented at the 1989 SPE/IADC Drilling Conference, New Orleans, U.S.A.

Biegler, M. W., & Kuhn, G. R. (1994, February). *Advances in prediction of stuck pipe using multivariate statistical analysis*. Paper IADC/SPE 27529 presented at 1994 IADC/SPE Drilling Conference. Dallas, Texas, U.S.A.

Brouse, M. (1982). How to handle stuck pipe and fishing problems. *World Oil, November*, 103-24.

Bushnell-Watson, Y. M., & Panesar, S. S. (1991, October). *Differential sticking laboratory tests can improve mud design*. Paper SPE 22549-MS presented at the SPE Annual Technical Conference and Exhibition, Dallas, Texas, U.S.A.

Castillo, H., & Wills, P. (2007, March). *Effective application of an innovative conveyance risk management methodology to reduce well intervention cost and risks in long/deep/tortuous wells*. Paper SPE 106562 presented at the SPE/ICoTA Coiled Tubing and Well Intervention Conference and Exhibition, The Woodlands, Texas, U.S.A.

Guzman, J. M., Khalil, M. E., Orban, N., Mohiuddin, M. A., Verna, J., & Ganda, S. (2012, April). *Stuck-pipe prevention solutions in deep gas drilling, new approaches*. Paper SPE 160875 presented at the SPE Saudi Section Technical Symposium and Exhibition, Al-Khobar, Saudi Arabia.

Haduch, G. A., & Samuels, D. A. (1994, February). Solution of common stuck pipe problems through the adaptation of torque drag calculations. Paper IADC/SPE 27490 presented at the SPE/IADC Drilling Conference, Dallas, Texas, U.S.A.

Harrison, C. G. (1982). Fishing decisions under uncertainty. *JPT*, February, 299-300.

Keller, P. D., Brinkman, P. E., & Taneja, P. K. (1984, May). *Economic and statistical analysis of time limitations for spotting fluids and fishing operations*. Paper OTC 4792 presented at the 1984 Annual Offshore Technology Conference, Houston, Texas, U.S.A.

Kemp, K., Elshahawi, H., Megat, A., Tollefsen, E., Kok, J., & Fey, S. (2011, May). *Analyzing the critical effects of time on station*. Paper presented at the SPWLA 52nd Annual logging Symposium, Colorado Springs, Colorado, U.S.A.

Love, T. E. (1983, February). *Stickiness factor - A new way of looking at stuck pipe*. Paper IADC/SPE 11383 presented at the IADC/SPE Drilling Conference, New Orleans, Louisiana, U.S.A.

Martins, A. L., Sansoni, U., Jr., de Camargo, J. W., & de Oliveira, S. F. (2000, November). *Guidelines for efficient spot fluid displacement in highly inclined and horizontal wells*. Paper SPE 64705 presented at the International Oil and Gas Conference and Exhibition, Beijing, China.

Robson, R. I. (2012, October). *Comparison of stuck-pipe and lost-in-hole outcomes by operator and well complexity in US land taken from a global bit run dataset.* Paper SPE 147555 presented at the SPE Annual Technical Conference and Exhibition, San Antonio, Texas, U.S.A.

Voghell, M., Mohanna, A., Hanley, C., Al-Khiriseh, C., Al-Mousa, A., & Al-Amri, A. (2013, March). *Downhole vibration analysis: Fishing agitation tool efficiency in stuck pipe recovery*. Paper SPE/IADC 163516 presented at the 2013 SPE/IADC Drilling Conference and Exhibition, Amsterdam, The Netherlands.

Wisnie, A. P., & Zhu, Z. (1994, September). *Quantifying stuck pipe risk in Gulf of Mexico oil and gas drilling*. Paper SPE 28298 presented at the SPE 69th Annual Technical Conference and Exhibition, New Orleans, Louisiana, U.S.A.

KEY TERMS AND DEFINITIONS

Differential Sticking: A situation in which the drilling assembly (pipe, drill collars and bottomhole assembly), casing or logging tool becomes embedded in filter cake that was previously deposited on a permeable zone and is held there by a significant amount of differential pressure.

Dogleg: A section of the hole that changes direction faster than anticipated or desired, increasing the overall friction to the drillstring, and the likelihood of getting stuck.

Filter Cake: The residue deposited on a permeable medium, such as filter paper, ceramic disc or permeable zones in the wellbore, when a drilling fluid is forced against the medium under a pressure.

Fishing: The application of tools, equipment and techniques for the removal of junk or debris from a wellbore.

Key Seat: A small-diameter channel, similar to the diameter of the drillpipe, worn into the side of a larger diameter wellbore as the result of a sharp change in direction of the wellbore (a dogleg), or if a hard formation ledge is left between softer formations that enlarge over time.

Mechanical Sticking: The limiting or prevention of motion of the drillstring while drilling caused by junk in the hole, key seats, wellbore geometry (deviation and ledges), formation instability, cement, or a buildup of cuttings in the annulus.

Pack Off: A situation in which the wellbore around a drillstring is pluged as a result of inadequate hole cleaning.

Sidetrack: A secondary wellbore drilled away from an original wellbore to bypass an unusable section of the original wellbore due to an irretrievable fish, junk in the hole, or a collapsed wellbore.

Spotting Fluid: A small volume or pill of fluid placed in a wellbore annulus to free differentially stuck pipe by covering the stuck region, breaking up the filter cake, reducing the stuck area and allowing pipe to be pulled free.

Stuck: An inability to remove the drillstring from the wellbore, lower it back into the wellbore, or rotate the drillstring in the wellbore.

Chapter 4
Lost Circulation

Nediljka Gaurina-Medjimurec
University of Zagreb, Croatia

Borivoje Pasic
University of Zagreb, Croatia

ABSTRACT

Lost circulation is defined as the uncontrolled flow of mud into a thief zone and presents one of the major risks associated with drilling. The complete prevention of lost circulation is impossible, but limiting circulation loss is possible if certain precautions are taken. Failure to minimize lost circulation can greatly increase the cost of drilling, as well as the risk of well loss. The risk of drilling a well in areas known to contain potential zones of lost circulation such as fractured, cavernous, or high permeability formations is a key factor in making a decision to approve or cancel a drilling project. The successful management of lost circulation should include identification of potential loss zones, optimization of drilling hydraulics, and remedial measures when lost circulation occures.

INTRODUCTION

Lost circulation or lost returns are the partial or total loss of mud to subsurface formations. It occurs when the openings in the formation allow the mud flow into the formation, resulting in losses ranging from 0.16 m^3/hr (1 bbl/hr) to a complete loss of returns. When circulation is lost, the drilling fluid costs rise considerably, along with expenses associated with rig and equipment delays. The industry spends millions of dollars a year to combat lost circulation and the detrimental effect it propagates. Potential problems associated

DOI: 10.4018/978-1-4666-4777-0.ch004

with lost circulation during drilling include: loss of mud, lost time, poor cement job, reduced safety, stuck pipe, additional casing string, failure to reach target depth, blowout and kill operations, downhole blowouts, environmental incident, and the abandonment of expensive wells. Lost circulation during cementing could result in reduced annular coverage, casing corrosion, poor zonal isolation and also reduced safety. In addition, during completion/workover operation, existing loss zone can cause the loss of completion fluid, lost rig time, formation damage and lost reserves of oil and gas. In partial lost circulation, mud continues to flow to the surface with some loss to the formation. Total lost circulation occurs when all the mud flows into a formation with no return to surface. If drilling continues during total lost circulation, it is referred to as blind drilling. This is not a common practice in the field unless the formations above the thief zone are mechanically stable, without production and the fluid is clear water. Blind drilling also may continue if it is economically feasible and safe.

The complete prevention of lost circulation is impossible because some formations, such as inherently fractured, vugular, cavernous, unconsolidated (high-permeability zones) or depleted low-pressure formations (usually sands), are not avoidable if the target zone is to be reached (Al Ubaidan et al., 2000; Algu et al., 2007; Bell et al. 1987; Davidson et al., 2000; Ferras et al., 2002). The problem of lost circulation was magnified considerably when operators began drilling deeper and/or depleted formations (Sanders et al., 2003; Suyan et al., 2009). The high hydraulic pressure forces mud to invade the depleted formation. If this situation exists, plans should be formulated to prevent lost circulation or stuck pipe from occurring in the depleted zone. Moreover, lost circulation has even been blamed for minimized production in that losses have resulted in failure to secure production tests and samples, while the plugging of production zones have led to decreased productivity. Special bridging agents and sealing materials should be used to form a good seal and filter cake on the depleted zone. The petroleum industry has invested a significant effort into understanding the mechanisms behind lost circulation, developing new tools to help locate the thief zone, and implementing new steps to minimize or eliminate this problem because prevention is more effective than remediation (Whitfill et al., 2007).

Whitfill (2003) recommends a fully engineered approach that incorporates a number of planning tools, including: borehole stability analysis, hydraulics modeling to estimate Equivalent Circulating Density (ECD), drilling fluid and Lost Circulation Materials (LCM) selection to help minimize effects on ECD.

Conventional lost circulation materials including pills, squeezes, pretreatments, and drilling procedures employing equivalent circulating density management often reach their limit in effectiveness and become unsuccessful in the deeper hole conditions where some formations are depleted, structurally weak, or naturally fractured and faulted (Wang et al., 2005). With continued drilling of deep water wells, High Pressure High Temperature (HPHT) wells, infill wells, etc., the mud-weight window is not only becoming much narrower, but also more uncertain. To address these issues, new lost circulation solutions and concepts such as borehole strengthening or Wellbore-Pressure Containment (WPC) have been developed. Proposed mechanisms behind various means proposed and used to enhance wellbore-pressure containment include sealing incipient fractures at the wellbore wall; propping open multiple short fractures at the wellbore wall, thus increasing compressive stresses around wellbore; and sealing fractures with various materials using hesitation-squeeze technology (Wang et al., 2005; Salehi & Nygaard, 2012).

Where unsustainable losses occcur and conventional circulation is no longer possible the Pressurized Mud Cap Closed Hole Circulation Drilling (CHCD) technique may be used to allow the continuation of drilling in circulation loss zones by controling the annulus pressure and injecting a sacrificial fluid into the thief formation. Masi et al. (2010) discuss the operational details of conducting CHCD as well as the risk-management approach, which involves the identification, evaluation, and mapping of all risks involved in each scenario (qualitative risk assessment). A probabilistic model is, therefore, developed to combine all the risks identified and to address their consequences within operational time.

LOST CIRCULATION ZONES AND CAUSES

Potential zones of lost circulation include highly-permeable formations, highly-porous formations, cavernous and vugular formations, fractured formations with natural or induced fractures, as well as depleted formations (Figure 1).

Highly-Permeable and Highly Porous Formations: Formations such as massive, unconsolidated or shalow sands, gravel beds, shell beds, reef and oyster beds, zones in carbonates (limestone or dolomite) and natural fractures, faults and transition zones in carbonates or hard shales can have sufficiently high permeability (in excess of 14 μm^2 (14 D)) for whole mud to invade the formation matrix, resulting in lost circulation. It can

Figure 1. Potential zones of lost circulation

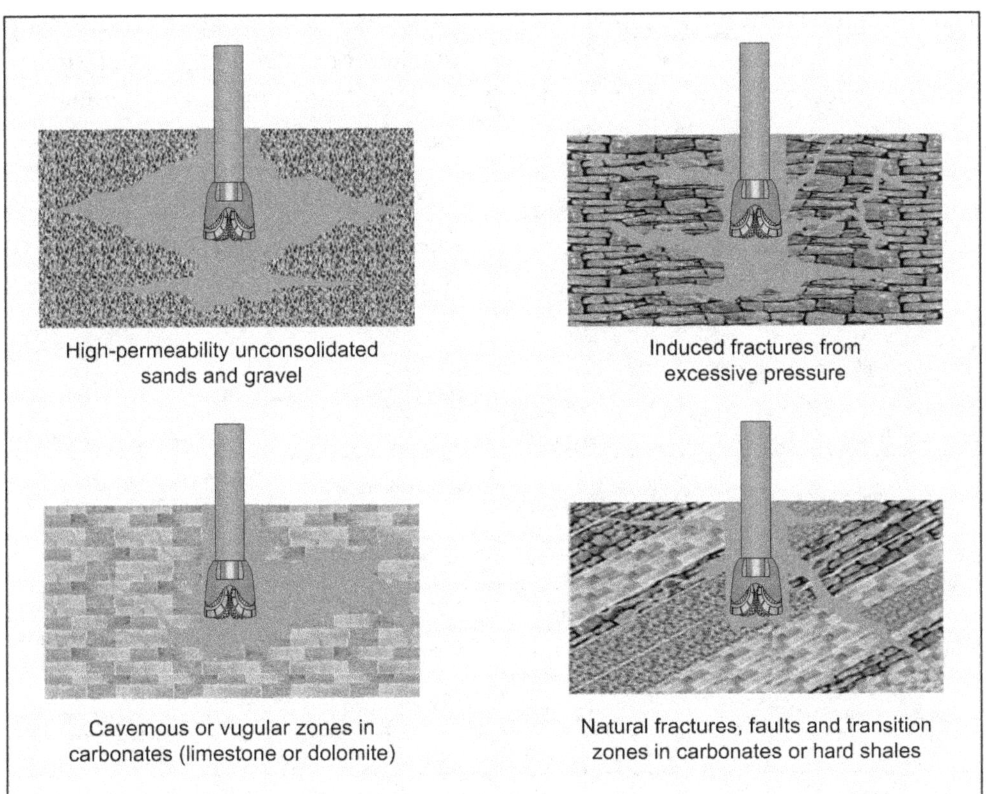

cause continuous gradual seepage losses and partial losses which may develop slowly and increase with penetration. The rate of loss is dependent on the degree of permeability. One important reason for preventing mud loss in shallow intervals is that it may cause these unconsolidated formations to wash out, forming a large cavity that is less stable which could cave in more easily from overburden and rig weight.

Cavernous and Vugular Formations: Formations such as limestones (with large caverns), dolomite, chalk, and other formations with secondary porosity can cause sudden and severe to complete loss of returns which may be accompanied by sudden erratic rates of penetration. Formations of this nature are generally predictable in areas with significant drilling histories. Cavernous or vugular zones are usually associated with low-pressure carbonate (limestone and dolomite) or volcanic formations. In limestone, vugs are created by the previous continuous flow of water that dissolved part of the rock matrix, creating a void space often later filled with oil. When these vugular formations are drilled, the drillstring may fall freely through the void zone and a rapid loss of mud is usually experienced. The volume and persistence of this kind of loss depends on the degree to which the vugs are interconnected. Similar vugs and caverns can develop during the cooling of volcanic magma or ash. Cavernous and vugular formations are often easily traceable from offset wells and predictable from mud logs and lithology.

Fractured Formations with Natural or Induced Fractures: Natural fractures exist in many cases, which may be impermeable under balanced pressure conditions. They may be indicated when losses occur during or immediately subsequent to rough drilling or sudden formation change. Mud loss also occurs in fissures or fractures in wells where no coarsely permeable or cavernous formations exist. These fissures or fractures may occur naturally, or may be initiated or extended by hydraulically imposed pressures. Losses may also occur at unsealed fault boundaries.

Induced Fractures: Are indicated where losses occur while tripping a pipe, breaking circulation, or raising mud weights. Hydraulic fracturing is initiated and lost circulation occurs when some critical fracture pressure is reached or exceeded (Ivan et al., 2002).

The major causes of induced fractures include excessive wellbore pressures (exceeding the fracture gradient) with excessive mud weights or high surge pressures and setting intermediate casing, especially in the transition zone, too high (Table 1). Once a fracture

Table 1. Excessive downhole pressures causes

Mechanical forces	Hole conditions	Mud properties
Excessive pump rates and velocities.	Sloughing shale or increased solids loading in the annulus.	Excessive viscosities and gel strengths.
		Buildup of drilled solids.
Raising or lowering the pipe too fast (surge/swab).	Accumulation of cuttings in a washed out portion of the hole or in the mud.	Thick filter cakes that reduce the hydraulic diameter of the wellbore.
Spudding bridges.	Cuttings beds or barite sag forming on the low side of a directional well, or possible slumping.	Excessive mud density or increasing mud density too fast.
Excessive Rate of Penetration (ROP).	Bridges.	Unbalanced mud columns.
Pipe whipping.	Kicks and well control procedures.	Barite sag.

is created or opened by an imposed pressure, it may be difficult to repair (heal) and it may never regain the original formation strength. Lost circulation may persist even though the pressure is later reduced. This is one reason why it is better to pretreat for, and prevent, lost circulation than to permit it to occur.

Excessive wellbore pressures are caused by high flow rates (high annular-friction pressure loss) or tripping in too fast (high surge pressure), which can lead to mud equivalent circulating densities.

Rapid movement of the drillstring will cause pressure surges that can induce fractures or reopen previously sealed loss zones. Unless a previously sealed loss zone has been pressure tested it should always be assumed this area remains a potential source of losses. The rate of pipe movement during trips and while reaming should be adjusted to compensate for this weakness.

Increasing pump rates too rapidly after connections and trips can put much higher circulating pressures on the formation due to the tendency of muds to thicken during trips. In addition, improper annular hole cleaning, excessive mud weight, or shutting in a well in high-pressure shallow gas can induce fractures, which can cause lost circulation.

Excessive Rate of Penetration (ROP) for a given flow rate will result in high cuttings concentration in the annular fluid causing a high ECD. If intermediate casing is set above the transition zone crossing from normal to abnormal pressures, the pressures exerted by the heavier mud (required to balance the increasing pressures) will often induce fracturing at the weak casing seat. Losses due to fracturing are most commonly near the previous casing seat, not at bit depth even if casing is properly set.

When dealing with induced fractures the problem is even more complicated than with natural fractures, as the shape and structure of induced formation fractures are always subject to the nature of the formation, drilling and mechanical effects, as well as geological influences over time (Sanders, 2003).

Depleted Formations. These formations (usually sands) are another potential loss zone. The problem of lost circulation was magnified considerably when operators began drilling deeper and/or depleted formations. Formation pressure in producing formations decreases due to the extraction of the formation fluids and may become subnormal. Drilling a new well in the same field, or general vicinity, requires adequate mud weights. However, in such a case, mud weights required to control other exposed formation pressures may be too high for the depleted low-pressure formation.

A type of loss is best determined by the lithology. The identifying features of four types of formations in which mud losses occur are presented in Table 2 (see also Figure 1).

PREVENTIVE MEASURES

Lost circulation, as the result of formation fracture or drilling into a porous formation, is largely proportional to the positive differential pressure of the circulating fluid over the formation pressure. Therefore, in order to prevent lost circulation, it is important to maintain density, gel strengths, yield point, and viscosity of the drilling fluid at the lowest levels that still allow for effective drilling, and in particular hole cleaning. High viscosities can increase thc ECD to a level that will break down the formation. Stopping and re-starting circulation take the fluid dynamics through many changes in turbulence (quantity and location) and, in unstable formations, inevitably erode the exposed formation wall, more than if circulation is continuous and steady. Continuous circulation provides the best conditions for establishing a minimum filter cake and

Table 2. The identifying features of thief formations in which mud losses occur

Type of formation	Identifying features
Shallow, unconsolidated formations where rock permeability can exceed 14 μm^2 (14 D).	• Offset records that identify the zone and its characteristics. • Gradual lowering of mud level in pits. • Loss may become complete if drilling is continued.
Carbonate formations that contain cavernous and vugular zones.	• Normally confined to limestone and dolomite. • Loss of returns is usually sudden and complete. • Drill string may drop from a few cm (inches) to a few dm (feet) just preceding loss. • Excessive torque may be experienced before loss.
Formations that have natural fractures such as limestone and some hard shale formations.	• Offset records and geological markers. • May occur in any type of hard, brittle rock. • Loss is evidenced by gradual lowering of mud in pits. • Loss may become complete if drilling is continued and more fractures are exposed.
Formations that have fractures induced through mechanical or hydraulic forces exerted on them.	• Offset records that include excessive annular hydraulics. • Could occur in any type of rock, but would be expected in formations with characteristically weak planes, such as soft shales. • Loss is usually sudden and complete. • Loss may follow any increase in mud weight or sudden surge in pressure. • Can be in competent or incompetent formations.

minimising erosion of the exposed formation. The steady pressure, throughout the drilling of the section, is most likely to minimise the loss of mud to the formation. A return flow rate was often determined by measuring the level of drilling fluid in the mud pit or by installing paddle-wheel meters in the return line. When the downhole flow rate is higher than the return flow rate, it is an indication of lost circulation. The ability to detect small differences between the downhole and the return flow rate can be hampered by the limited accuracy, sensitivity, and reliability of these two methods. Any variation in the return flow can be taken as an early sign of lost circulation or a kick and the response can be quicker and therefore more effective. The possibility of lost circulation can be limited by strengthening the wellbore, also. The most widely used method to increase wellbore strength involves utilizing specially-sized granular additives, others include rigid-plug forming treatments, chemical sealants and low-fluid loss muds. Wellbore-strengthening techniques are similar to lost-circulation techniques, but there are significant differences. Strengthening techniques still focus on avoiding loss, but also on increasing the apparent fracture gradient by sealing the fracture and isolating the fracture tip to keep it from further elongating, while lost circulation only mitigates losses of whole drilling fluid.

LOCATING LOSS ZONES AND DETERMINING THE SEVERITY OF LOSS

Locating the Loss Zone

Locating the exact point of lost circulation and restoring circulation as quickly as possible is very important since lost circulation increases mud cost and can cause other wellbore problems that result in additional operational costs. The probable location of the loss zone may be determined from knowledge available at the rig site such as operational information and information about drilled formations. Loss zone could be on bottom or off bottom of the hole. Losses are normally "on bottom" if: (a) they first occur while drilling ahead, (b) the loss is accompanied by a notable change in the rate of penetration (ROP), torque, or drilling vibration, and (c) the loss is due to natural fractures, faults, caverns, vugs or high permeability sands and gravels (well site geology data, drilling break with increase in torque and kelly free falls while drilling coupled with a instant loss in circulation). Losses are normally "off bottom" if: (a) they first occur while tripping, drilling fast or increasing mud weight, (b) they are the result of shutting the well in and killing the well, and (c) the annular loading is such as to increase the return apparent mud weight higher than the last casing shoe fracture gradient.

Contrary to the common belief, the majority of the losses do not occur at the bottom of the hole. It has been established that more than half of the losses occur just below the last casing shoe.

Several types of tools and methods are available for locating the thief point of lost circulation. These include (Power et al., 2003; Wang et al. 2005; Chilingarian & Vorabutr, 1981):

- Real-time geomechanical analysis,
- Pre-drill geological analysis,
- Pressure transducer surveys,
- Open hole logs,
- Hot wire surveys,
- Radioactive transducer surveys,
- Temperature surveys, and
- Spinner surveys.

The Real-Time Geomechanical Analysis: This process combines personnel, software tools, data synchronization and visualization in a process to transform all available data into actionable information to optimize drilling. It begins with data gathering as part of pre-drill planning and is updated in real time by measurements obtained while drilling. The program is a mechanism for deciding what information is needed to make decisions, developing a plan to acquire the information in the right time frame, interpreting results through tailored software tools and reporting the results and recommendations in a clear and precise manner (Power et al., 2003). A fracture prediction model can be included that employs an array propagation resistivity tool to reliably evaluate environmental effects such as hole enlargements or drilling induced fractures (Bratton et al., 2001). Two resistivity channels are monitored in real time. One real-time signal is a low-frequency phase shift resistivity that reads deep into the formation. It is insensitive to tool eccentricity and large boreholes when shale resistivity is low. The second signal, sent to the surface continuously, is the tool's shallowest measurement, the one most sensitive to eccentricity, hole size enlargement and fracturing. The separation of these two curves is an early indication of hole problems or induced fractures (Caughorn et al., 2002).

Pre-Drill Geological Analysis: This analysis can also be used for determining thief zone location.

- If the losses are experienced while drilling, it is likely that the loss zone is on bottom and is caused by natural fractures, caverns or highly permeable formations.
- Drilling into a sub-normally pressured, naturally fractured formation is usually indicated by a sudden high loss of returns accompanied by an increase in rotary torque. This is a reliable indication that the lost circulation zone is at a bit depth where no previous problems have been encountered.
- Induced fractures on bottom can be caused by a balled-up BHA or bit restricting the annulus.

- If losses are experienced while tripping or while increasing mud weight, then it is likely that the loss zone is not on bottom and is due to induced fractures. Recognizing losses while tripping back into the hole requires attention to the volume of fluid being displaced by the pipe. This volume can be determined by either visual observation or from regular examination of the pit level record.

Pressure Tansducer Surveys: This type of survey involves using a short cylinder open at the top and swaged at the bottom to restrict the flow of mud through the tube. A window with a neoprene diaphragm is fitted on one side of the tube. There is an electrode that moves back and forth between the two fixed electrodes on the diaphragm. As the pressure differential varies across the diaphragm, the potential varies in the electric circuit indicating the rate of flow of the mud and where the mud becomes static.This method appears to have certain advantages: (a) it is simple in construction and operation, (b) it is not easily clogged by lost circulation material (LCM), (c) it is workable in almost any type of mud, and (d) it can be used to locate a hole inside of casing.Two apparent disadvantages are: considerable mud flow is required and the equipment may not be readily available.

Open Hole Logs: Lost circulation problems occurring in weak formations far below the casing shoe can be located with logging data. Formation resistivity and annular pressure measurements, combined with time-lapse logging data, can be used to determine a more accurate fracture pressure, enabling cost-effective real-time drilling decisions (Akamine et al., 2003). Resistivity data are used to diagnose and precisely locate fractures. Pressure data are used to quantify fracture opening and closing. Early fracture diagnosis while drilling allows time to plan to reduce lost-circulation events. The resistivity and pressure data, viewed in both time and depth domains, contributes to a better understanding of fracture behavior. The abnormal real-time resistivity readings, suggest the initiation of fractures. It can be confirmed with time-lapse measurements made while tripping out of the hole. If the real-time resistivity data show elevated resistivities that can be indication of fracture growth. It can be confirmed with time-lapse measurements. The analysis provides the location of the problem zone, the formation type, and the wellbore pressure activating the fractures. The additional information provide a better understanding of the initiated fracture characteristics and enable drilling the section to total depth without mud losses.

Hot Wire Surveys: The hot wire tool is essentially a calibrated resistance wire that is sensitive to temperature changes. It is run to a desired point in the hole and the resistance is noted. Mud is then pumped into the hole. If the tool is above the point of loss, mud will flow by it, changing the resistance. If the resistance does not change, the tool is below the point of loss. The tool can be used in any kind of mud, but a large amount of mud is required while making the survey.

Radioactive Tracer Surveys: Radioactive surveys for locating the point of loss consist of making two gamma ray surveys. A base log is run before the introduction of radioactive material. A slug of mud containing radioactive material is then pumped down the hole and a new log is run. High concentrations of the radioactive material will be located at the point of loss. This method provides accurate data for locating the point of loss, but requires expensive equipment and additional deliberate loss of mud to obtain the desired information.

Temperature Surveys: The temperature survey depends on a subsurface thermometer for measuring the difference in the mud temperature and the formation temperature.

Figure 2. Lost circulation zone located by temperature survey (Goins & Dawson, 1953)

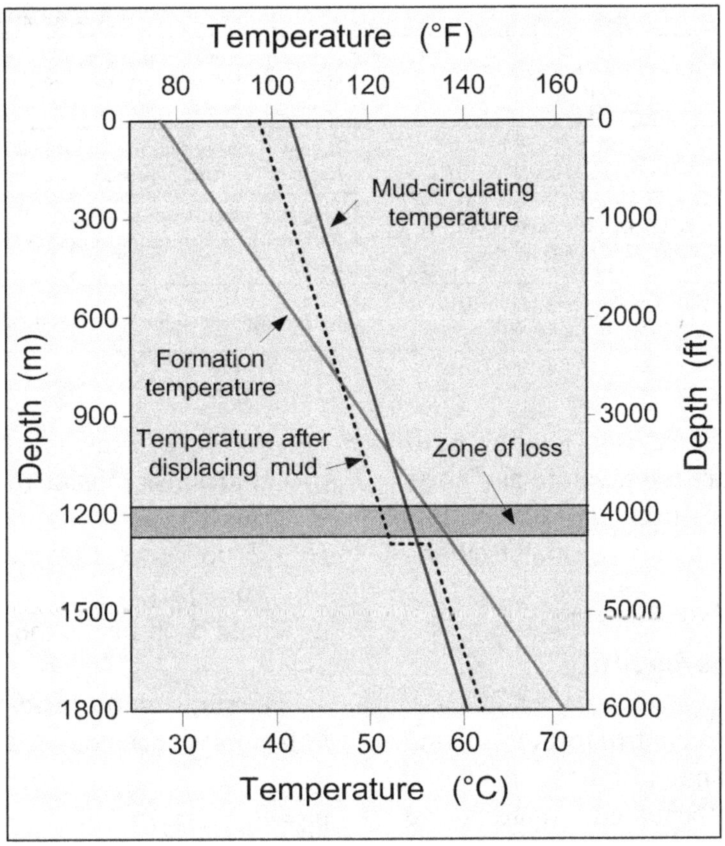

It is made by running a sensitive element in the hole that changes its resistance as the temperature changes. Two surveys are run. The first is run to establish the temperature gradient of the well after the mud has come to equilibrium with the formation. The second survey is run immediately after adding fresh cool mud to the well. A sharp temperature discrepancy will occur at the point of loss (Figure 2) (Goins & Dawson, 1953).

Spinner Surveys: The spinner survey is made by running a small spinner attachment into the well on a single conductor cable in such a manner that the vaned rotor will spin or turn if there is any horizontal motion of the mud. The rotor speed are recorded on film as a series of dashes or spaces. The rotor speed will be very slow until the point of loss is reached. There is a definite increase in the rotor's speed at the point of loss.There are two objections to this method: (a) it requires deliberate loss of large volume of mud, and (b) it is not effective where sealing material is already present in the mud.

Though it is a good practice to locate the thief zone, there are several reasons why surveys are not run more often. They are: (a) considerable time is spent getting the necessary equipment to the rig, and a deliberate loss of mud is required for these surveys, (b) the results of these surveys are sometimes difficult to interpret, and (c) conditions are not always such that the tools may be run because of abnormal subsurface pressure.

Table 3. Classification of mud loss according to its severity

Severity of loss	Rate of loss	Type of formation
Seepage losses	0.16 to 1.6 m^3/hr (1 to 10 bbl/hr)	Any permeable formation type when the bridging agents are not large enough to form a seal, or when there are no fine particles to complete the seal.
Partial losses	1.6 m^3/hr to 16 m^3/hr (10 to 100 bbl/hr)	Highly-porous or fractured formations, gravel, small natural fractures and barely opened induced fractures.
Complete loss (total loss of returns)	hole full to mud level at 61 to 152 m (200 to 500 ft)	Long, open sections of gravel, long intervals of small natural fractures, large natural fractures (caverns, vugs) or open induced fractures.
Severe complete loss	hole full to mud level at 152 to 305 + m (500 to 1,000+ ft)	Large, open natural fractures, caverns and open induced fractures.

The most commonly used logging tools for flow detection are temperature and noise tools. Oxygen activation and water flow measurements tools are also available (Wang et al., 2005).

Determining Loss Severity

The best approach to control lost circulation is to make an assessment of the severity of a loss zone and match the remedial material and technique to it in terms of both the size of the material and its function. The maximum allowable drilling fluid loss is of the order of 0.16 m^3/hr (<1 bbl/hr) (Chilingarian & Vorabutr, 1981). This loss can be measured in the pits at the surface. Preventive measures must be adopted in case the mud loss exceeds 0.16 m^3/hr. The severity of loss is best determined by the amount of loss and the static mud column height (Table 3).

In case of complete loss the static fluid level can be determined and then the pressure within the loss zone (P_z) is calculated as follows:

$$P_{lz} = \rho_m \cdot g \cdot (H_{lz} - h_m) \qquad (1)$$

where:

P_{lz} - pressure of the loss zone [Pa],
ρ_m - mud density inside the drill pipe [kg/m^3],
g - gravity [m/s^2],
H_{lz} - True Vertical Depth (TVD) of the loss zone [m],
h_m - mud drop inside the drill pipe [m].

The static mud density (ρ_{lz}) which the zone will support is calculated as follows:

$$\rho_{lz} = P_{lz}/(g \cdot H_{lz}) \qquad (2)$$

where:

ρ_{lz} - mud weight that the zone will support [kg/m^3].

LOST CIRCULATION MATERIALS

The standard practice when using conventional drilling techniques through loss circulation zones is to minimize losses by attempting to seal off the thief zone prior to continuing drilling. This is usually done by pumping *Loss Circulation Materials pills* (LCM's) and setting *gunks* and *cement plugs* such that the mud weight will overbalance formation pressure at all times (Masi et al., 2010).

The common Lost Circulation Materials (LCMs) that generally are mixed with the mud to seal loss zones may be grouped as *fibrous* (e.g. wood and mineral fiber, paper), *flaked* (e.g. cellophane, mica), *granular* (e.g. nut shells, graphite, calcium carbonate) and *a combination of fibrous, flaked, and granular materials* (Whitfill & Hemphill, 2003; Whitfill, 2008; Kefi et al., 2010). LCM slurries (pills) containing a blend of materials are most effective. These materials are available in fine, medium, and coarse grades for an attempt to seal low to moderate lost-circulation zones (Figure 3) (Howard & Scott Jr., 1951). The size and concentration of each LCM in the pill should be adjusted according to the severity of loss, and to the type of fluid in the hole (Ivan et al., 2003).

The size of the LCM material must be tailored to the severity of loss since the sizes of the openings taking the mud are seldom known. This logic is based on the knowledge that low rates of loss occur through small openings and higher rates of loss occur through larger openings. LCM materials that are too large will form a bridge at the face of the opening, while properly sized materials form a bridge in the opening where it is not as likely to be eroded away when circulation is restored (Figure 4).

LCM as blends of materials of different sizes, shapes and properties may be used one time or on a regular basis while drilling ahead. Particle size distribution of LCM should match the distribution of openings in rock (Bruton et al., 2001; Dick et al., 2000). They are mixed in mud from the circulating system or in freshly prepared, viscous, bentonit slurry. When selecting the size of LCMs to use, the size of the nozzles in the bit and other posible restrictions (e.g. Measurement While Drilling (MWD) tools, mud motors) should always be considered.

Figure 3. Effect of LCM material concentration and type on sealing fractures (Howard & Scott Jr., 1951)

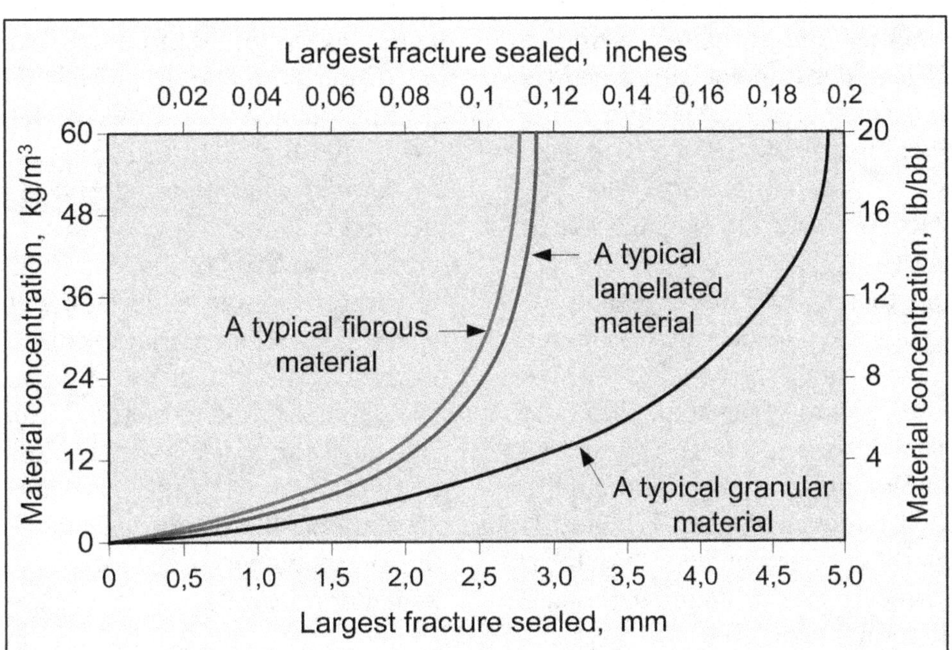

Figure 4. Improper (left) and proper fracture bridging (right)

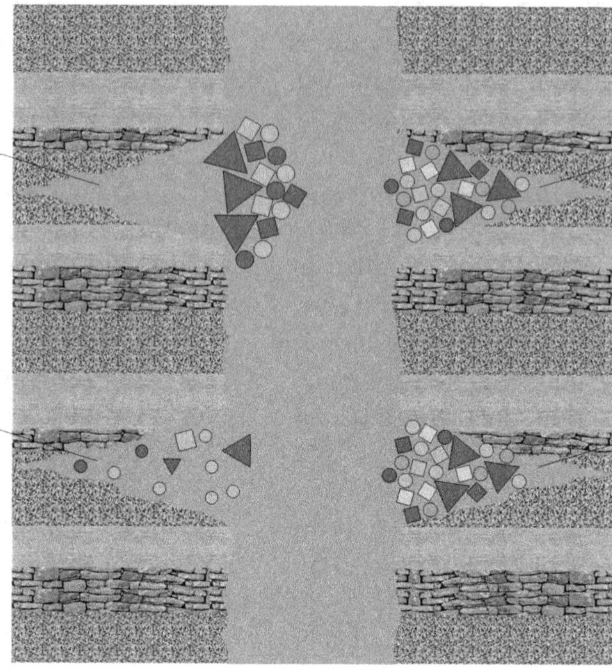

The control of seepage losses requires correctly sized LCM (fine granular material, fine fiber and cellophane flakes) and is ideally based on the knowledge of the fracture size or pore throat size in the seepage zone. Wellbore-imaging and formation samples are required in order to determine fracture/pore size and fracture density information. The Ideal Packing Theory (IPT) is then used for defining the right Particle Size Distribution (PSD) to eliminate the seepage losses (Power, 2003).

According to the IPT, the D_{90} value of the PSD (90% of the particles are less than a certain diameter) should be equal to the size of the fracture opening (Ivan et al., 2002). If the LCM particles are too large (Figure 3), the bridge builds up on the wellbore and is quickly eroded, while smaller LCM (Figure 4) goes through the open ing and does not form a bridge. Correct1y sized calcium carbonate and synthetic graphite have been highly successful in reducing and eliminating seepage losses, in both Water-Based Mud (WBM) and Synthetic-Based Mud (SBM). The synthetic graphite is inert and does not react with the mud, thereby not causing unwanted viscosity increases. One of the more significant properties of the synthetic graphite is its ability to deform under compression without breacking down, and the ability to heal fractures that are formed when the material is already in the mud system.

As the severity of the loss zone increases, only the size of the bridging material should be increased. Concentrations of LCMs above 60 kg/m^3 (20 lb/bbl) do not improve effectiveness. Coarse granular material, fine to coarse fiber and coarse flakes should be used against complete losses. In the case of severe lost circulations, the use of various plugs (gunk plug or cement plug) to seal the zone becomes mandatory (Ivan et al., 2003, Fidan et al., 2004; Romero et al., 2006). The gunk plug forms when water comes in contact with bentonite that has been placed in the hole as a concentrated suspension in diesel oil. Water causes the bentonite to swell to form a putty-like mass. When the gunk plug is placed through the drill pipe, care must be taken to have enough diesel oil precede and follow the bentonite suspension to serve as a barrier to prevent any mixing with water before the suspension has been displaced from the pipe. Typically a mixture of 825-1140 kg/m^3 (300-400 ppb) of bentonite in diesel or synthetic oil is pumped. The bentonite affinity for water causes rapid thickening (or hardening) that can seal a zone of loss. A 50:50 mixture of bentonite/oil and mud can plug open holes because it gives near instantaneous setting. The one drawback to this method of curing losses is that you are to be pretty sure that you know exactly what part of the hole the loss zone is in.

When total loss of returns is attributed to large vugular and cavernous formations a polymer plug may be spotted and allowed to hydrate and expand within the formation with time and temperature. Chemical reactive polymer pills consist of blends of loss circulation materials and a hydrateable cross-linked polymer that is spotted across the zone of interest and then squeezed at 1035- 1380 kPa (150-200 psi) (Bruton et al., 2001).

Polymer plugs are either surface activated as they are mixed and then chemically activated or cross-linked as pumping down hole begins or down hole activated when the pill contacts mud or formation fluid forming a plug. Down hole activated polymer pills are high concentrations of polymer suspended in an inert carrier such as diesel or mineral oil. These polymer plugs form hard very plastic plugs when the pill contacts the mud or formation fluid and some are used to control water flow on producing wells. Down hole activated polymer pills can be selected to work in various brines. Both surface activated and down hole activated pills react very much like bentonit diesel oil pills when contacting water (Figure 5).

High solid-content and high fluid loss squeeze pills are effective if carrying fluid leaks off into formation. Fluid leakoff requires formation permeability and a pressure differential. This type of treatment typically works well in depleted, highly permeable formations with water-based mud (Hang et al., 2005; Sanders et al., 2010). This treatment would hardly work in impermeable shale.

However, this is a long and often unsuccessful process because multiple leakoff zones may be encountered in sequence or because massively fractured zones cannot be cured. Furthemore, these methods have several disadvantages such as: offering uncertain results, being time consuming and therefore expensive (time and cost increase), and finally in hydrocarbon reservoirs these may seriously reduce well productivity (damages on well productivity). Major differences in permeability between shale, sandstone and vugular limestone show that one-size-fits-all solutions are not going to be effective or efficient (Whitfill & Wang, 2005).

Where unsustainable losses occcur and conventional circulation is no longer possible the Pressurized Mud Cap Closed Hole Circulation Drilling (CHCD) technique may be used to allow the continuation of drilling in circulation loss zones by controling the annulus pressure and injecting a sacrificial fluid into the thief formations, thereby minimizing the non-productive time and costs (large amounts of lost drilling fluid and non-productive rig time) typically required to implement the traditional methods of circulation loss control (Masi et al., 2010). The CHCD technique is not simple or an inexpensive system. It is not underbalanced drilling. The annulus is closed

Figure 5. The operational procedure of chemical sealant placement

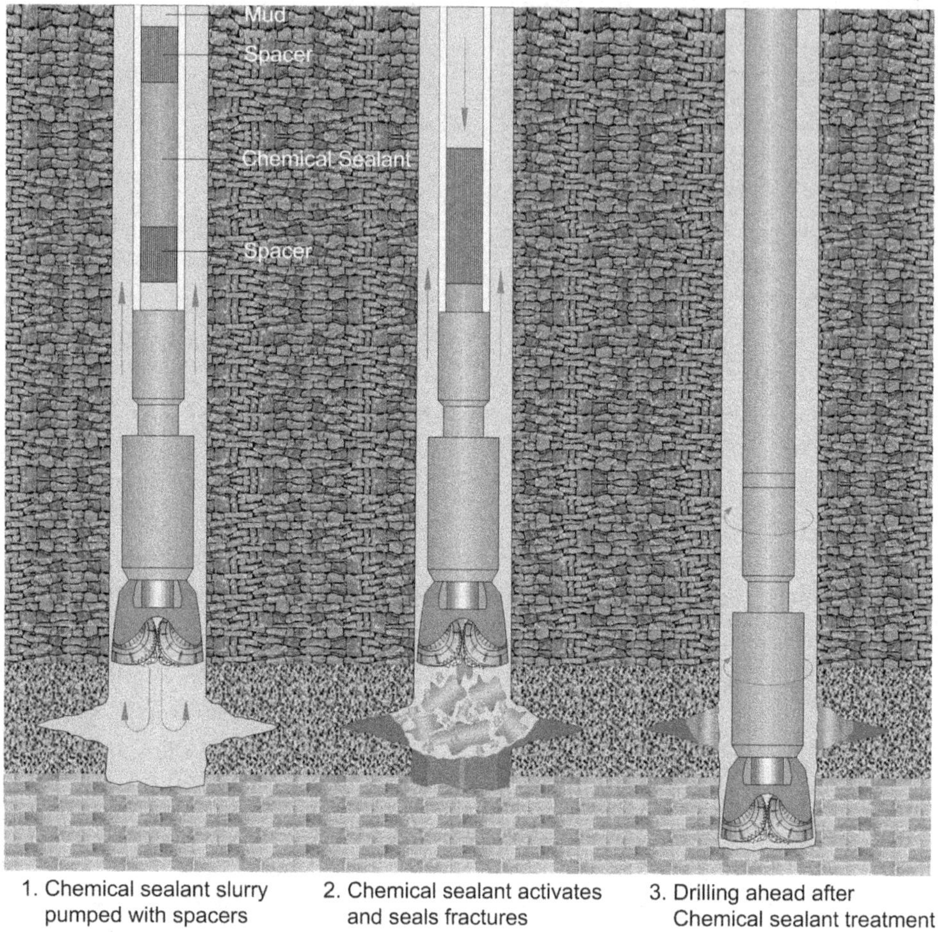

and no returns are circulated to the surface. It is used only when other systems for controling losses have proven to be ineffective. Large quantities of water and mud are required as well as additional equipment and specialized personnel. An accurate evaluation of several factors has to be performed. Drilling hazard is the major issue as the primary barrier to well influx is jeopardised, but also rig time, material availability and consumption have to be considered. Masi et al. (2010) developed a probabilistic model to combine all the risks identified and to address their consequences within operational time (quantitative risk assessment).

REMEDIAL PROCEDURES FOR LOSS CIRCULATION

When lost circulation is first noted, the conditions at the time the loss occurred should be accurately recorded and analyzed. The time of the occurrence (while drilling, circulating or tripping), the type of the loss (seepage, partial or complete) and the severity of the loss with respect to the exposed formations should be considered. This information will help determine why the loss occurred, where in the hole the loss occurred and the best remedy for the situation. The probable location of the loss zone may be determined from knowledge available at the rig site such as operational information and information about drilled formations. Failure to minimize lost circulation can greatly increase the already high cost of drilling, as well as risking loss of the well. Successful management of lost circulation should incorporate identification of potential loss zones, optimization of drilling hydraulics, and clear, easy to follow procedures should lost circulation be encountered. Many times when lost circulation is encountered, no log data or time for detailed log analysis is available. Parameters required for designing a treatment would rely on diagnostic methods. One of the major causes for a failed treatment is the loss zone identification. Accurate information is very important for a good definition of the loss problem and for selection of the best type of LCM and its placement procedure into the well (Table 4).

THE APPLICATION OF RISK ANALYSIS FOR LOST CIRCULATION

Risk analysis can be a useful tool in certain processes where uncertainty such as lost circulation is involved. The implementation of risk analysis involves three basic steps (Cunha, 2004): (1) identifying an opportunity (or event) where the tool can be applied, (2) quantifying the consequences of various possible decisions and (3) assessing, within the possible outcomes, the estimated best economic or operational result. The use of risk analysis tools requires methodical quantification of uncertainties and use of data from past operations that are not always easily available.

In order to have any method implemented, a careful analysis of existing data is necessary so that the quantification of uncertainties can be as realistic as possible. Historical data can be of great value as a base line. In

Table 4. Recommended treatments for different types of loss

Severity of loss	Recommended treatment
Seepage losses	Pull up and wait technique. Treat system with fine lost circualation material (LCM). Accompany LCM treatments with additions of a viscosifier to improve particle-size distribution. Avoid excessive circulation rates and/or penetration rates to minimize ECD and solids loading of the annulus. Reduce mud weight if possible.
Partial losses	Treat system with various sizes of LCM. Mixtures of different shapes are often quite effective. For severe losses, spot pill opposite the loss zone utilizing 86 to 114 kg/m^3 (30 to 40 lb/bbl) of various types of LCM. Pull up into casing and allow 6 to 8 hours. Squeeze the zone with an high-filtration slurry. The same techniques used in seeping losses should be used in partial losses.
Complete loss	Pull up and wait technique. Use high-filter-loss slurry squeeze technique. If returns are not regained with this technique, a hard plug such as a cement, cement-bentonite, cement-gilsonite or diesel-oil-bentonite-cement is recommended (see techniques for cement slurries). Reduce mud weight if possible.
Severe complete loss	Squeeze with either a high-filter-loss slurry squeeze or large amounts of diesel-oil-bentonite-cement slurries. Spot plug (e.g. Thermally Activated Oil Gelling Pill, Polymer-based Cross-linked Pills (PCP), Chemically Activated Cross-linked Pills (CACP), Reverse Gunk, Barite/Hematite Pills, Thixotropic Cement) If the severe, complete losses continue to occur as more open, natural fractures or caverns are penetrated, the technique of drilling blind or with aerated mud and setting casing should be considered. Reduce mud weight if possible.

areas where there is a lack of reliable data, information from other areas or fields with comparable situation can be used. The use of risk analysis was deemed as fundamental to minimize losses or to maximize the possibility of adopting, for a certain situation, the right decision. Particularly, for drilling operations there have been various articles where different aspects of the drilling process have been studied using risk analysis as an auxiliary tool on the decision-making process.

One characteristic of the use of the quantitative risk analysis is that normally, after implementation, its use is fairly simple and straightforward. The objective of risk analysis is to produce outputs that can be used to evaluate the nature and distribution of risk, and to develop appropriate strategies to manage the risk. Risk assessment methods can be qualitative or quantitative. Qualitative methods use descriptive terms to identify consequences and likelihoods of events and resultant risk as the product of consequence and likelihood. Quantitative methods identify likelihoods as frequencies or probabilities, and consequences in terms of relative scale (orders of magnitude) or in terms of specific values (for example estimates of cost). Monte Carlo simulation methods are often used to incorporate uncertainty into the quantitative estimates.

Once an event is identified and likelihood (incident occurrence probability) for all possible results have already been quantified, the last step will be to apply the model in order to have expected results (outcomes) for the various actions possible to be adopted.

Qualitative approaches to risk assessment are the most commonly applied, and the risk matrix can be used as a method to display the combinations of likelihood and the severity of consequences. They enable to separate risk events into risk ratings. There are simple and complex risk assessment matrices but all present categories of the likelihood that a particulart event will occur (incident occurrence probability) and the severity of consequences (the magnitude of harm or damage that could result).

The consequences of lost circulation can be as little as the loss of a few dollars of a drilling fluid, or as disastrous as a blowout and loss of life, so close monitoring of tanks, pits, and flow from the well, to quickly assess and control lost circulation, is taught and practiced. If the amount of fluid in the wellbore drops due to lost circulation, hydrostatic pressure is reduced, which can allow a gas or fluid being is under a higher pressure than the reduced hydrostatic pressure to flow into the wellbore. Another consequence of lost circulation is blind (or dry) drilling in which the drilling fluid is not returned to the surface. The effects of blind drilling range from as minor as destroying a bit to as serious as major damage to the wellbore requiring a new well to be drilled. Blind drilling can also cause severe damage to the drill string, including snapping the pipe, and the drilling rig itself.

Qualitative risk assessment methods use descriptive terms to define the likelihoods and consequences of risk events. Outputs from qualitative risk analyses are usually evaluated using a risk matrix format, such as the example in Table 5. Events with more significant consequences and likelihood are identified as higher risk, and are selected for higher priority mitigation actions in order to decrease the likelihood of the event occurring and/or reduce the consequences if the event was to occur.

Risk should be ranked so that priorities can be established. Each risk is categorized in accordance with the terms shown in Table 5. such as: Extreme (E), High (H), Moderate (M), and Low (L). Generally, remedial actions or acceptance for different risk category are:

Table 5. An example of a qualitative risk matrix

CONSEQUENCE		LIKELIHOOD				
		Rare E	**Unlikely D**	**Possible C**	**Likely B**	**Almost Certain A**
Severe	5	M	H	H	E	E
Major	4	M	M	H	H	E
Moderate	3	L	M	H	H	H
Minor	2	L	L	M	M	H
Negligibe	1	L	L	M	M	H
Risk rating: E- Extreme, H- High, M- Moderate, L- Low						

- **Extreme Risk (E):** Operation not permissible.
- **High Risk (H):** Remedial action to have high priority.
- **Moderate Risk (M):** Remedial action to be taken in appropriate time.
- **Low (L):** Risk is acceptable; remedial action is discretionary.

CONCLUSION

Lost circulation is one of the largest contributors to drilling non-productive time. It causes an increase in drilling fluid costs, and increase in expenses associated with rig and equipment. Especially with oil muds, losses can be very expensive, unsafe (underground blowout), and time consuming. While it is critical that losses be stopped once they occur, it is equally important that they be prevented whenever possible. For maximum success emphasis should be placed on the lost circulation assessment and a planning process to evaluate specific lost circulation problems and link them to existing products, systems and services. Lost circulation planning includes both prevention and remediation methods. Depending on the cause, there are several different engineering approaches to take for the prevention and treatment of lost circulation. An understanding of the cause and possible mechanisms of both the loss and the treatment is the key to determine the approach that includes both planning software and specialized materials. Managing the equivalent circulat ing densities (ECD) is a primary focus in the well plans. One important part of a preventive plan is the design of wellbore strengthening to improve the ability of the wellbore to contain higher pressures without losing circulation. The integrative preplanning process analyzes offset histories and formation data not only to identify risk zones but also to gather information on the exact fracture and pore size as well as fracture density. In an effort to control losses, a wide variety of lost circulation materials can be applied in thief zones (e.g. LCMs, pills containing sized solids, gunk squeezes, PCP squeeze, conventional cement squeezes, foamed cement, etc.). Therefore, a strategy for managing lost circulation must be an integral part of well planning and should include: best drilling practice, drilling fluid selection, Wellbore Strengthening Materials (WSM), and Lost Circulation Materials (LCM). The key to success is proper planning, risk assessments to identify potential high risk problems, quality materials and equipment, and experienced supervision.

REFERENCES

Akamine, A. Y., Bratton, T., Onyia, E., & Romanchock, M. (2003). *Apllication of real-time resistivity and annular pressure data in reducing lost-circulation events*. Paper presented at the SPE/IADC Drilling Conference. Amsterdam, The Netherlands.

Al Ubaidan, A., Esmail, W. J., Aborshaid, H., & Al Umair, M. (2000). *Successful isolation of a loss circulation zone of a short radius horizontal in a highly fractured environment*. Paper presented at the 2000 IADC/SPE Asia Pacific Drilling Technology. Kuala Lumpur, Malaysia.

Algu, D. R., van Oort, E., Landgrave, S., Grant, L., & Hillburn, D. (2007). *A systematic approach to drillling and delivering wells in severely depleted sands at ram powell (GOM)*. Paper presented at the 2007 SPE/IADC Drilling Conference. Amsterdam, The Netherlands.

Bell, R. J., Jr., & Davis, J. M. (1987). *Lost circulation chellenges: Drilling thick carbonate gas reservoir, natuna D-alpha block*. Paper presented at the 1987 SPE/IADC Drilling Conference. New Orleans, LA.

Bratton, T. R., Rezmer-Cooper, I. M., Desroches, J., Gillie, Y.-E., Li, Q., & McFayden, M. (2001). *How to diagnose drilling induced fractures in wells drilled with oil-based muds with real-time resistivity and pressure measurements*. Paper presented at the SPE/IADC Drilling Conference. Amsterdam, The Netherlands.

Bruton, J. R., Ivan, C. D., & Heinz, T. J. (2001). *Lost circulation control: Evolving techniques and strategies to reduce downhole mud losses*. Paper presented at the SPE/IADC Drilling Conference. Amsterdam, The Netherlands.

Caughron, D. E., Renfrow, D. K., Bruton, J. R., Ivan, C. D., Broussard, P. N., Bratton, T. R., & Standifird, W. B. (2002). *Unique cross-linking pill in tandem with fracture prediction model cures circulation losses in deep water of Gulf of Mexico*. Paper presented at the SPE/IADC Drilling Conference. Dallas, TX.

Chilingarian, G. V., & Vorabutr, P. (1981). *Drilling and drilling fluids*. New York: Elsevier.

Cunha, J. C. (2004). *Risk analysis application for drilling operatios*. Paper presented at the Petroleum Society's 5th Canadian international Petroleum Conferenc. Calgary, Canada.

Davidson, E., Richardson, L., & Zoller, S. (2000). *Control of lost circulation in fractured limestone reservoirs*. Paper presented at the 2000 IADC/SPE Asia Pacific Drilling Technology. Kuala Lumpur, Malaysia.

Dick, M. A., Heinz, T. J., Svoboda, C. F., & Aston, M. (2000). *Optimizing the selection of bridging particles for reservoir drilling fluids*. Paper presented at the SPE International Symposium on FormationDamage. Lafayette, LA.

Ferras, M., Galal, M., & Power, D. (2002). *Lost circulation solutions for severe sub-salt thief zones*. Paper presented at the AADE 2002 Technology Conference Drilling & Completion Fluids and Waste Management. Houston, TX.

Fidan, E., Babadagli, T., & Kuru, E. (2004). *Use of cement as lost circulation material – Field case studies*. Paper presented at the IADC/SPE Asia Pacific Drilling Technology Conference and Exhibition. Kuala Lumpur, Malaysia.

Goins, W. C., & Dawson, D. D. (1953). Temperature survey to locate zone of lost circulation. *Oil & Gas Journal*, *52*, 170–171.

Howard, G. C., & Scott, P. P. Jr. (1951). An analysis and control of lost circulation. *Journal of Petroleum Technology*, *3*(6). doi:10.2118/951171-G

Ivan, C. D., Bruton, J. R., Thiercelin, M., & Bedel, J. P. (2002). *Making a case for rethinking lost circulation treatments in induced fractures*. Paper presented at the SPE Annual Technical Conference. San Antonio, TX.

Kefi, S., Lee, J. C., Shindgikar, N. D., Brunet-Cambus, C., Vidick, B., & Diaz, N. I. (2010). *Optimizting in four steps composite lost-circulation pills without knowing loss zone width*. Paper presented at the IADC/SPE Asia Pacific Drilling Technology Conference and Exhibition. Ho Chi Minh City, Vietnam.

Masi, S., Molaschi, C., Zausa, F., & Michelez, J. (2010). *Managing circulation losses in a harsh drilling environment: Conventional solution vs. CHCD through a risk assessment*. Paper presented at the IADC/SPE Drilling Conference and Exhibition. New Orleans, LA.

Power, D., Ivan, C. D., & Brooks, S. W. (2003). *The top 10 lost circulation concerns in deepwater drilling*. Paper presented at the SPE Latin American and Caribbean Petroleum Engineering Conference. Port-of-Spain, West Indies.

Romero, S. N., Monroy, R. R., Johnson, C., Cardenas, F., & Abraham, G. A. T. (2006, September). Preventing lost circulation by use of lightweight slurries with reticular systems: Depleted reservoirs in southern Mexico. *SPE Drilling & Completion*, 185-192.

Salehi, S., & Nygaard, R. (2012). *Numerical modeling of induced fracture propagation: A novel approach for lost circulation materials (LCM) design in borehole strengthening applications of deep offshore drilling*. Paper presented at the SPE Annual Technical Conference and Exhibition. San Antonio, TX.

Sanders, M. W., Scorsone, J. T., & Friedheim, J. E. (2010). *High-fluid-loss, high-strength lost circulation treatments*. Paper presented at the SPE Deepwater Drilling and Completions Conference. Galveston, TX.

Sanders, W. W., Williamson, R. N., Ivan, C. D., & Powell, D. (2003). *Lost circulation assessment and planning program: Evolving strategy to control severe losses in deepwater projects*. Paper presented at the SPE/IADC Drilling Conference. Amsterdam, The Netherlands.

Suyan, K. M., Sharma, V., & Jain, V. K. (2009). *An innovative material for severe lost circulation control in depleted formations*. Paper presented at the SPE/IADC Middle East Drilling Technology Conference & Exhibition. Manama, Bahrain.

Wang, H., Sweatman, R., Engelman, B., Deeg, W., & Whitfill, D. (2005). *The key to successfully applying today's lost-circulation solutions*. Paper presented at the 2005 SPE Annual Technical Conference and Exhibition. Dallas, TX.

Whitfill, D. (2008). *Lost circulation material selection, particle size distribution and fracture modeling with fracture simulation software*. Paper presented at the IADC/SPE Asia Pacific Drilling Technology Conference and Exhibition. Jakarta, Indonesia.

Whitfill, D., & Wang, H. (2005). *Making economic decisions to mitigate lost circulation*. Paper presented at the 2005 SPE Annual Technical Conference and Exhibition. Dallas, TX.

Whitfill, D. L., & Hemphill, T. (2003). *All lost-circultion materials and systems are not created equal*. Paper presented at the SPE Annual Technical Conference and Exhibition. Denver, CO.

Whitfill, D. L., Wang, M., Jamison, D., & Angrove-Rogers, A. (2007). *Preventing lost circulation requires planning ahead*. Paper presented at the 2007 International Oil Conference and Exhibition. Veracruz, Mexico.

ADDITIONAL READING

Aldred, W., Hutin, R., Luppens, J. C., & Ritchie, G. (2008, March). *Development and testing of a rig-based quick event detection system to mitigate drilling risk*. Paper IADC/SPE 111757 presented at the 2008 IADC/SPE Drilling Conference, Orlando, Florida, U.S.A.

Arild, Ø., Fjelde, K. K., & Løberg, T. (2006, November). *An integrated approach to risk and hydraulic simulations in a well control planning perspective*. Paper IADC/SPE 103853 presented at the IADC/SPE Asia Pacific Drilling Technology Conference and Exhibition, Bangkok, Thailand.

Castillo, D. Ryles, P., & John, K. (2002, September). *The legendre field development and geomechanical program: from exploration to drilling to production*. Paper IADC/SPE 77255 presented at the IADC/SPE Asia Pacific Drilling Technology, Jakarta, Indonesia.

Collins, N., Kharitonov, A., Whitfill, D., Miller, M., & Kulakov, K. (2010, October). *Comprehensive approach to severe lost circulation problems in Russia*. Paper SPE 135704 presented at the 2010 SPE Russian Oil & Gas Technical Conference and Exhibition, Moscow, Russia.

Corbett, J. S. (2001, February). *An integrated and structured approach to the management of exploration and development risks*. Paper SPE 66511 presented at the SPE/EPA/DOE Exploration and production Environmental Conference, San Antonio, Texas, U.S.A.

Dear, S. F. III, Beasley, R. D., & Barr, K. P. (1995). Use of a decision tree to select the mud system for the Oso field, Nigeria. *Journal of Petroleum Technology*, (October): 909–912.

Dusseault, M. B., Maury, V., Sanfilippo, F., & Santarelli, F. J. (2004, June). *Drilling around salt: risks, stresses and uncertainties*. Paper ARMA/NARMS 04-647 presented at Gulf Rocks 2004, the 6th North America Rock Mechanics Symposium (NARMS), Rock Mechanics Across Borders and Disciplines, Houston, Texas, U.S.A.

El-Hassan, H., Abbas, R., Jarouj, H., & Munk, T. (2003, October). *Using a novel fiber cement system to control lost circulation: case histories from the Middle East and the Far East*. Paper SPE/IADC 85324 presented at the SPE/IADC Middle East Drilling Technology Conference & Exhibition, Abu Dhabi, U.A.E.

Fang, C. C., & Thaemlitz, C. (2005, March). *OSPAR-Compliant technologies for managing drilling-fluid lost-circulation events*. Paper SPE 94434 presented at the 2005 SPE/EPA/DOE Exploration and Production Environmental Conference, Galveston, Texas, U.S.A.

Fidan, E., Babadagli, T., & Kuru, E. (2004, June). *Use of cement as lost-circulation material: best practices.* Paper 2004-090 presented at the Petroleum Society's 5th Canadian International Petroleum Conference (55th Annual Technical Meeting), Calgary, Alberta, Canada.

Guan, Z., Ke, K., & Lu, Baoping (2010, June). *A new approach for casing program design with pressure uncertainties of deepwater exploration wells.* Paper SPE 130822 presented at the CPS/SPE International Oil & Gas Conferenece and Exhibition, Beijing, China.

Hettema, M. H. H., Bostrøm, B., & Lund, T. (2004, September). *Analysis of lost circulation during drilling in cooled formations.* Paper SPE 90442 presented at the SPE Annual Technical Conference and Exhibition, Houston, Texas, U.S.A.

Houtchens, B., Foster, J., & Tessari, R. (2007, February). *Applying risk analysis to casing while drilling.* Paper SPE/IADC 105678 presented at the 2007 SPE/IADC Drilling Conference, Amsterdam, The Netherlands.

Kok, K. H., Pruimboom, J., David, F. M., & Jit-Chiun, T. (2003, October). *Geosteer with resistivity forward modeling to prevent drilling into the lost circulation zone of a prolific carbonate reservoir.* Paper SPE/IADC 85306 presented at the SPE/IADC Middle East Drilling Technology Conference & Exhibition, Abu Dhabi, U.A.E.

Kristiansen, T. G., Mandziuch, K., Heavey, P., & Kol, H. (1999, March). *Minimizing drilling risk in extended-reach wells at Valhall using geomechanics, geoscience and 3D visualization technology.* Paper SPE/IADC 52863 presented at the 1999 SPE/IADC Drilling Conference, Amsterdam, Holland.

Malik, H. K., Litwin, R. J., & Jones, B. C. (2010, October). *Successful use of managed pressure drilling (MPD) to counteract loss circulation and HP gas zones: A case study from Red Rock South Wapiti field in Grande Prairie, Alberta, Canada.* Paper CSUG/SPE 137428 presented at the Canadian Unconventional Resources & International Petroleum Conference, Calgary, Alberta, Canada.

Marriott, T., Rogers, H., Lloyd, S., Quinton, C., & Tetrault, N. (2006, June). *Innovative cement plug setting process reduces risk and lowers NPT.* Paper 2006-015 presented at the Petroleum Society's 7th Canadian International Petroleum Conference (57th Annual Technical Meeting), Calgary, Alberta, Canada.

Morita, N., Black, A. D., & Fuh, G. F. (1990, September). *Theory of lost circulation pressure.* Paper SPE 20409 presented at the SPE Annual Technical Conference. New Orleans, Louisiana, U.S.A.

Nilsen, T. (2002, September). *Risk management for planning of challenging reservoir sections.* Paper IADC/SPE 77228 presented at the IADC/SPE Asia Pacific Drilling Technology, Jakarta, Indonesia.

Nilsen, T., & Karlsen, H. C. (2008, December). *Risk based decision support for the planning of a challenging HPHT drilling operation.* Paper IPTC 12526 presented at the International Petroleum Technology Conference, Kuala Lumpur, Malaysia.

Oakley, D., & Conn, L. (2011, March). *Drilling fluid design enlarges the hydraulic operating windows of managed pressure drilling operations.* Paper SPE/IADC 139623 presented at the SPE/IADC Drilling Conference and Exhibition, Amsterdam, The Netherlands.

Økland, D., Gabrielsen, G. K., Gjerde, J., Sinke, K., & Williams, E. L. (2002, October). *The importance of extended leak-off test data for combatting lost circulation*. Paper SPE/ISRM 78219 presented at the SPE/ISRM Rock Mechanics Conference, Irving, Texas.

Ostebo, R., Tronstad, L., & Fikse, T. (1991, March). *Risk analysis of drilling and well operations*. Paper SPE/IADC 21952 presented at the SPE/IADC Drilling Conference. Amsterdam, The Netherlands.

Rosato, M. J., & Courville, P. W. (2000, April). *Managing differential pressure during coiled tubing operation*. Paper SPE 60707 presented at the 2000 SPE/ICoTA Coiled Tubing Roundtable, Houston, Texas, U.S.A.

Sadlier, A., Wolfe, C., Reese, M., & Says, I. (2011, October/November). *Automated alarms for managing drilling pressure and maintaining wellbore stability - new concepts in while-drilling decision making*. Paper SPE 146298 presented at the SPE Annual Technical Conference and Exhibition, Denver, Colorado, U.S.A.

Sanad, M., Butler, C., Waheed, A., Engelman, B., & Sweatman, R. (2004, March). *Numerical models help analyze lost-circulation/flow events and frac gradient increase to control an HPHT well in the East Mediterranean Sea*. Paper IADC/SPE 87094 presented at the IADC/SPE Drilling Conference, Dallas, Texas, U.S.A.

Sheremetov, L., Batyrshin, I., Filatov, D., Martinez, J., & Rodriguez, H. (2008). Fuzzy expert system for solving lost circulation problem. *Applied Soft Computing*, *8*(1), 14–29. doi:10.1016/j.asoc.2006.11.003

Suyan, K. M., Dasgupta, D., Sanyal, D., Sharma, V., & Jain, V. K. (2007, November). *Managing total circulation losses with cross-flow while drilling: case history of practical solutions*. Paper SPE 109898 presented at the 2007 SPE Annual Technical Conference and Exhibition, Anaheim, California, U.S.A.

Tessari, R. M., Warren, T. M., & Jo, J. Y. (2006, October). *Drilling with casing reduces cost and risk*. Paper SPE 101819 presented at the 2006 SPE Russian Oil and Gas Technical Conference and Exhibition, Moscow, Russia.

Tranquini, N., Soto, J., Reveth, V., Prato, L., & Williams, H. (2007, April). *Proven synergistic approach in well cementing across thief zones: East Venezuela*. Paper SPE 107501 presented at the 2007 SPE Latin American and Caribbean Petroleum Engineering Conference, Buenos Aires, Argentina.

Wang, H. M., Sweatman, R., Engelman, B., Deeg, W., Whitfill, D., Soliman, M., & Towler, B. F. (2008). Best practice in understanding and managing lost circulation challenges. *SPE Drilling & Completion*, June, 168-175.

KEY TERMS AND DEFINITIONS

Equivalent Circulating Density (ECD): The effective density exerted by a circulating fluid against the formation that takes into account the pressure drop in the annulus above the point being considered.

Fiber Lost Circulation Material: A long, slender and flexible fiber of various sizes and lengths added to drilling fluid and placed downhole to help retard mud loss into fractures or highly permeable zones.

Flake Lost Circulation Material: A type of lost circulation material (e.g. mica flakes and pieces of cellophane) that is thin and flat in shape, with a large surface area. Its purpose is to seal off fluid loss zones in a well and help stop lost circulation.

Gunk Plug (Or Diesel-Oil Plug): A small batch of the slurry that consists of bentonite in diesel oil, and is pumped down a well that has lost circulation to seal the leaky zone.

Gunk Squeeze: The operation of squeezing a gunk plug into the loss zone.

Lost Circulation (Lost Returns): The reduced or total absence of drilling fluid flow up the annulus when fluid is pumped through the drillstring.

Lost Circulation Material: Fibrous, granular or flake material intentionally introduced into a mud and pumped down a well to reduce and eventually prevent the flow of drilling fluid into a weak, fractured or vugular formation or seal loss zones.

Lost Circulation Zone: A very permeable, highly fractured or cavernous rock layer that takes excessive amounts of drilling mud as it is being drilled.

Pill: Any relatively small quantity (less than 30 m^3 (200 bbl) of lost circulation material to plug a thief zone.

Seepage Loss: Loss of mud during drilling but with loss rate less than 3 m^3/hr [20 bbl/hr].

Thief Zone: A formation encountered during drilling into which circulating fluids can be lost.

Total Lost Returns: Situation during drilling process when drilling fluid is pumped through the drillstring but no fluid comes out of the annulus.

Vugular Rock: A rock containing cavities, voids or large pores that are commonly lined with mineral precipitates.

Chapter 5
Simultaneous Operations

Zdenko Kristafor
University of Zagreb, Croatia

ABSTRACT

Simultaneous operations can be defined as conducting independent operations in which the course of any operation may impact the safety of personnel, equipment, and environment of another operation. Operations are meant in terms of performing drilling, completion, workover, and construction or production activities simultaneously, mainly at offshore facilities. In general, they have an impact on the installation safety procedures and contingency planning program. Thus, for the purpose of performing simultaneous operations, they must comply with organizational safety and technological requirements. They are to be coordinated through joint planning efforts by supervisors who plan and direct activities. Typical chain-of-command as well as simultaneous operations decision making process flow diagrams are presented in this chapter. Basic steps to be regarded are performing risk assessment, assess and control risks, monitor the simultaneous tasks and communicate the control measures. Communication is the key to performing these tasks safely.

INTRODUCTION

There's large scope of both onshore and particularly offshore activities in the oil and gas and other marine industries where one activity can strongly affect performing another covered operation in a safe and risk acceptable manner. Simultaneous operations (SIMOPS) shall be thoroughly planned, analyzed and performed with the objective of limiting excessive risk imposed by multiple operations at the same time, comparing to the risk associated with

DOI: 10.4018/978-1-4666-4777-0.ch005

the execution of these operations individually. These operations can range from something as common as a crew change with a crew boat, food, fuel, material provision etc., to a much larger scale involving multiple, heavy construction assets and personnel (Ross, 2007). Is spite of the concept favorising opinion in the entering stage of project that "the avoided SIMOPS is the best SIMOPS", it is obvious that all simultaneous operations cannot be avoided. Relevant procedures for the control of simultaneous or parallel operations shall be developed and validated or assessed prior to commencement of operations.

Deepwater Horizon disaster prompted all those involved in the exploration and production of offshore oil and gas to reassess the safety of their industry. *The National Commission on the BP Deepwater Horizon Oil Spill and Offshore Drilling* (2011) made a series of recommendations, many of which have been implemented already. But, according to the Committee on Industry, Research and Energy of the European Parliament, the implications for Europe are less clear. Many of the recommendations for the U.S. regulatory regime reflect practices that have been applied and prevalent in parts of the E.U. for twenty years and more. One of the first recommendations in the National Commission's report states: "The Department of the Interior should develop a proactive, risk-based performance approach specific to individual facilities, operations and environments, similar to the "safety-case" approach in the North Sea" (Report to the President, p.252). In the North Sea area mature industries are operating in mostly well-developed fields under the oversight of well established and evolving regulatory regimes. "The "safety case" is a risk-based, site specific approach which requires operators to demonstrate to the relevant national health, safety and environmental authorities that all risks have been considered and controls implemented. Any residual risks should correspond to the ALARP (As Low As Reasonably Practicable) principle. The relevant authority should approve the safety-case, as well as any subsequent changes to it, before any operations can commence" (Draft report 2011/XXXX(INI)). In the U.S., Bureau of Ocean Energy Management, Regulation and Enforcement (BOEMRE) is a single agency where licensing, health and safety issues are consolidated.

The E.U. Commission on Safety of offshore oil and gas prospection, exploration and production activities formed the European Union Offshore Oil and Gas Authorities Group that shall be composed of Member States' authorities responsible for the regulatory oversight of offshore oil and gas activities and related policy issues. Different EU regulations have been proposed and adopted, among others, Proposal for a regulatory initiative on safety of offshore oil and gas prospection, exploration and production activities. In the context of the Directive 1994/22/EC the competent authority is required to consider the technical and financial risks, and where appropriate, the previous record of responsibility of applicants seeking exclusive exploration and production licenses. There is the need to ensure that when examining the technical and financial capability of the licensee the competent authorities thoroughly examine also its capability for ensuring continued safe and effective operations under all foreseeable conditions. Just to emphasize two chapters in this Regulation where "Prevention of major hazards related to offshore oil and gas activities" as well as "Preparation for and conduct of offshore oil and gas activities based on risk assessment" are treated. The operator's duty is to organize his activities around a best practice model, prepare a major hazards report and to submit this to the competent authority for assessment. In addition, the operator is to submit a notification for each well operation to the regulator. For both major hazard reports and well notifications, independent verification of the safety critical elements is required. An operator of an installation which is to be involved in a combined operation shall send to the competent authority a notification containing details of the combined operation in accordance with the requirements of Annex II, part 7 (EU Proposal, 2011).

International Association of Drilling Contractors issued HSE Case Guidelines for Offshore Drilling Contractors with enumeration of reference documents on international standards and industry guidelines (IADC, 2009). A detailed safety-case submission requirements and acceptance regime are explained with respect to referent regulators and legislation authorities in most leading countries involved in offshore operations and activities (Norway, U.K., Germany, Denmark, Australia, the Netherlands).

Several globally recognized associations, moreover most of major companies involved in offshore activities worldwide, provide operators and contractors with documents intended as a general guideline to assist them with better understanding and applying safe work practice expectations. A consistent simultaneous operations plan provides instructions in conducting safe and environmentally sensitive operations. The heightened potential for accidents when simultaneous operations are in progress as well as a need for attention to safety, communication and coordination necessary in these situations is recognized long ago. Offshore operating companies have developed their HSE plans, which also include simultaneous operations, based on risk assessment for offshore installations in the operational phase.

The basic steps in managing risk of simultaneous operations will be presented in this chapter, followed by life cycle model for SIMOPS, a typical chain-of-command flow diagram and SIMOPS general structure. To assist in decision making on whether a limited number of operations could be performed simultaneously and safely or not decision matrices were developed. In addition, a number of concisely structured SIMOPS manuals exist, but having in general very similar approach.

BACKGROUND

There is a wide range of definitions for simultaneous operations provided by professional associations, companies and entities involved in both onshore and offshore activities. These activities are not always recognized by all subjects involved. To safely perform certain tasks and activities that are running concurrently requires awareness of all the people involved in them. So, have we identified them as simultaneous tasks? "Identifying and controlling simultaneous operations is not currently a legislative requirement. However, the major companies are introducing this requirement into all operations" (http://EzineArticles.com, 2012).

Terms and Definitions

- SIMOPS is the performance of potentially conflicting operations at the same time, in the same location (www.slideshare.net/lifecombo/ngisimops). Both jobs may be done safely, but without coordination, one act could have a dangerous consequence to other.
- A term used mainly on offshore platforms, or installations with multiple wellheads, where more than one wellbore is being accessed, such as where a drilling rig, slickline unit or coiled tubing unit may be operating at the same time. Simultaneous operations gener-

ally have an impact on the installation safety producers and contingency planning processes (Schlumberger Oilfield Glossary, 2013).

- Simultaneous operations are defined as any two or more different sets of activities (macro activities) carried out, within a defined area, by different organizations (functional groups) or under different management systems that, because of their proximity or other critical factors, could interact adversely with each other (ENI, 2005).
- Simultaneous operations are described as the potential clash of activities which could bring about an undesired event of set of circumstances, e.g. safety, environment, damage to assets, schedule, commercial, financial etc. SIMOPS are defined as performing two or more operations concurrently. These activities typically include, but are not limited to the following:
    - A vessel undertaking a non-routine operation within an installation's 500 m zone;
    - Subsea umbilical, risers and flowline operations;
    - A field development with multi-vessel/contractor operations.

SIMOPS often involve multiple companies (owners, contractors, subcontractors, vendors), large multi-disciplined workforces and a wide range of daily, 24 hour, routine and non-routine construction and commissioning activities (IMCA, 2010).

- Woodhill Frontier limited (http://woodhill.co.uk) describe combined & simultaneous operations in terms of two or more installations, and combined operations such as construction and production, that need appropriate management systems and controls for effective use of resources and to minimize risks.
- Total Fina ELF (2001) states that construction, commissioning, utility systems running and production operations with potential interference between them should be forecasted to minimize any interference. Information relevant to simultaneous activities with a possible / potential impact to the other team should be given for every planned operation as well as for contingency and emergency situation.

From available sources different terms can be found for simultaneous operations. Terms such parallel, concurrent, multiple or combined are used, mostly in the same context as simultaneous. However, in some manuals or operational practices a subtle distinction among them is indicated. Anyhow, if two or more of the following operations are performed concurrently within the defined area for such activity, then these operations should be considered as simultaneous operations:

- Production operations or injection
- Live well operations
- Heavy construction
- Heavy lifts
- Rig operations or rig moves
- Mobilization and demobilization of equipment
- Anchoring the vessels
- Pigging operations exposing drilling and well area
- Hot work
- Drilling operations and workover
- Conductor driving and installation
- Seismic or geotechnical operations
- Injection to annulus
- Pumping through temporary lines
- Repair or change-out of Christmas tree
- Diving operations

- Helicopter landing or takeoff
- Emergency and/or spill response

First four quoted operations, some sources believe macro-activities. Any combination of two or more macro-activities can generate SIMOPS (if one of macro-activities is performed by using drilling rig) or parallel operations (in case without the use of drilling rig). Both require the same scheme in terms of basic steps to be shown later on. According to ENI (2005), if two different operations are carried out by using two drilling rigs, this can be identified as particular case of simultaneous operations, named concurrent operations – CONOPS. In most other considerations those terms are just synonyms.

Another way of looking at simultaneous operations in the offshore environment refers to the problems associated with dynamic positioning which is, unlike the back deck operations on the platforms, often considered as a secondary task. In this sense one can distinguish vessel SIMOPS as well as seafloor SIMOPS (Ross, 2007, IMCA 2010). Vessel simultaneous operations encompass two or more vessels undertaking a non-routine activity within the same area/close proximity to each other, or within an installation's 500 meters safety zone. Vessels may include a variety of barges, supply boats, diving support vessels, accommodation, pipe lay or cable lay, seismic, survey, heavy lift, ROV and other vessels operating in dynamic positioning mode. Installations include fixed and floating production platforms, drilling rigs, fluid production, storage and offloading unit/vessel (FPSO), fluid production units (FPU), and dynamically positioned production units.

When undertaking the seafloor activities, a number of operations in and around the trees, piles, manifolds, umbilicals, caissons, risers and flowlines must be thoroughly planned.

For special circumstances regarding Macondo well in the Gulf of Mexico, as a response to oil spill after Deepwater Horizon platform disaster, an overview of simultaneous operations is shown in Figure 1.

At the start of some offshore projects SIMOPS were not considered as a normal way of working, but as a downgraded mode of operation. Implementation of simultaneous operations was strictly exceptional and subjected to an acceptable risk assessment. However, in a later stage of field development, due to the necessity to reduce the delay between drilling and producing a well, to speed up the overall project by starting, for example, drilling on platforms not fully commissioned or built, and to allow routine maintenance, installation, lay-out of new pipelines and generally construction work on the field, simultaneous operations may become a normal way of working in a number of combinations drilling-production-construction to be conducted (Saintpère at all., 2005).

Norwegian standards, with the highest reputation in petroleum engineering, also include simultaneous operations. General safety requirements, barrier philosophy, preparation of program, risk analysis, application of consents, permits and approvals, HSE management system and simultaneous operations are elaborated in detail. As outlined in the scope, the NORSOK Standard for Drilling and Well Operations defines "functional and performance oriented requirements for well design and planning and execution of safe and efficient drilling and well operations in Norwegian waters" (NORSOK, 1998). It also stipulates an acceptable safety level. In larger extent it contributes in replacing individual oil company specifications, documents, guidelines and recommendations for existing and future oil industry development.

Figure 1. BP oil disaster simultaneous operations overview (http://arcanumdeepsecrets.wordpress.com)

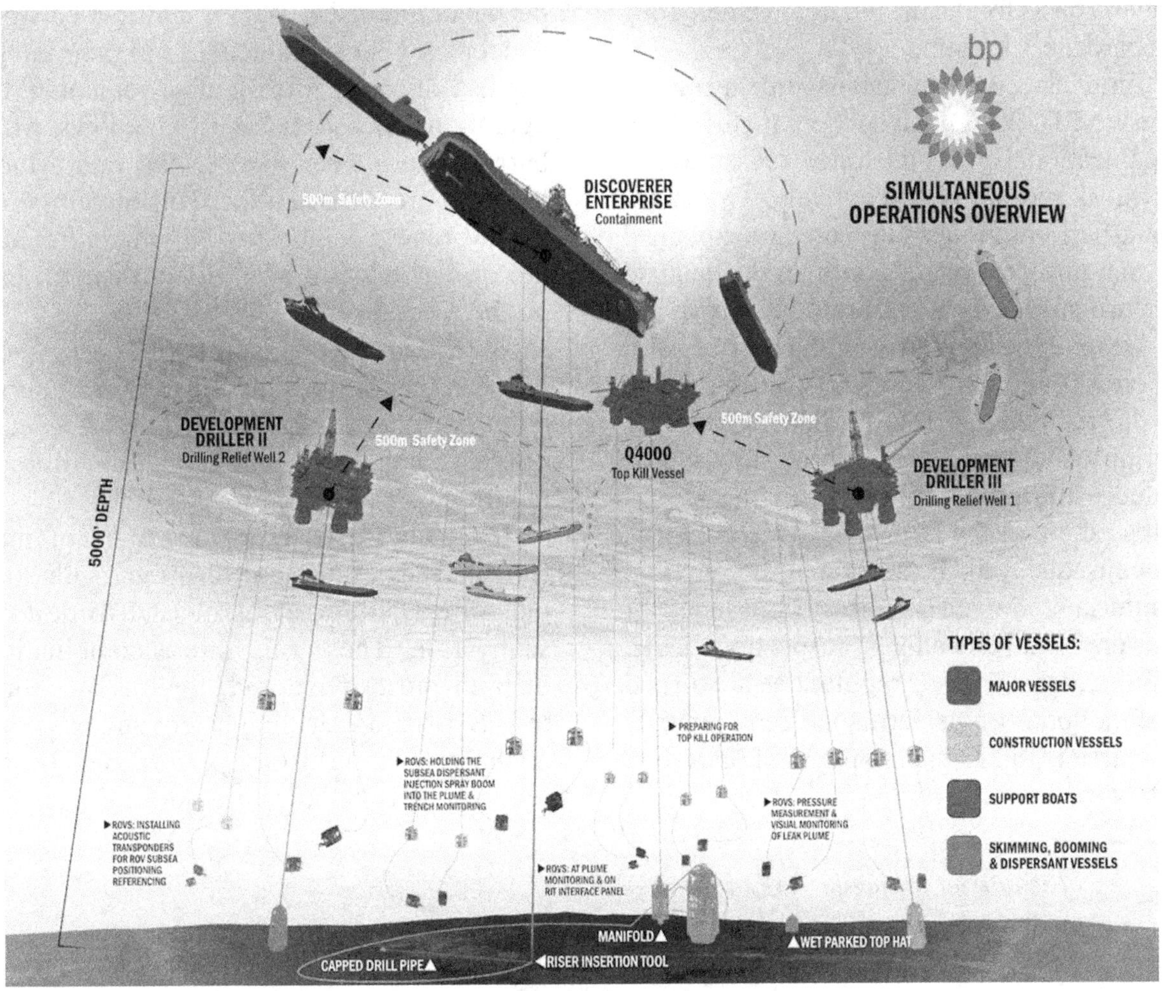

Concept of Barriers

Modern technologies have developed many safeguards and barriers to prevent major accidents and undesired events to happen. Better understanding and limiting their occurrence still represent challenges to oil industry. Most of operating companies have established and implemented the barrier philosophy in order to meet local and international legislative requirements, which shall also be respected and adapted by companies or personnel working for or behalf the operator. Vinnem at all. (2003) outline that quantitative risk assessment techniques have experienced wide application in Norwegian offshore oil and gas industry in the early eighties and came into force in Norwegian regulations in 1990. Pursuant to the regulations, in order to identify the hazardous events that may occur during the activities, the risk analysis is to be carried out. Later on, the Norwegian Petroleum Directorate (NPD) as a key authority, issued a regulation on safety barriers (on emergency preparedness) taking into consideration scenarios definitions, performance standards of emergency preparedness measures and analyses carried out as a basis for the design of the emergency preparedness system.

Emphasized by the Management regulations (NPD, 2002), the barriers include both organizational or administrative measures and physical measures respectively. Organizational barriers or safety measures are explained as measures to insure continuous adequate performance of technical and human barriers.

Despite the emphasis on the barriers philosophy (Figure 2), there is no strict definition in the regulations. ISO (2002) adopted the definition for barrier as "a measure, which reduces the probability of realizing a hazard's potential for harm and which reduces its consequence". Barriers may be physical (materials, devices, segregation, etc.) or non-physical (procedures, inspection, drills, training, etc.) Today's regulations are turned to functional or performance requirements for barriers rather than to physical ones (e.g. safety systems).

The protection against accidents in the petroleum industry rely on multiple barriers concept in which maintenance of their integrity is a key issue for risk management in the operational phase. It has to be known what barriers have been established and which function are designed for. This also implies the awareness on non-functioning or impairing of barriers and what effect this still has on the risk levels.

Another important aspect in the operational phase is that performing operations of short duration with the associated high risk level is often in combination with special conditions, requirements and premises.

In October 2010, U.S. Department of Interior enacted new offshore drilling regulation, including Drilling safety rule and Workplace safety rule. These rules are a combination of both prescriptive and, for the first time in U.S. offshore drilling safety and environmental regulation history, performance-based regulation. Both rules are addressed to the

Figure 2. The barrier principle

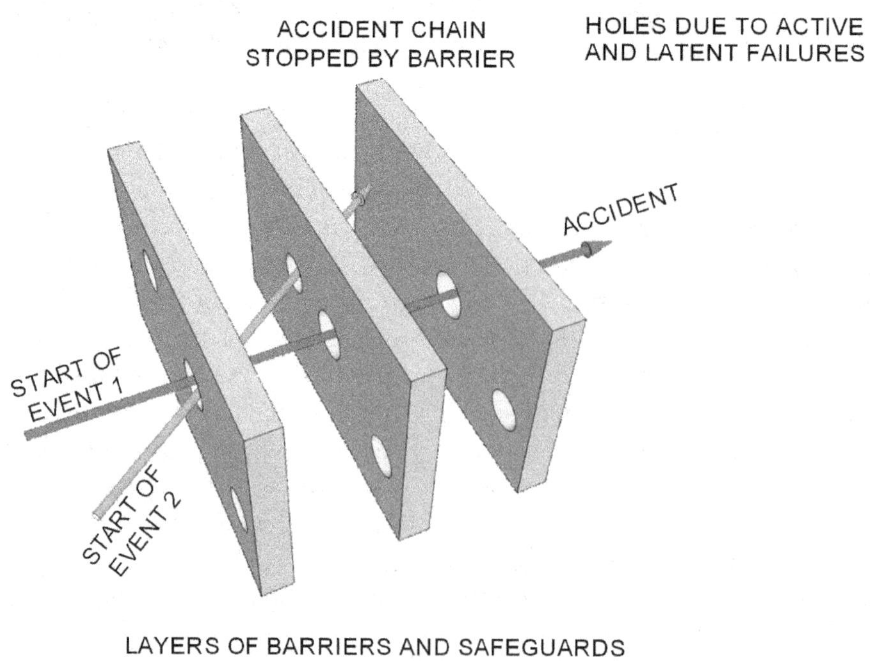

wellbore integrity and well control equipment respectively. Regarding drilling-completion-production activities, at least two independent and tested barriers across each flow path during the performance of high risk operations should be present at all times (also certified by a registered engineer). This request is a part of any operators´ company policy. If one of these barriers is in danger, all other operations must be warned and ready to suspend, until the problem is solved. When one barrier completely fails all simultaneous operations, (with the exception of those activities related to the barrier restoration) should be secured and suspended until the established number of barriers has been restored.

During drilling and workover operations a primary barrier is usually provided by hydrostatic fluid column, with blowout preventers available to be activated as a secondary barrier. While setting casing strings or liners, the casing itself and annulus cement shield isolate hydrocarbons from leaking, making the primary barrier. A shoe track with two types of mechanical barriers, cement in the shoe and the float collar valves or liner hanger seals, provide the secondary barrier. It is of the highest priority that no single failure of barrier or barrier element, whether caused by operational error or equipment failure, shall lead to loss of well control. That is the reason why two independent and tested barriers shall be available (NORSOK, 1998).

Simultaneous operations involving the potential leak of hydrocarbons, for example in live well operations or production operations and hot work activities in the vicinity of the wellhead area or of another potential source of hydrocarbon release, should be limited in number as much as possible, and be given particular attention in terms of risk assessment and operational procedures.

The Simultaneous Operations Process

A sequence of performing activities in an onshore or offshore area has its own routine. Their interaction and correlation makes the effectiveness of certain operations more complex. A lack in the skill to identify simultaneous operations sometimes represents a common difficulty. Performing some operations may strongly affect on success of others, increasing the risk of carrying them out. For this reason, when scheduling and planning activities, consideration should be given to bring the occurrence of simultaneous operations to the least possible extent. In managing risks of simultaneous operations the basic principles must be regarded. This process generally starts with identifying simultaneous operations. By identifying all of them it is enabled to manage risks. Consequently, full risk assessment must be performed prior to carrying out SIMOPS allowing to identify all risks introduced by the task as well as their level. Monitoring parallel activities allows to verify that the controls are in place and adequate. If not, work shall be stopped and control measures in the risk assessment shall be reviewed. While SIMOPS are being carried out, a single responsibility shall be assigned. At the outset of simultaneous operations communication shall be established and maintained on the daily basis, enabling all persons involved in tasks understand and implement control measures.

A literature and a field practice provide different approaches to explaining the basic steps of simultaneous operations, elaborated in short above, sometimes limited to specific conditions in which they occur. One of those attempts is unusual but very convenient because it hits the essence of simultaneous operations, using in its clarifying their acronym. According to Ross (2007) when working on offshore projects where a possible SIMOPS situation can occur, "we should think of SIMOPS in terms of **S**haring **I**nformation **M**aximizes **O**ur **P**ersonnel **S**afety. With this stated, a plan can be created by following a process outlined as SIMOPS - **S**tudy the project, **I**dentify possible SIMOPS, **M**eet with all parties involved, **O**rganize a **P**lan and **S**hare the plan/information". Timely identification of possible simultaneous operations for some project early on in the development phase will allow to distinguish situations that can cause costly delays, or even worse, cause serious harm or injuries to personnel involved in each operation.

The policy of a particular operator in relation to simultaneous operations can be summarized in few sentences. It is indicated that simultaneous operations, require suitable policies and procedures to be put in place that are effectively administered to ensure the control of hazards evolving from multiple activities, performed at the same time, and at the same place. Besides, the effective coordination of all activities and the definition of clear lines of responsibility is required. The operational personnel will be provided with the best practice with the fundamental principles needed to reduce the cumulative risk level of two or more high-risk activities occurring together to a tolerable level. In the same time it does not mean that all individual operations during SIMOPS are fully covered by procedures necessary to carry them out in a safe and effective manner. Each discipline is expected to have its own series of operational manuals and procedures.

In order to establish responsibilities and the decision-making structure all simultaneous operations shell be perceived and identified, defining whether individual activities can be performed simultaneously or not, and under what (additional) conditions. In that process, decision matrices, developed by most leading operators or contractors, are very helpful.

Decision Matrix

At the outset of simultaneous operations it is necessary to establish communication between all parties involved, maintained and discussed daily through all combination of conditions to be expected. This communication shall also be regularly documented. Based on these daily meetings a simultaneous operations plan is created in a written form, whose role is to:

- Designate the responsible parties for the various operations to be executed;
- Establish awareness on emergency response and contacts;
- Identify hazards and take precaution measures and actions for appropriate mitigation;
- Make communication between operations clear, and
- Establish a clear and consistent chain of command.

Operation premises and a detailed description of the planned tasks to be executed shall be included in specific operations programs for the operation in question. The program shall be subjected to an inter-discipline check involving the operator and the main contractor(s) prior to implementation. The program may be supplemented by more detailed guidelines and risk analyses. For example, with production and construction activities carried out concurrently only a permit-to-work (a system to manage and control work tasks and the personnel carrying them out) is required. But, depending on specific hazards and complexity of the project, insured by the proper authorization of designated work by the field management team, SIMOPS procedures may be required (Chevron, 2012). Any deviation from the program, which proved to be necessary to perform simultaneous operations must be formally identified and requested from the appropriate regulatory authority for approval. It should be noted that in some cases local regulations may require the approval of the SIMOPS Operating plan by competent statutory authorities before starting simultaneous operations.

To assist in decision making whether a limited number of operations, identified at the start of the project, could be executed simultaneously and safely or not, a decision matrix was developed. An installation-specific SIMOPS decision matrix, based on a detailed risk assessment study, should be prepared for every job, case by case. Therefore, such a matrix can be an integral part of some manual where a set of minimum requirements for each simultaneous operations scenario is described, within the scope of activities for some particular company. A decision matrix is usually presented as two-dimensional list of activities to be performed at the installation to each other. There is no limit placed on the number of operations to be carried out simultaneously. But, as stated by Gilbert (1991), a pair of operations have to be considered separately, ensuring that the implications of each activity are examined with respect to all other operations.

Addressing the objective of making SIMOPS decision matrices user-friendly, most of them are developed in a similar way enabling to make work schedule and to help manage the safe performing of work. The next example in Figure 3 shows a possible clash of activities at drilling and production installation.

When interpreting the advisability of conducting operations/activities concurrently, color or letter codes are used. At the intersecting point of a pair of activities the code indicates which activities can be executed without any restriction, which activities can be performed respecting some restrictions and which activities cannot be conducted at the same time.

During planning phase (i.e., rig mobilization/demobilization), as well as prior to and while carrying out operations simultaneously, a set of requirements is to be met. These requirements are usually explained in terms of "shall", "should" or "must" rules. In that sense, *shall* is a verbal form used to indicate requirements strictly to be followed in order to conform to the standard and from which no deviation is permitted, unless accepted by the Operator" (NORSOK, 1998). In other words, shall rules cannot be deviated from under any circumstances (Gilbert, 1991). *Should* is a verbal form used to indicate that among several possibilities one is recommended as particularly suitable, without mentioning or excluding others, or that a certain course of action is preferred but not necessarily required" (NORSOK, 1998). The should and must rules can be deviated from, provided special precautions are taken and the deviation is approved beforehand by the authorized person appointed through a typical chain-of command system.

As far as two or more vessels are operating in the same area in the vicinity to each other, the SIMOPS vessel hierarchy must be assigned. This makes the overall planning of operations smoother. Ross (2007) says, when assigning the hierarchy a great attention must be paid to the vessel(s) relying on advanced acoustic type of dynamic positioning (DP). These vessels (drill ships, semi-submersible platforms, heavy construction vessels, etc.)

Figure 3. The SIMOPS decision matrix (Coastal Energy, 2012)

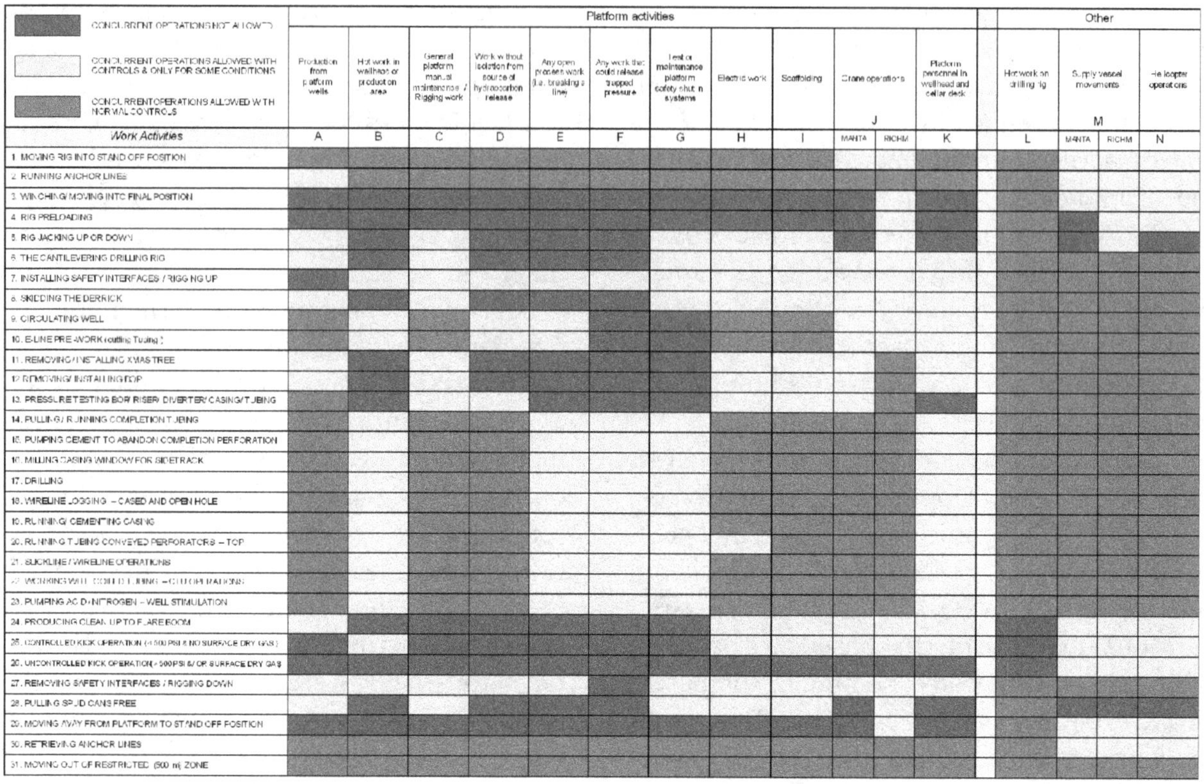

tend to be costly operations with high priority assigned, which other vessels can easily interfere with. There are many examples with hierarchy assignment, but when drilling in progress the DP drilling vessel will be the lead vessel during SIMOPS and this operation. During the developing phase of the project, entering to the other phase, this will and can be refined, but more than likely the drilling vessel will be ranked as the leading one.

Chain of Responsibilities

Depending on the scope of the individual company's activities, before carrying out simultaneous operations, a site specific SIMOPS operating plan has to be prepared. Besides the scope of activities, the operating plan should contain the resources required, elaborated system of responsibilities, and the schedule for the SIMOPS, based on all the strategies and company best practices that have to be taken in order to safely perform the activity. This plan will be used as a minimum. Supplemental plans or modifications may be developed to define all changes to the existing structures/installations /systems, as well as risk reducing measures, that need to be implemented before the safe execution of simultaneous operations. The main purpose of the simultaneous operation safety rules and procedures is to provide all personnel involved with simultaneous operations with a knowledge of the relevant safe practices and procedures.

In preparation of final document covering all the considerations outlined earlier, a clear management structure during carrying out simultaneous operations must be adopted. That is usually covered in chapters relating to the roles, responsibilities and coordinating procedures within specific operating manual/ guidelines. Mainly, "a single point responsibility" is declared as a key principle for the safe and efficient management of simultaneous operations.

Regarding the chain-of-command in place for a given simultaneous operations situation, clear direction must be provided. A full understanding of this decision-making practice has proved to be justified, greatly enhancing the capability of an emergency response organization. The various simultaneous operations situations may be addressed (platform drilling while producing at a hub platform, large commission work while producing at an outlying platform, etc.) and a flow chart describes the chain-of-command appropriate to each of them. Among many of flow diagrams available, one is shown in Figure 4.

The main person who has overall responsibility for the operations to be performed at the facility (installation) is simultaneous operations authority. When SIMOPS are being conducted the simultaneous operations authority (SOA) must be overseeing activities on the facility at all times, for both onshore and offshore operations. A person designated by the company is the highest in the SIMOPS hierarchy in which lower-ranked persons are responsible for coordinating with SOA one of the contributing simultaneous operations. In companies´ guidelines or manuals different names for the same hierarchical level are given, depending of the type of activity and their possible number. The most commonly used term world-wide for the person responsible for carrying out simultaneous operations is "person-in-charge" (PIC). Numerous op-

Figure 4. A characteristic chain-of-command (SIMOPS organization) flow diagram (adapted from ENI, 2005)

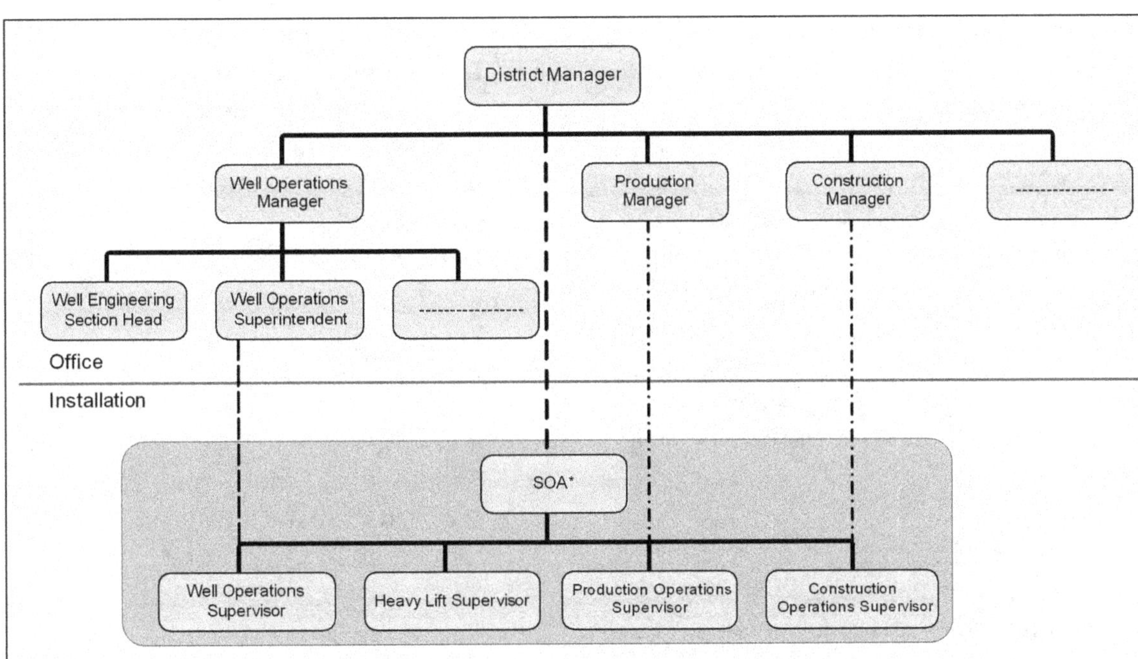

erators and contractors are using some alternate term as a synonym for person-in-charge, such as: SIMOPS Controller (Chevron, 2012), Simultaneous Operations Leader - SOL (Total Fina ELF, 2001), Simultaneous Operations Authority - SOA (ENI, 2005), Person-in-Charge or Lead Operator (Apache, 2013), etc.

For the role of the head of simultaneous operations (PIC/SOL/SOA...), a minimum knowledge is usually required:

- The ability to understand and control simultaneous operations;
- The requirements of the relevant safe work practices standards;
- Hazard identification and evaluation;
- The permit process;
- Awareness on impact their activity could potentially have on other operations;
- Emergency plan;
- The responsibility and authority to shut the operations down if something is observed to be unsafe (stop work authority);
- Establish clear communication between parallel operations;
- SIMOPS plan form, checklist and daily logs, and other.

Depending on type of operations, head of simultaneous operations may act as operation supervisor (OS), offshore installation manager (OIM), drill site manager (DSM) or well site manager (WSM). In parentheses the common acronyms are highlighted. The main duties of the person-in-charge for simultaneous operations are listed below (ENI, 2005; Chevron, 2012):

- The person-in-charge is the company designee who is responsible for coordinating and safe management of the simultaneous activities at the facility; he receives delegation for these responsibilities from the site manager (district manager, company representative);
- He liaises with all supervisors (production, construction, commissioning, maintenance, safety, etc. - also called SIMOPS representatives) and area /installation authorities;
- Participates in risk assessments, identifying potential hazards and proposing risk reducing measures;
- Reports to the operations manager;
- Should have the autonomy, authority and power to take decisions;
- Is responsible for defining priorities so that safety is maintained and downtimes reduced as much as possible;
- Should authorize all work permits in accordance with the permit-to-work (PTW) system, and will ensure countersignature by the respective supervisor(s) of the involved discipline(s);
- Is responsible for ensuring the compliance with rules and regulations, and the follow-up of the working condition of the safety systems;
- Is responsible for deciding the priority of simultaneous activities, and for making the final decision on general production shutdown in case of an emergency;
- Is responsible for approving the start-up of new wells or resumption of production from existing wells;
- Ensure that hand-over notes are prepared to allow for a smooth continuation of the operations at shift handover;
- Should ensure that emergency and all necessary drills are periodically performed in all work areas;

- Is responsible for the preparation and issue of the permit-to-work for production operations;
- Is the drill site manager or well site manager on all drilling and completion operations.

The head of simultaneous operations (PIC, SOA, SOL) should have sufficient technical background and knowledge related to the major activities being performed and should be trained/competent as to make clear and timely decisions. Although he may not be fully experienced in every aspect of all operations being executed, he will be formally informed of such activities and of their status and relevant program in order to be able to assess the levels of safety controls in place and give the final agreement for their start with respect to other activities in progress.

For unexpected events, requiring immediate intervention, that are implemented under the responsibility of the well/production/construction operations supervisor in charge (i.e. lower ranked persons), the head of simultaneous operations should be informed as soon as possible. Then, he should evaluate the impact of change and the measures taken on the other activities in progress, and should be ready to take any further decisions on the matter. Additionally, in establishing consistency with regard to the expectations of all personnel involved in dangerous or routine operations possibly causing an increase in risk, it is necessary (in written form) to specify conditions to be complied with and all precaution measures to be adopted before, during and after the carrying out simultaneous operations.

Life Cycle Model for SIMOPS

As stated earlier, the cycle of simultaneous operations goes through a series of interrelated and conditional steps. By their logic, interdisciplinary approach must include all procedures and measures to prevent obstructions or interference, to minimize constraints and to reduce risk of operations to the greatest extent possible. The SIMOPS program should establish a mechanism for review of proposed activities through ongoing communication, meetings and joint efforts of all parties involved. Emerging from this process, a decision matrix is derived to authorize simultaneous activities. The objectives of the SIMOPS program and procedures are to ensure their execution on controlled and safe manner. One can easily conclude that in offshore environments significantly more demanding, dangerous or risky situations are generated, comparing to onshore. Therefore, hereinafter described cycle mainly relates to offshore parallel activities, which does not mean that in number of specific circumstances cannot be applied to onshore situations.

Based on the large experience and representing offshore marine and underwater engineering companies, International Marine Contractors Association (IMCA) has become among leading authorities in providing structural approach to managing safety through risk assessment, technical interaction, supervision, trained personnel, management of change and feedback. Representing its members as well as offering guidance to industry their influence is beneficial to all industry parties, from operating companies, contractors and suppliers as well as to oil companies in general, governmental and regulatory bodies. For the purpose to provide guidance on simultaneous marine operations in support of offshore oil and gas

Figure 5. The SIMOPS process flowchart (based on IMCA, 2010)

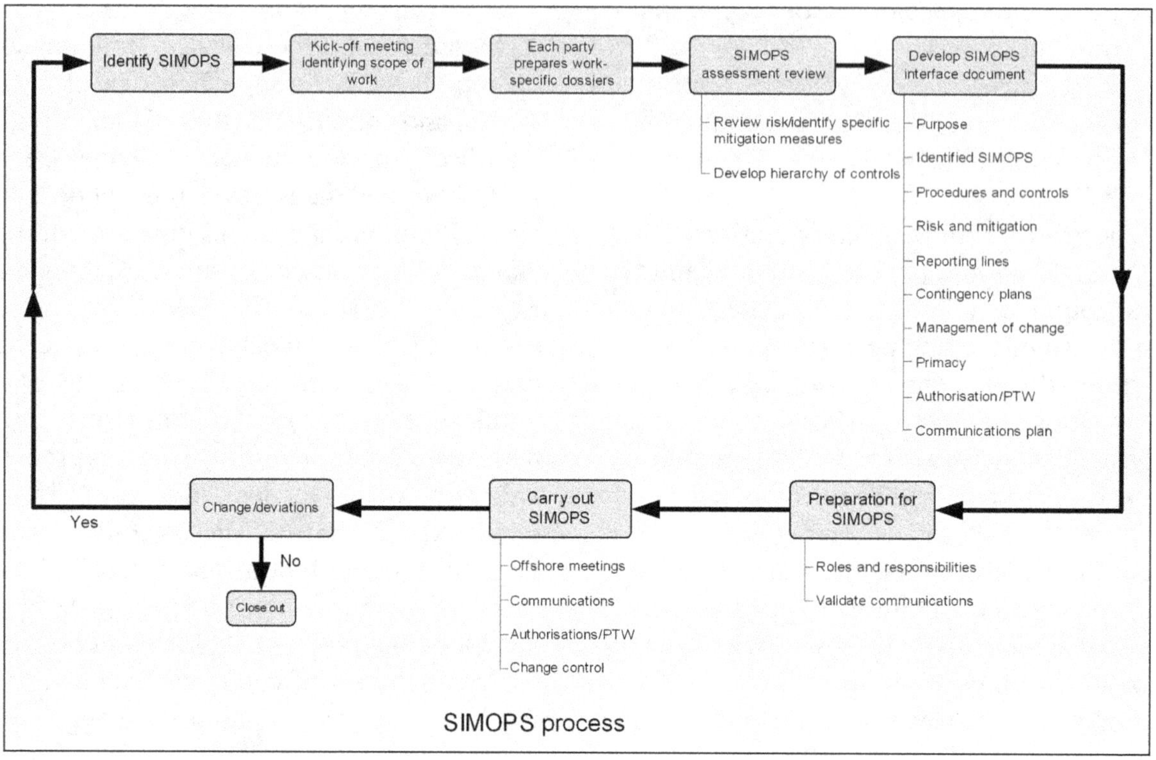

exploration and production or related construction and survey activities, the SIMOPS process is documented covering all its phases. Emphasis is made on the use of explanatory formats and schematics flowcharts.

Following the main SIMOPS premises, elaborated in previous sub-chapters, the explanation of basic steps (indicated in Figure 5) will be briefly elaborated.

- Identifying simultaneous operations

It is very important that simultaneous operations are identified at an early stage before the work starts. They can arise as the result of a variety of issues. Usually, there is overlapping of activities in the same area at the same, i.e. schedule clash. But also simultaneous operations may come about as a result of interference between platform operations and vessel activities, due to physical (loss of position, anchor patterns), environmental (weather limitations, currents, icebergs) and/or failure impacts (leakage, gas, explosion etc.). Furthermore, intervention in the area of 500 m around the existing infrastructure and other combined activities in the area of operation which could compromise project success.

- SIMOPS meetings

Prior to start-up of simultaneous activities all supervisors and authorities must be convened to a kick-off meeting by the site manager so that the scope of work can be composed and to ensure that all precautions have been implemented and reliable equipment is available.

The kick-off meeting should identify all the SIMOPS operations to be performed, time frame required for each part, as well as present the evidence that operations can start (according to the operational plan and authorization given by the operation manager). Emphasis will be placed on the operational steps to be taken in case of an emergency and the check list of corresponding mitigation measures should be drawn up. The meetings should be minuted and properly documented.

- Preparing documents (dossiers)

Based on the kick-off meeting and scope of work developed for the field operations a daily work coordination meeting is required, attended by simultaneous operations leader (SOL, SOA, PIC) and all supervisors involved in the simultaneous operations, in order to manage and plan the safe performing of the SIMOPS, in relation to the progress of ongoing activities. For each of the parties involved a work-specific dossier should be prepared containing the following (IMCA, 2010):

- A summary of work set out in a step by step procedure;
- A conceptual field sketch showing the surrounding offshore installations associated with the activities and restriction zones;
- A list of assets (platforms, vessels, jack-ups etc.) to be used in each part of the SIMOPS;
- A summary of constraints;
- Organization chart and a summary of the main hazards for each operation;
- Mitigation strategies and precautions proposed for each specific operation;
- As the operations proceed, significant deviations from the program shall be formally identified, recorded, and approved within the procedure, where applicable;
- Many other measures and limitations regarding whether, communication systems, assets allocation due to positioning systems used to avoid interference between vessels, subsea tools and equipment, remotely operated vehicles;
- Checklist of actions to be taken for each work-specific SIMOPS operation, and
- Contingency plans for all operations and measures listed previously.

- Performing SIMOPS assessment review and developing interface documentation

All the risks in conducting SIMOPS, contained in each party's dossier, should be identified and reassessed using methodology or tools such as hazard identification and risk assessment (HIRA), clash analysis and interdependency analysis.

During the SIMOPS assessment review it is important that the roles and responsibilities of each party involved are clearly identified and a hierarchy of control established for different phases of operations, covering in that way reporting lines in normal and emergency mode.

Depending on the scope of SIMOPS activities one interface document covering all the work could be required. Alternatively this can comprise several documents covering clearly identified specific SIMOPS activities. Typical SIMOPS interface document should contain all items as shown in Figure 5.

- Undertaking the simultaneous operations

If required by local regulations, the operational plan should be submitted to and approved by the relevant statutory authority before starting simultaneous operations. The plan should be updated or modified whenever significant deviation requests for already approved activities are planned, and a new approval should be obtained. Authorization to commence simultaneous operations is given by the operations manager once he has verified that the operational plan has been approved, when the requirements established in the operational plan are fulfilled, when an effective structure of management, supervision and control is in place and that the installations are suitable and prepared for the work to be carried out.

Once the SIMOPS has started daily calls or meetings (if feasible) should be held between the representatives of the involved parties for the purpose of monitoring progress of the current SIMOPS activity and to organize safe planning for the remaining work. As identified in the SIMOPS interface document, regular and emergency communications should be followed.

The person-in-charge should coordinate simultaneous operations, as identified in the interface document. The PIC should manage operations through single permit-to-work system. When deviations from the control limits are identified, the change should follow the agreed change control mechanism defined in the interface document.

- SIMOPS close out

Upon completed SIMOPS operations it is always helpful to undertake the process of their evaluation and closing either at the involved companies level or entire project level. The objective of the close out review is to examine what has been learnt in terms of the application of further SIMOPS activities.

CONCLUSION

Carrying out different operations concurrently is nowadays a common practice in a wide range of onshore and particularly offshore activities. Significant economic effect, which improves the efficiency of the process of exploitation of oil and gas in general, is achieved by their performing. Depending on the scope of work of particular companies involved, a large number of activities can take place simultaneously. Once simultaneous operations are identified, their planning and managing is based on the fundamental principles that include - the implementation of risk analysis before carrying them out, the hierarchy of responsibilities assigned and the permit-to-work system put in place during their execution, while ensuring all necessary action to meet HSE requirements. Communication is a key factor for the successful implementation of these basic principles while performing SIMOPS operations.

In order for simultaneous operations to be carried out in a safe and controlled manner, manuals or documents intended to provide guidance were developed by major companies and professional associations. The guidelines cover the entire simultaneous operations cycle, starting from their identifying, performing and completing. For the purpose of making a decision on performing SIMOPS, decision matrix significantly helps, which is mainly linked to the scope of activities within the company where it originates. Simultaneous operations are unavoidable and essential in offshore activities, but safety must never be compromised while performing. Therefore, the operational plan should be submitted to and approved by the relevant legislation authority, if so required by local regulations, before starting a simultaneous operation.

REFERENCES

Apache. (n.d.). Retrieved from https://sems-portal.apachecorp.com/

Arcanum Deep Secrets. (n.d.). Retrieved from http://arcanumdeepsecrets.wordpress.com/2010/07/01/bp-oil-disaster-simultaneous-operations-overview-infographic-diagram/

Coastal Energy. (n.d.). Retrieved from http://www.nucoastal.co.th/manuals_nct/SIMOP%20Manta%20RM/Manta%20to-20Richmond%20SIMOPS%20Risk%20Matrix%20FINAL.pdf

ENI S.p.A., E & P Division. (2005). *General best practice for well site simultaneous operations for onshore and offshore installation*. Author.

Ezine Articles. (n.d.). Retrieved from http://www.EzineArticles.com/?expert=Tina_Wallace

Gilbert, D. L. (2001). *New simultaneous operations guidelines enhance safety of offshore operations*. SPE 22802.

International Association of Drilling Contractors. (2009). *Health, safety and environment case guidelines for offshore drilling contractors*. Author.

International Marine Contractors Association (IMCA). (2010). *Guidance on simultaneous operations* (SIMOPS). Retrieved from www.Imca-int.com

ISO. (2002). *Petroleum and natural gas industry – Offshore production installations – Guidelines on tools and techniques for identification and assessment of hazards*. ISO 17776.

Manta-to-Richmond Simops Matrix. (n.d.). Retrieved from http://ebookbrowse.com/manta-to-richmond-simops-matrix-final-pdf-d459244888

National Commission on the BB Deepwater Horizon Oil Spill and Offshore Drilling. (2011). *Deep water - The gulf oil disaster and the future of offshore drilling*. Author.

Norwegian Petroleum Directorate (NPD). (2002). *Guidelines to regulations relating to management in the petroleum activities (the management regulation)*. Retrieved from http://www.npd.no/regelverk/r2002/frame_e.htm

Oilfield. (n.d.). Retrieved from http://www.glossary.oilfield.slb.com

EU Parliament. (2011). Draft report on facing the challenges of the safety of offshore oil and gas activities. *Committee on Industry, Research and Energy, 18*(8).

Parliament, E. U. (2011). *Proposal on safety of offshore oil and gas prospection, exploration and production activities, 2011/0309 (COD)*. Bruxelles: Author.

Petroleum. (n.d.). Retrieved from http://www.Standard.no/en/sectors/Petroleum

Ross, D. (2007). *Planning and communicating for conducting safe SIMOPS*. Paper presented at the Dynamic Positioning Conference. Houston, TX.

Saintpère, S., Faget, J.B., Payer, N., Vergnot, D., Baranthol, C. (2005). *Simultaneous drilling-production-construction operations: Managing safety vs. deliverability on the Peciko gas field*. Indonesia: SPE 96152.

Chevron. (2012). *Simultaneous operations*. Chevron.

Slideshare. (n.d.). Retrieved from http://www.slideshare.net/lifecombo/ngisimops

Standard, N. O. R. S. O. K. (1998). *Drilling & well operations, D-010, Rev.2*. Author.

Vinnem, J.E., Aven, T., Hundseid, H., Vassmyr, K.A., Vollen, R., & Ǿien, K. (2003). Risk assessments for offshore installations in the operational phase. In *Proceedings of ESREL 2003*. ESREL.

Woodhill. (n.d.). Retrieved from http://www.woodhill.co.uk/technical/os_cso.htm

ADDITIONAL READING

Bradford, T., & Cooke, R. (2001). *Probabilistic risk analysis: foundations and methods*. Cambridge, UK: Cambridge University Press. doi:10.1017/CBO9780511813597

Reason, J. (1997). *Managing the risks of Orginizational Accidents*. UK: Ashgate.

Schmidt, K. (2009, December). SIMOPS during construction, commissioning, startup, and operation of the world's largest LNG project, IPTC 13705, International Petroleum Technology Conference, Doha, Quatar.

KEY TERMS AND DEFINITIONS

Barrier: Administrative, organisational or physical measures that reduce the probability of realising hazard's potential for harm and which reduces its consequences.

Close Vicinity: Safety zone within 500 m.

Decision Matrix: The chart or any other graphic display to indicate which of the activities can be performed to each other.

Dynamic Positioning (DP): A vessel or a platform maintaining its position with a dynamic positioning system.

Permit-to-Work: A system to manage and control work activities and the personnel carrying them out.

Person-in-Charge: The person responsible for coordinating and safe managing the field operations.

Simultaneous Operations: Carrying out of potentially clashing activities at the same time, in the same location where the course of any operation may affect the safety of personnel, equipment and environment of another operation.

Chapter 6
The Macondo 252 Disaster:
Causes and Consequences

Davorin Matanovic
University of Zagreb, Croatia

ABSTRACT

The disaster of the Deepwater Horizon platform, while drilling the Macondo 252 well in the Gulf of Mexico in 2010 is for now the last of numerous tragedies, blowouts, and oil spills resulting from petroleum engineering activities. After the accident, several commissions, investigation groups, advisory committees, and company reports were prepared. They investigate causes and consequences of the disaster from different standpoints, but mainly come to the same conclusions. The "n^{th}" approach is presented in this chapter.

INTRODUCTION

The Macondo 252 well supposed to be an exploratory well in Macondo prospect onshore, at the water depth of about 1,500 m (5,000 ft). It was at the same time designed to serve as a production well if sufficient hydrocarbon reserves were proved. Unfortunately on April 20, 2010, after lot of bad decisions, human errors and safety barriers malfunctioning, gas that has found the way from the layer to the surface caused the explosion and fire. The consequences were dramatic. From the standpoint of families, the worst one was the loss of 11 lives and 17 people injured. As the efforts to fight the fire were unsuccessful, the platform sank in the morning of April 22, 2010 (the Day of the Planet Earth). All attempts to

DOI: 10.4018/978-1-4666-4777-0.ch006

close the well were also unsuccessful so the oil spill continues at the seabed level for 87 days more. About 210,000 gallons of oil was spilled every day and in only three days it covered over 580 square miles. Spreading to the coast it has influenced the coastal marine life as well.

SAFETY IN DRILLING OPERATIONS

To that time one of the best standards for elaboration of well integrity in drilling operations was NORSOK standard D-010 (2004) due the extensive activities offshore. It defines the requirements and guidelines related to well barriers with adequate schematics and tables determining features and acceptance criteria for primary and secondary barriers.

Drilling Concern

At the moment of the accident the Macondo 252 well was in phase of drilling and tripping with shearable drill string. According to adequate schematics primary well barrier has to be drilling fluid column. It must exert the pressure on the bottom of the well that will prevent any influx (kick) of formation fluid. Such pressure should be equal or adequate to control estimated or previously measured reservoir (pore) pressure with respect to surge or swabing pressure changes. The properties of the drilling fluid should be controlled. Pressure on the bottom should never exceed the formation fracture pressure. Control and verification is done by control of the fluid level/s in the well and spare tanks, the amount and intensity of fluid returns over the vibration screens and in the tanks. To maintain circulation all the time enough fluid should be prepared (about 20% of the fluid in circulation system at the moment). Secondary well barriers are casing, cement sheet in the annulus, wellhead, drilling blowout preventer system (BOP) and high pressure riser if installed. Casing design is one of the most important parts of the well design process. It determines casing material quality, wall thicknesses, joints and placement (from the bottom to the top of the well, or as a liner). The purpose of the casing/liner is to serve as the physical barrier against the well bore wall rocks (API Bull 5C2, 1999; Bull 5C3, 1994). All fluids inside the formation, formation pressure, and the weight of the casing are acting on the string all the time the drilling of the next section is in the process. Also casing must withstand those stresses through the well lifetime (API Spec. 5CT, 1995; API Spec. 5B, 1996). Casing design should be done on the basis of nominal dimensions and material quality. The ellipse of plasticity, showing allowed triaxial stresses is the best representation. The use of different design factors as is the practice in most of companies is not the best way because it can be misleading (Adams et al., 1993). To avoid uncontrolled fluid flow casing or liner should be leak tested after the placement. Usually casing/liner is cemented. That means to place cement slurry in the annular space and below the cementing shoe, wait for cement and continue with drilling if needed after tightness testing. Doing so the continuous and permanent assuming impermeable hydraulic seal is set along determined hole length in the annulus, to prevent any formation fluid flow and even more stabilize the formation rocks. The proper placement and quality is the primary demand for such kind of barrier. The predetermined length/casing to liner overlap, of the column should be achieved and proper cement sheath quality as well. Of great importance is the adequate "wait on cement" time that should be determined through testing on surface samples. The pressure in the annulus should be monitored regularly (not only through the drilling process). All casing strings are suspended over the wellhead that is constructed in the same manner that casing strings are landed in the well (API Spec 6A, 1999; API Spec 6A, 1999). Each wellhead spool should have annulus access ports with manometers (control of annular pressure change), seals and hanger. At the same time during the drilling process it is a basement for blowout preventers and risers mounting. Each section of the wellhead should be selected according to the maximum possible/anticipated well shut in pressure. The future purpose of the conduit should also be considered due the differences in fluid and pressures acting on the wellhead. Wing valves should be mounted and pressure and function tested to allow pressure control in the annulus or killing of the well if necessary. At the top of the wellhead for each segment of the well drilling, adequate BOP system should be used (API RP 53, 1997). Several preventers are used with different purposes; a) ram preventers for closing the annulus around the drill pipes or drill collars or to close the full cross section of the well, b) ram preventer capable to shear the drill pipe, c) annular preventer that closes around any pipe in the cross section and the full cross section area without pipes inside, d) rotational preventer if some under balance is needed or allowed. For drilling in deep water there should be more precautions, so the second annular preventer is used and double control and operating system too. One or both of them should be a part of the lower marine riser package. The dynamically positioned vessel must enable shearing of the full casing strings with sealing ability at the end; otherwise the liners should be run instead full casing string. High pressure riser is used as an extension of the BOP if they are positioned at different levels. The purpose is to prevent fluid flow from the well bore to the environment. It should be of the same pressure rating as the BOP used, with gas tight connections and sealing elements that can resist to fluids and temperatures that are acting on them. Each element of the BOP and riser system should be leak tested after the mounting or if reinstalled.

Drilling string is composed according to the needs (API Spec 7, 1997; API RP 7G, 1998; API Spec 5D, 1999). Usually it consists of the bit, drill collars, drill pipes, subs and according the way the string is rotated; kelly at the top (for rotary table connection) or downhole motor at the bottom, or nothing of that if top-drive is used. In a manner of the barrier it is important in fulfilling the purpose of providing the drilling fluid circulation without leaking in spite of any stress, temperature, corrosion or wear.

Stab in safety valve should be mounted and closed at the top of any open ended tubular joint of string suspended in the rotary table if string closure and sealing is needed. The influence of primary and secondary well barriers on risk must be determined through the analysis and understanding of hazards and accompanying events. They are considered as measures that are implemented or designed with the purpose to reduce likelihood of triggering potential of hazardous event and/or to reduce the consequences of such event (Holnagel, 2008). The performance of the safety systems in reducing the probability (likelihood) of the event to occur and their consequences can be represented according to Figure 1.

When using such system performance for drilling process, the sequence of usual events can be traced. Daily operations include drilling through different sequence of rocks/formations, with different pore pressures and fluids that have to be controlled because if not controlled appropriate they can lead to hazards. Prevention of unwanted events associated with high pressures and hydrocarbons is done by reduction of likelihood of kick (unwanted influx) by appropriate maintenance

Figure 1. Safety system performance with successive layers of defenses (Handal & Øie, 2013)

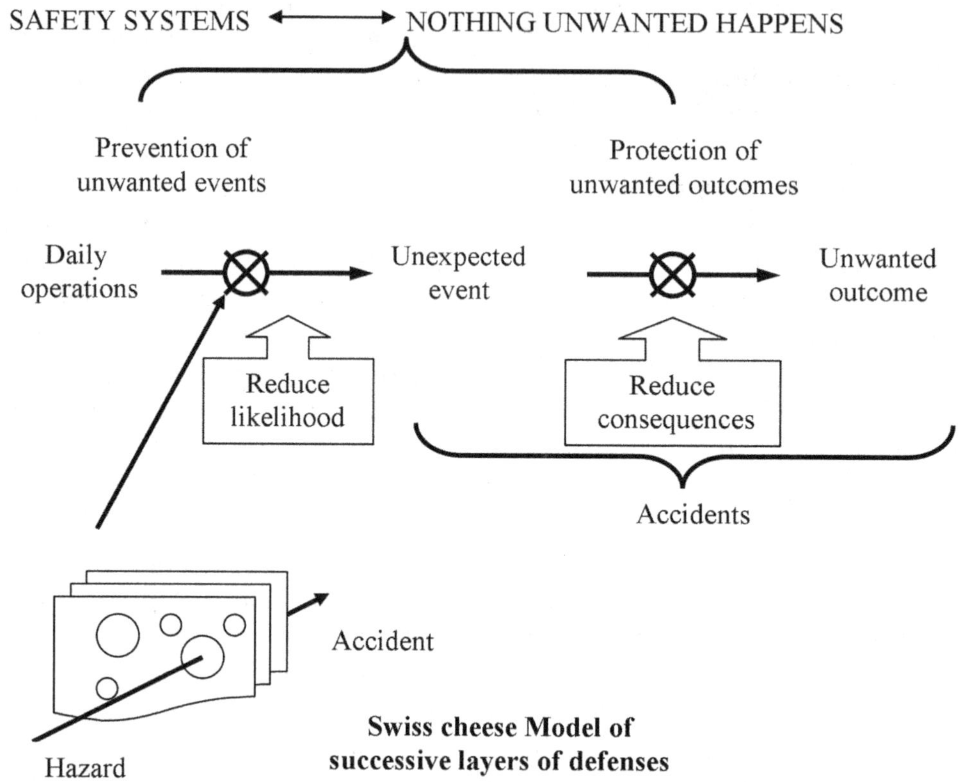

of the mud (primary well barrier). If for any reason primary barrier fails, the unexpected event can be experienced. Protection of unexpected event is achieved by reduction of likelihood and of unwanted outcomes by reduction of consequences. Reduction of consequences of primary barrier failure is reduced by activation of secondary barriers that must act all the time until the function of primary barrier is restored. Unvented outcome results when both barriers have failed. The quality of reduction measures can be represented with the number of slices of Swiss cheese (Reason, 1997). Such model represents the set of barriers where the accidents are prevented by successive protective systems, where holes represent the possible path for active failures and latent condition failures. Only when all of preventing systems fail to stop the hazard propagation of the accident is possible. Active failures result from human errors or technology malfunction that can lead to the accident (mud loss or excessive mud returns not detected, or BOP failure). Latent conditions represent the mistakes in process functioning and systems involved (cement slurry quality; wait on cement; casing or liner design). The model anticipates that any of barriers is capable to stop the accident sequence. In sequence, the next barrier must succeed if first has failed. They should all the time be independent in functioning. The as low as reasonable practicable approach is used to determine the number of necessary barriers. Single failures are not considered as the cause of a major accident. The "holes" that can lead to accident if correspond should be as small as possible and the parallelism avoided.

CAUSES OF THE MACONDO DISASTER

Serious of reports, articles and books have been written since the end of the struggling with the Macondo 252 well was successfully killed (put under control). The official reports are usually done by the authorities that create a comity of professionals from related topics (BOEMRE, 2010; BP, 2010; Lane et al., 2011; National Commission, 2011). The report should describe the sequence of actions or oversight that lead to the top event. Also the effectiveness of the control and emergency shut down systems should be reviewed. The impact of the technical staff and management must also be revised, because the universal truth in petroleum business is that about 85-90% of accidents are due to human factor (Yokel, F.Y. & Simiu, E., 1985). At the end some kind of conclusions or recommendations should appear in favor of prevention of future accidents of such type. They can lead to change of regulations, improvements in technology, equipment and managing or communication system. Going towards more and more challenging areas and situations it is obvious that the overall culture of the offshore industry must educe (Sutton, 2013). The hierarchy and responsibilities should be defined, and legislative in way of treaties. Much more the workers and their skills do not follow the challenges of new sophisticated equipment and technologies. The reasons for that are both, the fluctuation and deficiency of properly educated and skilled personnel and extreme advancement in other technologies implemented in offshore activities (Matanovic et al., 2001). To overcome the safety and environmental problems, safety management topics are elaborated in API RP 75 (2004). The national Commission Report (National Commission, 2011) has stated as the most important, that the Macondo disaster could have been prevented. In spite of worldwide known companies (British Petroleum, Halliburton and Transocean) the risk management was poor with lot of systematic failures. That has showed the real state in the industry involved in deepwater exploration. So in the future, not only regulatory oversight to ensure safety will be sufficient. The involved industry should also try to improve practices and work with the governments. At the end the scientific elaboration of environmental conditions, sensitivity and possible response is also necessary.

Overview of the Accident through the Critical Factors

The possible approach in determining the causes of the disaster is to use Root Cause Analysis (RCA). It can identify the root causes of the incident, but also direct corrective measures to minimize of problem to appear again (Garg & Gokavarapu, 2012). The method follows three steps: 1) first it defines the problem by defining what, when and where has happened, 2) than the problem is analyzed, and 3) the solution of the problem must be found. The answer to the first is that well integrity was not established satisfactory. Because of that hydrocarbons have entered in the well, they expand through the well uncontrolled. Reaching the surface with BOP not sealing, they spread over the platform, ignited and the platform exploded and sunk in fire.

The basis of the analysis is in the organization of the chronological order of events that lead to the disaster by building the timeline with description of events following one another. The starting event is the result of previous mistakes in well design, cementing and equipment functioning control. Because the *cement sheath in the annulus did not seal* the space and the *casing shoe and float collar failed* to prevent cement return. *Cement slurry design* was crucial for the first, because of the narrow margin between pore pressure and fracture gradient of the rocks. Because of that cement slurry was mixed with nitrogen to be at the downhole pressure and temperature foam cement. In such circumstances it is important to avoid any mixing of cement slurry with drilling mud or formation fluids. The independent laboratory that has prepared and tested such cement slurry has found that due the bad rheological properties the slurry was instable with potential for nitrogen break out. At the end due the long displacement it was possible that mixing the slurry with oil based spacer has occurred and some formation fluids should also penetrate to cement because no fluid loss additives have been used. All of that can result in nitrogen migration and formation of future way for gas encroachment.

Also due the overlap between casings there was the necessity of proper slurry placement. With small overlap there is an extreme necessity that casing is centralized inside the well bore and the walls of the hole cleaned of the mud cake. The *change in centralizer number* and positioning has happened. That has weakened the annulus cement barrier at the beginning.

The inside of the casing is protected of the formation fluids ingress by two sealing parts. Those are *casing shoe and the float collar* placed one length of casing above. Also the space between should be filled with cement. But findings show that they have also failed because of the contamination by nitrogen breakout and mud cake. Not confirmed, but possible, there was also some kind of float collar malfunction (check valves do not seal).

When running liner it is always suspended with the *casing hanger seal assembly* and placement must be seal tested. The findings have shown that unfortunately the seal assembly was not locked mechanically in the hanger, which can allow the seal assembly to be lifted up due the changes in pressures and temperature so forming another path for reservoir fluids penetration.

The *casing string* mentioned earlier as the one of the barriers has been selected and mounted properly so it did not contribute in leaking.

After the casing placement, cementing and after waiting on cement (about ten and a half hour) the *negative pressure test* was conducted (Omrcen, B. et al., 2011). That means to establish such conditions that the pressure inside the casing is below the reservoir pressure. In such conditions the three times grater fluid return that calculated was experienced, as an indication of communications flow paths with the reservoir. At the same time *pressure rise in drill pipes* was detected but misinterpreted by the well site leaders (showing their level of competencies). Providing negative pressure test was probably the trigger for gas breakthrough. Regardless that the main thing during all drilling processes is to control pressures on gauges mounted at appropriate places. Not only can the gauges be the indicators of the gas kick. The basic knowledge of the bachelor is to know basic kick indicators (Kristafor, et al., 1992; Matanovic & Gaurina-Medjimurec, 1993).

They are explained in Figure 2.

Not realizing the problem, mud in the well was displaced by the sea water that decreased the bottom hole pressure even more. The investigations have shown the *fluid gain* of about 6.2 m^3 (39 bbl) that was not detected by the crew nor automatic level controls. Some explanations tend to state that the reason was in end-of-well activities, and mud transfer to the supply vessel. About 30 minutes have passed from the displacement starting to the moment when differential pressure has been realized on the rig floor. About $38.6 \cdot 10^5$ Pa (560 psi) pressure rise was detected inside the drill pipes. The most important thing for the

Figure 2. Preliminary clues and indicators of the gas kick (Boyd, W. E., 1971)

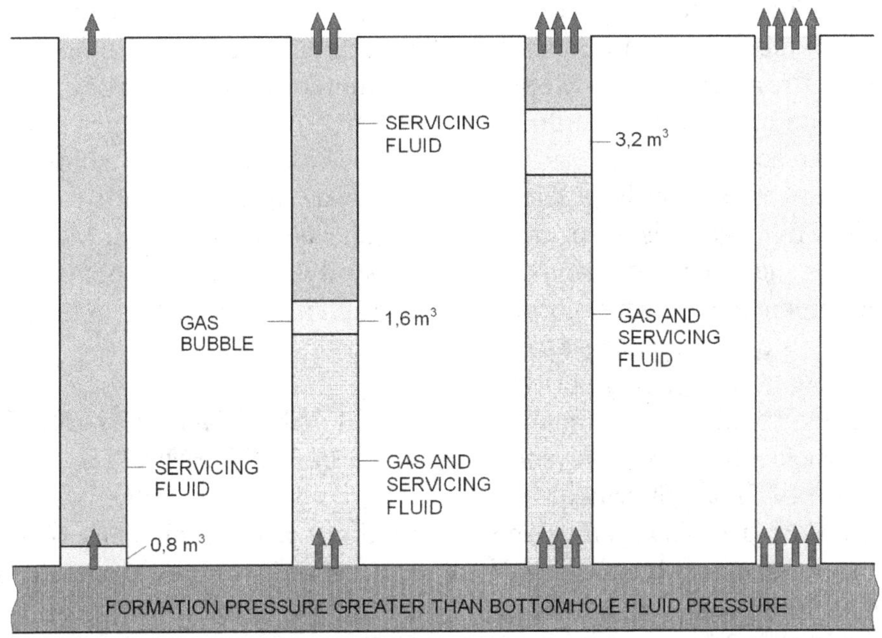

1. Pit level rises as intruding gas displaces mud. 2. Bottom-hole fluid pressure drops.	1. Pit level rises faster as rate of unloading increases. 2. Gas expands as it rises in the hole. 3. Bottom-hole fluid pressure drops. 4. More gas enters well bore.	Well flow continues and accelerates as drilling fluid unloads.	1. All mud out of hole. 2. Uncontrolled gas flow – BLOWOUT!

crew working offshore is to be certificated and trained according to the well control and survival. Unfortunately for the crew on the Deepwater Horizon that was not true. When mud started to flow uncontrolled over the rig floor, the crew tries to close annular preventer in the BOP system. Due the huge amount of the hydrocarbons in the riser (models have shown the amount of 48 m³ (300 bbl), and the fast expand of the gas when reaching the surface, the annular preventer did not fully seal around the drill pipes. Even than the new decision to divert the mud through the mud gas separator was wrong. In spite of that it would be better that crew has selected the diversion overboard through starboard diverter. So the majority of the gas from the mud was not vented securely. The following action of closing variable bore rams to seal the annulus caused the pressure rise in the drill pipes to about $395 \cdot 10^5$ Pa (5,730 psi). Modeling of gas dispersion over the rig floor and the platform has shown that also the secondary protective system has failed. The idea of those systems is to make impossible the gas to reach any ignition point and to ignite. They should also detect *the dispersion of gas beyond electrically classified areas.*

Opposite of that the gas has reached the supply air intakes for engine rooms, where the temperature and the concentration were high enough to start the fire and cause at least two explosions. The automatically shutdown system failed also to act.

Whenever the indicators show the possibility of gas kick, the first action is to securely close the well. For such purpose a number of preventers are mounted together to form a blowout preventer system. They all have to be controlled according to functioning periodically and testing registered and documented. The Macondo disaster had a blowout preventer malfunction as one of the eight critical factors for hazard to take place. The testing done previously has shown some problems in BOP activation but has been neglected. It must be stressed that offshore BOP stacks are more complex than those onshore. Also they are equipped with two pod systems. The fault tree for such systems can be used to indicate possible malfunctions and to be aware of consequences (Ostebo et al., 1991). The failure analysis using the fault tree representation is applicable for complex systems in a way of inclusion of each event contribution to the top event. At the same time it enables the better understanding of system failure behavior with easy overview, and possibility to calculate the probability of the critical event to occur in determined tome interval.

The system acting was investigated according to function prior and after the explosion. Before the explosion as was mentioned the annular preventer failed to seal around drill pipes and the variable ram preventer was activated successfully but too late.

After the explosion multiplex cables and hydraulic control lines were damaged. Losing the activation signal annular BOP has opened. With the rig and platform drifted from the location, the emergency disconnect system should be activated. The activation of automatic mode function should activate the blind shear rams to cut the pipes and loose free. Also the attempts with remote operating vehicle have succeeded to close them shearing was not complete and sealing was not achieved. Because of that oil and gas have flown at the sea bed all the time. The only way to kill the well was to drill so called relief wells. The first have been started in May 2010 and successfully finished with bottom kill maneuver in September 2010.

POST MACONDO CHANGES AND IMPROVEMENTS

Findings and conclusions of Commissions and Committees have lead to determine about possible solutions of the analyzed problems. Unfortunately the leadership and technical competencies that are required for effective performance in drilling operations has failed from the beginning. The "hunger" for money resulted with the accident (Cheng et. al., 2013).

Because the primary and secondary barriers (and third as well; unable to shear the pipes and seal the well in BOP) have failed the new guideline describing the barrier management has been issued (OGP, 2012). It describes barrier management as follows:

A comprehensive and common understanding from well design and construction through operations and abandonment of what constitutes barriers to flow (primary and secondary), and how barriers are verified, monitored and repaired. Understanding how well barrier elements make up a well barrier envelope and the importance of maintaining a minimum capability of two well barriers when the well is capable of discharging hydrocarbons or other fluids to the environment. In situations where two barriers are not feasible alternative control procedures may be adopted. (OGP, 2012, p.2)

Managed pressure drilling systems (MPD), and control equipment are introduced as the improvement of barrier analysis and functioning (Saeed et al., 2012; Falk et al., 2011). Three main steps illustrate such barrier analysis. Those are: (1) system and case description, (2) functional analysis of pressure control equipment, and (3) fault tree analysis with blowout as the top event. First step is the identification of the system with its main functions (mud as the primary barrier). All functions and sub functions of the barrier are determined through the hierarchy according how the function is accomplished. It goes as deep as the functioning of the barrier depends on (mud storage and preparation system, mud pumps functioning, mud cleaning systems; vibration screens, degassers etc.). The presentation is done through the function tree (Figure 3) that can be developed from both sides (Rausand & Høyland, 2004). The one mentioned, and the opposite by determining the lower sub-functions necessary for primary barrier to perform the task. All identified functions should than be listed with description of: (1) list of equipment performing specified function; (2) list of potential failures that may cause top event (kick; blowout); (3) what can be the potential consequences, with specified affect at the primary barrier; (4) which failures can be critical; and (5) list of potential recovery or redundancy measures. Parallel to that a technical hierarchy represented on Figure 3 is determined as well (Handal & Øie, 2013). Those two hierarchy lists can be considered like the first "slice of the cheese". Finally the fault three analyses are used to represent how each or combination of several system and failure events may cause the top event (blowout). Such representation can serve to identify holes in successive layers of defense with introduction of secondary barriers as well (Figure 4).

Figure 3. Presentation of technical hierarchy for primary well control

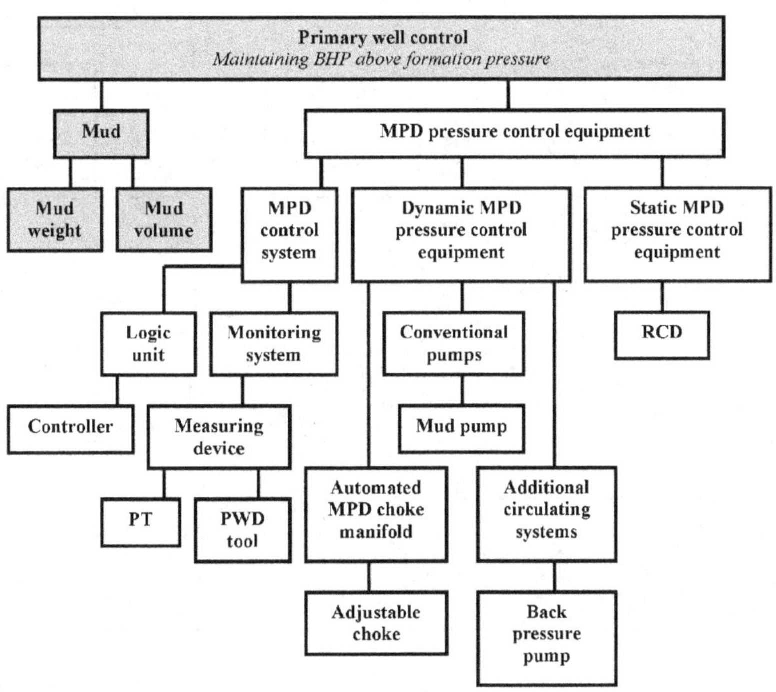

Figure 4. Fault tree analysis of loss of well control through the primary and secondary barrier

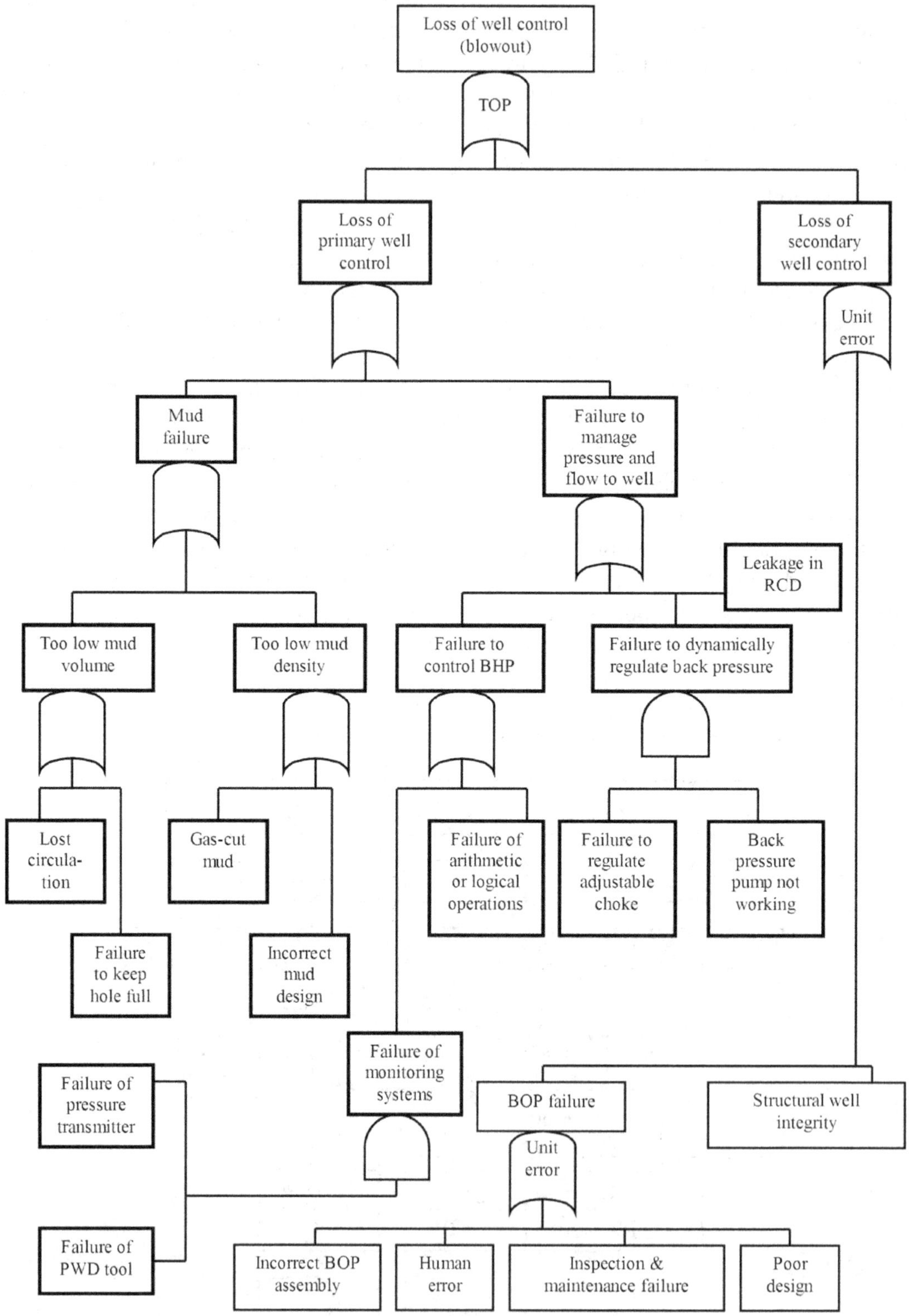

The MPD control equipment manages and controls flow and pressures in the well. That means to maintain all pressures within the operational pressure window. To do so it is equipped with logic units that serve as controllers (with different interfaces; dynamic and static pressure, hydraulics, mechanics, external well monitoring such as mud volume, drilling parameters control etc.). System is monitored through measuring devices such as; flow meters, pressure transmitters, temperature transmitters and other. Finally the pressure control equipment is used for dynamical adaptation of the pressure especially in the annulus. In that way the dynamical bottom hole pressure is controlled to avoid kicks or formation fracturing. Such dynamic pressure control equipment involve automated choke manifold with adjustable chokes, mud and cementing pumps, all kind of circulating system lines, check valves, annular and ram preventers, rotating preventer and rotating or no rotating control devices. All of these systems can be interpreted in the fault tree diagram (Figure 4). It helps to illustrate the possible relationships according to hazards. Unfortunately the lack of knowledge and experience of the personnel can be the mayor concern. It shows that even with several well barrier elements it is possible to experience kick and the blowout.

One of the proposed methods that should prevent unwanted spills is based on the use of warning indicators (Øien, & Nielsen, 2012). The Resilience based Early Warning Indicator (REWI) method that they have proposed maps the important underlying causes: (1) ineffective leadership, (2) deficient communication, (3) failure to provide timely procedures, (4) poor training and supervision, (5) ineffective management and oversight of control, (6) inadequate use of instrumentation and equipment, (7) poorly analyzed and respected risk, (8) focus on day rates not safety; to the direct causes mentioned earlier (Matanovic et al., 2001). The result of the method is obtained by mapping the issued REWI to the determined causes of the Deepwater Horizon (DWH) catastrophe. Such mapping is done based on the four questions asking: (1) if all of the contributing factors of the accident are included, (2) which links can exist between REWI and DWH causes, (3) is there a need for any other cause to be encompassed, and (4) the universal question of the possibility that the method prevents the accident. To method to be successful it is necessary to assure so called contributing success factors explained as the risk understanding, attention, response, decision support etc. Such factors have been developed through a number of workshops of professionals and scientists of different engineering background. Several general issues with appropriate indicators are stated from the beginning; even new ones can be added at any time. The goal is to provide a dialogue in direction that makes sense on the system and its behaviour. Doing so, it should be possible to prevent wrong cement and negative pressure test result explanations. Also the visible pressure changes that were lasting certain time, would not lead to the blowout if proper attention of adequately trained personnel existed. There is also a need for the biggest truth, that after the battle anyone can be a general. So even proved on the stated example the method should be developed and modified according to the challenge.

BOP Post-Macondo Upgrades

Two important and tragic things have been experienced in the Macondo accident, due the BOP malfunction. First one was the failure of annular preventer to close the annular space without leakage, and the second one is in inability to shear the pipes after the platform has sunk. The most important is the fact that BOP system is the final barrier to prevent the blowout. Different problems have been experienced with deepwater BOP control and activation. First one has appeared in 1980's when accumulator capacities have been recalculated for ideal gas to real gas (nitrogen) calculations. After that the problem rises with the depth of seabed. Than the double control lines have been placed (control and balance line). Nowadays after the Macondo disaster American Petroleum Institute (API) is changing the Recommended Practices for Blowout Prevention Equipment Systems for Drilling Wells to the Standard level (Sattler, 2013). The improvements and changes will be in direction of BOP and well compatibility and especially the shearing capability of the shear rams. Also due to the high costs of maintenance and original spare parts some of the Servicing Companies try to cut the costs. Because of that the BOP certification, increased equipment monitoring and specific maintenance requirements will be included in Safety and Environmental Management System (SEMS). The problem is in the possibility of individual perception as the reliability determination component, or even more the insurance companies' statistics. Due the frequent hydraulic control system leaks (especially control lines) the electronic control systems are implemented. The same problems that are experienced with intelligent completion systems are visible here so the improvement in components reliability is the main concert.

The oil spill from the destructed well had lasted about three months before the well was killed. It was due the inability to share the pipes inside shear rams. New upgrades of the system have been done in the use of depth compensated accumulators (Baugh et al., 2011). They enable to close the preventers with increased safety regardless the depth of the see. Much more the shearing ability and the emergency disconnect operations should be improved. The shear problem with drill pipe connection is more obvious with drill collars. Because of that the shareable drill collars have been proposed (outer thicker steel skin and inner thin steal skin, with lead centre wall). Also the other parts of the systems that are known and specified (API Spec 17D, 1992) can gain the level of control of the BOP system functioning.

CONCLUSION

The eight direct causes of the accident that were defined with the precise sequence should be prevented timely. The reason that has not happened lies in underlying causes. The last one stated is the "focus on time and costs rather than control of major accident risks" (Øien & Nielsen, 2012). Such inconsistency between time (money) saving but at the same time increasing risk at different stages of design and well construction and drilling process, have lead to the disaster. Total time saving achieved up to the moment of blowout, fire and explosions on Macondo 252 well were 170 hours (about 7 days). Change in casing design earns three days, using BOP in spite of obvious failures (one day), mud displacement before placing cement plug (one day), and the rest goes to the others (reducing number of centralizers, locking the sliding sleeve, eliminating cement bond and quality logging and inadequate pressure testing). Something that is not stated but should be also considered is the needed time to "wait on cement". That is the time needed that cement slurry solidifies and get proper mechanical properties. When talking about seven or eight (7 or 8) cemented annular spaces (barriers) the possibility of any one to leak is multiplied. The time graph (Omrcen et al., 2011) has shown the discrepancy between projected time schedule and that realy realized. The intervals for casing/liner placement, cementation, wait on cement and wellhead construction and testing have been shortened, which also has contributed to the accident.

REFERENCES

Adams, A. J., Parfitt, S. H., Reeves, T. B., & Thorogod, J. L. (1993). *Casing system risk analysis using structural reliability*. Paper presented at the SPE/IADC Drilling Conference. Amsterdam, The Netherlands.

API RP 14C. (2001). *Analysis, design, installation and testing of basic surface safety systems for offshore production platforms*. Washington, DC: American Petroleum Institute.

API RP 53. (1997). *Blowout prevention equipment systems for drilling wells* (3rd ed.). Washington, DC: American Petroleum Institute.

API RP 75. (2004). *Development of a safety and environmental management program for offshore opeartions and facilities*. Washington, DC: American Petroleum Insitute.

API RP 7G. (1998). *Recomended practice for drill stem design and operating limits* (16th ed.). Washington, DC: American Petroleum Institute.

Baugh, B. F. (1991). *RU (reusable) profile-second generation application*. ETTCE, ASME 91-Pet-32.

Baugh, B. F., & Rash, P. (1992, August). Frictionless stab plate engages new API 17D profile. *Ocean Industry*.

Baugh, B. F., Vozniak, J., & Schmidt, N. (2011). *Post-Macondo BOP safety upgrades*. Paper presented at the Offshore Technology Conference. Rio de Janeiro, Brasil.

Bellarby, J. (2009). *Well completion design*. Amsterdam, The Netherlands: Elsevier.

BOEMRE. (2010). *Report regarding the cause of the April 20, 2010 Macondo well blowout*. Retrieved from http://deepseanews.com/2011/09/boemre-final-report-on-deepwater-horizon/

Bower-White, G. (2013). *Demonstrating adequate management of risk: The move from quantitative to qualitative risk assessment*. Paper presented at the SPE European HSE Conference and Exhibition. London, UK.

BP. (2010). *Deepwater horizon accident investigation report*. Retrieved from http://www.bp.com/liveassets/bp_internet/globalbp/globalbp_uk_english/incident_response/STAGING/local_assets/downloads_pdfs/Deepwater_Horizon_Accident_Investigation_Report.pdf

Brickley, P., Coholan, P., Sharma, N., & Cadwallader, M. (2011). *When mother nature smiles – Multi-year analysis of currents at the Macondo well site*. Paper presented at the Offshore Technology Conference. Houston, TX.

Bull, A. P. I. 5C3. (1994). Bulletin on formulas and calculations for casing, tubing, drill pipe and line pipe properties. Washington, DC: American Petroleum Institute.

Bull, A. P. I. 5C2. (1999). Bulletin on performance properties of casing, tubing and drill pipe (21st ed.). Washington, DC: American Petroleum Institute.

Cheng, R., Wang, H., Shi, J., Ge, Y., Sun, Z., & Tian, H. (2013). *Drilling risk management in offshore China: Insights and lessons learned from the deepwater horizont incident*. Paper presented at the International Petroleum Technology Conference. Beijing, China.

Dhar, R. (2011). *Performance standards for safety critical elements – Are we doing enough?* Paper presented at the SPE European Health, Safety and Environmental Conference in Oil and Gas Exploration and Production. Vienna, Austria.

Falk, K., Fossli, B., Lagenberg, C., Handal, A., & Sangesland, S. (2011). *Well control when drilling with a partly-evacuated marine drilling riser*. Paper presented at the IADC/SPE Managed Pressure Drilling and Underbalanced Operations. Denver, CO.

Garg, T., & Gokavarapu, S. (2012). *Lessons learnt from root cause analysis of Gulf of Mexico oil spill 2010*. Paper presented at the SPE Kuwait International Petroleum Conference. Kuwait City, Kuwait.

Handal, A., & Øie, S. (2013). *Safety barrier analysis and hazard identification of blowout using managed pressure grilling compared with conventional drilling*. Paper presented at the IADC/SPE Managed Pressure Drilling and Underbalanced Operations Conference and Exhibition. San Antonio, TX.

Hollnagel, E. (2008). Risk + barriers = safety? *Safety Science*, *46*, 221–229. doi:10.1016/j.ssci.2007.06.028

Kristafor, Z., Golub, M., & Matanovic, D. (1992). *What else about sizing accumulator volume of BOP control systems*. Paper presented at the Mediterranean Offshore Conference. Tripoli, Libia.

Lane, G., Laleian, A., Marsh, G., Nadeau, P. H., Needoba, W., Nicholson, S., et al. (2011). *Final report on the investigetion of the Macondo well blowout*. Deepwater Horizon Study Group. Retrieved from http://ccrm.berkeley.edu/pdfs_papers/bea_pdfs/dhsgfinalreport-march2011-tag.pdf

Matanovic, D., & Gaurina-Medjimurec, N. (1993). *Some knowledge about killing gas production deep wells in Croatia*. Paper presented at the 22nd Petroleum Itinerary Conference and Exhibition. Tihany, Hungary.

Matanovic, D., Gaurina-Medjimurec, N., Rajkovic, D., & Kristafor, Z. (2001). *Economy and technology balance in well life cycle management*. Paper presented at the SPE 3rd International Symposium on Oilfield Scale. Aberdeen, UK.

McAndrews, K. L. (2011). *Consequences of Macondo: A summary of recently proposed and enacted changes to US offshore drilling safety and environmental regulation*. Paper presented at the SPE Americas E&P Health, Safety, Security and Environmental Conference. Houston, TX.

National Commission. (2011). *National commission on the BP deepwater horizon oil spill and offshore drilling*. Retrieved from http://www.oilspillcommission.gov/sites/default/files/documents/DEEPWATER_ReporttothePresident_FINAL.pdf

NORSOK Standard D-010. (2004). *Well integrity in drilling and well operations*. Lysaker, Norway: Standards Norway.

OGP. (2012). *Recommendations for enhancements to well control training, examination and certification*. Association of Oil and Gas Producers.

Øien, K., & Nielsen, L. (2012). *Proactive resilience based indicators: The case of the deepwater horizon accident*. Paper presented at the SPE/APPEA International Conference on Health, Safety and Environment in Oil and Gas Exploration and Production. Perth, Australia.

Omrcen, B., Rakic, L., Rajkovic, J., Aslanovski, N., & Omrcen, D. (2011). The gulf oil disaster – Lessons learned for present and future. *Naftaplin, 31*(63-64).

Ostebo, R., Tronstad, L., & Fikse, T. (1991) *Risk analysis of drilling and well operations*. Paper presented at the SPE/IADC Drilling Conference. Amsterdam, The Netherlands.

Rausand, M., & Høyland, A. (2004). *System reliability theory: Models, statistical methods and applications* (2nd ed.). Hoboken, NJ: John Wiley & Sons.

Reason, J. (1997). *Managing the risks of organizational accidents*. Burlington, UK: Ashgate Publishing Company.

Saeed, S., Lovorn, R., & Knudsen, K. A. (2012). *Automated drilling systems for MPD – The reality*. Paper presented at the IADCE/SPE Drilling Conference and Exhibition. San Diego, CA.

Sattler, J. P. (2013). *BOP performance-development and consequences in a post-Macondo world*. Paper presented at the SPE/IADC Drilling Conference and Exhibition. Amsterdam, The Netherlands.

Somers, T., & Baugh, B. F. (1993). ROV capabilities for TLP support, including API 17D tooling system. In *Proceedings of Energy Technology and Exhibition Conference (ETCE)*. ASME.

Spec, A. P. I. 17D. (1992). Specifications for subsea wellhead and Christmas tree equipment. Washington, DC: American Petroleum Institute.

Spec, A. P. I. 5CT (metric). (1995). Specification for casing and tubing (metric units) (5th ed.). Washington, DC: American Petroleum Institute.

Spec, A. P. I. 5B. (1996). Specification for threading, gauging and thread inspection of casing, tubing and line pipe threads (14th ed.). Washington, DC: American Petroleum Institute.

Spec, A. P. I. 7. (1997). Specification for rotary drill stem elements (39th ed.). Washington, DC: American Petroleum Institute.

Spec, A. P. I. 5D. (1999). Specification for drill pipe (4th ed.). Washington, DC: American Petroleum Institute.

Spec, A. P. I. 6A. (1999). Specification for wellhead and Christmas tree equipment (17th ed.). Washington, DC: American Petroleum Institute.

Springett, F., & Franklin, D. (2007). *Advancements in subsea BOP stack equipment and controls*. Paper presented at the IADC Well Control Conference. Singapore.

Sutton, I. (2013). *Summarize the deepwater horizon/Macondo reports*. Paper presented at the Offshore Technology Conference. Houston, TX.

(1971). Well servicing and workover. Boyd, W. E. (Ed.), *Control of formation pressure, lesson 9*. Austin, TX: Petroleum Extension Service, The University of Texas at Austin.

Yokel, F. Y., & Simiu, E. (Eds.). (1985). *Application of risk analysis to offshore oil and gas operations – Proceedings of an international workshop*. Gaithersburg, MD: National Bureau of Standards.

ADDITIONAL READING

Bower-White, G. (2013, April). *Demonstrating Adequate Management of Risk: The Move from Quantitative to Qualitative Risk Assesment.* Paper SPE 164969 presented at the SPE European HSE Conference and Exhibition, London, UK.

Fitzgerald, B., Breen, P., & Patric, J. (2012, September). *Making the Safety Case Work – Post Macondo and Montara.* Paper SPE 158199 presented at the SPE/APPEA International Conference on Health, Safety, and Environment in Oil and Gas Exploration and Production, Perth, Australia.

Freudenburg, R., & Gramlin, R. (2011). *Blowout in the Gulf, The BP Oil Disaster and the Future of Energy in America.* London, UK: The Massachusetts Institzte of Technology Press.

Greve, J. F. (2011, October-November). *A New Approach to Deepwater Drilling Data Analysis Offers Enhanced Real Time Capabilities in a Post Macondo World.* Paper SPE 146624 presented at the Annual Technical Conference and Exhibition, Denver, USA.

Holland, P. (1997). *Offshore Blowouts, Causes and Control.* Houston, USA: Gulf Publishing Company.

Hull, M. E., Rees, J. C., Botto, A., Deegan, F. J., & Whooley, A. (2011, May). *Risk-Based Case Study of Floating Facility Drilling Riser Design Concepts in Deepwater Gulf of Mexico.* Paper OTC 21873 presented at the Offshore Technology Conference, Houston, USA.

Hurzeler, K., & Scheuermann, R. (2012, May). *The Development and Maintenance of an Evergreen Spill Response Capability in the Gulf of Mexico.* Paper OTC 23691 presented at the Offshore Technology Conference, Houston, USA.

Lantz, J. G., Little, P. E., Nadeau, J. P., & Reynolds, J. D. (2012, May). *Lessons Learned following Macondo – Safety Enhancement on the U.S. Outer Continental Shelf.* Paper OTC 23463 presented at the Offshore Technology Conference, Houston, USA.

Moomijan, C., Jr. (2013, March) *Macondo Litigation and its Impact on the Offshore Industry – What Every Operator, Driller, Service and Supply Company Needs to Know.* Paper SPE/IADC 163483 presented at the SPE/IADC Drilling Conference and Exhibition, Amsterdam, The Netherlands.

Øien, K., Massaiu, S., & Tinmannsvik, R. K. (2012). *Guideline for implementing the REVI method, Resilience based Early Warning Indicators. SINTEF report A22026.* Trondheim, Norway: SINTEF, Technology and Society.

Øien, K., Utne, I. B., & Herrera, I. A. (2011). Building Safety Indicators: Part 1 – Theoretical foundation. *Safety Science, 49*(Issue 2), 148–161. doi:10.1016/j.ssci.2010.05.012

Ostebo, R., Kallestad, O., & Grytdal, I. (2001, April/May). *Subsea Reliability – Success factors for sustainable deepwater concept development.* Paper OTC 13290 presented at the Offshore Technology Conference, Houston, USA.

Paltrinieri, N., Cozzani, V., Øien, K., & Grøtan, T. O. (2011). Prevention of atypical scenarios through the use of resilience based early warning indicators. Proceedings of the European Safety and Reliability Conference, pp. 42-50,Taylor&Francis, London, UK.

Saad, S., Lovorn, R., & Knudsen, K. A. (2012, March). *Automated Drilling Systems for MPD – The Reality.* Paper IADC/SPE 151416 presented at the IADC/SPE Drilling Conference and Exhibition, San Diego, USA.

Sklet, S. (2006). Safety barriers: Definition, classification, and performance. *Journal of Loss Prevention in the Process Industries, 19*, 494–506. doi:10.1016/j.jlp.2005.12.004

Størseth, F., Tinmannsvik, R. K., & Øien, K. (2009, September). *Building safety by resilient organization – a case specific approach.* Paper presented at the European Safety and Reliability Association Annual Conference (ESREL), Prague, Czech Republic-

Visser, R. C. (2011, May). *Offshore Accidents, Regulations and Industry Standards.* Paper SPE 144011 presented at the SPE Western North American Regional Meeting, Anchorage, USA.

KEY TERMS AND DEFINITIONS

Blowout: An uncontrolled out flow of oil gas or other reservoir or well fluids into the environment at the surface or into other underground formation. It usually occurs when formation pressure is greater than pressure applied by the hydrostatic pressure of the servicing fluid column in the well.

Blowout Preventers: Stack of several sealing devices installed at the top of the well (on the wellhead) which can close annular place around the pipes in the well or entire cross section of the well without pipes in it.

Casing: Steel pipe placed in the well to prevent the walls of the borehole to cave inside, than to prevent fluid communication among different formations when cemented.

Macondo 252: The Macondo 252 well supposed to be an exploratory well in Macondo prospect onshore, at the water depth of about 1,500 m (5,000 ft). Unfortunately on April 20, 2010, after lot of bad decisions, human errors and safety barriers malfunctioning, gas that has found the way from the layer to the surface caused the explosion and fire. The consequences were dramatic. From the standpoint of families, the worst one was the loss of 11 lives and 17 people injured. As the efforts to fight the fire were unsuccessful, the platform sank in the morning of April 22, 2010 (the Day of the Planet Earth).

Managed Pressure Drilling Systems (MPD): And control equipment are introduced as the improvement of barrier analysis and functioning.

Mud: The liquid (suspension of clay in water) that is circulated through the drill string, bit and annular space along the well to clean the bottom and carry the cuttings from the well, exert adequate pressure at the bottom and the walls of the hole among other functions.

Shear Rams: The component in a blowout preventer system that has the purpose to cut the drill pipe body when needed and continue to seal the area; especially important offshore when quick move of the rig from the position is necessary.

Resilience Based Early Warning Indicator (REWI) Method: That they have proposed maps the important underlying.

Wait on Cement: The time needed that after the placement of cement slurry in the well it can harden sufficiently.

Wellbore: Hole (borehole) made by drilling or coring in the ground.

Chapter 7
Risk Analysis of Completion and Production Systems

Davorin Matanovic
University of Zagreb, Croatia

ABSTRACT

A variety of definitions are available through the literature, but the universal one is in defining the well completion as the optimal path for the reservoir fluids to be produced. That means to achieve a desired production with minimal costs. Wells represent the greatest part of expenditure when developing the reservoir. For a long period of time it was defined to be simple, reliable, and safe with enough flexibility to allow future operations. Nowadays, so called "intelligent completions" appear to give more financial benefits, flexibility, and control. The reliability of system components is essential for long-lasting production. In addition, the differences according to natural flowing well risk and artificial lift are given.

INTRODUCTION

The optimal well completion minimizes initial and overall cost according to production completion, while providing the optimal path for reservoir fluid to reach the surface. When designing the completion of an production well many circumstances should be encountered. The completion must balance the factors related to (Patton & Abbott, 1979) environment, constraints, and resources. The aim of the design is to decide about intervals of the reservoir to be completed, completion method, number of completions in the

DOI: 10.4018/978-1-4666-4777-0.ch007

well, casing and tubing configuration and size, perforating considerations etc. So the optimum production (injection) performance include safety by providing the means of production interruption or control restoration in controlled or emergency situations (Peden & Leadbetter, 1986). The integrity of the system and reliability of the components must allow long lasting production with little need for intervention. Because everything is subject of the economic evaluation, the associated costs of initial completion, production costs, and workover and remedial operations and stimulations should be encountered. To obtain optimal completion design there is a need of well performance prediction, than the perforating procedures and equipment are selected, servicing fluid specified and production string and its components as well, and completion sequence and running procedures elaborated. Completion design is influenced by several parameters. They are related to the determined well purpose, dependent on the environment (e.g. land, off-shore), drilled hole diameter, reservoir properties, production characteristics and possibilities and available completion techniques. Much more it is necessary if desired to enable borehole wall stability, selective production from different layers, create minimal pressure drop due the fluid flow, enable the adjustment of flow rate, and allow needed operations and measurements without a need for workover and over all to ensure the well safety. To achieve optimal completion design the prediction of well performance is of interest (now and changes in the future) (Economides, 1998). The well servicing fluid with concern to compatibility with formation rocks and formation fluids is the next factor in the loop. Specification of perforating equipment (retrievable, non-retrievable, tubing conveyed), procedures (underbalanced or overbalanced) and perforations density and spacing impact the inflow performances. Tubing and other down-hole components have to satisfy according the flow optimization, mechanical properties (three-axial stresses) and overall reliability. The specification of the completion sequence and the running procedures are also implemented in the iteration design process, to be in position to consider all eventualities. The design quality depends on the available data. Such data are of two kinds: (1) quantitative data are based on reservoir condition and well parameters and changes can be recalculated and (2) qualitative data are based on the individual or company experience. Data can be grouped according to the scope. Those talking about well objectives and reservoir parameters include well path data, casing program, reservoir depth and thickness, fluid contact zones, reservoir initial pressure and temperature, maximal production rate, well fluids composition and properties and flowing bottom-hole and tubing head pressure. Initial completion and workover data should include the well history, equipment failure records, perforating conditions and servicing fluid characteristics. Completion equipment data specify tubing properties, perforating equipment weight and gun outer diameter and length due the tubing clearance. All other parts of the down-hole equipment should also satisfy material manufacturing requirements, with inner diameters (due to flow restrictions and mandrel seating) and setting position (depth, sequence etc.). Due the change of formation fluids properties when flowing up the tubing, there is also the necessity of determining the depth of wax, hydrate or scale formation if any. Corrosion problems are also strongly related to changes with depth and temperature. Finally clay swelling and wettability changes can appear due the completion or workover works.

From the production point of view, safety is a top concern. Whatever the way of production is applied (naturally flowing well or some kind of artificial lift used) the designer must be aware of changes in the reservoir pressure with rise of cumulative production. Depending on the level of wellhead pressure and the fluid produced (oil or gas) the pressure rating of the controlling or safety elements must be selected. That means e.g. double wing valves on the Christmas tree and subsurface safety valve on wells that produce gas.

SAFETY IN COMPLETION AND PRODUCTION

The critical point in selection the completion or production equipment is related to safety. It should be the same in relation to installation or operation process. The completion design means at the same time the optimal path for fluids up and down through the well and at the same time the barriers that are controlling high pressures and direct fluid flow. There are different approaches in various companies, but there is a general overall rule that there have always to be at least two means of well control. Whatever they are they have to be tested according to function needed. Very good and reliable approach can be found in NORSOK Standard D-010 (2004). It defines well integrity systems in drilling and well operations. They should include two inevitable barriers all the time of well activities or production takes place. That also means that there is a pressure difference that can cause uncontrolled outflow from the wellbore to the environment on the surface. Any well barrier should be designed and selected to withstand the maximal differential pressure that can occur. It is always tested according to reliable and leak proof functioning. It should be placed at the appropriate place that location is well known. Barriers can be defined as primary and secondary (Bellarby, 2009).

Completion Concern

When completing the well the primary barrier is represented by servicing fluid column and installed and pressure tested casing/liner and isolating valve or plug. The secondary barrier at that time is the wellhead connected to casing with mounted blowout preventer system. When installing downhole equipment the blowout preventer system should enable the shearing of any tool that will pass the barrier and than to seal the cross section of the wellhead (Fowler & Roche, 1993). Also the sealing around any tool that can be inside the preventer during system installation must be enabled. Any of the barriers should be verified through the leak testing according to the predetermined differential pressure. Also the operational functioning (close/open) should be performed periodically. The leak testing is performed immediately after the installation, always after replacement, when suspicion on leak exists and when the expected rate of differential pressure that the barrier will be exposed to in near future is close to pressure rating of the system. The testing pressure is applied from the flow direction side. Two tests are performed. First the low pressure one with 1.5 to 2 MPa lasting five minutes because the leak will rather start at low pressures. The second one, high pressure leak test is done up to the pressure rating of the system part tested and must maintain the same for 10 minutes. The cross-section of well completed for production, with specified elements used as barriers or safety and control systems is presented in Figure 1.

Completion covers all activities to install the tubular with appropriate downhole equipment and to suspend it at the surface allowing the fluid to flow from the reservoir through the production system to the gathering system. Optimal downhole production completion

Figure 1. The cross-section of well completed for production, with specified elements used as barriers or safety and control systems

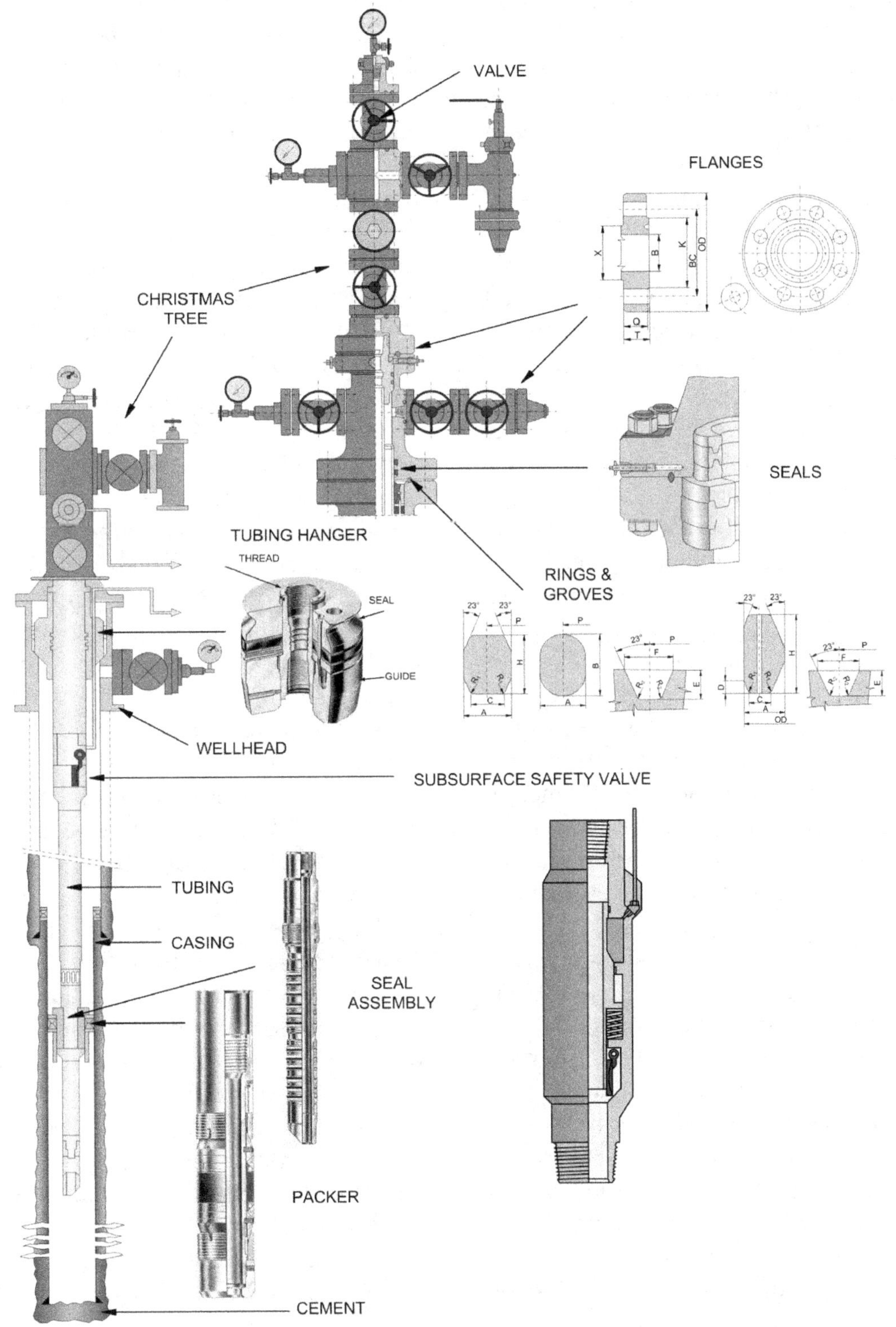

equipment includes tubing (string of production pipes connected with appropriate threads/connections), with sealing device (packer) that seals the annular space between the casing and tubing and so forcing the fluid to flow inside the tubing and other parts of the string. Some kind of flowing device (sliding sleeve or sliding doors) are used to enable the communication between those two areas if needed. All gas wells and also all wells with reservoir pressure high enough to lift hydrocarbons to seabed level should be equipped with some kind of subsurface safety valve. The usual kind is that controlled from the surface by the control line/s. Also some kind of annular safety valve is used in gas lift completions or with potential of hydrocarbon flow through the annulus.

Before the completion starts the well is off course drilled, production casing set in the borehole and the annular space cemented to the top (usual practice in Croatia). All casing strings (if more) are suspended in adequate spool and sealed. Casing strings can be positioned from the bottom of the well to the top, or as liners (just overlapping previous casing string with length of 50 to 100 meters or more if needed). Its function is to serve as a support for the surrounding rocks and to prevent uncontrolled fluid flow between the wellbore and back-side of casing.

The worst up today example for catastrophic destruction of wellhead/casing systems was due the sand production and formation rocks compaction at Wilmington field resulted with surface subsidence of up to 10 m. Several earthquakes have been registered and about 300 producing oil wells have been damaged with complete loss of 120 more (Suman, 1974). Usually the compaction of the reservoir increases axial loading and shortening of casing or slotted liner. If loads exceed elastic limit of the steel, pipe can be plastically deformed or slots can be bulged or distorted. It is improved with injection of cement slurry. The changes in the downhole producing area depend on the amount of cumulative sand volume that has been produced. The worst case, catastrophic sand production is the result of high rate sand influx that can collapse production casing or tubing, or fill up parts of the producing system and terminate production (Veeken et al., 1974).

To fit the purpose the casing strings have to withstand any pressure or load that can appear during the lifetime of the well (API Bulletin 5C2, 1984; API Bulletin 5C3, 1985; API Spec. 5B, 1996; API Spec. 5CT (metric), 1995). When designing casing strings the service-life model should be used instead the conventional casing design based on design factors (Klementich & Jellison, 1986). Loads imposed during initial running and cementing (high differential burst and collapse pressures) differ from those after cementing (significant changes in temperatures). They have stated that the performance life of the casing begins with cementing phase, and than when cement has set additional forces and stresses influence the casing servicing life.

The probabilistic reliability method applied to casing design was used to enable lowering the overall costs of such tubular that was estimated to be about 16% of average years drilling expenditures (Payne & Swanson, 1990). They have realized that the expected load and given casing rating for particular design were defined by the load capacity of the tubular and the maximum load expected through the well's life. The objective was to ensure that all that time tubular load capacity is greater than maximum load expected. Also the quantitative risk analysis (Adams et al., 1993) has given the assign in dependence of design factors and probability of failure according to the acceptable probability of failure. Casing strings are always designed using some kind of design factors (Adams et al., 1993). Regardless the design factor used the blowout rate due the structural failure alone was through the investigated period in the range of 10^{-4}, what means 1 failure (blowout/leakage) in 10,000 wells. At the same time failures due the human error have been recorded in ten time higher range. So the overall blowout rate was 1 in 1,000 wells.

The cementation can be realized as one of the most important stages in well completion process. When cement slurry is set down it cures and forms the cement sheath that stabilizes and supports the casing and avoids fluid migration so providing a zonal isolation between different layers in formation. Bad cementation is a great problem and can be if detected solved with technical and financial additional expenses. One way is the use of squeeze cementing. The annular space when cemented provides a continuous (when cemented to the top) and impermeable hydraulic seal where it is placed. That is a permanent state so cement must resist pressures from all sides, prevent uncontrolled fluid flow and support casing or liner string for the well life. The quality and the position of the cement in annulus must be verified according to the strength, cement bond logs and variable density tool logs. The problems can arise if there are leakage pathways or debonding from casing or the well walls. Two types of mechanisms can influence the cement integrity. As first that is mechanical degradation due to high compressive or tensile loads that overcame its strength (Goodvin & Crook, 1992). Second one is the chemical degradation that propagates slowly unless leakage pathways have been formed before (Barlet-Gouédard et al., 2006). The worst situation is the simultaneous functioning of both effects. The sheath integrity can change due the volume variations through the hydration process (shrinkage/expansion and heat production), contraction or dilation of the casing due the change of mud density and temperature, cement volume decrease with pore collapse, and due the thermal cycling (Bois et al., 2012). The bulk shrinkage is the result of the decrease in pore pressure as a result of cement hydration. That means that the volume of reaction product (cement) is smaller compared to the volume of reactants (water and cement powder). It has been shown (Thiercelin et all, 1998) that the bulk shrinkage is dependent on the boundary conditions and initial conditions. The most important is the initial pore pressure that if high enough prevents the cavitations of water. Experience in the field has also indicated the problems of gas leakage through cement due the pressure testing of the casing (radial deformation of casing and so the cement sheath). The reason can also be the reduction of the hydrostatic pressure inside casing after the cement has set (Jackson & Murphey, 1993). It can be due the change of high density mud with the light servicing fluids (completion and packer fluid). The special consideration must be undertaken when dealing with high pressure gas wells or the wells that are used for gas storage purpose or CO_2 storage wells. In such cases the long period of seal ability is needed and the problem results from changing pressures in the wells and in the formation in contact (Teodoriu et al., 2010). If unexpected pressure rise in the annulus is registered that my be the result of wellhead seal or casing leak, or improper cementation or cement channeling. Because of that surface casing and casing head are also influenced by the corrosion. The oxygenated water or moisture positioned in the annulus will tend to corrode materials in contact (Singh et al., 2012).

As the casing strings are set down, the wellhead is constructed. All changes that affect the casing string are in some way affecting the wellhead to. Because of that it is necessary to monitor the fluctuation in pressure changes registered in different parts at the wellhead. Wellhead is constructed by adding spools as the well is drilled. Each casing string is suspended and sealed in adequate spool (API Spec. 6A, 1999; API RP 6AR, 1994). Sealing between spools is provided by the use of metal rings. Once mounted and sealed wellheads can serve the purpose for a long period of time. Problems can arise due the mechanical and thermal induced stresses. Mechanical stresses are connected to settling of overburden layers resulting of formation compaction as it becomes depleted. Almost the same problem is one mentioned connected with catastrophic sand production. Very good analysis of resultant casing/ wellhead load distribution can be found in Stahl and Baur (1983) for offshore platform casing design, and Kocian et al. (1990) for surface casing onshore wells. The analyses have been focused on the differences among single string design and multistring systems joined at the wellhead. The focus was on the axial and radial response of the wellhead and casing system under combined axial loading, thermal changes (casing expansion or shrinkage) and annular pressure buildup. They have realized that for hot production operations the most intensive buckling is at the tubing and inner casing, because of their greatest free length. Likely such changes have no significant impact on wellhead displacement. Working in direction of cooling the system (killing or injection with cold fluid) the surface casing will be influenced mostly. Static loads and their changes are mainly distributed onto the outer casing. But because it is short and in most cases almost always cemented to top there will also be little impact on overall wellhead displacement.

When landing the production string at the top of the wellhead blowout preventers are mounted. The mud in the well is replaced with clean servicing fluid compatible with the formation rocks and fluids, with the density adequate to formation pressure. When completing the well with open end completion string several elements are used as the barrier. The primary barrier is the servicing fluid (exerts a hydrostatic pressure that is the product of the fluid density, column height and gravity) that must overcome the formation pressure. It must be adjusted not to exceed the formation fracture pressure when circulated or due the surge and swab pressures while tripping. The control is done by verifying the stable fluid level, density and rheological properties. Secondary barrier represents casing string usually cemented to the top of the annulus and suspended in the wellhead. The casing hanger holds the casing in place and primary and secondary seals seal the annulus. For emergency tubing cutting purpose and closing of full area or area around the pipe inside the well, blowout preventers are used (API RP 53, 1999). Two of them are always used. They are determined by the expected wellhead maximal pressure including safety margin for killing operation. They can be of ram type or spherical type. Shearing is possible only with ram type preventers. Also the number of pipe rams must be equal to the number of possible string outer diameters. Kill and choke lines are connected on the wellhead connectors.

String of production pipes (tubing) is designed according to algorithm presented in Figure 2 (Peden & Leadbetter, 1986). The design of the completion string is done with regard to flow optimization with regard to available reservoir energy. Tubing outer diameter is restricted with inside casing diameter. So the optimal inside diameter must be determined through the nodal analysis. It is recommended always to determine real tubing properties according to stresses. They are changing with change of state in the well; shut-in well, production or injection. Pipes

Figure 2. Tubing design procedure

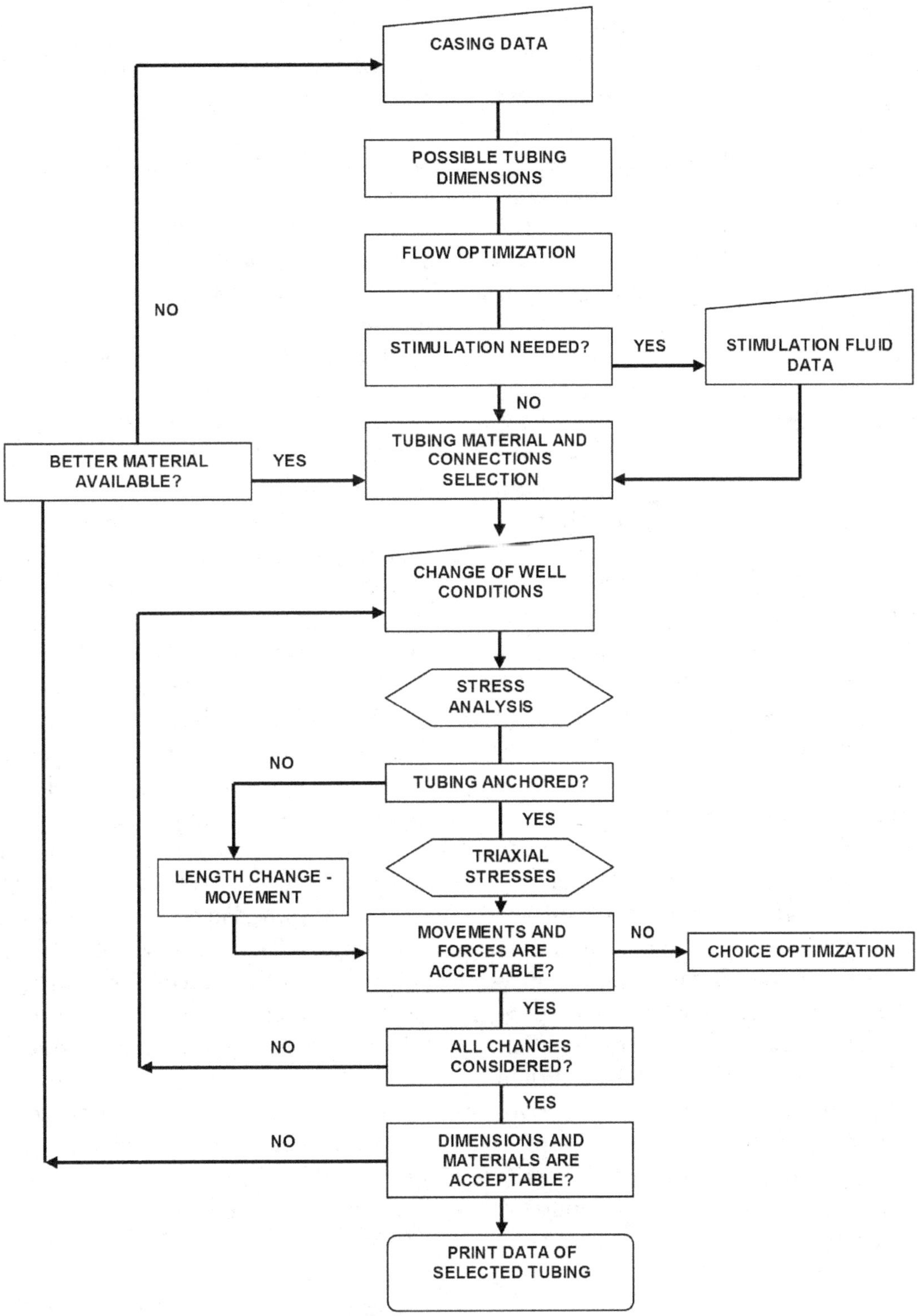

and joints are checked for axial tension, or compression, burst pressure or collapse pressure, and triaxial stresses. Ovality is also one part of concern especially when determining axial tensile stresses combined with collapse pressure. Design factor can be always used according the company policies. Because tubing is connected with packer, possible movements and forces have to be determined too. The temperature is the factor that strongly affects material properties or changes. Strength of the material decreases with temperature increase. Young's modulus of elasticity lowers with stresses at the plastic deformation range. The critical parts of the string are connections especially gas tight connections. There is strong difference in threads and make up control when dealing with steel tubing for oil wells and corrosion resistant alloy joints for sour gas completions (Bannon et al., 1991; Payne et al., 1995). Premium threaded connections have been analyzed through the finite element analysis and full-scale testing (Hilbert & Kalil, 1992). Full-scale testing can only determine if the connection is leaking or not. Finite element analysis gives insight in stresses from makeup alone and then changes under loading conditions (Sato et al., 1985). Seal integrity is achieved with the metal-to-metal contact. There is also a great influence of surface finish and compound grease to galling and material deformation. Grease is used to fill the gap between threads. Such way it helps to transfer loads from one surface to the other and whole joint acts as a solid element. When to much grease is trapped in the gaps there is a significant pressure built up in that area when making connections (Tsuru et al., 1990). Landing nipples can be positioned along the string to enable mandrels placement. They can than hold plugs, check valves etc.

Production Concern with Naturally Flowing Wells

On the other hand for the naturally flowing well the primary barrier is the Christmas tree and the tubing packer system, and the secondary consists of the casing (cemented or no cemented), wellhead and tubing hanger and for gas wells the subsurface safety valve. Production through wells with artificial lift system depends on the system and is usually controlled by Christmas tree, master valve, wellhead and casing as primary barrier, and artificial lift system shut-down as the secondary one. They should be independent of each other, but sometimes they rely one upon another.

After the well is completed and casing and cement shield perforated the production stage begins. It concludes with any workover, well operation, plugging or well abandonment. In fact for naturally flowing wells production is controlled eruption. The flow is controlled through the Christmas tree and can be justified or interrupted if needed by the body valve, master valve or wing valves. Operating production parameters should the entire time be in the boundaries specified for completion design. Well integrity during this period depends also on primary and secondary barriers. The primary is combination of production packer and tubing with surface or subsurface controlled subsurface safety valve (if needed). Secondary are casing (cemented, suspended and sealed in the wellhead spool), tubing hanger (connected with tubing by threaded connection and sealed inside tubing spool) and production (Christmas) tree. Also any annular space is equipped with access line, valve and if possible gauge to detect pressure change.

Packer is the sealing and anchoring element that consists of a body, cones, slips and sealing elements. When activated it divides the inside of the tubing and the annular space between casing inner diameter and tubing outer diameter above packer sealing element. It is necessary that the functioning of the packer is permanent regardless the changes in pressures and forces acting with the change of conditions in the well. The slips should hold packer in place while sealing assembly and polished bore extensions should enable tubing movement according the changes from one state to the other. The barrier verification is according to the allowed differential pressure usually in the direction of flow.

At the top production string (tubing) is suspended and sealed in tubing hanger that is pressure tested with other elements of the wellhead.

Subsurface safety valves (SSV) purpose is in fact to stop the flow of fluids up the tubing in situations when there is an erratic change in flow line pressure if the well head is damaged, Christmas tree or connected lines have been destroyed or malfunction for whatever reason. Surface controlled or differential subsurface safety valves (API RP 14B, 1996) consist of a body with threaded connections, close/open sealing mechanism that allows or stop the fluid flow. When controlled from the surface they are connected with the hydraulic control unit by the hydraulic control line/s. In such conditions closing pressure is the function of control fluid gradient. In other cases it depends on the surrounding temperature and pressure conditions with regards to hydrates forming or wax or scale deposition. The other thing to be considered is the position of kick-off point, to be below it in case of well collision. It must be tested initially by applying low and high differential pressure. Also periodical testing in regular intervals (each month, every three months, and every six months) with test duration of 30 minutes must be performed. They can be tubing retrievable (when landed as a part of a tubing string) or wireline retrievable (when mounted inside the landing nipple or tubing retrievable one). When analyzing liability or reliability it is optimal to consider the use of tubing retrievable one with the possibility to repair method by wireline tools manipulation inserting the wireline retrievable valve inside opened tubing retrievable one. If than both fail the tubing can be retrieved at the end (Busch et al., 1985). In addition surface controlled valves are mostly in use. That means that the control line/s is used to force the closing element to be opened. If the pressure in control line is lost (destroyed, bleed-of) the valve closes. Control lines can be the reason of valve malfunction especially in deep see completions and productions. Than dual control lines are usually used. The reliability of closing element (ball or flapper) has been detected to be about same (Medley, 1978; Engen & Rausand, 1982; Gunt, 1976). Even so the experience in the field has shown that there are more problems in opening ball valves than flapper valves if control line fails. Also the problem with wire line tools manipulation through the flapper valve can result in valve closure when pulling out because of possible hitch.

Production tree (API Spec. 6AV1, 1996; API Spec. 6D, 1994) can be placed on the surface or as a sub-sea one. In general it consists of a housing (flanged spools or as a solid block) with bores that are fitted with valves (swab, master, kill and flow). The purpose is to provide the controlled path for the fluid coming up the tubing to the surface lines of gathering system. Also it mast be able to stop the flow by closing some of valves mentioned.

Production Concern with Artificial Lift

At some point of the production the pressure decline comes to the value when some kind of artificial lift is need to maintain the production from the reservoir. That means that the pressure in the reservoir is to low to raise the hydrocarbons from the depth of the reservoir to the surface. Several systems can be used than. The broadly used is the beam pumping, than come gas lift, electrical submersible pumps, Moineau pumps and hydraulic lift. For all of them the pressure on the surface in the producing part of the well will be the function of the pressure exerted by the driving force of the artificial system used. That means when the system drive is shut off there will be no pressure at the surface. That is true for all systems except the gas lift system. Regardless the gas lift system used there is the part of the well under the pressure of the injected gas. That means that such amount of the gas in the container can bleed if some kind of leaking is possible. For such purpose the annular safety valves (ASV) are used. They should seal the annulus between tubing and the casing, so protecting the surface facilities and workers from the gas trapped in the event of wellhead integrity lost (Morris & Pringle, 1988).

Reliability Concept

The main item in risk or reliability analysis is the equipment failure rate. So the unavailability or unreliability of the system can be predicted, and so can be the risk (Shafagi, 2008). The same can be applied to completion and production systems. The blowout as the worst detrimental impact or the leak of with less impact is considered to influence the personnel, production systems and environment. So the prediction of possible blowout or leakage is essential to determine the risk.

When evaluating the production system the analysis applied by Woodyard (1980; 1981) can be applied. The fault tree and reliability analysis have been used to combine the consequence or the risk to be avoided with the component or the entire system reliability. To estimate reliability it is necessary to determine the failure rate of the component. It is well known from the tribology approach that the failure rate is greatest in the start up period and when wear out period is reached. In between there is a long useful life period of minimal and stable failure rate. That stresses out the importance of elimination of early failures through proper control of materials and production of the items and also on time replacements when necessary. To calculate the reliability of the system Equation 1 can be used:

$$R(t) = e^{-\lambda \cdot t} \tag{1}$$

where $R(t)$ is the probability that the part of the system will operate without failure for the period (t) for stated operating conditions; (λ) expressed in hours^{-1} or years^{-1}, is the component failure rate; and (t) is the application period expressed in hours or years.

At the other hand the probability of the failure during the same period of time will be expressed by Equation 2:

$$P(t) = 1 - e^{-\lambda \cdot t} \tag{2}$$

where $P(t)$ is the probability that the part of the system will fail within the period of time (t).

Some other analysis terms can be used. One of them is equipment mean-time-between failures (*MTBF*). It represents the average time between occurrences of a specific failure mode of the part of the system and mathematically it is the inverse of the equipment failure rate (Equation 3):

$$MTBF = \frac{1}{\lambda} \qquad (3)$$

Sometimes there is of interest to know the mean-time-to-restoration (*MTTR*). That is the average time needed to restore the equipment operation (time to detect failure, diagnose problem, and repair and again start the production). It and can be expressed by Equation 4:

$$MTTR = \frac{1}{\theta} \qquad (4)$$

where (θ) is the equipment restoration rate usually represented in hours, because (*MTTR*) is very small compared to (*MTBF*).

Component reliabilities can be obtained from historical data (depending on availability they can also be very limited), controlled experiments that are not practical and engineering judgement that can be broad but of varying quality. Some intend also to use the insurance company documentation that can also lead to false conclusions.

In production systems the most important is the sealing function, so just ranking the seals and their reliability can be enough to compare different systems with the same or different sealing elements. The best seals according to the reliability are metal-to-metal seals and wellhead packing, when installed at the surface and tested. Next, ranged as good are all other static seals installed at the surface and permanent packer sealing elements as well. Fair reliability enable dynamic seals on machined parts, retrievable packer and static packer bore sealing elements and downhole wireline installed seals. The worst are the dynamic packer bore seals regardless installed on the surface or downhole. The overall reliability of the system can be calculated if reliabilities of the components are known. Data on mean times between failures can (*MTBF*) be obtained from literature. For example, the tubing joints failure survey (metal-to-metal gas tight) has shown the *MTBF* to be in a range of $2.52 \cdot 10^8$ hours. Because it represents the best seal quality any other sealing element below that class will have lover range of *MTBF*. Surveys with subsurface safety valves have been done by different companies and institutions (Brill, 1974; Lawson & Brill, 1975; Engen & Rausand, 1982; Busch et al., 1985). The quality and the reliability is improved with each subsequent survey. The mean value of *MTBF* obtained was about $1.82 \cdot 10^5$ hours for valves closing element and $1.84 \cdot 10^6$ hours for control systems. Broader analysis has been presented by Grassick et al. (1990) and Grassick et al. (1992). They have covered gas-lift completions, through the analysis of wellhead tree, tubing, SSSV, single gas-lift completion, dual, hybrid and concentric. Results were presented as the average blowout frequency against part reliability as *MTBF*. Results have shown that there is the greatest reliability of the dual string gas-lift completion with *MTBF* of about $1.75 \cdot 10^5$ hours. Tubing if properly selected is very reliable part lasting up to forty years when continuously installed, which means almost double compared to the overall gas-lift installation. Unfortunately due the uncontrolled production it is possible to complete the well with new tubing and change it in a year. The wellhead and Christmas tree can be realized as the most reliable part of the system because it can withstand even more than tubing string if installed properly and not disconnected.

The Fault Tree approach presented in Figure 3 represents components associated

Figure 3. Fault Tree for a simple completion of gas production well

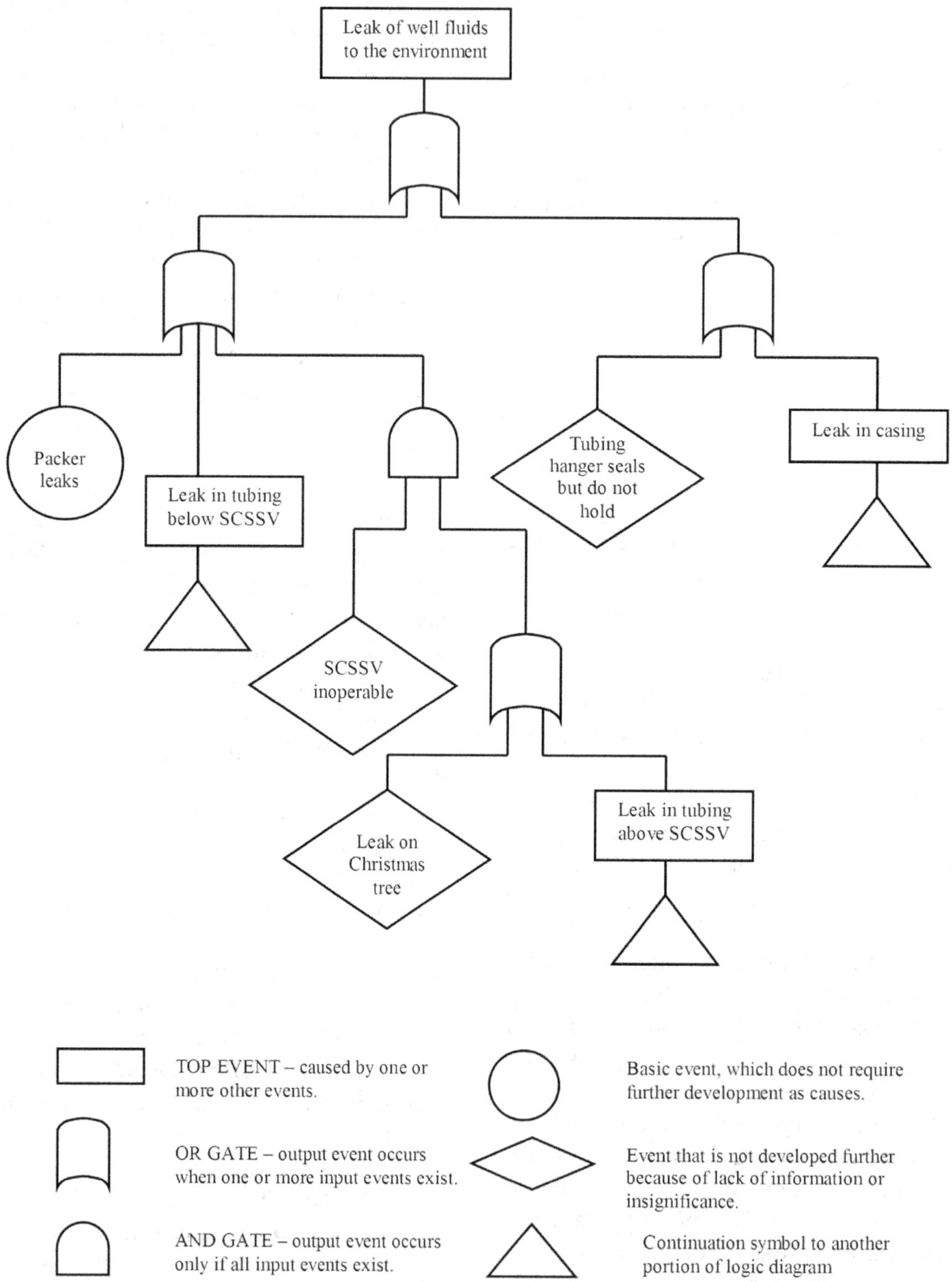

with gas production. The probability of the leak in the system can occur only if the casing and tubing hanger seals fail and tubing or packer leaks at the same time. The surface controlled subsurface safety valve (SCSSV) can influence the system only if the tubing leaks above it.

Intelligent Completion Reliability

Intelligent completion comes out of today's more complex technical demands with developmental scenarios that incorporate economical and environment protection aspects. That is particularly important in off-shore activities or in communications with multiple layers through only one well. Concept of "intelligent completion" basically means to lower the basic control equipment down hole regardless that is producing or injection well. Intelligent completion (Vachon et al., 2006) includes two primary concepts: (1) instantaneous surveillance– that means the possibility to measure and gather data of changes in the layer pressure and changes in flow at layer level; and (2) instantaneous control (real-time control) – the possibility to control and change flow conditions by remote steerage of down hole control equipment. Data collection is provided by the use of different sensors, and data transfer enable electrical and optical cables, while driving of downhole equipment is possible through hydraulic control lines. Figure 4 illustrates basic differences between conventional (a) and intelligent (b) completion.

Each project is evaluated according to two criterions: (1) risk or reliability with regard to technical integrity of the system and (2) by its return. The reliability of the system (Mathieson et al., 2003) must be in accordance with requirements set up in the project. Modeling of intelligent completions are verified with accelerated testing procedures, and come-at-able data enable creation of framework scheme for correct decision making during designing process, if:

- The incremental value derived from implementation of intelligent well technology is defined,
- Available and applied adequate equipment according to project requirements and functionality,
- Ensured that all factors that enable proper determination of earning of extra profit through the time of intelligent completion implementation exist.

So the reliability becomes to be a key requirement for all systems that permanently and remotely control inflow and monitor the changes in reservoir. The reliability lifetime has in the past been mostly determined from the history of the available data. The thorough analysis has shown that it corresponds to the people survival function, due the same mathematical behavior over time (Veneruso et al., 2000). The failure of the system elements regardless they are electrical or mechanical are a function of operational time. At the beginning an initial period exists where failures result from manufacturing and installation reasons. In the middle is a long period of constant failure rate, and at the end, there is a wear out period with increased failure intensity due the wear, fatigue or aging. Through the twenty years of experience with permanent downhole installations the improvements in testing and engineering processes ad the quality to system reliability and risk reduction. The main reasons for failures appear to be numerous environmental parameters such as temperature and pressure changes (cycles),

Figure 4. Basic differences between conventional (a) and intelligent (b) completion

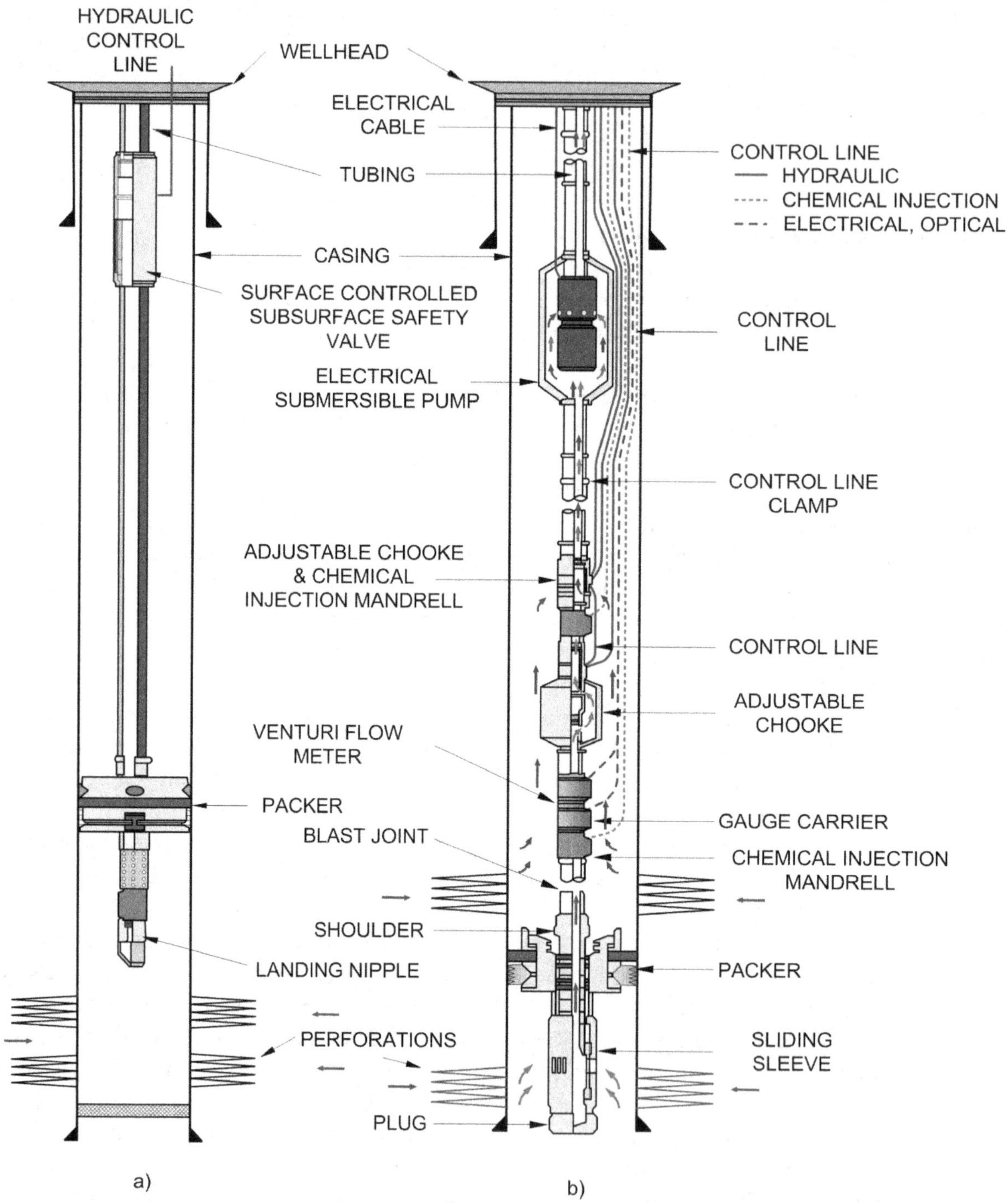

flow rates, erosion due the sand production, corrosion due the well fluids chemistry and vibrations as well (Drakeley et all., 2001). To increase elements reliability an active reliability program has been established. It starts with product design phase by definition of the mission profile (the set of fundamental boundaries for the system, functionality, life expectations and environment). In the testing phase Failure Modes Effects and Criticality Analyses (FMECA), Failure Mode Testing (FMT), Fault Tree Analysis (FTA) and Accelerated Life/Stress Testing (QAL/ST) are performed (Williams &.Holbrook, 2001). Much more the Weibull reliability analysis method enables to analyze failure rates that are not constant and vary with time. It can be written as:

$$R = e^{-\left(t/\alpha\right)^{\beta}} \qquad (5)$$

where probability (R) is the function of time (t), Weibull time parameter (α) and Weibull shape parameter (β) that can be greater than 1 when the failure rate increases with time, less than 1 when it decreases with time and equal to 1 when it is constant over time (Veneruso et al., 2003).

CONCLUSION

General conclusion of the analysis of blowouts or well leaks (Roland, 2000) connected with completion and production is related to duration and released amounts of hydrocarbons. Well leaks have short duration (up to 10 minutes). Also the amount of released fluid is less than 100 m^3 per leak. From at the moment, broadest and available database (Holand, 2006) it is obvious that the risk of the blowout or well leak during completion is 0.00077 incidents per completion and 0.000047 incidents per well year during production.

REFERENCES

Adams, A. J., Parfitt, S. H., Reeves, T. B., & Thorogod, J. L. (1993) *Casing system risk analysis using structural reliability.* Paper presented at the SPE/IADC Drilling Conference. Amsterdam, The Netherlands.

API RP 14B. (1994). *Design, installation, repair and operation on subsurface safety valve system* (4th ed.). Washington, DC: American Petroleum Institute.

API RP 53. (1999). *Recommended practice for blowout prevention equipment systems for drilling wells* (3rd ed.). Washington, DC: American Petroleum Institute.

API RP 6AR. (1994). *Recommended practice for repair and remanufacture of wellhead and Christmas tree equipment.* Washington, DC: American Petroleum Institute.

Barlet-Gouédard, V., Rimmelé, G., Goffé, B., & Porcherie, O. (2006). *Mitigation strategies for the risk of CO_2 migration through well-bores.* Paper presented at the IADC/SPE Drilling Conference. Miami, FL.

Bellarby, J. (2009). *Well completion design.* Amsterdam, The Netherlands: Elsevier.

Bois, A-P., Garnier, A., Galdiolo, G., & Laudet, J-B. (2012, June). Use of a mechanistic model to forecast cement-sheat integrity. *SPE Drilling & Completion,* 303-314.

Brill, J. P. (1974). *API activities for improving subsurface safety valve reliability.* Tulsa, OK: The University of Tulsa.

Bulletin, A. P. I. 5C3. (1994). Bulletin on formulas and calculations for casing, tubing, drill pipe, and line pipe properties (6th ed.). Washington, DC: American Petroleum Institute.

Bulletin, A. P. I. 5C2. (1999). Bulletin on performance properties of casing, tubing and drill pipe (21st ed.). Washington, DC: American Petroleum Institute.

Busch, J. M., Policky, B. J., & Llewelyn, D. C. G. (1985, October). Subsurface safety valves: Safety asset or safety liability. *Journal of Petroleum Technology,* 1813–1818.

Drakeley, B. K., Douglas, N. I., Haugen, K. E., & Willmann, E. (2001) *Application of reliability analysis techniques to intelligent wells.* Paper presented at the Offshore Technology Conference. Houston, TX.

Economides, M. J., Watters, L. T., & Dunn-Norman, S. (1998). *Petroleum wells construction.* Chichester, UK: John Wiley and Sons Ltd.

Engen, G., & Rausand, M. (1982). *Reliability of downhole safety valves used in north sea.* Paper presented at 14th Annual OTC. Houston, TX.

Fowler, J. H., & Roche, J. R. (1993). *System safety analysis of well control equipment.* Paper presented at the 25th Annual OTC. Houston, TX.

Goodvin, K. J., & Crook, R. J. (1992). Cement sheath stress failure. *SPE Drilling Engineering, 7*(4), 291–296.

Grassick, D. D., Kallos, P. S., Dean, S., & King, S. D. J. (1992, May). Blowout risk analysis of gas-lift completions. *SPE Production Engineering,* 172–180.

Grassick, D. D., Kallos, P. S., Jardine, I. J. A., & Deegan, F. J. (1990, November). Risk analysis of single- and dual-string gas-lift completions. *JPT, 1374-1376,* 1418–1419.

Gunt, W. G. (1976). *Subsurface safety valves for gas storage wells*. Paper presented at Eastern Regional Meeting of the Society of Petroleum Engineers of AIME. Columbus, OH.

Hilbert, L. B., Jr., & Kalil, I. A. (1992). *Evaluation of premium threaded connections using finite-element analysis and full.scale testing*. Paper presented at the IADC/SPE Drilling Conference. New Orleans, LA.

Holand, P. (2006). *Blowout and well release characteristics and frequencies*. Trondheim, Norway: SINTEF Technology and Society, Safety and Reliability.

Jackson, P.-B., & Murphey, C. E. (1993). *Effect of casing pressure on gas flow through a sheath of set cement*. Paper presented at the SPE/IADC Drilling Conference. Amsterdam, The Netherlands.

Klementich, E. F., & Jellison, M. J. (1986, April). A service-life model for casing strings. *SPE Drilling Engineering*, 141–152.

Kocijan, E. M., Melford, R. N., Hilbert, L. B., & Kalil, L. A. (1990). *Compressive load casing design*. Paper presented at the SPE/IADC Drilling Conference. Houston, TX.

Lawson, J. D., & Brill, J. P. (1975). *Improving surface safety valve reliability – A progress report on API sponsored research*. Tulsa, OK: University of Tulsa.

Mathieson, D., Rogers, J., Dajagopalan, S., & McManus, R. (2003). *Reliability assurance, managing the growth of intelligent completion technology*. Paper presented at the SPE Annual Technical Conference and Exhibition. Denver, CO.

McSpaden, A. R., & Glover, S. (2009). *Analysis of complex wellhead load events for conductor and surface casing strings*. Paper presented at the SPE/IADC Drilling Conference and Exhibition. Amsterdam, The Netherlands.

Medley, E. L. (1978). *Experience with surface controlled sub-surface safety valves*. Paper presented at European Petroleum Conference and Exhibition. London, UK.

Morris, A. J., & Pringle, R. E. (1988, August). Ultra-high-pressure tubing-retrievable annular safety valve. *SPE Production Engineering*, 361–365.

NORSOK Standard D-010. (2004). *Well integrity in drilling and well operations*. Lysaker, Norway: Standards Norway.

Patton, D., & Abbott, W. A. (1979, June). Well completion and workover: Part 3 selecting the best well completion. *Petroleum Engineering International*, 64-74.

Payne, M. L., & Swanson, J. D. (1990, December). Application of probabilistic reliability methods to tubular designs. *SPE Drilling Engineering*, 299–305.

Payne, M. L., Zerbi, J. L., & Sims, D. C. (1995). *Tubular design optimization for ARCO China's Yasheng 13-1 development*. Paper presented at the International Meeting on Petroleum Engineering. Bejing, China.

Peden, J. M., & Leadbetter, A. (1986). *Rationality in completion design and equipment selection in the north sea*. Paper presented at the SPE European Petroleum Conference. London, UK.

Roland, T. H. (2000). *New risk assessment approach shows significant reduction in oil blowout risk.* Paper presented at the SPE International Conference on Health, Safety, and the Environment in Oil and Gas Exploration and Production. Stavanger, Norway.

Sato, T., Hori, M., & Ichiara, T. (1985). *Finite element analysis of premium connections for tubing and casing.* Paper presented at the 8th Energy-Sources Technology Conference & Exhibition. Dallas, TX.

Shafagi, A. (2008). Equipment failure rate updating – Bayesian estimation. *Journal of Hazardous Materials, 159*, 87–91. doi:10.1016/j.jhazmat.2008.01.042 PMID:18294769

Singh, S. K., Subekti, H., Al-Asmkh, M., & Al-Samarraie, L. (2012). *An integrated approach to well integrity evaluation via reliability assessment of well integrity tools and methods: Results from Dukhan Field, Qatar.* Paper presented at SPE International Production and Operations Conference and Exhibition. Doha, Qatar.

Spec, A. P. I. 6D. (1994). Specification for pipeline valves (gate, ball, and check valves) (21st ed.). Washington, DC: American Petroleum Institute.

Spec, A. P. I. 5CT (metric). (1995). Specification for casing and tubing (metric units) (5th ed.). Washington, DC: American Petroleum Institute.

Spec, A. P. I. 5B. (1996). Specification for threading, gauging and thread inspection of casing, tubing and line pipe threads (14th ed.). Washington, DC: American Petroleum Institute.

Spec, A. P. I. 6AV1. (1996). Specification for verification test of wellhead surface safety valves and underwater safety valves for offshore service. Washington, DC: American Petroleum Institute.

Spec, A. P. I. 6A. (1999). Specification for wellhead and Christmas tree equipment (7th ed.). Washington, DC: American Petroleum Institute.

Stahl, B., & Baur, M. P. (1983, November). Design methodology for offshore platform conductors. *JPT.*

Suman, G. O., Jr. (1974, November). Sand control, part 1. *World Oil.*

Teodoriu, C., Reinecke, K. M., Fichter, C., & Wehling, P. (2010). *Investigations on casing-cement interaction with application to gas and CO_2 storage wells.* Paper presented at the SPE EUROPEC/AEGE Annual Conference and Exhibition. Barcelona, Spain.

Thiercelin, M. J., Baumgarte, C., & Guillot, D. (1998). *A soil mechanics approach to predict cement sheath behavior.* Paper presented at the SPE/ISMR Eurock '98. Trondheim, Norway.

Tsuru, E., Maruyama, K., & Yazaki, Y. (1990, March). Influence of compound grease on the performance of premium connections. *SPE Drilling Engineering*, 71–78.

Vachon, G., Vega, P., & Lee, J. (2006). *Closing the well-centric production optimization loop using intelligent completions.* Paper presented at the First International Oil Conference and Exhibition. Cancun, Mexico.

Veeken, C. A. M., Davies, D. R., Kenter, C. J., & Koojima, A. P. (1991). *Sand production prediction review: Developing an integrated approach.* Paper presented at 66th Annual Technical Conference and Exhibition. Dallas, TX.

Veneruso, A. F., Hiron, S., Bhavsar, R., & Bernard, L. (2000). *Reliability qualification testing for permanently installed wellbore equipment.* Paper presented at the SPE Annual Technical Conference and Exhibition. Dallas, TX.

Veneruso, A. F., Kohli, H., & Webster, M. J. (2003). *Towards truly permanent intelligent completions: Lifelong system survivability through a structured reliability assurance process.* Paper presented at the SPE Annual Technical Conference and Exhibition. Denver, CO.

Williams, B., & Holbrook, J. (2001). *Adopting reliability practices from the space industry into intelligent wells.* Paper presented at the Offshore Technology Conference. Houston, TX.

Woodyard, A. (1980). *Risk analysis of well completion systems.* Paper presented at the 55th Annual Fall Technical Conference and Exhibition of the DPE of AIME. Dallas, TX.

Woodyard, A. (1981, April). Determining blowout risks for producing wells. *World Oil,* 87-91.

ADDITIONAL READING

Aasen, J. A., & Aadnoy, B. S. (2004, September) *Multistring Analysis of Well Growth.* Paper SPE/IADC 88024 presented at the IADC/SPE Asia Pacific Drilling Conference, Kuala Lumpur, Malaysia.

Adams, A. (1991, October) *How to Design for Annulus Fluid Heat-Up.* Paper SPE 22871 presented at 66th Annual Technical Conference and Exhibition of the Society of Petroleum Engineers, Dallas, USA.

Aggrey, G. H., & Davies, D. R. (2007, September) *A Rigorous Stohastic Coupling of Reliability and Reservoir Performance When Defining the Value of Intelligent Wells.* Paper SPE 107197 presented at the Offshore Europe, Aberdeen, Scotland, UK.

Ahammed, M., & Melchers, R. E. (2006). Gradient and parameter sensitivity estimation for systems evaluated using Monte Carlo analysis. *Reliability Engineering & System Safety, 91*(5), 594–601. doi:10.1016/j.ress.2005.04.005

Ale, B. J. M., Baksteen, H., Bellamy, L. J., Bloemhof, A., Goossens, L., & Hale, A. et al. (2008). Quantifying occupational risk: the development of an occupational risk model. *Safety Science, 46*(2), 176–185. doi:10.1016/j.ssci.2007.02.001

Bane, D. E., & Peoples Brock (2009, May) *Subsurface Safety Valve Control System for Ultra deepwater Applications.* OTC 19870, Paper presented at the Offshore Technology Conference, Houston, USA.

Bois, A. P., Garnier, A., Rodot, F., Saint-Marc, J., & Aimard, N. (2011) How to prevent Loss of Zonal Isolation through a Comprehensive Analysis of Microannulus Formation. *SPE Drilling & Completion, March 2011,* pp. 13-31.

Bradley, W. B. (1975, September) *The Effect of Casing Wear on the Burst Strength of Casing: Part 1 – Joint Leakage.* Paper ASME 75-PET-11 presented at the Annual Petroleum Mechanical Engineering Conference, Tulsa, USA.

Bradley, W. B. (1975, September) *The Effect of Casing Wear on the Burst Strength of Casing: Part 2 – Statistical Burst Strength of Worn and Unworn Casing Strings.* Paper ASME 75-PET-11 presented at the Annual Petroleum Mechanical Engineering Conference, Tulsa, USA.

Bradley, W. B. (1975) Here's How Casing Wear Affects Joint Leakage. Oil&Gas J., December 25, pp. 170-173.

Cornelussen, K., Sørli, F., Brandager Haga, H., Teonold, E., Menezes, C., Grimbert, B., & Owren, K. (2007, November) *Well Integrity Management System (WIMS) – A Systematic Way of Describing the Actual and Historical Integrity Status of Operational Wells.* Paper SPE 110347 presented at SPE Drilling Annual Technical Conference and Exhibition, Anaheim, USA.

Fanoiki, J. B. (2002, September) *Sustaining Surface Casing Integrity in a Saline Environment.* Paper SPE 77244 presented at: IADC/SPE Drilling Conference, Jakarta, Indonesia.

Goodman, M. A., & Halal, A. S. (1993, October) *Case study: HPHT casing design achieved with multistring analysis.* Paper SPE 26322 presented at SPE Annual Technical Conference and Exhibition, Houston, USA.

Halal, A. S., Mutchell, R. F., & Wagner, R. R. (1997, March) *Multi-String Casing with Wellhead Movement.* Paper SPE 37443 presented at the SPE Production Operations Symposium, Oklahoma City, USA.

Imbò, P., & Gandini, G. (2011, March) *Electro Magnetic Wireline Retrievable - Surface Controlled Subsurface Safety Valve: A New Backup for Surface Controlled Subsurface Safety Valve to Avoid Workover.* Paper presented at 10th Offshore Mediterranean Conference and Exhibition, Ravena, Italy.

Jellison, M. J., & Klementich, E. F. (1990, June) *An Expert System for Casing String Design.* Paper SPE 20328 presented at 5th SPE Petroleum Computer Conference, Denver, USA.

Johnson, F. W. (1984, September) *Microcomputer Finite Element Analysis of Tubular Connections.* Paper SPE 13070 presented at the 59th Annual Technical Conference and Exhibition, Houston, USA.

Johnson, R., Jellison, M. J., & Klementich, E. F. (1987). Triaxial-Load-Capacity Diagrams Provide a New Approach to Casing and Tubing Design Analysis. *SPE Drilling Engineering,* (September): 268–274.

LeBoeuf, G., Adams, S., Pittman, A., & Dodd, P. (2008, May) Case History: *New Design in Surface-Controlled Subsurface Safety Valves Resolves Valve Problems in Subsea Completions in the Gulf of Mexico.* Paper OTC 19620 presented at the Offshore Technology Conference, Houston, USA.

Levine, D. C., Thomas, E. W., & Tolle, G. C. (1979, September) *Annular Gas Flow after Cemmenting: A Look at Practical Solutions.* Paper SPE 8255 presented at the SPE Annual Fall Technical Conference and Exhibition, Las Vegas, USA.

Matanovic, D., Cikes, M., & Moslavac, B. (2012). *Sand Control in Well Construction and Operation.* Berlin, Germany: Springer-Verlag. doi:10.1007/978-3-642-25614-1

Millet, F., Petit, H., Wallez, G., Lalanne, P., Ducasse, A., & Barzu, E. (2008, September) *Improving Well Safety and Maximizing Reserves Using an Innovative Surface Controlled Subsurface Safety Valve (SC-SSSV).* Paper SPE 113829 presented at the SPE Annual Technical Conference and Exhibition, Denver, USA.

Ostebo, R., Kallestad, O., & Grytdal, I. (2001, April/May) *Subsea Reliability – Success factors for sustainable deepwater concept development.* Paper OTC 13290 presented at the Offshore Technology Conference, Houston, USA.

Ravi, K., Bosma, M., & Gastebed, O. (2002, February) *Improve the Economics of Oil and Gas Wells by Reducing the Risk of Cement Failure.* Paper SPE 74497 presented at the IADC/SPE Drilling Conference, Dallas, USA.

Samuel, G. R., & Gonzales, A. (1999, October) *Optimization of Multistring Casing Design with Wellhead Growth.* Paper SPE 56762 presented at the SPE Annual Technical Conference and Exhibition, Houston, USA.

Stewart, R. B., & Shouten, F. C. (1988). Invasion and Migration in Cemented Annuli: Causes and Cures. *SPE Drilling Engineering, 3*(1), 77–82. doi:10.2118/14779-PA

Thiercelin, M. J., Dargaud, B., Baret, J. F., & Rodrigez, W. J. (1998) Cement Design Based on Cement Mechanical Response. *SPE Drilling & Completion, December 1998,* pp. 266-273.

Weiner, P. D., & True, M. E. (1969, March) *A Method of Obtaining Leakproof API Threaded Connections in High Pressure Gas Service.* Paper 69-040 presented at the API Spring Meeting, Southern District Div. of Production, USA. Drilling and Production practice 1969, pp. 40-47 http://www.onepetro.org

KEY TERMS AND DEFINITIONS

Blowout: An uncontrolled out flow of oil gas or other reservoir or well fluids into the environment at the surface or into other underground formation. It usually occurs when formation pressure is greater than pressure applied by the hydrostatic pressure of the servicing fluid column in the well.

Casing: Steel pipe placed in the well to prevent the walls of the borehole to cave inside, than to prevent fluid communication among different formations when cemented.

Casing Head: Flanged steel fitting connected to the top of the casing in the well (introduction casing) or other spools mounted one over another. It provides space for slips (casing hanger) and packing assembly landing so to seal the annular space.

Christmas Tree: Composed or monoblock steel element with control valves, pressure gauges and chokes, mounted on the top of wellhead, to control the flow of oil or gas when produced naturally.

Packer: Element of the downhole equipment which purpose is to seal the annular space between the tubing and casing or wellbore walls. That enables to control the fluid flow through the packer body and continue through tubing. Elements of the packer can be only the body with connections and the sealing element, but also there can be slips, cones, drag blocks, "J" lock, equalizing valve and etc.

Packer Fluid: Some kind of servicing fluid (e.g. oil, salt water or mud) that is placed in the annular space with the density near to killing fluid density. It should not change properties with the time and must be noncorrosive.

Safety Valve: A device that closes or opens automatically when an abnormal condition occurs. Surface safety valve is mounted in the Christmas tree assembly, to stop the fluid flow from the well if assembly is damaged. Sub-surface safety valve is mounted in gas wells below the surface deep enough to prevent the valve freezing, hydrates forming or wax or scale deposition.

Tubing: Is steel or reinforced (glass or kevlar fiber) polymer pipe that is run into a well and serves as a conduit for reservoir fluid production or fluid injection from the surface to the reservoir.

Tubing Hanger: Consists of slips and packing elements and is used to suspend and seal tubing inside the tubing head.

Tubing Head: Is a spool with flanged fitting that holds tubing string over the tubing hanger and at the same time seals the space between tubing and casing holding the pressure. The upper flange enables the connecting and support for a Christmas tree.

Wellbore: Hole (borehole) made by drilling or coring in the ground.

Chapter 8
Risk and Remediation of Irreducible Casing Pressure at Petroleum Wells

Andrew K. Wojtanowicz
Louisiana State University, USA

ABSTRACT

Oil well cement problems such as small cracks or channels may result in gas migration and lead to irreducible pressure at the casing head. Irreducible casing pressure also termed, Sustained Casing Pressure (SCP) is hazardous for a safe operation and the affected wells cannot be terminated without remedial operations. It is believed that even very small leaks might lead to continuous emissions of gas to the atmosphere. In the chapter, the author describes physical mechanisms of irreducible casing pressure and qualifies the associated risk by showing statistical data from the Gulf of Mexico and discussing the regulatory approach. This chapter also introduces a new approach to evaluate risk of casing pressure by computing a probable rate of atmospheric emissions from wells with failed casing heads resulting from excessive pressure. Also presented is a new method for assessing potential for self-plugging of such wells flowing wet gas as the gas migration channels could be plugged off by the condensate.

DOI: 10.4018/978-1-4666-4777-0.ch008

ANNULAR AND SUSTAINED CASING PRESSURES

Petroleum wells are usually constructed such that their casing–casing annuli do not experience abnormal pressures. The exception occurs in gas lifted wells where gas is injected into the production tubing–casing annulus. Annuli differ from other well components in that they are usually not the result of purposeful design. Rather, they are a consequence of the design of tubulars and the well construction process. Therefore, the ability of an annulus to withstand loads that occur on its components is (or should be) evaluated at the end of the design process.

The right figure in Figure 1 shows different kinds of annuli in a well bore (JIP, 2001). The primary annulus (Type I, or A) is formed by the production tubing and casing. It is bounded on the top and bottom by the wellbore seal assembly and completion hardware (including packers and seals) respectively. In addition, there may be an annular safety valve, gas lift valves and related equipment depending on the nature of the well.

The secondary (outer) annuli (B, C, etc.) can be of two kinds- Types II and III. The Type II annulus is formed by two adjacent casing strings. It is bounded at the top by the wellhead seal assembly and at the bottom by the cement. The cement top in this instance is above the shoe of the outer string of the annulus. The type III annulus is essentially similar, except that its bottom is open to the formation. The cement top lies below the shoe of the outer casing string, either by design or accident.

By definition (as well as design) an annulus is a sealed volume, and there should be no flow paths that cause migration of fluids into (from) the annulus from (into) its surroundings. In

Figure 1. Simplified well schematic and types of annuli in a wellbore (Detail "A")

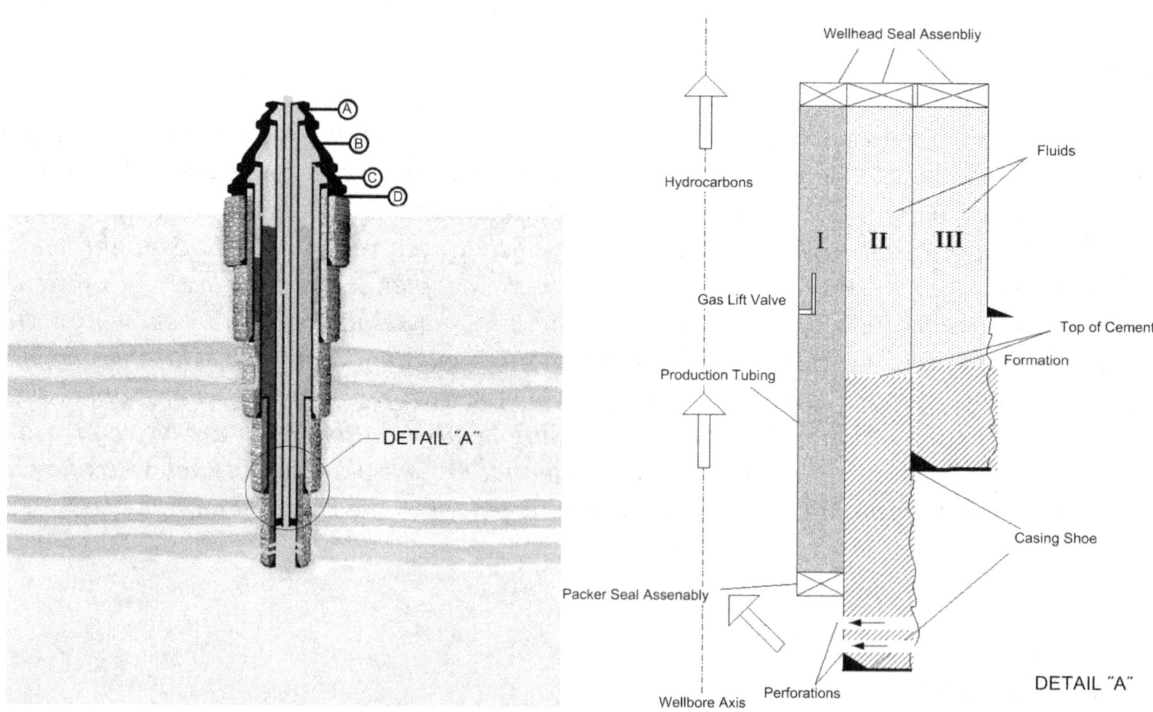

principle, given the annular configuration, all leak paths that can compromise its integrity should be identified. This identification is an integral part of the design process and is the basis for the diagnosis and management of sustained casing pressure (SCP), later in the life of the well.

The left figure in Figure 1 is a typical well completion showing the placement of cement to seal off the interior of various casing strings from the subsurface formations exposed by the drill bit (Bourgoyne et al., 1999). Ideally, the well should have pressure only on the production tubing. Pressure gauges on all of the casing strings should read zero if:

- The well is allowed to come to a steady-state flowing condition, and
- The effect of any liquid pressurization due to heating of the casing and completion fluids by the produced fluids is allowed to bleed off by opening a needle valve at the casing top.

Only a small volume of fluid would be bled off in order for the casing pressure to fall to atmospheric pressure if the pressure was caused by thermal expansion effects.

If the needle valve is closed and the well remains at the same steady-state condition, then the *annular casing pressure* should remain at zero. If the annular casing pressure returns when the needle valve is closed, then the casing is said to exhibit *sustained casing pressure* (SCP). In some cases the pressure can reach dangerously high values.

SCP Problem Occurrence

Persistent, unintended pressure on closed casing/casing annuli (SCP) and leakage of formation fluids from open casing/casing annuli are a significant problem in many oil and gas producing areas. The impetus for this chapter has primarily been the SCP problem in the Gulf of Mexico (GOM). However, the "surface casing vent leakage" problem with gas wells in Alberta has essentially the same downhole causes. It has received substantial attention via regulation by the Alberta Energy and Utilities Board and prevention and remediation efforts by the industry. Serious problems resulting from unintended pressure on casing/casing annuli have also been reported in the San Juan Basin of New Mexico, in South Louisiana, in India, Norway, Poland, and in Tunisia. Hydrocarbon intrusion into drinking water aquifers has occurred in the San Juan Basin and in Alberta, and its potential for occurrence should be a major concern in any onshore producing area.

Occurrence of SCP was early reported in technical papers by Attard (1991) and Michel (1991). Attard's paper describes the repeated occurrence of SCP in the primary and secondary annuli of wells in the Hutton field, offshore North Sea. This paper discusses the engineering basis for management and remedial actions that were necessary following the discovery of SCP in the wells. The paper by Michel deals with operational aspects of diagnosing SCP and isolating causes with specific details pertinent to wells on the North Slope of Alaska.

In the late 1990s and early 2000s, Louisiana State University (LSU) studies (Bourgoyne et al., 1999), revealed that:

- 50% of the casing strings exhibiting sustained casing pressure are production casing strings;
- 10% of the casing strings exhibiting sustained casing pressure are intermediate casing strings.
- 30% of the casing strings exhibiting sustained casing pressure are surface casing strings;

- 10% of the casing strings exhibiting sustained casing pressure are conductor casing strings.

The study also concluded that poor cementing was the major source of SCP in the outer annuli.

Also analyzed was a GOM field most severely affected by the SCP problem (Wojtanowicz et al, 2000). Of the total 26 wells in this field, 22 wells (85%) displayed SCP. Distribution by casing type in this field was quite different to that for the whole GOM;

- 30.8% of the casing strings exhibiting sustained casing pressure are production casing;
- 65.4% of the casing strings exhibiting sustained casing pressure are intermediate casing strings;
- 34.6% of the casing strings exhibiting sustained casing pressure are surface casing strings;
- 15.4% of the casing strings exhibiting sustained casing pressure are conductor casing strings.

Shown in Figure 2 is a summary of the occurrence of SCP on the outer casing strings in the Gulf of Mexico Outer Continental Shelf (OCS) region by block. The occurrence of SCP on production casing was assumed to be caused by tubing, packer, or wellhead seal leaks and thus not related to geology. For this reason, the reported occurrence of SCP on production casing is not included in the data of Figure 2. The left side of Figure 2 gives the occurrence of SCP on outer casing strings as a percentage of the wells completed in the various GOM areas. The right side of Figure

Figure 2. Sustained Casing Pressure by area in GOM (Bourgoyne et al., 1999)

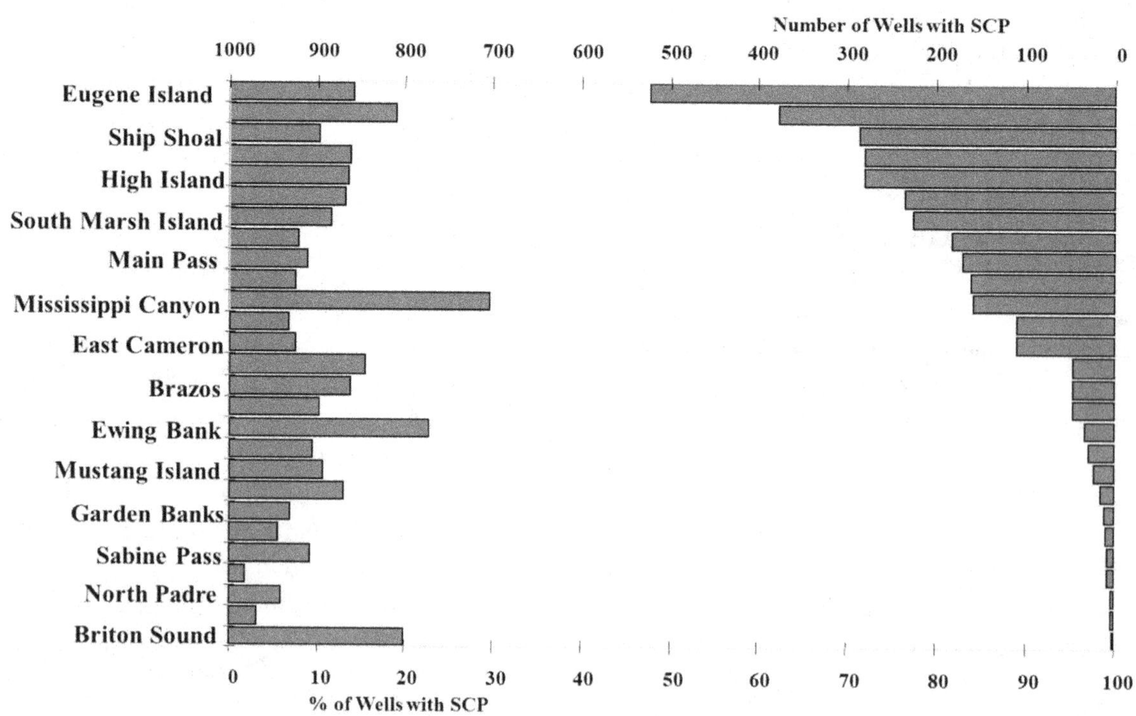

2 gives the total number of wells in which SCP was reported on one or more outer casing strings. Note that the occurrence of SCP is widespread and affects a large number of wells.

Figure 3 shows the frequency density and the cumulative frequency distributions of SCP for all casing strings by pressure for all wells in GOM with SCP. These plots show that about 90% of sustained casing pressures observed are less than 1,000 psi ($70{\cdot}10^5$ Pa) in magnitude.

Discussion of SCP Accidents in GOM

During the period of 1980 to 1990, the oil and gas industry in the GOM Outer Continental Shelf (OCS) experienced several catastrophic accidents as a direct result of high SCP. Four such case histories are described in more detail and discussed, below (Bourgoyne et al., 1999; JIP, 2001).

Figure 3. Sustained Casing Pressure value distribution in GOM

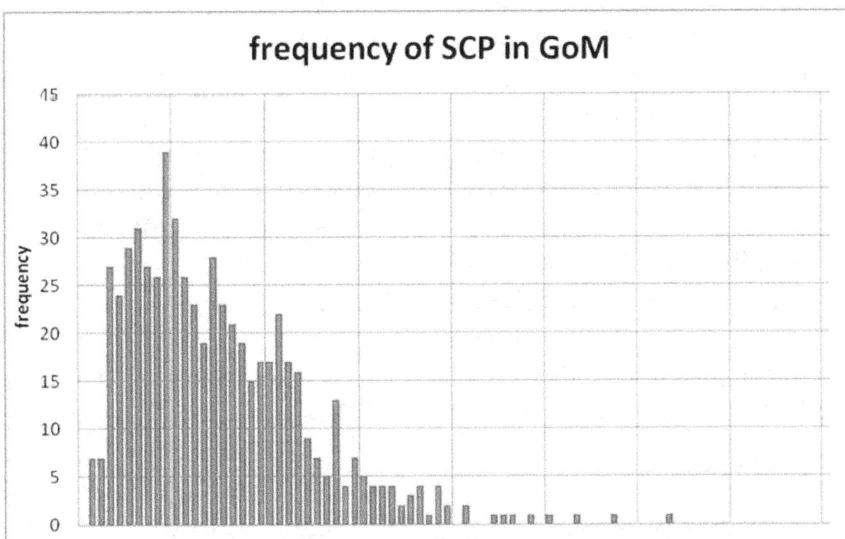

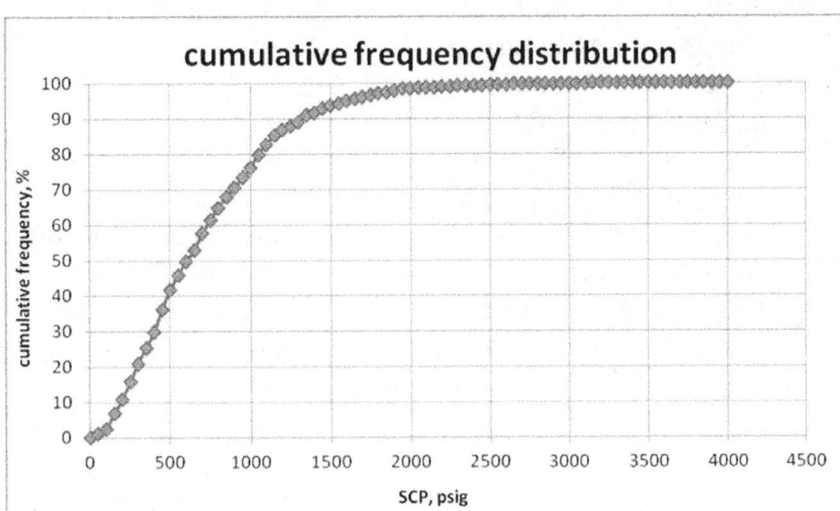

Case 1

Two wells on a platform developed SCP on the production casing about six years after the wells were completed. The operator reported that the shut-in casing pressure was about 3,400 psi ($234 \cdot 10^5$ Pa) and the minimum internal yield on the casing was about 6,900 psi ($475 \cdot 10^5$ Pa). Thus, the operator argued that a safe operation could be maintained. The well continued to be produced. Two years later, the well began blowing out from the annulus between the production casing and surface casing. The well was out of control for 46 days and released an estimated 600 MMSCF ($17 \cdot 10^6$ m^3) of gas and 3,200 Bbl (509 m^3) of condensate during this period. Pollution washed up on about 4 miles of beach. The well cratered and the platform tipped over. The blowout was killed using a relief well and the platform and wells had to be abandoned and removed. The wells were plugged and cut off below the mudline. It is believed that the production casing became pressurized through tubing or packer leaks. Failure of the production casing led to pressure on the outer strings through which the blowout occurred.

Discussion

This case describes a well that cratered due to SCP of 3,400 psi on the production casing. The only detail pertaining to the well parameters is the minimum internal yield point (MIYP) of the production casing (6,900 psi). The case history concludes by stating that "It is believed that the production casing became pressurized through tubing or packer leaks. Failure of the production casing led to pressure on the outer strings through which the blowout occurred." Though the outcome cannot be disputed, it is difficult to make definite conclusions in the absence of a design basis, post-mortem or even a well schematic. For instance, why did the production casing fail at 3,400 psi, if its pipe body yield is 6,900 psi? What was the load state on the production casing (i.e., what was the differential pressure on it?). What kinds of connections were used? Is it possible that the blow-out was a consequence of pressure build up (or failure of cement) in the outer casings, and was independent of the fact that the production casing also saw pressure? The only clue to probable mismanagement of the problem during initial stages comes from the fact that the operator applied for a departure when the SCP was measured at 3,400 psi (which is nearly 49% of the MIYP). Unless the basis for granting departure and the diagnostic results prior to departure grant are known, among other things, it is difficult to re-construct the events leading to the explosion.

Case 2

Five years after a well was put on production, SCP was measured on the production casing and the well was allowed to continue production for a period of one year. At the end of this year, the operator requested permission to continue production, reporting that the SCP on the production casing ranged from 1,400 psi ($96.5 \cdot 10^5$ Pa) to 1,800 psi ($124 \cdot 10^5$ Pa).. Mineral Management Service (MMS) granted the renewal with a diagnostic monitoring program in place to periodically bleed down the pressure to determine the rate of pressure buildup. About six months later, the SCP began fluctuating and bubbles were observed below the platform. The underground blowout was confirmed in one of the wells that had been sidetracked during drilling operations because of a stuck drill string. The foundation below one of the platform legs eroded, and the platform began to shift and settle. This platform leg failed below the mudline. All of the wells on the platform were temporarily plugged, and work proceeded on repairing the platform and killing the underground blowout. A relief well was needed to kill the blowout. This work was eventually successful, with about 250,000 cubic yards (191,150 m^3) of fill sand being needed to fill the crater around the platform leg. The wells were returned to production after about two years of blowout control and remediation work. Holes in the production tubing at and below about 1500 ft (457.2 m) were found during the remediation work. The next two outer casing strings had also failed in the vicinity of the tubing leak.

Discussion

Case 2 describes a well with an underground blowout which eventually caused platform settling at the mudline. Holes in the tubing were discovered at 1500 ft. Once again, as in case 1, the cause and effect sequence is not clear from the data presented.

Case 3

About four years after the well was drilled, the well began to flow mud, gas, and water from the annular space between the surface casing and the conductor casing. Some of the wells had SCP on the production casing. All of the six wells within the leg of the platform containing the flowing well were killed with mud. It was noted that the flow stopped when the SCP on an adjacent well was bled down from about 700 psi ($48 \cdot 10^5$ Pa). The flow path was thought to be from the production casing of one well, into a shallow water sand, and up the surface casing/conductor casing annulus. A number of wells had to be abandoned and replacement wells drilled as a result of this problem.

Discussion

Case 3 is not clearly relevant to the effect of migration from well's deeper section the incident of SCP was limited to the surface casing/conductor annulus, and the cause-effect analysis is not fairly clear from the reported history.

Case 4

Soon after the well was completed and put on production, the operator reported SCP on the intermediate casing of 4,600 psi ($317 \cdot 10^5$ Pa), which was about 46% of the minimum internal yield point and requested permission to continue production. Initially, it was thought that the casing pressure could have been due to thermal expansion. Eighteen months later, work was done on the well to determine why the production rates were lower than expected. Temperature and Thermal Decay Time (TDT) logs were run and a bottom hole pressure (BHP) survey was made. It was determined that an underground blowout was in progress through holes in the tubing, production casing, and intermediate casing. Flow was exiting into salt water sand below the surface casing. About two months were required to kill the underground blowout, and the well was plugged and abandoned. Some damage was done to the platform foundation, and some settlement of the platform occurred.

Discussion

It was believed that the primary (tubing-casing) annulus played important role in this case. The cause of the SCP was assumed to be thermal pressure buildup. Again, it is not clear what the pressure on the primary annulus was when the intermediate casing recorded 4,600 psi. Thermal pressure buildup effects are felt also in the primary annulus. However, the primary annulus pressure was not reported to be considered in analysis of this case history. Based on what eventually happened, this case seems to demonstrate wrong diagnosis, since an underground blowout was the cause of the problem.

The reported case histories demonstrate a very important point that SCP could have (and has had) disastrous consequences if it is poorly diagnosed or managed. They emphasize the issue of SCP as a situation that must be addressed by regulatory authorities and operators alike. The most important conclusions from the LSU report are:

- There is a large incidence of sustained casing pressure in one or more annuli in GOM wells,
- If poorly managed, SCP has the potential for dangerous consequences.

REGULATORY CONTROL AND INDUSTRY STANDARDS FOR SCP

In the US, Mineral Management Service has systematically streamlined and tightened technical requirements for operators of wells with casing pressure. The principal requirement derives from the US Code of Federal Regulations 30 CFR 250.517 (2010) that addresses tubing and wellhead equipment with intent to achieve and maintain pressure control of wells. MMS has interpreted 30 CFR 250.517 to mean that no SCP is to be maintained on any annulus of an OCS well. However, with thousands of wells affected in the GOM, immediate elimination of all SCP was impractical. Thus, MMS has developed a step-wise strategy to remove SCP in wells that represent the greatest hazard, and establish a tiered-monitoring system for the remaining wells, while working towards elimination of the problem. The strategy has been reflected in Letters to Lessees (LTL) and Notices to Lessees (NTL) on August 5, 1991 and January 13, 1994, and in the draft NTL published November 9, 2001.

Regulations of MMS in the Gulf of Mexico require that an operator may continue production (i.e. become "self-approved") if:

- Casing pressure remains at less than 20% of internal yield rating of casing; and
- Casing pressure bleeds to zero during diagnostic tests.

If casing pressures are greater than 20% of internal yield, a departure from the MMS regulations may be applied for. The granting of a departure allows the well to continue producing without elimination of SCP.

Normally, departures are granted for producing wells with casing pressures that bleed to zero and demonstrate a relatively slow subsequent 24-hour build-up rate. (However, for wells that have been temporarily abandoned or permanently abandoned MMS requires that all casing pressures remain at zero which means elimination of SCP is mandatory.) To date, the LTL's have permitted departure from mandatory elimination of casing pressure specified by 30 CFR 250.517 for thousands of wells with small or non-persistent casing pressures. In 2000, for example, the MMS processed 632 requests (of 672 total received) of which 455 (72%) received the departure status (238 wells qualified for self-approved departure and 217 wells were granted departure) (Bourgoyne et al., 1999).

However, the proposed Notice to Lessees of November 9, 2001 would further reduced operator eligibility for being granted a departure. It introduced one-year fixed-term departures for some producing wells, eliminates departures for non-producing wells, and requires operators to eliminate SCP on temporarily abandoned wells. Also, the proposed regulation requires operators to document their plans for SCP removal thus making operators actively responsible and prepared for future removal of SCP in all wells. In conclusion, there is an undeniable trend in the MMS regulatory strategy to require remedial treatments of SCP rather than tolerate the SCP problem.

Operators are stakeholders in the SCP problem since they are being affected by the potential impacts of SCP on personnel and facility safety, and loss of reserves, as well as the regulatory strategy of MMS. The petroleum industry is trying to preserve productivity of wells with SCP and comply with regulatory requirements while controlling operational costs. The industry has accumulated considerable experience in monitoring and diagnosing SCP. It has improved cementing technology to prevent SCP. Also, it has developed several new techniques for control and removal of SCP. There is a need, therefore, to combine all this experience in a systematic set of guidelines.

In response to the proposed NTL of November 9, 2001, the petroleum industry, through American Petroleum Institute (API), and Offshore Operators Committee (OOC) submitted comments and proposed an industry-developed Recommended Practice (RP) on SCP. This new API RP 90, published in August 2006, addresses the monitoring, diagnostics, and remedial actions that should be taken when SCP occurs (API RP-90, 2006). The RP objective is to summarize and standardize the consensus industry knowledge for dealing with the SCP problem in a set of performance-based procedures. Moreover, as the development was supported by MMS, it is expected that the RP will be ultimately incorporated in the MMS regulations.

Evaluation of Wells with SCP

Regulators require that all casing head pressures, excluding drive or structural casing, must be reported to the agency that, presently for the GOM – is Bureau of Safety and Environmental Enforcement (BSEE). Diagnostic tests are required on a well when sustained pressure is first discovered on one or more casing strings among other data, the diagnostic test records must contain:

- Identification of the casing annulus;
- SCP value and initial pressures on tubing and all other casing strings at beginning of test;
- Pressure bleed-down and pressure buildup charts showing time required to bleed pressure down to zero and the return pressure recorded for 24-hour period.

The pressure bleed-down and build-up charts or tables should identify and record pressures on strings not being bled as well as the casing string being bled to detect possible communication between annuli.

Monitoring of casing head pressure (P_{csg}) is different in fixed-platform versus subsea wells. For fixed platform wells, each non-structural casing string is equipped with gauge and the pressure in each annulus can be monitored monthly from taps or flanges installed directly on the wellhead. For subsea wells, pressure in the innermost tubing-casing annulus can be monitored. However, other annuli are hydraulically isolated after the casing strings have been landed in the wellhead. Thus it is a technical challenge to monitor the pressures in subsea well-heads.

In the USA the BSEE regulations require that casing pressure in the fixed platform wells must be monitored on a regular basis. A bleed-down/ build-up (B-B) test must be performed if (P_{csg}) is greater than 100 psig ($6.9 \cdot 10^5$ Pa) (MMS 30 CFR Part 250). In Canada, Energy and Utilities Board (EUB) regulates SCP using the flowing bleed-down pressure and the increase of (P_{csg}) during the shut-in period (BCO&GC, 2008). If flowing pressure is greater than $14 \cdot 10^5$ Pa, or increases more than 42 psig ($2.9 \cdot 10^5$ Pa) during test shut in period, the SCP is considered to constitute high risk. In Norway, NORSOK Standard D-010 Well Integrity in Drilling and Well Operations (NORSOK, 2004) regulates SCP using an arbitrary sub-surface failure criterion. If (P_{csg}) is greater than $70 \cdot 10^5$ Pa for any intermediate casing, SCP is considered high risk.

The B-B test is performed by bleeding off the wellhead pressure through a one-half inch needle valve, followed by a 24 hour shut-in period. Based on the outcome, the environmental risk is categorized as none, small or high. If the pressure cannot be bled off within 24 hours, the risk is considered high. Else if it is bled to zero but builds back up when shut in, the risk is considered small. If no build up is observed, the (P_{csg}) is not considered due SCP constituting no risk. The three cases are shown on a qualitative B-B test chart in Figure 4.

The API Recommended Practice 90 identifies risk of SCP based on the magnitude of (P_{csg}) and its comparison with the maximum allowable well-head operating pressure (MAWOP) (API RP 90). If any (P_{csg}) is greater than 100 psig ($6.9 \cdot 10^5$ Pa) or exceeds the casing's minimum internal yield pressure (MIYP), a B-B test must be performed. A flowchart demonstrating the risk-rating logic is shown in Figure 5.

The logic of B-B test analysis can be summarized as follows:

1. If the pressure bleeds to zero psig and doesn't build back up within 24 consecu-

Figure 4. Possible outcomes from Bleed-down/Buildup (B-B) test

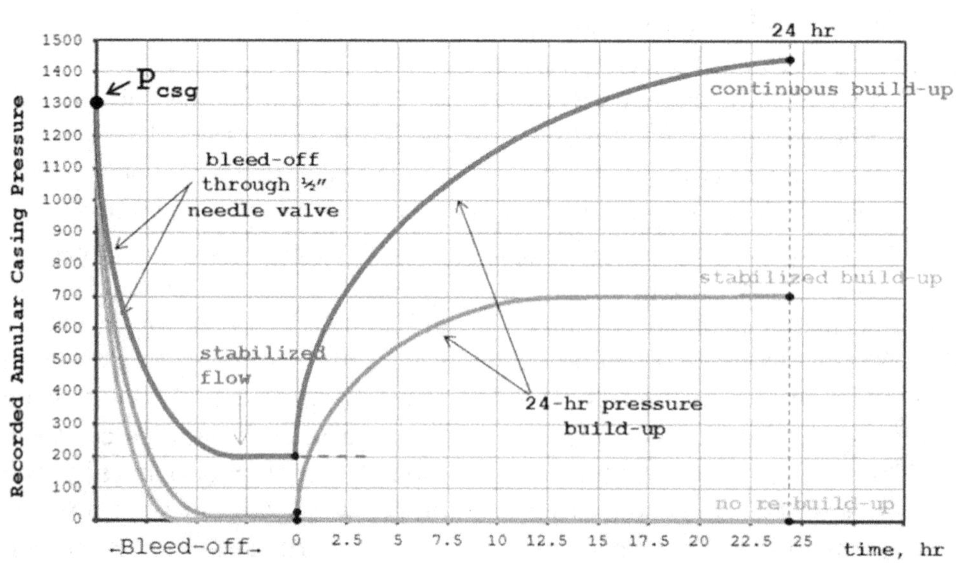

Figure 5. Current evaluation of SCP (Kinik, 2012)

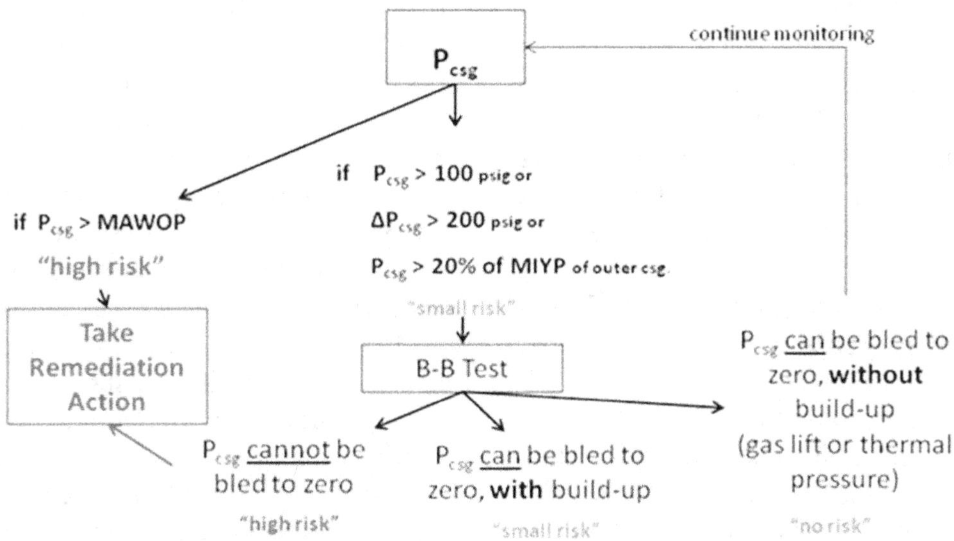

tive hours, then the annulus in question does not have SCP. The pressure source was either thermal in origin or resulted from a leak of very low rate.

2. If the pressure bleeds to zero through a 1/2-in. (12.7 mm) needle valve at a low differential pressure and builds back up to original pressure within 24 consecutive hours, then the annulus in question has a small leak. The leak rate is considered acceptable and the barriers for pressure containment are considered adequate. This well will need to be monitored for changing conditions. An increase in the annulus pressure is not necessarily an indication of an increase in leak rate. This annulus will need to be re-evaluated periodically.
3. If the pressure bleeds to zero psig and builds back up to a lower pressure within 24 consecutive hours, then the annulus in question has a small leak. This assumes that the well is able to bleed to zero psig through a 1/2-in. needle valve at a low differential pressure. The reasons for the pressure not building to its original value within 24 consecutive hours may include the following:
   a. The leak rate is very small.
   b. There is a large gas cap at the top of the annulus.
   c. A portion of the original pressure was caused by thermal effects.
   d. The initial pressure build-up after the bleed-down has a full column of fluid, and higher pressure will develop later as small gas bubbles slowly migrate to the top of the annulus.
4. If the pressure does not bleed to zero psig within 24 cumulative hours through a 1/2-in. needle valve, then the barrier to pressure containment may have partially failed and, in some cases, the leak rate may be unacceptable. This condition may indicate that the leak rate is greater than what may pass through the orifice of the 1/2-in. needle valve at a low differential pressure.

a. If this condition is on the "A" annulus, further investigation is needed to determine the leak path and leak source. Repair plans may also need to be developed.
b. If this condition exists on the outer annuli, it is recognized that options for correction are very limited.

Wells with annular pressure that do not bleed to zero psig should be evaluated further on a case-by-case basis.

It is important to note that SCP may develop in time and, if not remediated, may not remain constant. It may increase over time as a result of factors such as deterioration of the cement sheath, damage to primary cement caused by mechanical shock impacts during tripping, thermal cracking, or dissolution of cement by acidic formation brines. Several studies have reported cases of the problem-free wells developing sustained casing pressure over time (Xu, 2002).

RISK OF AIR POLLUTION DUE SCP – MAXIMUM EMISSION RATE

Sustained casing pressure brings about a risk of well integrity loss resulting in emissions to the environment (Kinik & Wojtanowicz, 2011, and Kinik, 2012). Currently, the risk is qualified by the value of annular surface pressure that may cause failure of casing head. The qualification is made by comparing casing pressure with maximum available wellhead operating pressure (MAWOP) to size the problem and decide on the course of further action (MMS 30 CFR Part 250, 2010 and API RP 90, 2006). The value of MAWOP is set arbitrarily by considering the collapse failure of the inner tubular and burst failure of the outer tubular. The MAWOP value equals either 50% of MIYP of the pipe body for the casing being evaluated, or 80% of MIYP of the pipe body of the next outer casing, or 75% of collapse rating of the inner tubular pipe body, whichever is smaller. For the outermost casing, MAWOP is the lesser value of 30% of MIYP of the pipe body for the casing or production riser being evaluated or 75% of inner tubular pipe body collapse rating. The analysis based on MAWOP addresses only risk of casing head failure without considering its consequences – emissions to the atmosphere.

When the well's external integrity fails, emissions of hazardous substances and criteria pollutants into the ambient air are difficult to quantify without special monitoring equipment. A source test is the most accurate way to measure the type and quantity of the emission, which requires proper sampling, equipment and training of the personnel (Wilson & Frederick, 1999). Methods have been published to calculate or estimate the emission rates for specific equipment and processes. However, there is no quantitative methodology for possible air emissions from wells with SCP. Such a procedure has been recently developed using the LSU model and method. Calculation of emission rates required a correct modeling of uncontrolled gas flow in the well annulus open to the atmosphere (Kinik & Wojtanowicz, 2011).

In a well with sustained casing pressure (SCP), the wellhead pressure (P_{csg}) results from the system equilibrium. As the gas from the source formation is charged into the cement, accumulation of the gas above the mud column brings the system to the equilibrium (Bourgoyne et al., 1999). A schematic of vertical well with (P_{csg}) is shown in Figure 6.

When the well's annulus is In equilibrium, formation pressure (P_f) is balanced with (P_{csg}), hydrostatic pressure created by the mud column (P_{hyd}) and hydrostatic pressure of the fluids inside the cement leak. If the cement leak is filled with gas, the pressure balance or pressure at the top of cement (P_{TOC}) can be simplified as (Xu, 2002),

$$P_f \cong P_{TOC} = P_{csg} + P_{hyd} \quad (1)$$

In case of the casing well-head failure, P_{csg} =0, i.e. an instant pressure imbalance is formed, which is the driving force for the gas flow. The resultant flow rate in such case depends on the total pressure drop downstream from the gas source as,

$$q_{gas} = f\left(P_f - \Delta P_{total}\right) \quad (2)$$

where,

$$\Delta P_{total} = \Delta P_{cement} + \Delta P_{mud} + \Delta P_{gas} \quad (3)$$

and,

ΔP_{cement} - frictional pressure loss through the cement sheath,

ΔP_{hyd} - mud column hydrostatic pressure, and

ΔP_{gas} - frictional pressure loss through the gas column.

Figure 6. Vertical well with equilibrium sustained casing pressure

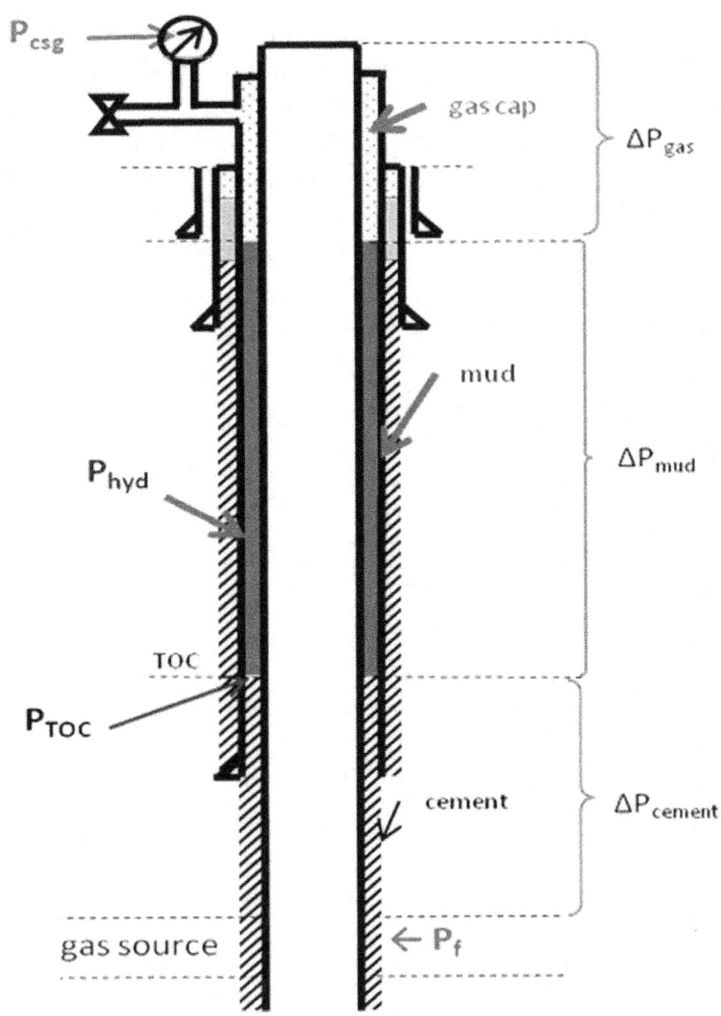

In Equation 3 each term is a complex function of the model parameters controlling the flow mechanism. Therefore, calculation of the maximum gas rate requires mathematical definition of each component and coupling the components at the cement top using Nodal analysis.

The well flow system comprises four nodes shown in Figure 6: gas formation, cement, mud, and well-head. Graphical representation of the flow system performance is presented in Figure 7.

Performance of the overall flow system can be expressed as two nodes coupled at the cement top. The bottom node is that the formation responses to pressure drop by delivering flow, and the upper node is that the pressure drop from the top of cement to the atmosphere. The two nodes represent flow in the cement sheath and gas migration in a stagnant mud column, respectively. The approach is similar to the widely accepted IPR-TPR well performance analysis in gas well production design (Lee & Wattenbarger, 1996). However, the complexity added by the flow in mud column requires a different mathematical modeling approach and input data obtained from the B-B test analysis (Xu and Wojtanowicz,2001; Wojtanowicz et al, 2000).

The Cement Top Inflow Performance (CTIP) represents gas flow in the cement sheath and gas source formation. It depends solely on cement leak size and the reservoir pressure of gas bearing formation. Gas well testing theory provides mathematical description of flow from the formation to the top of cement. The flow is a combination of radial and linear flow in series. In the LSU model, the reservoir pressure (P_f) is assumed constant.

The Cement Top Outflow Performance (CTOP) represents gas migration upwards from the cement top through the mud column and the liquid-free annulus above free level of liquid. (When liquid unloading occurs, at high gas rates a narrow annulus and a liquid-gas mixture with higher average density could result in a significant pressure gradient that would add to the flowing pressures at the top of cement (TOC). At low gas rates, however, contribution of the pressure drop due to frictions above mud level become insignificant.

Figure 7. System performance of SCP well

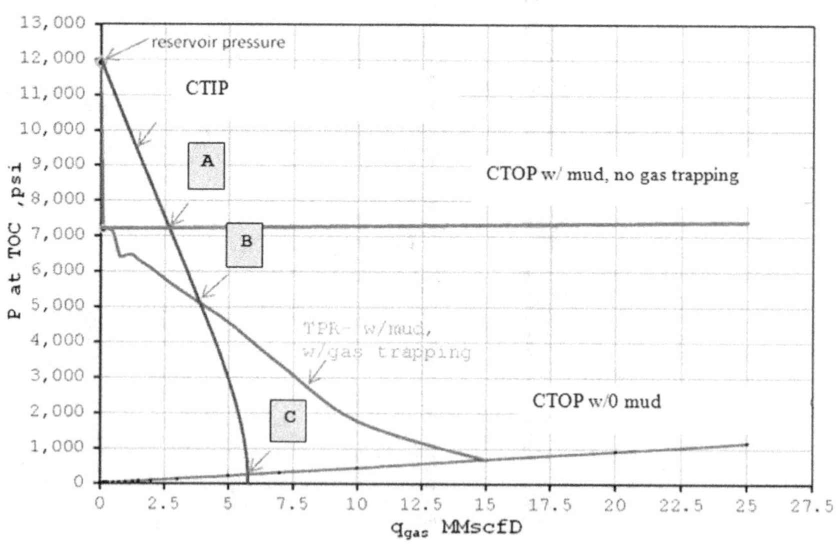

Additional restriction to flow is the failed well-head. The restriction behavior of this component depends on the case by case well-head failure incident, thus in this study the well-head is assumed to form no restriction to flow.

Mathematically, the maximum steady-state gas flow rate (q_g) is the common solution at the cement top. Graphically, the solution is the intercept of the CTIP and CTOP curves. Top cement inflow performance and cement top outflow performance curves are shown in Figure 7.

A Maximum Emission Rate (MER) model and software, developed at Louisiana State University (LSU), provides numerical and graphical solutions for the two concurrent mechanisms with a coupling criterion. The coupling criterion is the common value of pressure at the top of cement (PTOC). The graphical output gives superimposed plots of the CTIP and CTOP relationships. The numerical output gives a single convergence value of the equilibrium gas rate resulting from an iterative solution of the equation,

$$P_{CTIP}\left(q_g\right) - P_{CTOP}\left(q_g\right) = 0 \qquad (4)$$

where, (P_{CTIP}) and (P_{CTOP}) are the system inflow and outflow performance relationships, respectively.

There are several possible scenarios of gas flow and emissions for this annular flow system depending on the well configuration, condition of the mud and the cement sheath. For example, an expansion of the mud volume due gas cutting may trigger liquid unloading from the annulus causing reduction in PTOC. Thus, for different scenarios the gas flow may stabilize at various rates as shown by the points A, B and C in Figure 7.

More complexity is added by the liquid unloading phenomenon. Pressure differential between the reservoir pressure and PTOC determines the flow rate in the cement sheath. Hydrostatic pressure created by the stagnant mud column controls PTOC. As more gas is released from the TOC into the mud column, the mud volume expands and if its length exceeds the distance to the surface, some liquid get permanently removed at the top casing head and the annulus gets "unloaded" that, in turn, further reduces pressure at the cement top, PTOC, and the gas flow rate. Thus, liquid unloading may set out an irreversible domino effect resulting in maximum gas rate corresponding to the absolute open flow potential of the system, AOF.

Other possible scenarios are as follows. If the mud was not trapping any gas, and allowing the gas bubbles to migrate to the top with zero gas cutting, the mud column would not expand. With no expansion and unloading, (q_{gas}) would be defined at point A in Figure 7. Hence, mud rheology becomes the critical parameter since gas trapping is primarily controlled by the residence time of the gas in the mud. Gas residence time is a direct function of gas rise velocity in mud.

If gas trapping occurs, certain volume of annular fluid (mud) would be " unloaded" from the annulus before the system comes to a steady state gas flow, which yields the equilibrium rate shown at point B. Depending upon the leak size and formation pressure, complete unloading may happen - the worst case scenario. In such a case, the gas emission rate, (q_{gas}), indicated by point C, is at maximum.

SELF-PLUGGING POTENTIAL OF SCP WELLS

From the regulatory standpoint, occurrence of SCP disqualifies wells for safe operation. In principle, wells with large SCP cannot be produced and the wells with any SCP value cannot be terminated (Plugged and Abandoned, P&A) without remedial operations. At present, the SCP removal is either capital intensive, as it requires well's workover with a rig, or less expensive but little reliable with rig-less techniques such as CARS or Bleed-and-Lube. However, gas migration in some wells may cease with little intervention when the migrating gas is wet and flow channels in the cement small. Particularly, for the gas condensate wells gas migration can be stopped due the self-plugging effect.

The effect of plugging off the gas migration derives from the analogy between the permeability damage around gas wells and the leak conductivity "damage" in the cemented annular (Duan & Wojtanowicz., 2005). The self-plugging effect should work in wells where SCP sources are the wet-gas or condensate gas formations. When the wellhead pressure drops after opening the casing head needle valve, the pressure in the cement channel may decrease below dew point. Hence, condensate liquid would fill-up the channel and possibly stop the gas flow, as shown in Figure 8. This outcome is called "self-plugging" by condensate liquid.

Conditions for Self-Plugging

After pressure in the cement channel drops below the dew point, condensation begins. The condensate liquid may initially form a film around the inside passage. The retrograde liquid saturation in tiny channels will increase until the throats of the passage are blocked and the gas flow ceases. The process may take some time and involve cycles of plugging and unplugging some throats until the column of condensate liquid develops sufficient hydrostatic and capillary pressure to stop the flow.

Obviously, the self-plugging effect is only applicable to SCP wells with a wet/condensate gas source as the hydrostatic pressure balance of liquid phase in the cement channel is essential for complete plugging. Natural hydrocarbon fluids are classified as oil, wet gas and dry gas by drawing critical temperatures and pressures line on the phase envelope, shown in Figure 9. Condensate gas exists for temperatures above critical point temperature and below the maximum dew temperature point. Outside the envelope, there is no liquid phase. Inside the envelope, the isolines represent different concentrations of gas in the mixture of gas and condensate. Thus, gas changes into liquid (condensate oil) as pressure reduces. This is a reverse phenomenon comparing to normally increasing of the gas phase as pressure decreases. It is called "retrograde gas condensation". A rich condensate gas may retrograde more liquid with heavier components than poor gas.

The amount of condensate liquid depends on gas composition, shape of the envelope (gas behavior) and flow conditions. In general, the richer the condensate gas is, the greater the retrograde liquid saturation becomes. Also, the volume of condensate is proportional to cumulative volume of gas flow. The latter is important for this method since the filling -up of cement channel with condensate liquid may require keeping the casing head open to atmosphere for some time.

Figure 8. Wellbore sketch of cement leak plugged with condensate liquid

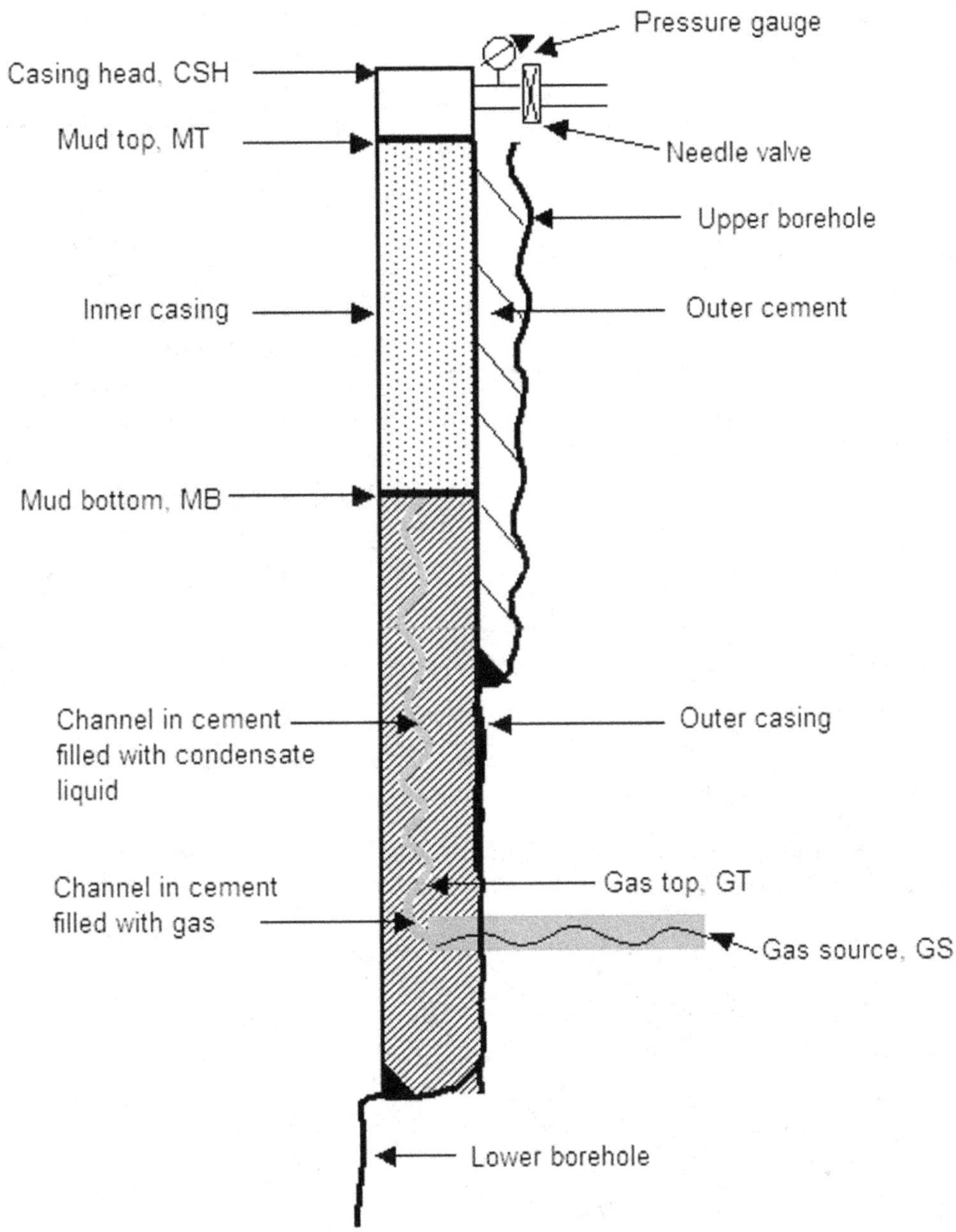

As shown in Figure 8, after the casing head opens to atmosphere, pressure in cement drops, the gas begins to flow, and condensation begins. As the flow continues a condensate liquid film lines up the flow channels and builds up the hydrostatic pressure while the gas may still flow in the channels. In Figure 9, the cement section BM-PT_{100} represents the two-phase column of gas and condensate. Note that point PT_{100} represents mathematically the intercept of the annular hydrostatic pressure and the phase envelope contour plots. In addition, the intercept is located at depth of the gas-condensate liquid contact. Analytically the pressure at the intercept point p_{100} is a solution to equation,

$$p_{100}(T) = p_a(T) \tag{5}$$

where,

$p_{100}(T)$ - gas phase envelope contour plot

$p_a(T)$ - plot of annular hydrostatic pressure in flowing well.

The solution is,

$$p_{PT100} = p_{CSH'} + \left(\frac{dp}{dz}\right)_m * z_{BM'} + \left(\frac{dp}{dz}\right)_l * \left(z_{PT100} - z_{BM}\right) \tag{6}$$

where,

$z_{PT100} = z(T_{PT100})$, from the $T=T(z)$ correlation – depth of intercept point at phase envelope

When the well stops flowing, the point of intercept corresponding to the bottom of the condensation zone moves to its final (equilibrium) position – PT_{100}. Note that Equation (6) gives the value of hydrostatic pressure at point PT_{100}' when the well is still flowing gas and the condensate liquid has already filled up some channels in the well cement. The flow will stop when capillary pressure component adds to the hydrostatic pressure and brings the system to equilibrium. For the equilibrium conditions, pressure at the casing head ($p_{CSH''}$), becomes zero. The intercept point pressure for the plugged well is

$$p_{PT100} = \left(\frac{dp}{dz}\right)_m * z_{BM} + \left(\frac{dp}{dz}\right)_l * \left(z_{PT100''} - z_{BM}\right) \tag{7}$$

Figure 9. Well annular pressure distribution for casing head closed (CSH) and opened (CSH') superimposed over gas phase envelope (Duan & Wojtanowicz, 2005)

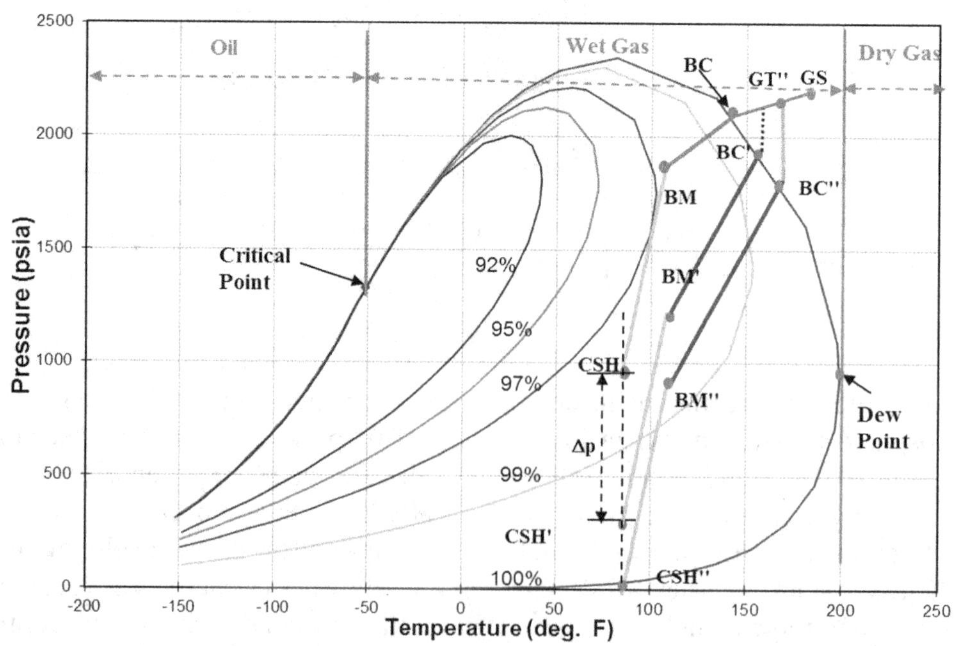

It is obvious that computation of Equations (6) and (7) can only be performed by trial-and-error using a gas phase envelope software.

Capillary Pressure

Capillary pressure develops when condensate liquid accumulates in cracks or channels of the annular cement. Geometry of these conduits is complex and has been simplified in Figure 10. Conductivity of the cement leak may be represented by a cluster of tiny sinuous tubes (matrix model) or a microannulus.

For a cluster of tubes having diameter d_c, capillary pressure is,

$$p_c = \frac{4\sigma \cos\left(\theta\right)}{d_c} \tag{8}$$

For a microannulus, having a slot size of y, shown in Figure 10, capillary pressure is,

$$p_c = 2\sigma \cos\left(\theta\right)\left(\frac{1}{y}\right) \tag{9}$$

After condensation, some liquid drops may lump at narrow spots in the cement channels and block the gas flow. Hence, capillary pressure plays an important role in stopping the gas migration. As the cement voids surface is hydrophilic, droplets of condensate entering internal cement pores and throats are resisted. Thus, capillary pressure would oppose the gas flow.

Pressure Balance Calculation

After gas condensation in the well cement begins and the gas flow continues, hydrostatic pressure is gradually developed in the growing condensate liquid column. The system comes to equilibrium when the gas source formation pressure is balanced by the hydrostatic and capillary pressures as

$$p_{GS} = \left(\frac{dp}{dz}\right)_m * z_{BM} + \left(\frac{dp}{dz}\right)_l * \left(z_{PT100"} - z_{BM}\right) + \left(\frac{dp}{dz}\right)_g * \left(z_{GS} - z_{GT}\right) + p_c \tag{10}$$

where, $(dp/dz)_m$, $(dp/dz)_l$, and $(dp/dz)_g$ are pressure gradients of mud, condensate liquid and gas, respectively.

Figure 10. Cement leak models: A – microannulus, B – matrix

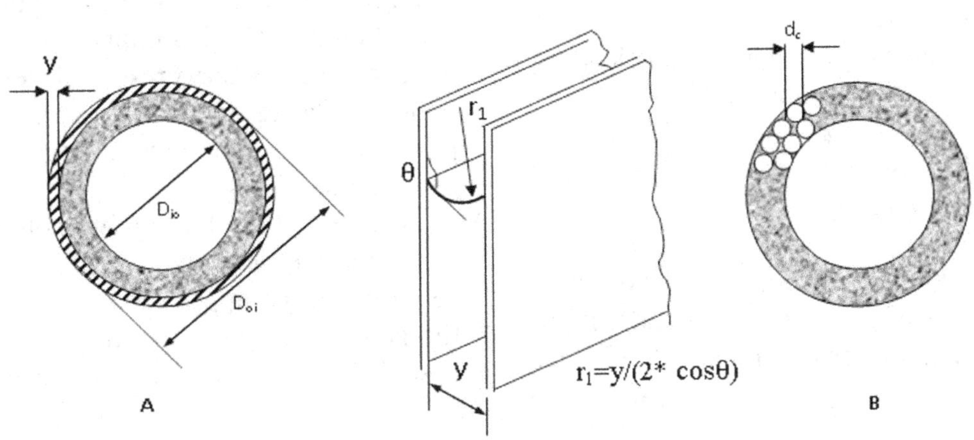

Most of the data in Equation (10) is readily available from the well/cementing records and a laboratory analysis of the gas and condensate liquid. However, two pieces of the data capillary pressure, p_c, and the depth of gas condensation (z_{BC}), need to be discussed. Analysis of Figure 10 indicates that Equation (10) specifies minimum value of capillary pressure (p_c), needed to reach pressure equilibrium and stop the gas flow. As evidenced by Equations (6) and (8), an estimation of the cement leak size is needed to determine the capillary pressure. Such an estimate could be made from the analysis of SCP bleed-down and build-up (B_B) test. An improvement to the current testing procedures is needed to provide the data (Xu & Wojtanowicz, 2001; XU, 2002).

Assessment of Well's Self-Plugging Potential with Condensate

Assessment of the self-plugging process involves several steps and is based on the following assumptions:

1. Gas phase envelope is constant and not affected by the process.
2. The liquid condensate in cement channel becomes a continuous phase that generates hydraulic pressure.
3. Frictional measure loss of flowing gas is neglected.
4. Flow rate of gas during liquid loading of the cement leak is constant.
5. There is no unloading of liquid phase by flowing gas in the channels due to very low gas velocity.

The last assumption was verified by comparing estimated gas velocities in the leaking cement with those corresponding to the Turner *et al.*'s critical rate. The velocity of migrating gas proved to be much smaller than critical.

The required Input data is listed in Table 1A in the Appendix. The table also shows possible sources of input data: SCP testing, well records and gas measurements. At first, capillary pressure is computed from the estimated channel size based on the leak conductivity (K_a), from the B_B test. Since the size of sinuous tubes in the leak matrix model is unknown, their number (N), is computed from the matrix model (see Appendix) and their size is given as,

$$d_c = 774.6 * \frac{A^2}{(\pi / 2 + \sqrt{3}) * q_g} * \sqrt{\frac{C * (p_{GS} - p_{BM})}{\mu * (z_{GS} - z_{BM}) * \overline{T}_c}} \quad (11)$$

For micro-annulus model, the gap size is,

$$y = \frac{\pi * D_{oi} - \sqrt{(\pi * D_{oi})^2 - 128\pi * K_a / A}}{2 * \pi} \quad (12)$$

By substituting (11) to (8) and (12) to (9) we derive formulas for capillary pressure in the cement leak represented either by micro-tubes (matrix) or micro-annulus, respectively.

If the value of capillary pressure satisfies Equation (10), the SCP well has a potential for self-plugging. The potential for self-plugging requires that capillary pressure develops at the bottom of the liquid-loaded section of cement-point "BC" and its value is large enough to reach point "GT".

A SCP well with plugging potential may still continue flowing if capillary pressure develops in the upper section of cement, or the actual leak geometry deviates from the model. Thus, it is important to determinc the maximum time of the well flow testing. If the flow does not cease by this time, the SCP could not be removed with this method. It happens that the best estimate of this time is the cement leak fill-up time calculated together with the flow rate as follows.

Volume of condensate liquid needed to fill up the cement flow passage is computed using estimated capacity of the passages. For the micro-tube model of leaking cement the capacity is,

$$v_l = \pi * N * d_c^2 * (z_{BC} - z_{BM}) \tag{13}$$

Volume capacity of the cement micro-annulus is

$$v_l = \pi * y * (D_{oi} - 0.5 * y) * (z_{BC} - z_{BM}) \tag{14}$$

As the gas-to-liquid ratio of wet gas is very large, a considerable amount of gas should pass through the cement leak to generate the fill-up condensate liquid. The values can be determined from the phase envelope section BM-PT_{100}. The gas volume is

$$v_g = v_l * \frac{\overline{T_s}}{T_c} * \frac{\overline{p_c}}{p_s} * f \tag{15}$$

The rate of gas flow during liquid loading of the cement leak is,

$$q_g = 0.11924 * \frac{K_a}{T_w * \mu_g} * \frac{(p_{GS}^2 - p_{BM}^2)}{z_{GS} - z_{BM}} \tag{16}$$

The fill-up time estimate is,

$$t_l = v_g / q_g \tag{17}$$

Example of Self–Plugging Assessment

According to the U.S. Energy Information Administration (EIA), about 20% wells in the Gulf of Mexico (GOM) were producing condensate gas in 2003 (EIA). Thus, we may roughly assume that 20% of wells with 'SCP" in GOM will be affected by wet gas, which makes this method potentially applicable to the area.

To demonstrate the method, properties of a wet gas from another area are used here as an example (Duan & Wojtanowicz, 2005). The formation pressure is 3,650 psi, and the formation temperature gradient is 2.5 F/100 ft. Average reservoir temperature is 255 F. The depth of condensate liquid was calculated based on Equation (7).

In this well, the fill-up volume is 37.2 cu ft and the gas migration channels are small enough to generate high capillary pressure value, 757 psi, to balance the formation pressure (see Appendix).

The gas flow plugging effect is evaluated using the phase envelope graph in Figure 11. Similar to Figure 9, the plot in Figure11 indicates capillary pressure value large enough to balance gas formation pressure (section PT_{100}-GT). The plot also shows the continuity of the pressure line (lower plot) from the gas formation (GS) to the casing head (CSH") open to atmospheric pressure.

The results in Table 1A reveal that bleeding gas from the casing head would give the emission rate (if the gas is not diverted) of about 272 scf/d and the plugging off the cement leak should be completed in 23 days. It is also expected that the well should stop flowing at this time. If not, the casing head should be closed and other methods considered for SCP removal.

Figure 11. Example of SCP well self-plugging with condensate; pBC=pGT (Duan & Wojtanowicz, 2005)

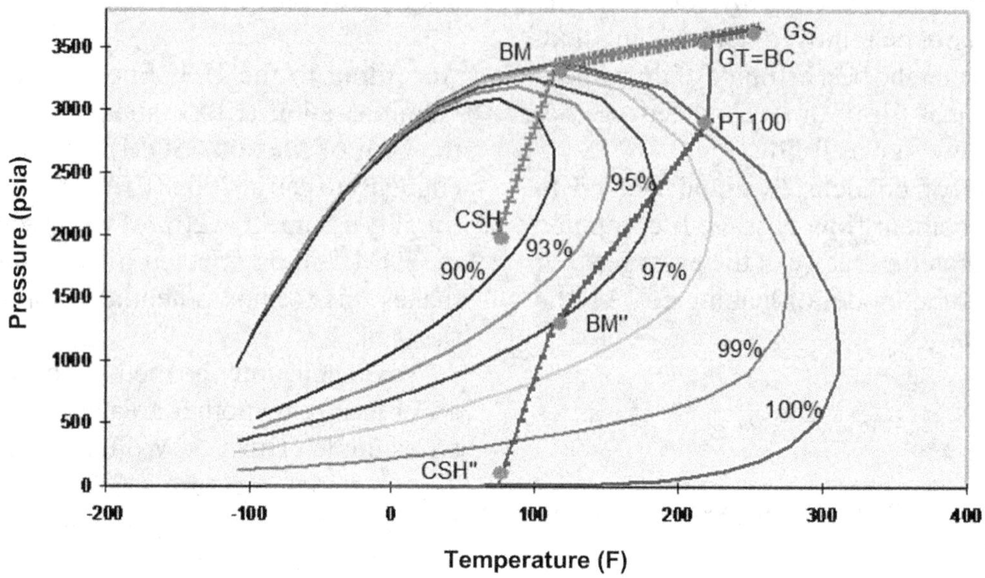

Also, if the flow cannot be diverted but just released and flared, the total volume of gas 6.2 Mscf, is too small to cause serious air pollution problem.

Discussion of Self-Plugging Potential

The theoretical study shows physical merit of SCP well self plugging and demonstrates feasibility of new method for self-plugging potential evaluation. Wells with SCP problem resulting from wet gas migration may be sealed-off with gas condensate liquid if their casing head remains open for a sufficient time period. The proposed technique is very cheap and requires no mechanical intervention in the well.

A simple method, developed in this work, evaluates well's potential for plug-over based on static balance of pressure in the well annulus opened to atmospheric pressure of the casing head. The method determines length of the top section of well cement that would be "loaded" with condensates liquid from the flowing gas. It also estimates the value of capillary pressure opposing the gas flow.

Capillary pressure is essential for condensate plug-over but it can only be estimated from the value of hydraulic conductivity of leaking cement. Hydraulic conductivity (and other data needed for this method) could be computed from SCP bleed-off/build up testing. Present testing procedures, however, do not quantify the data and need to be improved.

Since the self-plugging potential is merely an estimate, the promising well needs to be tested by continuous bleeding enough gas that cement is loaded up with condensate liquid, which may (or not) result in cessation of flow and removal of SCP. The bleeding time can be estimated using the proposed method. Moreover, the study also demonstrates that opening casing head to gas flow may result in a very small flow rate and the total volume of gas released during the flow test may be also small resulting in minimal air pollution.

Nomenclature

A = cross-sectional area of annulus, in^2
K_a = annular conductivity, md-in^2
BC = bottom of condensate
BM = mud column bottom – initial
BM' = mud column bottom – final
C = coefficient for low gas flow in pipe, ft/s
CSH = casing head – initial
CSH' = casing head – final
d_c = pore diameter, m
D_{oi} = inside diameter of outer casing, in
D_{io} = outside diameter of inner casing, in
f = average gas to condensate ratio, v/v
GS = gas source
GT = gas top
μ_g = gas viscosity, cp
N = number of micro-tubes in matrix model
p = pressure, psi
p_c = capillary pressure, psi
$\overline{p_c}$ = average pressure in cement leak, psi
p_S = surface pressure, psi
SCP = sustained casing pressure, psi
PT_{100} = intercept point at phase envelope
q_l = condensate flow rate, ft^3/d
Q = flow rate, ft^3/d
r = radius of a tube, m
r_1 = radius of curvature of o/g meniscus, m
T = average temperature, F
T_S = surface temperature, K
T_w = average well temperature, F
t_1 = fill-up time, days
$\overline{T_c}$ = average cement temperature, K
y_l = condensate volume, scf
y_g = total gas volume for cement leak fill-up, scf
z = depth, ft
γ_g = gas gravity
σ = surface tension, dynes/cm
θ = contact angle, degree

REFERENCES

API RP 90. (2006). *Annular casing pressure management for offshore wells*. Washington, DC: American Petroleum Institute.

Attard, M. (1991). *The occurrence of annulus pressures in the North West Hutton Field: Problems and solutions*. Paper presented at the Offshore Europe Conference. Aberdeen, UK.

BCO&GC. (2008). *Surface casing vent flow/gas migration testing, reporting and repair requirements*. Vancouver, Canada: British Columbia Oil and Gas Commission.

Bourgoyne, A. T., Jr., Scott, S. L., & Manowski, W. (1999). *A review of sustained casing pressure occurring on the OCS*. Retrieved from ftp://www.mms.gov/TubingStudy/

Bourgoyne, A. T., Jr., Scott, S. L., & Regg, J. B. (1999). *Sustained casing pressure in offshore producing wells*. Paper presented at the Offshore Technology Conference. Houston, TX.

Duan, S., & Wojtanowicz, A. K. (2005). *Surface induced removal of sustained casinghead pressure by sealing the well's leaking cement with condensate from flowing gas*. Paper presented at the Petroleum Society's 6th Canadian International Petroleum Conference (56th Annual Technical Meeting). Calgary, Canada.

Joint Industry Project Report (JIP). (2001). *Best practices for prevention and management of sustained casing pressure*. Houston, TX: Stress Engineering Services, Inc.

Kinik, K. (2012). *Risk of well integrity failure due sustained casing pressure*. (Master Thesis). Louisiana State University, Baton Rouge, LA.

Kinik, K., & Wojtanowicz, A. K. (2011). *Identifying environmental risk of sustained casing pressure*. Paper presented at the HSSE Conference. Houston, TX.

Lee, J. W., & Wattenbarger, R. A. (1996). *Gas reservoir engineering*. Richardson, TX: SPE.

Michel, C. M. (1991). *Methods of detecting and locating tubing and packer leaks in the Western operating area of Prudhoe Bay field*. Paper presented at the 1991 Production Operations Symposium. Oklahoma City, OK.

MMS 30 CFR Part 250. (2010). *Annular casing pressure management for offshore wells, code of federal regulations, title 30, chapter ii, minerals management service*. Washington, DC: United States Dpt. of Interior.

NORSOK Standard D-010. (2004). *Well integrity in drilling and well operations*. Lysaker, Norway: Standards Norway.

Shoham, O. (2006). *Mechanistic modeling of gas-liquid two-phase flow in pipes*. Richardson, TX: Technology & Engineering, Society of Petroleum Engineers.

Wilson, M. J., & Frederick, J. D. (Eds.). (1999). *Environmental engineering for exploration and production activities*. Richardson, TX: Society of Petroleum Engineers.

Wojtanowicz, A. K., Nishikawa, S., & Xu, R. (2000). *Diagnosis and remediation of sustained casing pressure in wells*. Retrieved from http://www.bsee.gov/

Xu, R. (2002). *Analysis of diagnostic testing of sustained casing pressure in wells*. (PhD Dissertation). Louisiana State University, Baton Rouge, LA.

Xu, R., & Wojtanowicz, A. K. (2001). *Diagnosis of sustained casing pressure from bleed-off/buildup testing patterns*. Paper presented at the SPE Production Operations Symposium. Oklahoma City, OK.

APPENDIX

For the matrix model, assuming equal size of the micro-tubes, the number of the tubes is

$$N = \frac{A}{(\frac{\pi}{2} + \sqrt{3}) * \frac{d_c^2}{4}} \quad \text{(A-1)}$$

For laminar gas flow in a vertical micro-tube with known pressure drop (p_{GS} - p_{BM}), we have,

$$(\frac{q_g}{A * N})^2 = \frac{37500 * C * (p_{GS} - p_{BM}) * d_c^2}{\mu * (z_{GS} - z_{BM}) * \overline{T}_c} \quad \text{(A-2)}$$

Substituting Equation (A-1) in (A-2)

$$\frac{q_g^2}{(\frac{A^2}{(\pi / 2 + \sqrt{3})} * \frac{4}{d_c^2})^2} = \frac{37500 * C * (p_{GS} - p_{BM}) * d_c^2}{\mu * (z_{GS} - z_{BM}) * \overline{T}_c} \quad \text{(A-3)}$$

Diameter of a single micro-tube is,

$$d_c = 774.6 * \frac{A^2}{(\pi / 2 + \sqrt{3}) * q_g} * \sqrt{\frac{C * (p_{GS} - p_{BM})}{\mu * (z_{GS} - z_{BM}) * \overline{T}_c}} \quad \text{(A-4)}$$

For gas flow in a micro-annulus, the conductivity is

$$K_a = A * (D_{oi} - y) * \pi * \frac{y}{32} \quad \text{(A-5)}$$

Solving (A-5) for size, y, gives

$$y = \frac{\pi * D_{oi} - \sqrt{(\pi * D_{oi})^2 - 128\pi * K_a / A}}{2 * \pi} \quad \text{(A-6)}$$

where,

$$A = \frac{\pi}{4}(D_{oi}^2 - D_{io}^2)$$

Table 1A. Input and output data for example

Input Data		**Unit**	**Data Source**
Test Data			
Casing head pressure(SCP)	2004	psi	*
Flowing rate	260	ft^3/day	*
Annular conductivity	939.53	md*in^2	*
Gas specific gravity	0.677	(air=1)	*
Gas source depth	7200	FT	*
Gas source pressure	3650	psi	*
Temperature gradient	2.5	F/100FT	*
Reservoir temperature	255	F	*
Well and Formation Data			
Outter casing diameter (ID)	13.625	in	**
Inner casing diameter (OD)	8.125	in	**
Mud depth	1500	FT	**
Mud gradient	0.85	psi/FT	**
Physical Properties			
Gas viscosity	0.01	cp	***
Water gas surface tension	50	dynes/cm	***
Contact angle	170	degree	***
Condensate liquid gravity	0.87	(water=1)	***
Condensate liquid gradient	0.41	psi/FT	***
Calculated Data			**Block Status**
Bottom mud pressure	1275	psi	
Cement area	93.95	in^2	
Phase envelope intercept point pressure	2909	psi	
Condensate liquid column	1634	psi	
Size of microannulus	0.038	10^{-6}m	
Capillary pressure	757	psi	
Permeability	10.00	md	
Fill-up capacity of microannulus	37.19	ft^3	
Cumulative fill-up gas	6172.2	ft^3	
Gas flow rate	271.7	ft^3/d	
Output Data			
Blocking gas flow rate	271.7	ft^3/d	
Blocking flowing time	22.7	day	
Cumulative gas volume for blocking	6172.2	ft^3	

* *SCP Diagnostic Test; ** Well record; ***Laboratory testing of gas*

Chapter 9
Risk Analysis in the Process of Hydraulic Fracturing

Sonja Koščak Kolin
University of Zagreb, Croatia

Marin Čikeš
University of Zagreb, Croatia

ABSTRACT

This chapter focuses on risk to the environment from hydraulic fracturing operations, starting with transport of materials and ending when the well is routed to the production facilities. The initial assumption for the fracturing risk analysis is that the well is new and was constructed correctly so that all producible formations are securely isolated behind the barriers of casing and competent cement. The justification for this assumption is that the vast majority of fracturing is the first major stimulation in a well and occurs immediately after completing a new well. Although many well development problems are blamed on fracturing, there are only excluded problems that are real and worthy of the discussion to help define boundaries of the fracturing risk (King, 2012).

INTRODUCTION

While hydraulic fracturing is a well-developed technology that has been used for more than 60 years in petroleum engineering, its widespread use for coal-bed natural gas and shale gas development has raised questions about the appropriate regulatory approach to ensure that groundwater resources and surface water are protected (Arthur, 2011). Millions of fracs have been pumped (SPE - Society of Petroleum Engineers estimate 2.5 million fracs world-

DOI: 10.4018/978-1-4666-4777-0.ch009

wide). The technical literature on hydraulic fracturing is extensive, addressing nearly every aspect of oil and gas development, with over 550 papers in shale fracturing. The shale papers in the past several years alone are from more over 70 universities, over a dozen state, federal and international agencies and more then a hundred energy and service companies (King, 2010; Arthur, 2009).

However, recent exposure of the general public to this technology, particularly as practiced with large volume of water, has resulted in considerable fear and misunderstanding of what is occurring downhole. Fortunately, the industry has been studied the problem of fracture height growth for several decades and has been monitoring fractures with tiltmeters for two decades and with microseismics for over one decade. All data of investigation has shown that hydraulic fracturing in typical reservoirs is not threat to fracture into and contaminate or otherwise disturb groundwater (Warpinski, 2011). In engineering terms, fracturing concerns a precise stimulation activity, limited to the fluid action in initiating and extending cracks in the rock; while, for many concerned citizens, bloggers and environmentalists, fracturing has come to represent nearly every phase of the well development cycle from drilling to production.

First, scale drawings of the distance from the surface and near-surface fresh water supplies (nearly all fresh water formations are within the first 300 meters from the surface) are needed to show the physical distance between the surface and the completion interval of interest (pay zone).

Fracture height, predicted by computer models and confirmed by microseismics monitoring during fracturing, post-frac tracer flows, temperature logs and even by mine-back experiments (Warpinski, 1985), show most vertical effective fracture growth at 100 meters or less. Fracture height growth in most formations is known to be effectively limited by barriers and fluid leakoff (loss of fluid to the rock). Fracture heights limited by these physical and active barriers will simply not reach into fresh water sands.

Although the fracturing process is more than thousand meters away from the water table, methane is showing up in residential water wells, but this contamination in water wells is caused by other natural and manmade causes. Part of the reason is the natural occurrence with biogenic methane forming from shallow decay of organic materials and natural seeps of thermogenic methane (gas formed deep in the earth) that have been coming to the surface for millions of years, particularly in regions with shale and coal outcrops or shallow formations that share the water table with fresh water wells.

However, part of the increasing methane content in water well may be coming from near-by improperly constructed gas or oil well. These older wells predate the invention of hydraulic fracturing and most predate any significant well construction regulations. Deeper formations are at higher pressures and exposure of these zones without sufficient cementing isolation will allow gas seepage that will build up a higher pressure that was customary at shallower depths. This type of methane leak is usually low volume and is made possible by poor or incomplete cementing practices. The incident can be noticed soon after drilling and can start before the well is fractured.

In a fracturing risk analysis, that is main focus of the chapter, any local conditions will influence the conclusions. For regional or area studies the impact and the occurrence information, from traffic accidents to fractured well incidents should be from the area, but wider studies of other factors may be necessary for purposes of well population and incident review. Estimating risk at both regional and area levels allows identification of risk areas and refinement of that risk through knowledge of local conditions. Different companies operating in different areas of the country may use the same risk estimation approach but reach different values in roads, well construction requirements, infrastructure, engineering experience, local geology, regulations and other factors.

The level of permeability in a rock holding oil and gas dictates whether the reservoir must be hydraulically fractured (Figure 1). At lower permeabilities or where oil viscosity is high or reservoir pressure is low, the flow of fluids toward the wellbore can be assisted by fracturing, which creates a flow path of much higher permeability. Stable fractures offer a flow path with average permeability of 100 to over 1000 times the permeability of the formation (Gaskari, 2006). At lower permeabilities, such as shale, most wells will not flow economic quantities of fluids without extensive hydraulic fracturing.

The oil and gas containing shale featured are classified on the basis of the size of the very fine particles or grains that make up the rock. They are actually very fine grained sandstones, often with similar mechanical properties to the sandstones that comprise conventional gas and oil producing reservoirs. As shown in Figure 1, the difference is that conventional sandstones may have permeabilities in the range of 0.5 to 20×10^{-3} μm^2, while this gas shale may have permeabilities of 0.000001 to over 0.0001×10^{-3} μm^2. Hydraulic fracture treatments in wells from unconventional reservoirs are required to ensure economic production rates for a long period of time (Čikeš, 2000).

HYDRAULIC FRACTURING PROCESS

Fracturing Design and Chemicals

Fracturing design is normally made by computer, specifying fracturing volume, rate and other factors to achieve goals of fracture height, fracture width and fracture length or fracture complexity. Typical fracture geometry (height, h_f, width, w, and length, L) is shown on the Figure 2.

The goal is to design a fracture that will stay in the pay zone, develop the maximum pay or producing formation contact and achieve maximum flow of hydrocarbons and minimum flow of produced water.

Transport and storage of fresh or salty fracturing water, chemicals and equipment fall under the general heading of transport, and along with well construction, have been identified as a potential source of pollution. Chemical spill risk can be reduced by using

Figure 1. Permeability range of producing formations (King, 2012)

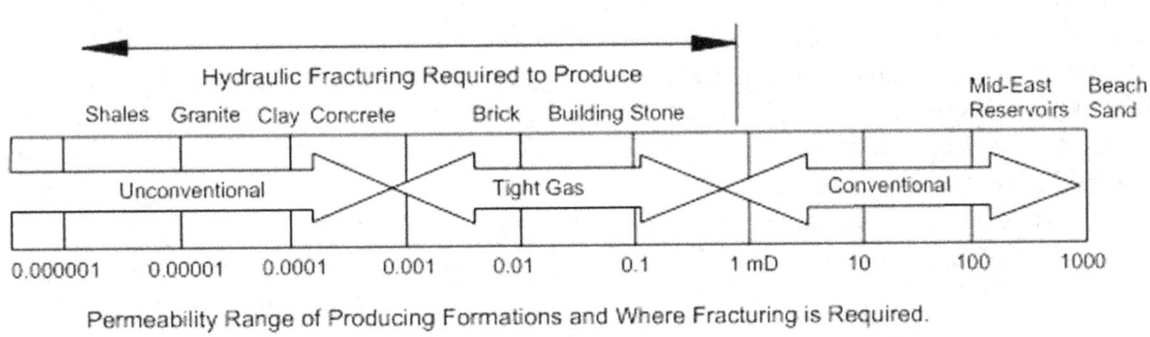

Figure 2. Fracture geometry, PKN (Perkins-Kern-Nordgren) model (Economides, 2000)

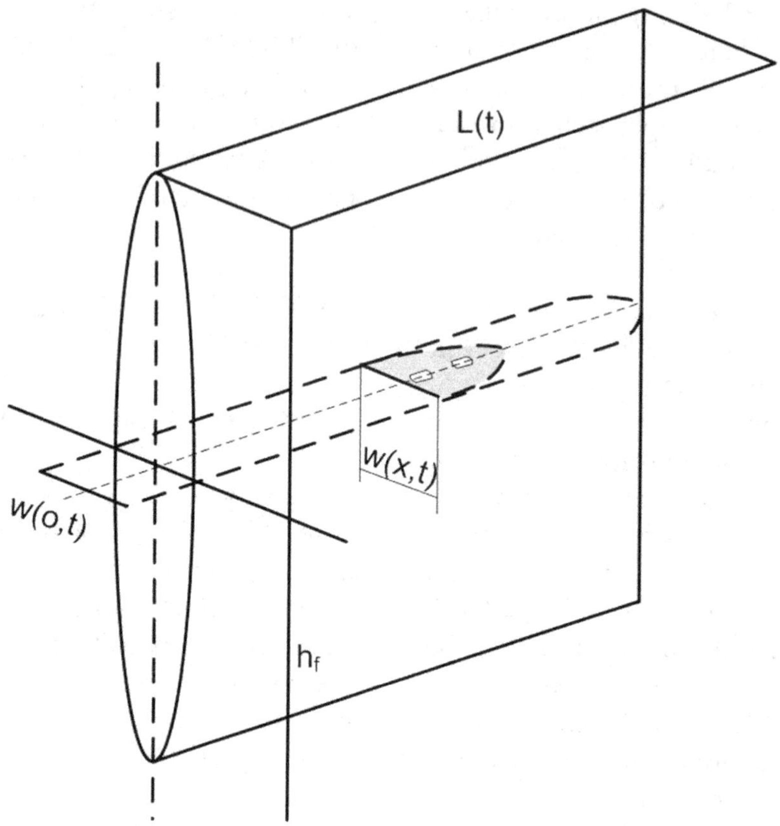

double wall containers, collision-proof totes and/or use of dry additives. Surface storage vessel leaks and spills can range from less than a few liters during connections in fracturing fluid lines to the very rare leak of a truck load (20 m^3) or the highly unlikely leak of a full fracturing tank (80 m^3) of water.

Leak impact can be reduced by container mats underneath pipe connections, portable tank containment berms and tank monitoring to immediately spot leaks. Impact of fracturing base fluid leaks is usually minor if the leaking fluid is fresh water, since most fracturing chemicals are mixed into the fluid only as it is pumped into the well. Safe transport, storage and handling of chemical concentrates are major concerns. These risks are sharply reduced when non-toxic or even foodgrade additives replace traditional chemicals.

Mixing and pumping of the frac increases risk of leaks and spills as stored fracturing fluid is pumped, first to the chemical addition trailer and then the blender where sand is added, before going to the high pressure pumps and down the well.

Chemicals, or more precisely, the lack of disclosure of chemical identities, have probably received and deserved the most vitriolic attacks in the "anti-frac" literature. There are very few chemicals used in fracturing and definitely not the "hundreds of toxic chemicals". Independent laboratory analysis of surface water sources used for fracturing do show a variety of chemicals at trace concentrations below EPA (Environmental Protection Agency) limits, that are not added as part of the fracturing process, but instead come from agricultural sources (herbicides, pesticides, fungicides, etc.), that carry over into ground water runoff. These same chemicals are found in the raw water feed into drinking waters in nearly all areas, whether or not oil and gas operations are present. BTEX (benzene, toluene, xylene and ethylbenzene), diesel, fluorocarbons, etc., have been removed from fracturing in the past several years as a caution or by regulations based on fear of somehow contaminating fresh water supplies with fracturing fluids. Most of the fracturing treatments used in shale are water with a friction reducer and no significant gelling agents – called a slick water frac. The shale responds very well to these inexpensive fracture treatments. Basic slick water formulation (Fontaine, 2008) for shale may include:

- **Water:** About 98% to 99% of total volume - commonly fresh water (<500 mg/kg salinity), but increasingly containing treated produced water.
- **Proppant:** About 1% to 1.9% of total volume – usually sand or ceramic particles carried by the fracturing fluid into the fracture to keep the fracture open when hydraulic pressure is released.
- **Friction Reducer:** About 0.025% of total volume – the non-acid form of polyacrylamide used as adsorbent in baby diapers and as a flocculent in drinking water preparation, and reduces friction pressure of water flowing through the pipe during high rate pumping, thereby reducing required pump power output and air emissions from the pumps.
- **Disinfectant (Biocide):** About 0.005% to 0.05% of total volume – common biocides of glutaraldehyde (a common antimicrobial used in hospitals and even municipal water treating systems) or quaternary amine (drinking water disinfectant and common over-thecounter skin antiseptics). Disinfectants are used to control the growth of certain kinds of microbes that would destroy gelled fracture fluids, or, in unusual cases, may create a sour gas generation (H_2S or hydrogen sulfide) problem in the reservoir.
- **Surfactant:** That modify surface or interfacial tension, break or prevent emulsions, and perform other specific actions are used at 0.5 to 2 liters per one cubic meter in all or part of the fracture fluid volume in some cases.
- **Gelation Chemicals:** (Thickeners) such as guar gum and cellulose polymers are not frequent additions to slick water frac formulations, but may be used in hybrid fracturing (that use a ungelled water to initiate the fracture and a gelled water to carry some of the proppant or sand). These gelling materials are common food additives, do not break down into toxin and are not considered of concern (Hoeman, 2011).
- **Scale Inhibitors:** Rarely to commonly used depending on the specific shale, prevent mineral scale precipitates, eliminating the potential for concentrating problem ions in scale and blockage of tubing and equipment.
- **Hydrochloric Acid:** (Same material used in swimming pools and to clean brick and other masonry projects) May be used in some cases to reduce fracture initiation pressure (the downhole pressure needed to create the first small crack in the rock).
- **Corrosion Inhibitor:** One of several organic compounds that may be toxic, is used at 0.2% to 0.5% in only the acid (total inhibitor volume per fracturing is 30 to 40 liters and only used if acid is used.

Chemicals returning from a well after a fracturing treatment are at a fraction (usually 20% or less for chemicals and about 40% for polymers) of what was pumped down the well (King, 1988). Polymers decompose quickly at temperature, biocides are spent on organic demand and degrade, surfactants are adsorbed on rock surfaces and scale inhibitors precipitate and come back slowly at 10 to 15 mg/kg for several months. If these chemicals are selected to have minimum impact (low/no toxicity, full biodegradation, etc.) then impact from initially added chemicals in a spill of recovered water is minor. Fluids returning from the formation are best used as a resource for pressure maintenance in oilfield enhanced recovery operations. Deep well disposal of produced fluids (including frac flow back), is common, but there are other oil field uses for this resource including treatment for re-use as saline fracturing water. Surface release of any water that is high salinity and has chemicals above the specific safe limit is not a viable alternative to oilfield reuse or disposal.

Utilizing alternate sources of water for drilling and fracturing to displace fresh water use is a major work area in the industry. The water argument is a complicated issue that requires the oil and gas industry to examine alternate sources. As these have been investigated, a strong economic argument has emerged in many areas for switching from fresh water to salt water based fracturing fluids.

Water used for shale gas developments includes water for drilling, cementing and fracturing. While fresh water for near surface drilling and cementing is advantageous since the typical leak-off volumes of 1 to 3% will be fresh water, deeper drilling and all fracturing may be able to take advantage of salt water as a base to prevent formation interactions.

Ideally, the salinity content of any water used in an oil or gas reservoir should roughly match the salinity of the fluids already there. Most shale of marine origin has water salinities near that of the sea water when deposited, usually about 35,000 mg/kg total salinity. Formations in contact with waters in evaporite beds may be higher salinity (Blauch, 2010). Use of large volumes of fresh water for fracturing in arid areas causes severe problems. By switching to a water source in the saline to brine water range and/or adopting closed-loop fracturing water supply where possible, the pressure on fresh water supplies could be greatly reduced.

Pumping

Pumping of a fracturing stage may last from 20 minutes to about 6 hours, depending upon the design and intent of the fracturing. This period of high pressure operation is probably the only time most wells will experience pressures above a level that will force reverse fluid flow (into the formation). Fracture vertical growth may extend up to a hundred meters above the pay zone in a few cases where there are no natural upper rock fracture barriers immediately over the pay zone, but more likely, the fracture will be quickly limited by one of dozens of rock barriers above and below the pay zone. The fracture is also limited by increasing loss of fracture fluid as it leaks into permeable formations as more rock is contacted by the fracturing fluid. Driving a fracture upwards through thousand of meters of rock is simply not possible, given the limits imposed by natural fracture barriers, fluid leakoff and the natural stresses of the formation and rocks above the pay zone.

Hydraulic fracture height growth limits in vertical fractures from horizontal wells can be mapped with passive microseismics monitoring that triangulates the location of sounds made by rock breaking during shear fracturing. Microseismics measurements are made with a 60 to 120 m long set of microphones in a well within a hundred meters of the well being fractured or, in some cases with several sets of microphones at the surface. Rock breaking during pumping will deflect slightly due to relaxation of built-up stress in the rock. In-situ rock stress is the result of historical uplift, creep of salt formations and other tectonic forces. In nature, these forces will build up until they reach the point of rock failure. Evidence of this rock stress relaxation reaction over geologic time is the presence of numerous small faults and a few very large faults.

The energy released by rock breaking during pumping has a magnitude about one hundred thousand times smaller than the magnitude of the slightest "felt" earthquake (magnitude ~3.0) (USGS – U.S. Geological Survey calculator). Hydraulic fracturing related creation of a moderate magnitude earthquake (magnitude greater than 5) in an unknown fault system is a very remote possibility because of the generally low incidence of quakes after or during fracturing and the ease of spotting larger faults on 3D seismics surveys. Hydraulic fracturing may be able to create small fault slippage and thus very small tremors that are at or usually well below the "felt" threshold. Gas storage reservoirs that exposed to fast charge and discharge rates (movement rates of more than millions of standard cubic meters per day) have had long-term studies of rock movement and micro-tremors (Deflandre, 2002).

The actual act of fracturing in a properly designed and constructed wellbore is the lowest risk action involved in the well development process, especially in wells more than about 600 meters deep. The minimum depth at which fracturing is practical and safe is set by state regulators with knowledge of local fracturing experience and geological hazards such as faults and karsts, and effective fracture barriers (rocks that resist fracture initiation or penetration). State regulators may also set specific limits on well depth, fracture volume, rate, type or fluid. If well construction is not properly done, then communication may be possible through the wellbore annulus (the area between the un-cemented casing and the wellbore rock wall): this is a pollution risk. The special case of fracturing in very shallow wells, particularly those at depths less than about 600 m or with fresh water within 300 m of the hydrocarbon containing formation is cause for concern and very careful evaluations are required.

Fluid backflow, or flowback, is the activity of cleaning up or flowing the well after fracturing to recover part of the fracturing fluid and initiate the gas production. Recovery of fluid depends on the formation and how much water that the formation adsorbs and absorbs into its structure. The amount of water in a formation, or its percent of water saturation, depends on the composition and form of the minerals such as clays at the microscopic level. If the formation minerals do not have sufficient water in their structure, they will trap and hold water from any available source until the minerals reach an irreducible water level. Water trapped in this manner may dry out again over geologic time through dry gas evaporation of the bound water, but is not likely to move during years of production. Water removed by dehydration will not transport chemicals, which will remain trapped in the rock (Woodroof, 2003).

Flowback of fracturing fluids during the first two to three weeks after a shale fracturing may experience fluid recovery rates of 0.5 to as much as 1 m^3/min for a few hours, often dropping to 150 m^3/day (100 liters per minute or about 8 times what a garden hose will flow) within 24 hours and then quickly decreasing over several days to the less than 50 m^3/day, by the end of the second or third week. This is followed by a gradual decrease to a few hundred liters per day within a few weeks. Modeling the flowback behavior has benefits in optimizing well production operations (Gdanski, 2010). Methane production is generally absent as the first water is produced. Water rate usually drops quickly as gas production starts. Although methane venting during cleanup is still done in exploratory mode (to test rates in the first few wells in an area for pipeline sizing), development wells can be turned to the production system within hours of gas flow start.

Fracture Monitoring

The crack that develops during fracturing, which produce a break in the rock, is narrow, usually 2 to 3 mm in width (1/10 to 1/8 in.) and grows outward, upward and downward widening slightly until a barrier is encountered or there is sufficient fluid leakoff into side fractures or permeable formation to stop the fracture from growing. Even at an injection rate of 16 cubic meters per minute, the secondary fractures and permeable streaks will soon absorb enough liquid from the main fracture to limit outward and upward fracture growth.

When a fracture grows in a planar style, as is common with sandstones and limestones, it conforms to a general direction set by the stresses in the rock, following what is called "fracture direction" or orientation, perpendicular to the plane of least principle stress (stresses are created by tectonic forces generated by uplift and other natural forces). The fractures are most commonly vertical and may extend laterally several hundred meters away from the wellbore and usually grow upward until it contacts a rock of different structure, texture or strength to stop the fracture. These "seal" or fracture barrier rocks, which stop the fracture upward or downward growth, are very common in every environment.

A second control method that stops fracture growth is the loss of fracture fluid, called fluid leakoff, into the reservoir rock. Fracturing fluid under pressure is required to extend the fracture, which also increases contact area with the formation. However, as the fluid increases fracture contact area with the formation or invades natural fractures, part of the fracturing fluid is lost to the formation, decreasing the amount of fracturing fluid left to drive fluid into the formation. With the large contact of thousands of natural fractures present in shale and the low viscosity of slick water fracture fluids, the rapidly increasing loss of fluids to the formation quickly matches the maximum fracturing injection rate and the fracture stops growing. This is seen repeatedly in microseismics event replays of monitored fracturing with a very rapid fracture growth at the start of the fracturing that very often decreases to no growth during the last 10% of the job.

The formation contact area that the fracturing fluid creates with the pay zone in a naturally fractured formation is normally very large, often accounting for several thousands to a hundred thousand square meters in densely naturally fractured shale well, all within a total extent of a few hundred meters away from the wellbore.

Downhole camera pictures in a well with an open-hole completion are shown in Figure 3. The camera was built and run by Amoco Research in the late 1960's through the 1990's to study and map hydraulic fracture growth in a variety of formations and compare performance of fracture variables (Smith, 1982). The pictures here have a total view, (width and height), of about 5×3.8 cm (2×1.5 in.). Depth is about 1400 m, with pressure in bar and time of day in 24 hour time.

Photo "A" shows narrow fractures stopping and starting again at a 6 mm (¼ in.) thick shale break that interrupts the development of vertical fracture growth. Thicker shale lay-

Figure 3. Downhole camera pictures of hydraulic fractures (King, 2012)

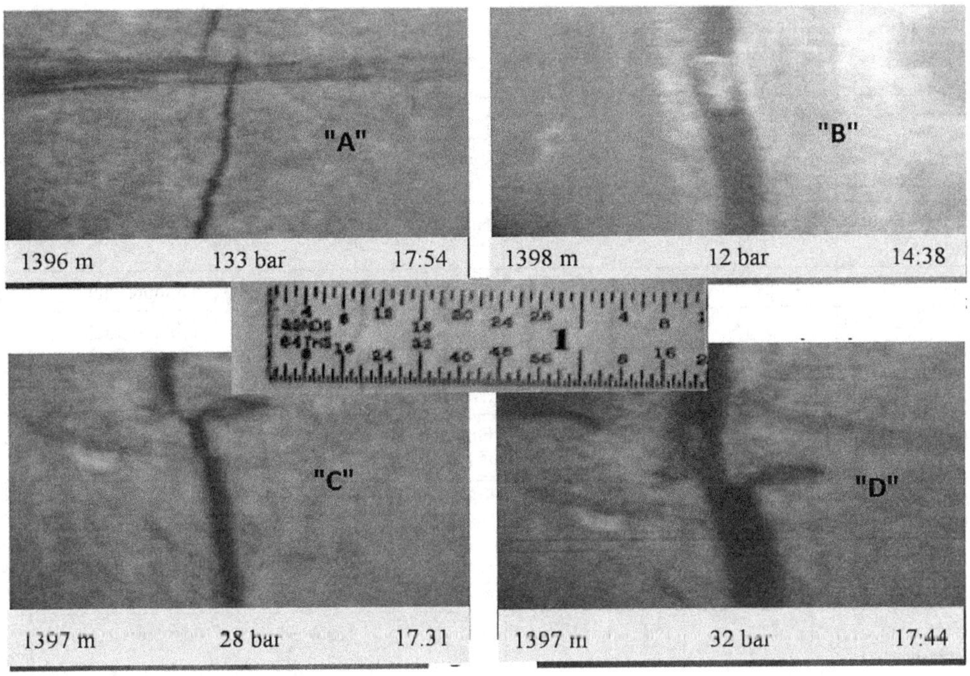

ers will effectively stop the fracture growth unless pressure can reapply itself above the barrier such as happens in this open wellbore. If pressure is reduced and cannot be reapplied, the fracture will stop growing.

Photo "B" is a fully developed fracture about 6 mm (¼ in.) wide with a camera-confirmed height of approximately 6 m. This fracture was in a limestone. The rock in the fracture is a piece of formation debris swept into the fracture after the crack was widened sufficiently to accept particles.

Figures "C" and "D" are of the same fracture. Photo "C" was after the fracture was initiated but while the wellbore was at low pressure. Photo "D" was of the same section of fracture after pressure was applied. The camera was a rare, side-looking camera that used close-ups of fractures in open-hole completions to help categorize fracture development from the wellbore. One very unusual experiment with the camera captured creation of a vertical fracture while pressuring up on a formation and pumping at about 0.8 m^3/min. The vertical fracture grew past the camera with a recorded creation rate of about 30 cm per second.

Cameras are only good for viewing the surface of an exposed wellbore in clear fluid and most only work at low pressures. Only a limited amount of information on fracturing can be deduced unless the well is an open-hole completion. Downhole cameras are commonly used in the industry to help diagnose well conditions. Other technologies using electrical resistance, sound, impression blocks and measurements of temperature, pressure and sound are also common diagnostic tools (Barree, 2002; Salamy, 1991; Cipolla, 2009).

Primary monitoring methods for describing fracture orientation, length, height, width and placement factors include microseismics, tilt meters, tracers, data frac and several other technologies, as shown in Table 1.

Table 1. Fracture monitoring technique (King, 2012)

Fracture monitoring technique	Fracture location in well	Fracture direction	Fracture height	Fracture length	Investigation timing	Specific fracture flow
Surface pressure	No	No	Indirect	No	Real time	No
Downhole pressure	No	No	Indirect	No	Post frac or real time	No
Fiber optics (temperature, pressure, sound)	Yes	No	No	No	Real time	Yes
Microseismics	Yes	Yes	Yes	Yes	5 minutes delay	No
Tilt meters	Yes	Yes	Yes	Yes	5 minutes delay	No
Temperature logging	Yes	No	Only in vertical wells	No	Post frac	Yes
Tracer tagged proppant	Yes	Transverse or longitudinal	Only in vertical wells	No	Post frac	No
Chemicals tracers in fracturing fluid	No	No	No	No	Post frac	Yes
Production log	Yes	No	Only in vertical wells	No	Post frac	Yes
Pressure build up, production tests and interference tests	No	Indirect	Indirect	Indirect	Post production	No

These technologies are most actively applied during the exploratory drilling and early development, where fracturing treatments are refined and observations can be made about local geologic impacts that optimize wellbore and fracture placement.

Once the well construction, lateral placement and fracture dynamics are understood in an area, the development phase is entered and well costs can be optimized with learning generated on the first few wells. Microseismics and tracers are rarely used in later stages of development.

Fracture modeling by computer depends on the accuracy and range of a large number of variables (Britt, 2009; Cipolla, 2009; Meyer, 2010, Smith, 1982). Formation property assumptions on permeability, porosity, in-situ stresses, mineralogy, fracture barrier locations and a dozen or more other variables with wide ranges, make the first estimates of computer modeling less than ideal. The first wells in a development are opportunities to learn. Combining known regional geology with local geology findings generated during drilling are the first solid inputs. As fracture monitoring techniques add accuracy to the assumptions, fracture models will be able to improve accuracy. Most early jobs are kept small until a more optimized fracture design is constructed. Even with a more polished fracture design, improvements and innovative changes will still be part of the optimization process. Local knowledge on fracturing is a very good starting point in any area.

Data gathering from hydraulic fracturing jobs can also be used to examine fracture growth and risk properties on a local or regional basis.

A FRACTURING RISK ANALYSIS

Risk is generated by uncertainty of the outcome of an action (Holton, 2004). The elements of risk consist of probability (belief that a detrimental event can occur and that a frequency can be predicted for that event) and severity (or impact) of that event. Probability that the fracturing materials spill can occur during the transport to the well is simple to determine using insurance tables on wreck frequency, fuel spills, truck roll-overs and some equipment failures where reliability data exists.

Potential pollution from the specific act of fracture pumping is not an easy prediction to make since the bulk of the evidence for actual cases of pollution is negative and the documentation of claimed detrimental results are absent or discounted by scientific and engineering investigation. This creates significant uncertainty around the unmeasured or un-measurable part of any impact of pollution from the specific act of fracturing. Unfortunately, it is the best that can be done with the available scientific information.

The main fracturing activity discussed in the risk analysis is slick water hydraulic fracturing in shale formations. The pollution potential include examination of pollution routes for fracturing fluids, chemicals, returned fluids, unreturned fluids, seismic disturbances, natural seep alteration, penetration of aquifers by fractures and emissions from fractures. Water sources specifically protected from contamination include fresh and brackish water aquifers; sources that are suitable for economic levels of treating to make drinking water. The objective is to identify high-risk elements of fracturing activities and to assign appropriate risk reduction actions.

The first action is to separate well construction issues from fracturing issues. Well construction issues, specifically the design, supervision, application, testing and monitoring of well construction, are influenced by a company's design philosophy and commitment to a sound design. Any well activity that begins with a sub-standard well will likely fail, regardless of the quality of the well activity. The root cause of problem would be the well construction, not the well activity.

Fracturing issues, defined here as the transport of fracturing materials to the well, the specific act of fracturing, the recovery of fracturing fluids from the well prior to production and the transport of fracturing materials from the well, offer similar base-line impact and probability, regardless of the operator, given adherence to lawful and safe operational practices. Reduction in base-line impact and probability comes from a company's commitment to use and optimize practices that reduce the impact of the event and/or the probability of the event.

Variances in local transport incident occurrence and local geology necessarily make a hydraulic fracture risk evaluation a local analysis. There cannot be a single detailed over-all fracturing risk analysis. General analyses can be constructed but the user must specify a range of occurrence suitable for the areas to be compared. Many of the factors discussed will be further influenced by the operator's experience, geological and engineering skill, ethics and projection of well design life and well functions during that life (King, 2012).

Local Variances Included in the Risk Analysis

1. **Traffic Wreck Incidence Frequency:** From state-to-state and road types from 2-lane rural to controlled access interstate.
2. **Well Construction:** Overpressure related incident – these are fairly constant across a region with the lower frequency being in areas where there is a high incidence of fracturing with common knowledge of adequate fracturing application for fracturing in the area. Prior to the fracturing job, well casing is tested to the pressure expected in fracturing

plus a safety factor. Casing ruptures are very rare and are immediately seen by annulus pressure monitoring.

3. **Depth of Intervals:** Fracture height growth is limited by natural rock barriers and leakoff of fracturing fluid into permeable rock. As demonstrated by microseismics monitoring during fracturing, fractures rarely reach a vertical growth of a hundred meters above the pay zone, particularly when fracturing rates are tailored to specific intervals. Shallower zones, nominally those less than 600 meters would have the highest potential for hydraulic fracturing communication into a fresh water sand, although even for this potential event, there is no documented proof. The deeper the well, the less likely such an event would be. Deeper zones have more separation, more fracture barriers and more permeable rock that would sap fracturing fluid away from the stimulation. Most commercial shale reservoirs are at 1000 m to 3500 m deep, with an average depth of about 2000 m.
4. **Fracture Intersection:** Fractures intersecting other fractures or other wells at the same depth in a pay zone are likely as the wellbore spacing (distance between wellbores at the depth of the pay zone) is diminished. Shale wellbores are intentionally set close together because fracture penetration is dominated by extensive natural fractures that limit overall fracture growth. Fracture intersection at depth and below the main cap rock or seal does not create a problem unless the intersected well is open to a fresh water zone.

   Fractures intersecting an abandoned well are a rare, but known occurrence. If the well is properly abandoned, there is little or no risk of fresh water contact or contamination. If the well is an old, improperly abandoned well, the casing strength cement quality and the pipe strength must be considered low unless proved otherwise by testing. The low pressure of these old wells would very likely create an immediate pressure drop in any fracturing treatment that entered the old well, prompting investigation. Even when intersected, a properly abandoned well will withstand fluid movement towards surface. No documentation could be found for this type incident, but estimates from field personnel active in fracturing application over several decades indicates the figure is about one intersection or less in 10.000 fracturing jobs.
5. **Other Factors:** These are fracturing incidents that may or may not have an impact on fresh water pollution but many have a detrimental effect (King, 2012).

FUTURE RESEARCH DIRECTIONS

It is shown in the risk analysis that fracture treatments do not penetrate fresh water supplies in properly constructed well, which is derived from a preponderance of evidence, such as monitoring fracture growth, absence of fracturing chemical components outside the pay zone and a 60 years history documenting fracturing containment in numerous geologic settings. However, future research should include more evidence on fracture pumping and not innocently verdict the entire well development process.

Risk analysis also showed that the potential for chemical contamination of underground or surface sources of fresh water in adequately constructed wells is arguably less than one in a million fracs. As it comes exclusively from the road transport of fracturing components, onsight storage and surface mixing, in the future it can be sharply reduced by more attention to the root cause of these spill or leak events.

Although the impact of spill and leak events are generally low, they can be decreased further by reducing number, amount (concentration and/or activity), toxicity and environmental permanence of chemicals used in fracturing. Chemical rating systems that focus on these issues should be a part of the planning for any fracturing treatment.

For targeted hydrocarbon pays of less than 600 m depth, state regulators with knowledge of local geological systems may need to set specific limits on well depth, fracture volume, rate, or type of fluid. The special case of fracturing in very shallow wells, particularly those at depths less than about 600 m or with fresh water within 300 m of the hydrocarbon containing formation, is cause for concern and additional evaluations of geology, fracturing rates and volumes are required.

Methane presence is commonly recorded in water wells and may predate any drilling or fracturing in the area. Methane may increase during a water well's life as water is produced, as a result of liberating methane that was adsorbed in and onto organic materials in the sediments. This type of methane increase is particularly active in fresh water containing coal seams and high clay content shale. As the fresh water aquifer levels are drawn down, the methane adsorbed on the organics will desorb and overall methane content in the water well will increase. Avoiding coal seams and high organic source rocks by proper cement isolation in the fresh water wells is required to minimize this problem (King, 2012).

CONCLUSION

Drawing conclusions from this analysis involved an initial separation of well construction issues (where a company's design philosophy in a local geologic area can be a major risk differentiator) from the specific rock fracturing risk (where every company faces near-identical risks).

The primary conclusion that the potential for even a small amount of chemical contamination of underground or surface sources of fresh water from the specific act of fracturing, applied in adequately constructed wells with pay zone depth of greater than 600 m, is arguably less than one in a million fracs due to the self-limiting nature of fracturing leakoff and the numerous frac barriers found in every deeper formation sequence.

The height of fracture growth in deep wells is usually a few hundred meters above the targeted hydrocarbon zone but thousands of meters below the deepest fresh water sands, which is documented by downhole microseismics, tilt meters, tracers, logging and other methods (Fisher, 2011).

With proper well construction, there was no documented case located of fracturing chemical migration to a fresh water aquifer or to the surface from a zone deeper than 600 m. Cases of suspected contamination by chemicals in shallower zones are known, with many, if not all, linked to poor isolation of the well during the well construction phase.

REFERENCES

Arthur, J. D., Bohm, B., Coughlin, B. J., & Layne, M. (2009). *Evaluating implications of hydraulic fracturing in shale gas reservoirs.* Paper presented at the SPE Americas Environmental and Safety Conference. San Antonio, TX.

Arthur, J. D., Hochheiser, H. W., & Coughlin, B. J. (2011). *State and federal regulation of hydraulic fracturing: A comparative analysis.* Paper presented at the SPE Hydraulic Fracturing Technology Conference and Exhibition. The Woodlands, TX.

Barree, R. D., Fisher, M. K., & Woodrood, R. A. (2002). *A practical guide to hydraulic fracture diagnostics technologies.* Paper presented at the SPE Annual Technical Conference and Exhibition. San Antonio, TX.

Blauch, M. E. (2010). *Developing effective and environmentally suitable fracturing fluids using hydraulic fracturing flowback waters.* Paper presented at the SPE Unconventional Gas Conference. Pittsburgh, PA.

Britt, L. K., & Schoeffler, J. (2009). *The geomechanics of a shale play: What makes a shale prospective!* Paper presented at the SPE Eastern Regional Meeting. Charleston, WV.

Čikeš, M. (2000). Long-term hydraulic-fracture conductivities under extreme conditions. *SPE Production and Facilities, 15*(4), 255–261.

Cipolla, C. L. (2009). Modeling production and evaluating fracture performance in unconventional reservoirs, SPE 118536. *Journal of Petroleum Technology*, (9): 84–90.

Cipolla, C. L., Lolon, E. P., Mayerhofer, M. J., & Warpinski, N. R. (2009). *Fracture design considerations in horizontal wells drilled in unconventional gas reservoirs.* Paper presented at the SPE Hydraulic Fracturing Technology Conference. The Woodlands, TX.

Deflandre, J. P., & Huguet, F. (2002). Microseismic monitoring on gas storage reservoirs: A ten-year experience. In *Proceedings of Worldwide Gas Production, Monitoring and Storage, Forum* (pp. 441-445). World Petroleum Congress.

Economides, M. J., & Nolte, K. G. (2000). *Reservoir stimulation.* West Sussex, UK: John Wiley & Sons.

Fisher, K., & Warpinski, N. (2011). *Hydraulic fracture height-growth: Real data.* Paper presented at the SPE Annual Technical Meeting and Exhibition. Denver, CO.

Fontaine, J., Johnson, N., & Schoen, D. (2008). *Design, execution and evaluation of a typical marcellus shale slickwater stimulation: A case history.* Paper presented at the SPE Eastern Regional/AAPG Eastern Section Joint Meeting. Pittsburgh, PA.

Gaskari, R., & Mohaghegh, S. D. (2006). *Estimating major and minor natural fracture pattern in gas shales using production data.* Paper presented at the SPE Eastern Regional Meeting. Canton, OH.

Gdanski, R. D., Fulton, D. D., & Johnson, B. J. (2010). *Returns matching reveals new tools for fracture/reservoir evaluations.* Paper presented at the SPE Tight Gas Completions Conference. San Antonio, TX.

Hoeman, K., Klasner, S., & Karcher, A. (2011). *Method for identification and analysis of polysaccharide-based, hydraulic-fracture flowback fluid*. Paper presented at the SPE International Symposium on Oilfield Chemistry. The Woodlands, TX.

Holton, G. A. (2004). Defining risk. *Financial Analysts Journal*, *60*(6), 19–25. doi:10.2469/faj.v60.n6.2669

King, G. E. (2010). *Thirty years of gas shale fracturing: What have we learned?* Paper presented at the SPE Annual Technical Meeting and Exhibition. Florence, Italy.

King, G. E. (2012). *Hydraulic fracturing 101: What every representative, environmentalist, regulator, reporter, investor, university researcher, neighbor and engineer should know about estimating frac risk and improving frac performance in unconventional gas and oil wells*. Paper presented at the SPE Hydraulic Fracturing Technology Conference. The Woodlands, TX.

King, G. E., & Lee, R. M. (1988). Adsorption and chlorination of mutual solvents used in acidizing. *SPE Production Engineering*, *3*(2), 205–209. doi:10.2118/14432-PA

Meyer, B. R., Bazan, L. W., Jacot, R. H., & Lattibeaudiere, M. G. (2010). *Optimization of multiple transverse hydraulic fractures in horizontal wellbores*. Paper presented at the SPE Unconventional Gas Conference. Pittsburgh, PA.

Palmer, I. D., King, N. S., & Sparks, D. P. (1991). *The character of coal fracture treatments in oak grove field, black warrior basin*. Paper presented at the SPE Annual Technical Conference and Exhibition. Dallas, TX.

Salamy, S. P., Aminian, K., Koperna, G. J., & Locke, C. D. (1991). *Pre- and post-stimulation well test data analysis from horizontal wells in the devonian shales*. Paper presented at the SPE Eastern Regional Meeting. Lexington, KY.

Smith, M. B., Rosenberg, R. J., & Bowen, J. F. (1982). *Frac width – Design vs. measurement*. Paper presented at the SPE Annual Conference and Exhibition. New Orleans, LA.

Warpinski, N. R. (1985). Measurement of width and pressure in a propagating hydraulic fracture. *SPE Journal*, *25*(1).

Warpinski, N. R. (2011). Hydraulic fracture height in gas shale reservoirs. In *Proceedings of SEG Annual Meeting*, (pp. 3705-3706). San Antonio, TX: SEG.

Woodroof, R. A., Asadi, M., & Warren, M. N. (2003). *Monitoring fracturing fluid flowback and optimizing fluid cleanup using frac tracers*. Paper presented at the SPE European Formation Damage Conference. The Hague, The Netherlands.

ADDITIONAL READING

Arthur, J. D., Bohm, B., & Cornue, D. (2009, October). *Environmental Considerations of Modern Shale Development*. SPE 122931. Paper presented at the SPE Annual Technical Meeting, New Orleans, LA USA.

Asadi, M., Woodroof, R. W., Malone, W. S., & Shaw, D. R. (2002, September). *Monitoring Fracturing Fluid Flowback With Chemical Tracers: A Field Case Study*. SPE 77750, Paper presented at the SPE Annual Technical Conference and Exposition, San Antonio, TX USA.

Barrufet, M. A., & Mareth, B. C. (2009, May). *Optimization and Process Control of a Reverse Osmosis Treatment for Oilfield Brines*. SPE 121177. Paper presented at the Latin American and Caribbean Petroleum Engineering Conference, Cartagena de Indias, Colombia.

Blauch, M. E., Myers, R. R., Moore, T. R., & Houston, N. A. (2009, September). *Marcellus Shale Post-Frac Flowback Waters – Where is All the Salt Coming From and What are the Implications?* SPE 125740. Paper presented at the SPE Regional Meeting, Charleston, WVA USA.

Britt, L. K., & Smith, M. B. (2009, September). *Horizontal Well Completion, Stimulation Optimization, and Risk Mitigation*. SPE 125526. Paper presented at the SPE Eastern Regional Meeting, Charleston, WVA USA.

Burnett, D. B., Yu, O., & Schubert, J. J. (2009, March). *Well Design for Environmentally Friendly Drilling Systems: Using a Graduate Student Drilling Class Team Challenge to Identify Options for Reducing Impacts*. SPE 119297. Paper presented at the SPE/IADC Drilling Conference and Exhibition, Amsterdam, The Netherlands.

Carman, P. S., & Cawiezel, K. E. (2007, January). *Successful Breaker Optimization for Polyacrylamide Friction Reducers Used in Slickwater Fracturing*. SPE 106162, Paper presented at the SPE Hydraulic Fracturing Technology Conference, College Station, TX USA.

Cramer, D. D. (2008, February). *Stimulating Unconventional Reservoirs: Lessons Learned, Successful Practices, Areas for Improvement*. SPE 114172. Paper presented at the SPE Unconventional Reservoirs Conference, Keystone, CO USA.

Dao, B., & Vijn, P. (2002, March). *Environmentally Acceptable Cement Fluid Loss Additives*. SPE 74088, Paper presented at the SPE International Conference on Health, Safety and Environment in Oil and Gas Exploration and Production, Kuala Lumpur, Malaysia.

DeMong, K., Fahlman, J., & Schnell, R. (2010, October). *Coping With Surface and Downhole Interference on Tightly Spaced Completions Pads in the Horn River*. CSUG/SPE 138026, Paper presented at the Canadian Unconventional Resources & International Petroleum Conference, Calgary, Alberta Canada.

Edwards, K. L., Weissert, S., Jackson, J., & Marcotte, D. (2011, January). *Marcellus Shale Hydraulic Fracturing and Optimal Well Spacing to Maximize Recovery and Control Costs*. SPE 140463, Paper presented at the SPE Hydraulic Fracturing Technology Conference, The Woodlands, TX USA.

Enzien, M. V., Yin, B., Love, D., Harless, M., & Corrin, E. (2011, April). *Improved Microbial Control Programs for Hydraulic Fracturing Fluids used during Unconventional Shale-Gas Exploration and Production*. SPE 141409. Paper presented at the SPE International Symposium on Oilfield Chemistry, The Woodlands, TX USA.

Fisher, M. K., Heinze, J. R., Harris, C. D., Davidson, B. M., Wright, C. A., & Dunn, K. P. (2004, September). *Optimizing Horizontal Completion Techniques in the Barnett Shale Using Microseismic Fracture Mapping*. SPE 90051. Paper presented at the SPE Annual Technical Conference and Exhibition, Houston, TX USA.

Friedmann, F. (1986). Surfactant and Polymer Losses During Flow Through Porous Media. SPE 11779. *SPE Reservoir Engineering*, *1*(3), 261–271. doi:10.2118/11779-PA

Gupta, D. V. S., & Hildek, B. T. (2009, January). *Frac Fluid Recycling and Water Conservation: A Case History*. SPE 119478. Paper presented at the SPE Hydraulic Fracturing Conference, The Woodlands, TX USA.

Holt, S., Sanders, J., Rodrigues, K., & Vanderhoof, M. (2009, April). *Biodegradable Alternatives for Scale Control in Oilfield Applications*. SPE 121723. Paper presented at the SPE International Symposium on Oilfield Chemistry, The Woodlands, TX USA.

Houston, N., Blauch, M., Weaver, D., Miller, D. S., & O'Hara, D. (2009, September). *Fracture Stimulation in the Marcellus Shale-Lessons Learned in Fluid Selection and Execution*. SPE 125987. Paper presented at the SPE Regional Meeting, Charleston, WVA USA.

Howard, P. R., Mukhopadhyay, S., Moniaga, N., Schafer, L., Penny, G., & Dismuke, K. (2009, May). *Comparison of Flowback Aids: Understanding Their Capillary Pressure and Wetting Properties*. SPE 122307, Paper presented at the 8th European Formation Damage Control Conference, Scheveningen, The Netherlands.

King, G. E., & Haile, L. Shuss, J., & Dobkins, T. A. (2008, November). *Increasing Fracture Path Complexity and Controlling Downward Fracture Growth in the Barnett Shale*. SPE 119896. Paper presented at the SPE Shale Gas Production Conference, Fort Worth, TX USA.

Kundert, D., & Mullen, M. (2009, April). *Proper Evaluation of Shale Gas Reservoirs Leads to a More Effective Hydraulic-Fracture Stimulation*. SPE 123586. Paper presented at the SPE Rocky Mountain Petroleum Technology Conference, Denver, CO USA.

Palmer, I. D., Sparks, D. P. (1991). Measurement of Induced Fractures by Downhole TV Camera in Black Warrior Basin Coalbeds, SPE 20660, *JPT Journal of Petroleum Technology, No. 3*, 270-278.

Perkins, C. K. (2008, October). *Logistics of Marcellus Shale Stimulation: Changing the Face of Completions in Appalachia*. SPE 117754. Paper presented at the SPE Eastern Regional/AAPG Eastern Regional Section Joint Meeting, Pittsburgh, PA USA.

Rimassa, S. M., Howard, P. R., & Blow, K. A. (2009, June). *Optimizing Fracturing Fluids From Flowback Water.* SPE 125336. Paper presented at the SPE Tight Gas Completions Conference, San Antonio, TX USA.

Schweitzer, R., & Bilgesu, H. I. (2009, September). *The Role of Economics on Well and Fracture Design Completions of Marcellus Shale Wells*. SPE 125975. Paper presented at the SPE Eastern Regional Meeting, Charleston, WV USA.

Suarez-Rivera, R., Deenasayalu, C., & Yang, Y. (2009, May). *Unlocking the Unconventional Oil and Gas Reservoirs: The Effect of Laminated Heterogeneity in Wellbore Stability and Completion of Tight Gas Shale Reservoirs*. OTC 20269. Paper presented at the OTC, Houston, TX USA.

USGS. (2012). *Earthquake Facts and Statistics*. Retrieved March 27, 2013, from http://earthquake.usgs.gov/earthquakes/eqarchives/year/graphs.php

KEY TERMS AND DEFINITIONS

Coal-Bed Methane: Methane contained in coal seams. A coal seam is a layer or stratum of coal parallel to the rock stratification.

Flowback: After the hydraulic fracturing procedure is completed and pressure is released, the direction of fluid flow reverses, and water and excess proppant flow up through the wellbore to the surface. The water that returns to the surface is commonly referred to as flowback.

Hydraulic Fracturing: The process of using high pressure to pump sand along with water and other fluids into subsurface rock formations in order to improve flow of oil and gas into a wellbore.

Permeability: Ability of rock to transmit fluid through pore spaces.

Proppant: A granular substance (sand grain and synthetic material as sintered bauxite and ceramics) used to hold the walls of the fracture apart to create a conductive path to the wellbore after pumping has stopped and the fracturing fluid has leaked off. Placing the appropriate concentration and type of proppant in the fracture is critical to the success of a hydraulic fracturing treatment.

Shale: A fine-grained sedimentary rock composed mostly of consolidated clay or mud. Shale is the most frequently occurring sedimentary rock.

Tight Sands: A geological formation consisting of low permeability and low porosity sandstones.

Chapter 10
Workover Impact on Accidental Risk

Bojan Moslavac
University of Zagreb, Croatia

ABSTRACT

Certain risks are very common to any application of a downhole control system during workover while others will be field-specific or at least increased or decreased by the given well conditions. Workover anomalies may be caused by erosion, corrosion, mechanical errors, temperature effects on electronics, wear and tear on the dynamic seals, or seizure of moving components. Obviously, the simpler the system and the fewer moving parts, the fewer components are available to fail. Procedures and supporting control software must be developed to ensure optimum system use. Blowouts are not very common today as most blowout preventing systems serve the purpose. However, occasionally, these systems may fail due to either technical problems on the wellbore equipment or human error. Both reasons can cause enormous damage on wellbore equipment, reservoir, surface facilities, or even human loss. The right accidental risk assessment approach can reduce the chances of repeating bad scenarios that happened in the past.

INTRODUCTION

In general, workover refers to any kind of well intervention or remedial treatment on oil or gas well involving different methods and techniques common in drilling and completion department. In most cases it implies the wireline or slickline operations, tubing replacement, coiled tubing services, snubbing unit services etc. Reasons for such treatments or interventions are numerous (Jahn et al., 2008)- mechanical damage of downhole equipment (corroded or damaged tubing string, stuck wellbore tools), reservoir productivity impairment, flow restriction due to sand production or scale deposition, water

DOI: 10.4018/978-1-4666-4777-0.ch010

or gas breakthrough, cross flow in the wellbore or behind casing, downhole control system failures, wellhead completion parts failures and other. Usually, if a workover operation is needed, a workover rig is required as well, except in cases of coiled tubing or subbing unit operations.

Perrin et al. (1999) described some of the equipment failures and modifications urging for intervention. That would be *wellhead equipment failures* (leaks at the lower master valve, tubing hanger or tie-down screws, damaged back pressure valve seat, leak or failure at the controllable subsurface valves control line outlets, wellhead penetrator failures etc.), *downhole equipment failures* (tubing or wireline retrievable safety valves are faulty, landing nipple or wireline retrievable safety valve, packer, locator, slip joint or circulating sleeve are leaking, control line failure, annular safety system failure, gas-lift valves stuck, fish not processed properly via wireline, downhole gauges failures) and many other problems met during the life of a well.

Rig site safety is considered to be a significant factor in the oilfield for many years now and continues to improve its worthiness. In spite of all safety precautions undertaken during a certain operations performed in the well accidents continue to occur with the results mentioned earlier. What can we do to further improve safety on the rig is a question all the oilfield companies try to answer investing a lot of assets in it. Term rig safety firstly refers to human life preservation and then wellbore integrity and equipment maintenance. As everything starts with well integrity no human life or wellbore equipment will be preserved without it.

WELL INTEGRITY

Well integrity represents risk reduction of formation fluids uncontrollable releases. Reasons for well integrity loss are either of technical nature, as well completion or surface equipment failure or in a form of human error. Well control loss, and thus blowout, is a result of the two mentioned. Blowouts can lead to disastrous events with the result of a loss of human life, environmental impacts or equipment damage. Fatality risk in exploration activities is partially expressed by blowout occurrence. According to Corneliussen (2006) measure for the fatality risk is demonstrated as FAR (fatal accident rate), which is a frequency rate defined as the expected number of fatalities per 10^8 hours of exposure. When elaborating the environmental pollution, workover blowouts are more likely to cause severe pollution than the drilling blowouts because of the cased wellbore all down to the production zones. If the well blows out in such a completed well, the fluid blowing out can be water, gas, oil or condensate. Drilling blowouts mainly result with gas blowout.

Workover blowout is the end result of well integrity loss due to loss of the well barriers, primary and secondary ones, described in detail by Holand (1997) and Bellarby (2009). *Primary barrier* is defined as a barrier which has a task to stop unpredictable fluid flow (examples: wellbore fluid, tubing, x-mass tree valves). *Secondary barrier* presents a primary barrier backup (examples: downhole safety valve, tubing hanger, wellhead). Secondary barriers have to be independent of the primary barriers in case of failure of the latter. Swabbing, low fluid column weight, trapped gas, low cement preflush weight or bullheading are considered to be the main causes for the primary barrier loss within the wellbore operations like pulling the wellbore equipment out of the hole, installing the equipment, perforating, pressure testing, circulation, snubbing operations or well abandonment. Secondary barrier is mainly lost within unavailability or failure of string safety valves, kelly stabbing failure, BOP activation failure, insufficient frictional back pressure and annulus valve or casing head failure (Holand, 1997). Therefore, occasional pressure testing of wellbore equipment is essential if barrier losses are to be avoided.

Well barriers during the workover operations are basically the same barriers that exist during the drilling processes. But, there are some differences. Workover fluids, for example, are usually solids-free and they are free to escape into formation in open hole production systems without the mud cake formed on the wellbore wall, which is not the case in drilling operations.

Well Integrity Management System (WIMS) is one of the risk assessment approaches used to describe the status and handle the well integrity issues (Corneliussen, 2007; Al-Ashhab et al., 2004; Tarmoom et al., 2007). Generally, there are three basic elements of the system- data gathering, risk review and ranking and well integrity database. Data gathering considers identification of comprehensive data set. Second element implies risk assessment methodology of reviewed and ranked data after the analysis completion. Third element presents an organization of the processed data into the huge database.

Bullheading

Bullheading is most often used when normal well killing methods with conventional circulation do not work and if there is a possibility of hydrogen sulfide appearance in formation fluids, but is also a phrase for any kind of fluid force pumping downhole (acid treatment or scale removal for example). In case of gas kick, when pumping kill fluids down the string, the gas is compressed and the wellbore pressure exceeds the reservoir pressure and gas leaks off into formation. Kill fluid should be pumped until all the gas is displaced from the well and tubing string filled up with kill fluid having a sufficient density to balance the formation pressure (Oudeman et al., 1994).

As the bullheading is not a routinely performed well control method and considering the fact that very high pressures are sometimes involved a special attention should be addressed to operation execution. Exceeding the pressure rating of the used equipment can lead to a burst with the end result of equipment damage and well kill failure which has a blow out potential. There are some operational considerations for bullheading that have to be followed to avoid kick occurrence: a) Detailed job design before the operation execution- especially on the equipment pressure rating, b) Additional pumping unit available at the rig site for better well control, c) Large pumping fluid volumes available in case of major losses, d) Possibility of annulus area isolation in case of kill line failure.

There are some numerical models (Oudeman et al., 1994; Dykhno et al., 2005) developed to simulate a bullheading operation which is very helpful for multiphase flow performance and analysis. In bullheading operations, when using sophisticated software, associated blowout risk can be reduced to a minimum.

Swabbing - Surging

In the oilfield swabbing is a term for a pressure reduction below the pipe string when moving upward. Adversely, pressure increase below the pipe string when moving downward corresponds to surging. Swabbing is one of the most frequent blowout causes in workover and drilling operations(Holand, 1997; Nielsen & Karlsen, 2008). It appears mainly if wrong pull out rate is applied not allowing fluid column to settle down fast enough below the string. The end result is a state of downhole pressure underbalance (negative differential pressure) with formation fluids entering the wellbore. Bottom line, whatever the reason for string pull out is, it has to be done with extreme caution taking care about pull out rate concerning the tubular hydraulics factors (casing/tubing diameters and clearances, wellbore fluid weight and viscosity etc.). Surging may cause formation fracturing and fluid leak off, but this is elaborated within another chapter dealing with fracturing operations.

Low Fluid Column Weight

Normally, bottom hole pressure in static conditions should be greater than the reservoir pressure (primary barrier) to prevent formation fluid influx. If the hydrostatic bottom hole pressure decreases below the formation pressure point, fluids start to flow into the well endangering the well integrity with possible well blowout. In these situations, the most important thing is to act promptly and close in the well. Of course, not every fluid influx is a potential for well closure. It depends on the intensity and volumes entering the wellbore, so a good influx detection device or method is extremely important. In workover operations low fluid column weight situation can appear after the cement squeezing job, pumping the cement preflush, perforation or simply while circulating water down the tubing and out the annulus when the well starts to flow (Holand, 1997). Like in conventional drilling techniques, high pressure zones can be discovered during the coiled tubing drilling process as well. If the rig crew does not respond promptly, an abundant gas influx may initiate serious well control measures.

Trapped Gas

Trapped gas often refers to a gas that is trapped under the production equipment, a tubing string anchored with a packer. If the completion is to be replaced, when pulling the string to retrieve the packer trapped gas can rupture the tubing, enter the area of lower pressure, expand, push the workover fluid and damage the wellbore equipment if shear ram or annular preventer is not closed quickly enough. Trapped gas can also form beneath the closed subsurface safety valve. In that case, if a careless blind back off is performed, pressure can force the tubing out of the well.

COILED TUBING OPERATIONS

Coiled tubing (CT) as a reeled flexible type of carbon steel, titanium or polymer tubing is applicable in various wellbore operations (Baker Hughes, 2003) like *wellbore cleaning operations* (i.e. formation sand or proppant cleaning out, wellbore debris catching, underreaming the tubing or other completion accessories, scale or cement milling, high pressure jet washing), *fishing* (i.e. completion parts recovery with various fishing tools, removing and recovering obstructions- valves for instance, pipe cutting with mechanical, chemical or explosive cutters, milling operations), *zone isolation* (i.e. through tubing zonal isolation, wellbore accessories set up and retrieval), *well stimulation and fracturing* (hydraulic fracturing and acidizing operations), *sand control* (through tubing sand control operations), *flow management* (velocity strings, production/ injection logging, downhole equipment shifting), *wellbore abandonment* (bridge plug or cement plug placement)and *sidetracking* (re-entry wellbores drilling, slimhole drilling). It can be deployed in pressurized or dead well.

Obviously, such a huge variety of CT applications sometimes imposes certain complications within the operational execution. Some leading reasons for coiled tubing failures are (Crabtree, 2008) corrosion, mechanical damage, manufacturing flaws and defects, handling errors, H_2S cracking, erosion, fatigue and overload (Figure 1).

Corrosion and Cracking

Carbon steel coiled tubing, as any other tool or accessory made of steel, is prone to corrosion. Unfortunately, installed coiled tubing systems face produced water and CO_2/H_2S laden natural gas as well (Martin & Van Arnam, 2003). CO_2 is very aggressive corrosive gas which mainly corrodes downhole equipment by localized pitting at a certain penetration rate depending on the partial pressure of the CO_2 in the water phase, which represents CO_2 content. Partial pressure greater than $2.068 \cdot 10^5$ Pa (30 psi) almost ensures corrosion appearance. A typical 44.45 mm (1 ¾ in) OD coiled tubing with 2.77 mm (0.109 in) wall thickness is predicted to corrode at 100 mpy- mils per year (1 mpy equals 0.0254 mm per year). If it is expected to be used in over a year, there should be no hazard threat. But, in this case, if it is used for more than a year, an extensive corrosion inhibition plan has to be applied (Martin & Van Arnam, 2003).

Some evaluations (Metcalf et al., 2003; Nasr-El-Din & Metcalf, 2004) of the impact of H_2S on coiled tubing performance have already been performed. According to those, preferred maximum tensile strength of coiled tubing for corrosion avoidance in H_2S sour environment is $551.58 \cdot 10^6$ Pa (80 kpsi). Maximum concentrations of possibly harmful elements within wellbore fluids should be: 0.002 wt% (sulfur), 0.025 wt% (phosphorus) and 0.6 wt% (carbon). Corrosion aspects and types such as sulfide stress cracking (SSC), stress corrosion cracking (SCC), hydrogen induced cracking (HIC) and stress oriented hydrogen induced cracking (SOHIC) are often used to describe coiled tubing failure cause if a string overtensioning is not likely to be the damage cause. From the data above it can be seen that a proper corrosion inhibition and operation performance within safe tension margins is

Figure 1. Reasons for coiled tubing failures (Crabtree, 2008)

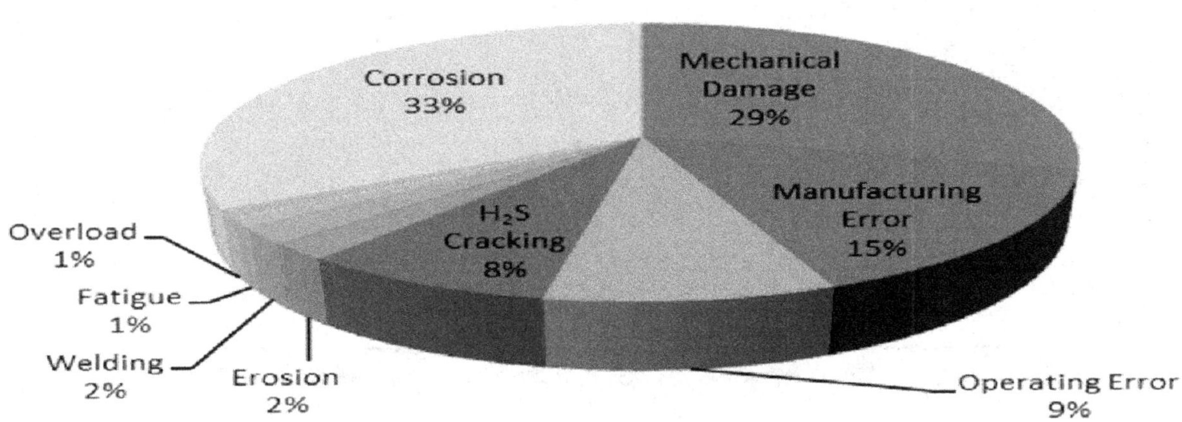

very important if the wellbore integrity is meant to be maintained. With 41% portion of total coiled tubing failure causes (Figure 1), corrosion and H_2S cracking are the most accounted.

Erosion

As different fluids and particles are being pumped through the coiled tubing string, erosion effects start to change the flow characteristics and reduce the wall thickness of the string. It will affect the string capacity to contain pressure and apply load, reduce tensile strength and increase fatigue in the coiled tubing. Field experience implies different wall thickness loss in coiled and straight section. During the frac packing treatments, for example, certain amount of proppant is pumped into the well to support fractures and prevent migration of fines into the well (Ghalambor et al., 2009). According to Rispler et al. (2001) and Rosine et al. (2008), CT wear (wall thickness loss due to erosion) increases exponentially with the weight of proppant pumped. Coiled tubing with 31.75 mm (1 1/4 in) OD or less has a potential for total damage by erosion if more than 700 tons of proppant are being pumped within the operation considering the wall thickness of 2.03 mm (0.08 in). If the wall loss is expressed by millimeters per 100 tons of proppant pumped, which is correlated with distance of measuring spot from the CT reel core, the data obtained by Rosine et al. (2008) corresponds to Figure 2. Abnormal increase in wall loss is observed in reeled section of the coiled tubing and at the tubing guide. In this case, first 400 m of the coiled tubing is a critical section if huge amounts of proppant are being pumped down the string. A thorough operational hazard analysis with worst scenarios is mandatory if the technical risk level is to be reduced to a minimum.

Figure 2. Measured coiled tubing wall loss (Rosine et al., 2008)

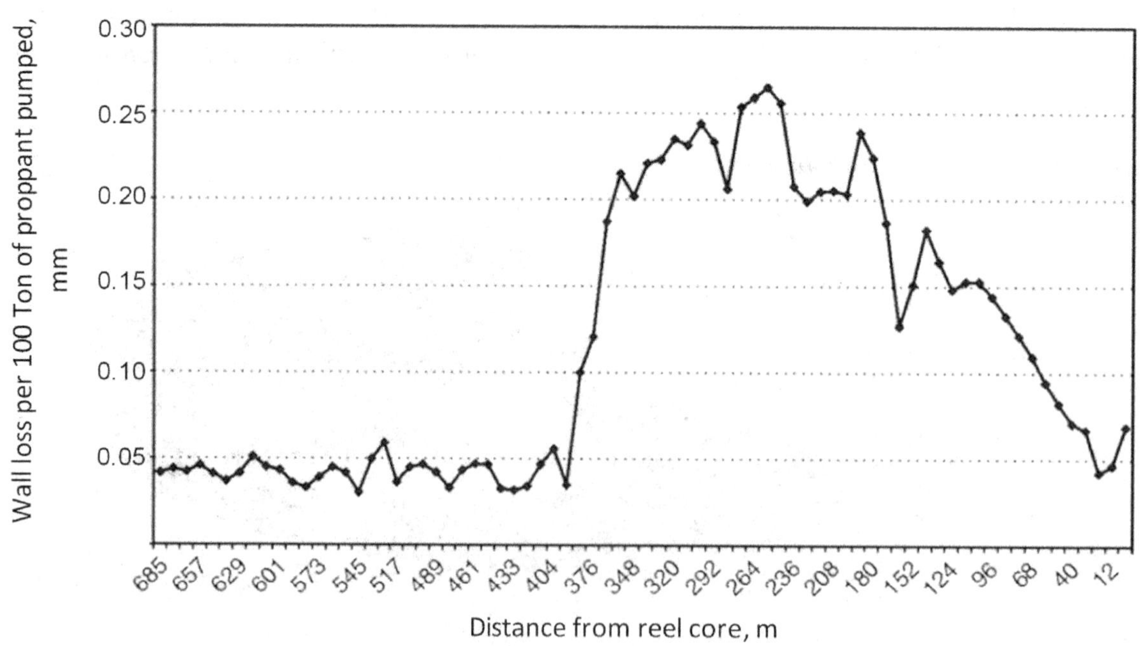

Mechanical Damage

Coiled tubing is used in rough and dynamic environment that renders it delicate to mechanical damage. As Crabtree (2008) and Padron et al. (2007) state, mechanical damage of coiled tubing is quite a big problem in the oilfield (accounts for 29% of total failure causes). It includes running in or out damage of the tubing body and surface equipment damage (reel, gooseneck or other). This type of damage can affect the coiled tubing fatigue life by creating concentration of stresses and strains or by causing micro cracking leading to premature fatigue failures. Coiled tubing can be repaired by cutting out the damaged tubing section and replacing it with a butt-weld (welding technique used to connect parts which are nearly parallel and do not overlap) or by removing the damage by grinding and polishing to a smooth surface finish. Fatigue failure process can be divided into five stages (Padron et al., 2007): a) Cyclic plastic deformation prior to fatigue crack initiation, b) Initiation of micro cracks, c) Propagation of micro cracks to form more of them, d) Propagation of one or more micro cracks and e) Final failure.

One of the ways to reduce the likelihood of coiled tubing mechanical damage failures and improve reliability of the operations is to understand and estimate the fatigue life. Experimental work on coiled tubing has to be done for that matter, and that implies cycle fatigue test methods and full scale bending tests (Sisak & Crawford, 1994). The test results depend on the damage geometry- depth, width, length and shape (ellipsoidal, cylindrical or conical); tubing characteristics (diameter, wall thickness and grade) and operating parameters (internal pressure and bending radius).

Operating and Manufacturing Errors

Manufacturing defects are not very common in coiled tubing installations but are possible to occur (Figure 1 – 15% of total failures). Detected defects are either due to a problem with the steel manufacturing or the tube manufacturing processes (Crabtree, 2008). Welds to make a tube shaped pipe (seam or bias welds) can be poorly performed. Seam welding problems dominate by inclusions in the weld and lack of fusion. Inclusions are usually a result of the slitting process where the master coil is cut into the widths required for the strips later formed into coiled tubing.

Operating the coiled tubing unit is not an easy job to perform handling the tubing run in and out. Tubing overloading and thus buckling and parting is sometimes a result of wrong operating practices or carelessness. Such problems are solved by a frequent personnel education and stimulation. Operating errors account for about 9% of total failures.

Well Control

Wellbore integrity during the coiled tubing operations has to be established, no matter dead or live (pressurized) well. Typical well control equipment scheme for coiled tubing operations consists of stripper (stuffing box) which provides primary seal between wellbore fluids under high pressure and the surface under atmospheric pressure during the coiled tubing operations and BOPs. Several stripper designs are available today: *1) Side door stripper*, *2) Radial stripper* (applies a radial force directly to the stripper element) or *3) Inaccessible sealing elements stripper* with coiled tubing in place (early model).

Blowout preventers in a role of well barriers prevent wellbore fluids from escaping to the surface. Like the conventional BOPs, coiled tubing BOPs are rated according to maximum expected wellhead pressure during the operational phase plus the safety factor. BOP stacks used are mainly rated to $344.74 \cdot 10^5$ Pa (5,000 psi), $689.48 \cdot 10^5$ Pa (10,000 psi) and $1{,}034.21 \cdot 10^5$ Pa (15,000 psi) incorporating blind, shear, slip and pipe ram preventer types (Sas-Jaworsky et al., 2006). Each BOP serves as a wellbore barrier for high pressure confinement if needed closing around the pipe (pipe and slip rams) or full wellhead inner diameter profile (blind and shear rams). High pressure causing fluid blowout can appear either from the tubing string or tubing/casing annulus. Downhole safety valves like back pressure valve, flapper check valve or ball valve are used to control the wellbore from the tubing side. On the other hand, BOP's control the annulus or a full wellhead profile (CTES, 2005).

Like any other operation, job planning is an important aspect of wellbore operations executed by coiled tubing. Although a pre-job safety meetings and discussions are very useful and can minimize risk of blowout or other hazards and accidents by analyzing operational steps in detail, some failures cannot be avoided by applying a thorough safety toolbox talk. Leaks due to string corrosion, poor connection sealing or valve malfunctions, cement break-downs, collapsed coiled tubing string, well fluid return system failure, circulating system or pumping failure- they can all appear by virtue of reasons shown in Figure 1.

According to NORSOK standard D-010 (2004) there is a well barrier acceptance criterion that has to be followed to safely perform coiled tubing operations and minimize technical risk occurrence. Steps to carry out the aforesaid are:

- Well control equipment is in place, which consists of CT stripper, BOP stack and CT safety head;
- Metal-to-metal sealing should be applied to all connections from the production tree to CT BOP stack using manually and remotely operated valves;
- Pressure rated line should be connected to the kill inlet of the surface well control equipment;
- If subsurface safety valve is leaking, coiled tubing safety head should be installed and tested prior to rigging up the well control equipment;
- Wellbore and surface tanks fluid level should be monitored at all times;
- If the bottom hole assembly (BHA)that cannot be cut is used, a special break out joint use is mandatory;
- In offshore environment shear/seal ram on the lower riser package (LRP) should be the upper closure device in the secondary well barrier when running the coiled tubing in completed well. Also, the same ram inside the drilling BOP stack should be the upper closure device in the secondary well barrier whenever a drilling BOP stack is installed when running the coiled tubing in the well;
- A two well barrier situation is maintained if the coiled tubing is damaged above the stripper and it is verified that there is no influx of well fluids in the CT string nor the check valves are leaking.

SNUBBING OPERATIONS

Snubbing unit is a special type of hydraulic workover intervention unit operating in live well conditions where the fluid column is not able to balance the formation pressure (Figure 3). It allows tubular to be run with specialized handling and sealing equipment under the pressure. Each pipe is made up or broken up while running in or pulling out of the well. To be able to do that dual slip system is required- stationary and travelling slips. Both slip systems are operated sequentially to grip the pipes while they are snubbed into the well. The stroke unit is simply mounted on top of the x-mass tree with BOP stack and stripper below the handling equipment. Stripper has sealing elements to contain wellhead pressure while running in or pulling out the string. To run the snubbing unit a hydraulic power unit is required. A diesel engine drives main pumps for the jacks and an auxiliary pump for the BOPs, equalizing and bleed valves, slips and the winch (Leggett et al. 1991, Perrin et al. 1999).

Safety Considerations

Safety wise, during the pipes snubbing process into a live well, they are compressed due to the wellbore pressure, so the risk of pipe buckling is present. To avoid the possibility of buckling appearance, the length of unguided pipe (from stripper to the travelling slips) has to be limited to critical buckling length. It greatly depends on the pipe outer diameter and wellhead pressure. Critical moment in such an operation is at the beginning of the running in phase. Force acting from below the pipe end is the greatest when there are no other pipes below to confront the wellhead pressure force. Some safety margins on maximum allowable pipe diameter and corresponding length of the unguided pipe if the wellhead pressure is about 10 MPa, were given by Perrin et al. (1999). Such a pressure can be handled by 48.3 mm (1.9 in) OD pipe with about 4 m

Figure 3. Snubbing unit

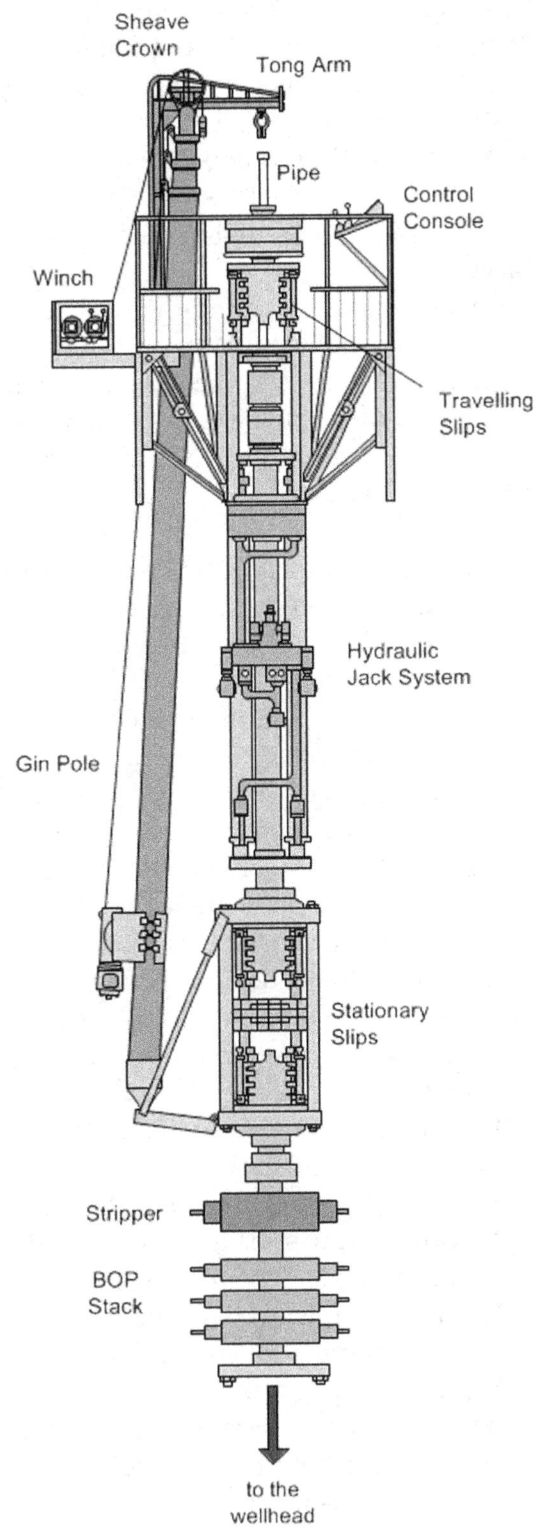

of unguided pipe length or by 88.9 mm (3 ½ in) OD pipe with about 7 m of the mentioned length. An increase in length of the unguided part of the pipe may impose severe equipment damage or even worse. Stripper rubber sealing elements are designed to last more than 3 000 m of string run in without any problems. After that round trip a rubber change might be necessary.

To maintain safety on the rig, during the snubbing operations well barriers have to be established as well. Some guidelines and requirements on safely performed snubbing operations are presented by NORSOK standard D-010 (2004):

- Well control equipment should consist of a stripper, dual stripper sealing elements (rams), snubbing BOP stack and a safety head;
- All connections from the production tree to the top of the snubbing unit should be flanged or clamped with metal-to-metal sealing;
- Valve inlets and outlets should have flanged or clamped connections. The inner valve (shall not be used as a working valve) should be bi-directional with metal-to-metal seal in the bore and one of the two valves should be remotely operated;
- Pressure rated line should be connected to kill inlet of the well;
- Safety head should be installed and pressure tested prior to rigging up the well control equipment;
- Fluid level in the well should be monitored at all times when deploying long BHA in an open well;
- If the bottom hole assembly (BHA) that cannot be cut is used, a special break out joint use is mandatory;
- Tool and components that cannot be cut by safety head should be identified;
- Wireline conveyed bridge plug for setting inside the pipe being snubbed should be available;
- At least two pump down plugs should be available on site;
- The length from the swab valve to the upper stripper ram should be normally the well barrier limit when defining the lubrication length.

There is a wide array of failure types with regard to snubbing operations and each of them should be taken to consider when planning the operation execution. Power unit shut down and loss of hydraulic circuits often first comes to mind when operating a unit depending on it. Prevention of this kind of problem lies in a good maintenance and engineering solutions. Further on, a slip bowl or stripper rubber failure may occur if the operating hours on stripper and slips have exceeded the tolerable limits. An occasional slip elements and stripper rubber change is required as mentioned previously. Leakage from the BOP rams is possible if they are damaged or inadequately designed by means of installation of wrong pressure rated BOP. Possibility of unintentional closing of shear rams on the BOP stack exists if the handles or buttons on the BOP control panel are not secured or accidentally engaged. This is a good example of a human error accident type. External and inside leaks during the wireline jobs through the snubbing unit, or a choke system failure, can also appear and are matters of well control system failure. Workstring inside blowout, intentional and unintentional dropping of the workstring or its shearing greatly impact the wellbore integrity which is very much endangered when operating in high pressure snubbing operations.

WIRELINE OPERATIONS

In the oilfield wireline mainly refers to operations in producing or injection wells running a steel cable, a slick line, and setting or retrieving wellbore tools. Wireline jobs can be classified into three main groups (Perrin et al., 1999): a) Checking and cleaning the downhole tubular or the bottom hole (tubular calibration, corrosion detection, clogging, sediments top detection and other), b) Downhole measurements (recordings of bottom hole temperature and pressure, fluid sampling, interfaces location, production logging and other) and c) Running or retrieving wellbore tools (setting and pulling subsurface safety valves, choke valves, two- or one-way-check valves, gas lift valves, shifting circulating sleeves, fishing, perforating and other). To be able to perform such operations, besides the cable line and tools meant for run in, surface logistics and wellhead equipment is needed as well. Standard surface equipment consists of a unit with the winch and a drum where the wire is spooled up. The whole system is driven by an engine. Stuffing box and the lubricator (a pipe that allows tools to be run into the well under pressure) are situated above the wellhead with bleed off valve located on the lubricator used for pressure release. Figure 4 shows a typical wireline installation with all belonging parts.

During the wireline operations like deployment of various electrical and mechanical downhole tools on electrical cable, braided cable or slickline, well integrity is maintained through the primary and secondary well barriers. Through the well life time different barrier systems exist in wireline operations. So, primary barriers most often include ce-

Figure 4. Wireline installation

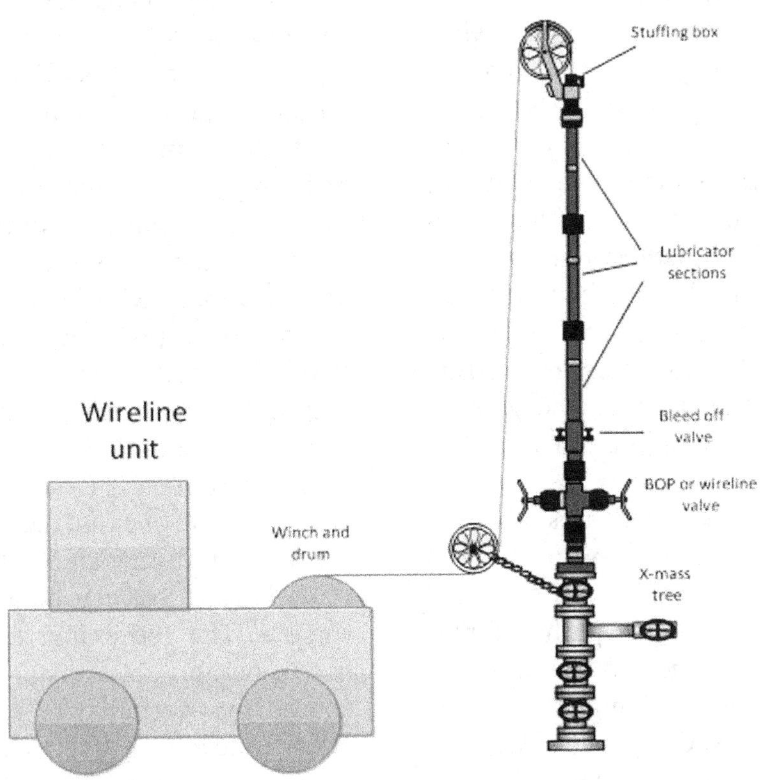

ment, casing, production packer, completion string, tubing hanger, production tree, BOP, lubricator or stuffing box, while the secondary barriers mainly include wireline safety head, wellhead and sometimes already mentioned primary barriers.

As wireline operations are frequently performed during workovers, drilling and production, blowouts may occur in such events. According to NORSOK standard D-010 (2004) there is a wide array of incident scenarios that may happen in wireline operations. They include loss of air supply during the operation execution; loss of winch power or mechanical failure; external leaks in lubricator, BOP stack or below the safety head; broken cable blown out of the wellbore; leak in the stuffing box, BOP cable ram, surface production tree or lubricator bleed off valve; external leaks in landing string; emergency situations on the rig; fluid influx during the wireline logging etc. The *most frequent* operations and activities in which blowouts occasionally occur are wireline running in and out activities. Main causes for such events are primary and secondary barrier loss due to wireline lubricator failure (wellbore seal off failure), subsurface safety valve failure, X-mass tree failure or running the wireline without the lubricator (Holand, 1997). Wireline blowouts are typically released through the tubing with release point above the X-mass tree. The most critical release points are the lubricator and stuffing box, so these elements should be given a particular care including pre-job tests and proper certification.

Besides a typical blowout risk, there is a cable break down risk as well. Normal wireline operations consider uniform tool movement applying risk free pulling forces within allowed boundaries. When an unexpected situation occurs requiring extra tension to be applied (tool sticking and dragging) problems appear when the whole system is underdesigned. To prevent the before said an extensive simulation of possible scenarios has to be run for appropriate cable, winch and drum selection in particular operational conditions (wellbore depth, anticipated drags and maximum tension values). High tension operations planning is extensive and sometimes time consuming. The wireline conveyance has to be properly selected, maintained and operationally evaluated risk wise with defined prevention and mitigation actions (Sarian et al., 2013).

PRODUCTION STRING CHANGE OPERATIONS

Installation and retrieval of the tubing string from the wellbore should be a routinely performed job without significant problems during the run in. But, there are some problems that may occur on the well which needs a tubing string change (Heriot-Watt, 2000): a) Mechanical failure of the tubing string- collapse or a leak, b) Inability to conduct tubing wireline equipment replacement through the string due to internal diameter limitations, c) String replacement due to an alternative design to allow the installation of other options or d) String replacement for reservoir performance optimization. To retrieve the string a certain pull has to be applied for packer deactivation. If a straight pull does not work, various tubing cutting techniques and tools can be used (mechanical cutters, chemical cutters, jet cutters, plasma cutters etc.). If the whole string cannot be retrieved in one run, the retrieval operation can be partial by means of retrieving segments of the string that can be accessed from the surface. The retrieving, fishing equipment has to be able to withstand tensile stresses caused by overpulls for production packer release and to face tubing head pressure at the surface in live wells. A suitable unit for live well operations is a snubbing unit mentioned in one of the previous subchapters. Reasons for the production string failures are almost the same like in case of coiled tubing.

RISK ASSESSMENT

It is very well known fact that the oilfield business investments bring uncertain outcomes. Wellbore hardware malfunctions may cause additional expenses to appear endangering oilfield projects economically. Unfortunately, at the rig site accidents of any type used to happen very often which has considerably lessened nowadays. The main reason for such reduction is a huge well history database, rigor safety precautions, better quality of the materials involved in wellbore tools and surface hardware make, personnel education etc. Assessing the outcomes and assigning probabilities of occurrence and associated values is the way to deal with risk management. Risk quantification is the end result all the uncertainty methods try to form. Modeling and quantification of uncertainty is a prelude to risk mitigation which focuses on unacceptable ranges of possibility. Generally speaking, risk mitigation is a kind of protection from unwanted situations (Lake & Warner, 2006). Therefore, different risk assessment methods and approaches are involved to simplify the operational phases execution and make it more safe. Risk analysis is usually compiled by using several different models and risk input data as both the models and the data are incomplete. When performing one of the well-known risk assessment approaches like *quantified risk analysis* (QRA), there are three important aspects required to ensure a satisfactory quality of the analysis (Holand, 1997):

- A thorough understanding of the system and its operational characteristics,
- The proper use of risk analysis methods, and
- Risk input data of high quality.

Risk and decision analysis software is as diverse as the analysis methods doing *Monte Carlo simulations* and *decision trees*. In order to quantify the fault trees, input data is required for the related base events. It includes activity frequencies, equipment failure probabilities, conditional probabilities for base events of a more operational nature and well characteristics (Edmondson & Hide, 1996).

After Corneliussen (2006) main well risk factors in the operational phase are divided in two main groups: a) Frequency factors and b) Consequence factors. Frequency factors include well barrier design as number of well barriers present in a certain operational phase; well barrier system availability factors (functionality, reliability, detection and repair of failures) and external hazards like weather and climate, human errors, fires and explosions. They are very much influenced by decisions made on well system level in the design phase. Consequence factors are mainly controlled by decisions made on installation level or by reservoir properties and include installation type (surface, subsea etc.), installation design (subsea wellheads, templates, safety functions, living quarters placement etc.), installation activity type (manned or unmanned), fluids leakage type (hydrocarbons, water or other), fluids leakage characteristics (leak rates, quantity, leak duration), fluids leakage exit points (casing, formation, wellhead, x-mass tree or other).

According to before said, in workover operations the main focus is on the two main safety functions of the well: a) To prevent uncontrolled leakage of well fluids from the well to the environment (this function is usually referred to as *well integrity* and is a *continuous* safety function that may fail at any instant of time) and b) To shut in the well flow in case of a dangerous incident on the downstream side of the x-mas tree. The well shut-in function is an on demand function activated in a random critical situation. Some basic steps in related risk assessment are:

- **Acquire a Risk Assessment Model:**
  - Set of algorithms or rules that use available information and data relationship to measure the levels of risk.
- **Collect and Prepare Data:**
  - Processes that result in data sets that are ready to be read into and used by risk assessment model.
- **Devise and Implement a Segmentation Strategy:**
  - It is important to segment the system into sections with constant risk characteristics.
- **Assess the Risks:**
  - After a risk model has been selected and the data have been prepared, risks can be assessed. This is a process of applying the rules to the collected data.
- **Manage the Risks:**
  - This step consists of determining what actions are appropriate given the risk assessment results.

FUTURE RESEARCH DIRECTIONS

As for the future research, most probably it will go in direction of new risk analysis software development advancing history database with much more important data to deal with. Software input variables should be more certain and direct by means of accurately determining well characteristics, failure probabilities and activity frequencies. It will be achieved if the oil business companies continue to cooperate and share the valuable experience.

REFERENCES

Al-Ashhab, J., Afzal, M., & Emenike, C. O. (2004). *Well integrity management system (WIMS)*. Paper presented at the International Petroleum Exhibition and Conference. Abu Dhabi, UAE.

Baker Hughes. (2003). *Coiled tubing solutions- Solve downhole problems with reliable, cost-effective technology*. Houston, TX: Author.

Bellarby, J. (2009). Well completion design. *Developments in Petroleum Science, 56*.

Corneliussen, K. (2006). *Well safety- risk control in the operational phase of offshore wells*. (Doctoral Thesis). Department of Production and Quality Engineering, The Norwegian University of Science and Technology, Trondheim, Norway.

Corneliussen, K., Sorli, F., Haga, H. B., Tenold, E., Menezes, C., Grimbert, B., & Owren, K. (2007). *Well integrity management system (WIMS)- A systematic way of describing the actual and historic integrity status of operational wells*. Paper presented at the Annual Technical Conference and Exhibition. Anaheim, CA.

Crabtree, A. R. (2008). *CT-failure monitoring: A decade of experience*. Paper presented at the Coiled Tubing and Well Intervention Conference and Exhibition. Dallas, TX.

CTES. (2005). *Coiled tubing manual*. Conroe, TX: National Oilwell Varco-CTES.

Dykhno, L. A., Hatton, G. J., & Nisbet, W. J. (2005). *Bullheading optimization for hydrate management in deepwater wells*. Paper presented at the Offshore Technology Conference. Houston, TX.

Edmondson, J. N., & Hide, D. K. (1996). *Risk assessment of hydrocarbon releases during workover and wireline operations on completed wells on offshore platforms*. Paper presented at the European Petroleum Conference. Milan, Italy.

Ghalambor, A., Ali, S. A., & Norman, W. D. (Eds.). (2009). *Frac packing handbook*. Richardson, TX: Society of Petroleum Engineers.

Halliburton- Boots & Coots. (2012). *Hydraulic workover services- pressure control*. Online free purchase manual.

Heriot-Watt University. (2000). *Production technology I*. Edinburgh, UK: Author.

Holand, P. (1997). *Offshore blowouts- Causes and control*. Houston, TX: Gulf Publishing Company.

Jahn, F., Cook, M., & Graham, M. (2008). *Hydrocarbon exploration and production* (2nd ed.). Amsterdam, The Netherlands: Elsevier.

Lake, L. W., & Warner, H. R. Jr., (Eds.). (2006). *Petroleum engineering handbook, emerging and peripheral technologies*. Richardson, TX: Society of Petroleum Engineering.

Leggett, R. B., Griffith, C. A., & Wesson, H. R., Jr. (1991). *Snubbing unit applications in potentially high-rate gas wells: A case study of the Anschutz Ranch East Unit, Summit County, Utah*. Paper presented at the 66th Annual Technical Conference and Exhibition of the SPE. Dallas, TX.

Martin, J. R., & Van Arnam, W. D. (2003). *The development of a corrosion-resistant coiled tubing product*. Paper presented at the Coiled Tubing Conference. Houston, TX.

Metcalf, S., Delorey, J., & Allen, S. (2003). *Coiled tubing acid corrosion protection: Laboratory vs. field performance*. Paper presented at the Asia Pacific Oil and Gas Conference and Exhibition. Jakarta, Indonesia.

Nasr-El-Din, H. A., & Metcalf, A. S. (2004). *Workovers in sour environments: How do we avoid coiled tubing failures?* Paper presented at the 1st International Symposium on Oilfield Corrosion. Aberdeen, UK.

Nilsen, T., & Karlsen, H. K. (2008). *Risk based decision support for the planning of a challenging HPHT drilling operation*. Paper presented at the International Petroleum Technology Conference. Kuala Lumpur, Malaysia.

NORSOK Standard D-010. (2004). *Well integrity in drilling and well operations*. Lysaker, Norway: Standards Norway.

Oudeman, P., terAvest, D., Grodal, E. O., Asheim, H. A., & Meissner, R. J. H. (1994). *Bull heading to kill live gas wells*. Paper presented at the European Petroleum Conference. London, UK.

Padron, T., Luft, B., Kee, E., & Tipton, S. M. (2007). *Fatigue life of coiled tubing with external mechanical damage*. Paper presented at the Coiled Tubing and Well Intervention Conference and Exhibition. The Woodlands, TX.

Perrin, D., Caron, M., & Gaillot, G. (Eds.). (1999). Well completion and servicing- oil and gas development techniques. Paris, France: Ed.s Technip, Institut Français du Pétrole Publications.

Rispler, K., McNichol, J., Matiasz, K., & Rheinlander, M. (2001). *Using an artificial neural network to develop a wall-loss model for coiled tubing fracturing operations*. Paper presented at the Coiled Tubing Roundtable. Houston, TX.

Rosine, R. S., Blanco, I. L., & Bailey, M. (2008). *Comparison of computational fluid dynamics of erosion in coiled tubing to field and test data*. Paper presented at the Coiled Tubing and Well Intervention Conference and Exhibition. Woodlands, TX.

Sarian, S., Varkey, J., Protasov, V., & Turner, J. (2013). *Polymer-locked, crush-free wireline composite cables reduce tool sticking and HSE risk in emerging deepwater reservoirs*. Paper presented at the North Africa Technical Conference & Exhibition. Cairo, Egypt.

Sas-Jaworsky, A., Blount, C., & Tipton, S. M. (2006). Coiled –tubing well intervention and drilling operations. In *Petroleum engineering handbook: Drilling engineering* (pp. 687–742). Richardson, TX: Society of Petroleum Engineering.

Sisak, W. J., & Crawford, D. J. (1994). *The fatigue life of coiled tubing*. Paper presented at the International Association of Drilling Contractors/Society of Petroleum Engineers Drilling Conference. Dallas, TX.

Tarmoom, I., Bin Thabet, H., Samad, S., Chishti, K., Hussain, A., & Arafat, M. (2007). *A comprehensive approach to well-integrity management in Adma-Opco*. Paper presented at the Middle East Oil & Gas Show and Conference. Bahrein International Exhibition Centre, Kingdom of Bahrein.

ADDITIONAL READING

Al Zahrani, A. R., Al-Nasser, R. H., Collen, T. W., & Khade, S. (2009, May). *Case Study: First Successful Offshore ESP Project in Saudi Arabia*. Society of Petroleum Engineers paper No. 126066 presented at the Technical Symposium and Exhibition, Alkhobar, Saudi Arabia.

Aven, T. (2008). *Risk Analysis- Assessing Uncertainties Beyond Expected Values and Probabilities*. West Sussex, England, UK: John Wiley & Sons.

Berg Anderson, L., Aven, T., & Maglione, R. (1996, June). *On Risk Interpretation and the Levels of Detail in Quantitative Blowout Risk Modeling*. Society of Petroleum Engineers paper No. 35967 presented at the International Conference on Health, Safety & Environment, New Orleans, Louisiana, USA.

Busch, J. M., Policky, B. J., & Llewelyn, D. C. G. (1985). Subsurface Safety Valves: Safety Asset or Safety Liability? *Journal of Petroleum Technology*, (October): 1813–1818.

Cooper, S. R. (2008, October). *People and Parts – Transforming the Risk Drivers*. Socicty of Petroleum Engineers paper No. 116375 presented at the Asia Pacific Oil & Gas Conference and Exhibition, Perth, Australia.

Grant, R. (1994, June). *Eliminating Workover Risk Though the Use of Pumping Well Buildup Tests*. Petroleum Society of CIM & AOSTRA paper No. 94-55 presented at the 45th Annual Technical Meeting, Calgary, Canada.

Grassick, D. D., Kallos, P. S., Jardine, I. J. A., & Deegan, F. J. (1990). Risk Analysis of Single- and Dual-String Gas-Lift Completions. *Journal of Petroleum Engineering, November*. (pp. 1374-1419).

Kumamoto, H. (2007). *Satisfying Safety Goals by Probabilistic Risk Assessment*. London, England, UK: Springer-Verlag London Limited.

Lake, L. W. (Ed.), & Clegg, J. D. (Ed.) (2007). Petroleum Engineering Handbook. Volume IV, Production Operations Engineering. Richardson, Texas, USA: Society of Petroleum Engineering.

Landgren, K. M., Das, J. A., & Abraham, P. (2013, March). *Improving Operations Through Next Generation Analytics and Mobility*. Society of Petroleum Engineers paper No. 163715 presented at the Digital Energy Conference and Exhibition, Woodlands, Texas, USA.

Lang, K. (2006, April). *Innovations in Task-Based Assessments*. Society of Petroleum Engineers paper No. 98641 presented at the International Conference on Health, Safety and Environment in Oil and Gas Exploration and Production, Abu Dhabi, U.A.E.

Mavroudis, D. (2001). *Downhole Environmental Risks Associated With Drilling and Well Completion Practices in the Cooper/ Eromanga Basins. Report Book 2001/00009*. Australia: Department of Primary Industries and Resources South Australia.

Molnes, E., & Strand, G. O. (2000, October). *Application of a Completion Equipment Reliability Database in Decision Making*. Society of Petroleum Engineers paper No. 63112 presented at the Annual Technical Conference and Exhibition, Dallas, Texas, USA.

Munday, K., Trim, S., & Turna, E. P. (2006, September). *Want to Make Tree Operations Safer? Why Not Use the DHSV as a Barrier?* Society of Petroleum Engineers paper No. 96337 presented at the Annual Technical Conference and Exhibition, San Antonio, USA.

Murchie, S., Billingham, M., Evans, S., & Bautheas, M. (2012, March). *Risk Reduction and Efficiency Gains in Well Intervention Operations Through the Application of Real-Time Digital Slickline Services*. Society of Petroleum Engineers paper No. 154439 presented at the Coiled Tubing & Well Intervention Conference & Exhibition, Woodlands, Texas, USA.

Murtha, J. A. (2001). *Risk Analysis for the Oil Industry*. A Supplement to Hart's E&P.

Nilsen, T., Sandøy, M., Rommetveit, R., & Guarneri, A. (2001, March). *Risk-Based Well Control Planning: The Integration of Random and Known Quantities in a Computerized Risk Management Tool*. Society of Petroleum Engineers paper No. 68447 presented at the Coiled Tubing Roundtable, Houston, Texas, USA.

NORSOK standard D-002 (2000). *System Requirements Well Intervention Equipment*. Online free purchase, Norwegian Technology Center. Oslo, Norway.

Peball, U., & Sorin, D. (2011, February). *Addressing MAH (Major Accident Hazards) – From Assessment to Actions*. Society of Petroleum Engineers paper No. 140879 presented at the European Health, Safety and Environment Conference in Oil and Gas Exploration and Production, Vienna, Austria.

Pietersen, C. M., & Engelhard, W. F. J. M. (1991, November). *Offshore Safety Assessment*. Society of Petroleum Engineers paper No. 23508 presented at the First International Conference on Health, Safety and Environment, The Hague, The Netherlands.

Qureshi, A. R. (1993). Safety Case for High Hazard Potential Installations – The Way Forward. *Proceedings of the Third International Offshore and Polar Engineering Conference* (pp. 631-637). June. Singapore.

Rabia, H. (1985). *Oilwell Drilling Engineering- Principles & Practices*. Newcastle, UK: University of Newcastle upon Tyne, Graham & Trotman.

Rachmat, E., Suzanne, J. G., & Boibien, C. G. (1993). Light Workover. *Proceedings of the Society of Petroleum Engineers from Asia Pacific Oil & Gas Conference & Exhibition* (pp. 415-425). February. Singapore.

Rae, P., Di Lullo, G., & Ahmad, A. bte (2002, October). *Eliminating Environmental Risks in Well Construction and Workover*. Society of Petroleum Engineers paper No. 77812 presented at the Asia Pacific Oil and Gas Conference and Exhibition, Melbourne, Australia.

Rasmussen, J. (1997). Risk Management in a Dynamic Society: A Modelling Problem. [Great Britain: Elsevier.]. *Safety Science*, *27*(2/3), 183–213. doi:10.1016/S0925-7535(97)00052-0

Roald, T. H. (2000, June). *New Risk Approach Shows Significant Reduction in Oil Blowout Risk*. Society of Petroleum Engineers paper No. 61255 presented at the International Conference on health, Safety and the Environment in Oil and Gas Exploration and Production, Stavanger, Norway.

Shah, S. N., Jain, S., & Zhou, Y. (2004, March). *Coiled Tubing Erosion During Hydraulic Fracturing Slurry Flow*. Society of Petroleum Engineers paper No. 89479 presented at the Coiled Tubing Conference and Exhibition, Houston, Texas, USA.

Solli, T. E. (2011, October). *Workover/Well Intervention and Regulatory Challenges*. Society of Petroleum Engineers paper No. 22586 presented at the Offshore Technology Conference, Rio de Janeiro, Brazil.

Sonnier, B., & Sackash, M. (2008, October). *Intervening With Coiled Tubing Through the Tubing-Casing Annulus to Restore Well Control*. Society of Petroleum Engineers paper No. 113756 presented at the Coiled Tubing and Well Intervention Conference and Exhibition, Dallas, Texas, USA.

Vefring, E. H., Rommetveit, R., & Børge, E. (1991, June). *An Advanced Kick Simulator Operating in a User-Friendly X-Window System Environment*. Society of Petroleum Engineers paper No. 22314 presented at the Petroleum Computer Conference, Dallas, Texas, USA.

Wakama, M., Eta, E., & Adeniyi, A. (2004, August). *Well Integrity Management in Shell Nigeria*. Society of Petroleum Engineers paper No. 88934 presented at the Annual International Technical Conference and Exhibition, Abuja, Nigeria.

Wiggins, M. L., & Zhang, X. (1994). Using PC's and Monte Carlo Simulation to Assess Risk in Workover Evaluations. Society of Petroleum Engineers Computer Applications, June. pp. 19-23.

Woodyard, A. H. (1982). Risk Analysis of Well Completion Systems. *Journal of Petroleum Technology*, (April): 713–720.

Worth, D. J., Crepin, S., Alhanati, F., & Lastiwka, M. (2008, October). *Risk Assessment for SAGD Well Blowouts*. Socicty of Petroleum Engineers/Petroleum Society/Canadian Heavy Oil Association paper No. 117679 presented at the International Thermal Operations and Heavy Oil Symposium, Alberta, Canada.

KEY TERMS AND DEFINITIONS

Bullheading: Well kill method used when normal well killing with conventional circulation does not work. It is also a phrase for any kind of fluid force pumping downhole.

Coiled Tubing: Reeled flexible type of carbon steel, titanium or polymer tubing.

Corrosion: A process of gradual destruction of metals by chemical reaction with its environment.

Risk Assessment: Determination of quantitative and qualitative value of risk related to certain situations in operational phases.

Snubbing Unit: A type of hydraulic workover intervention unit operating in live well conditions where the fluid column is not able to balance the formation pressure.

Well Integrity: Application of technical, operational and organizational solutions to reduce risk of uncontrolled release of formation fluids throughout the life cycle of a well.

Wireline: Specific operations in producing or injection wells running a steel cable, a slick line, and setting or retrieving wellbore tools.

Workover: Any kind of well intervention or remedial treatment on oil or gas well involving different methods and techniques common in drilling and completion department.

Chapter 11

Gathering Systems and Processing Facilities Risk Analysis

Svijetlana Dubovski
INA Plc, Croatia

ABSTRACT

Gathering system is defined as one or more segments of pipeline, usually interconnected to form a network that transports oil and natural gas from the production wells to one or more production facilities, gas processing plant, storage facility, or a shipping point. There are two types of pipeline networks: radial and trunk system. Produced well fluids are often complex mixtures of the liquid hydrocarbons, gas, and some impurities that can have detrimental effects on the integrity of the gathering pipelines. It is necessary to eliminate most of the impurities before oil and natural gas can be stored and sold. Complexity of the processing facility depends on the treated fluid composition. Environmental impacts during the oil and gas transportation and processing phase will cause long-term habitat changes. To minimize that, it is very important to implement appropriate activities across the designing, construction, operational, and decommissioning phases.

INTRODUCTION

The evolution of today's gathering systems and surface facilities actually began several thousand years ago. Seepages of asphqaltic bitumen in Mesopotamia around 3000 B.C. were the raw material for the "petroleum" industry that flourished for about three thousand years. It primarily produced mastic and caulk used in construction. The production system was crude and involved merely the recovery of hydrocarbon in jars and casks. Surface facilities for the production at the time involved a crude type of distillation (JPT, 1999).

DOI: 10.4018/978-1-4666-4777-0.ch011

Today, dramatic changes in the range of new technologies available ensure a new environmental and safety approach. Awareness of the importance of environmental issues has become increasingly vital to our society. Designer's understanding of all processes, safety and environmental tasks of newly developed technologies is essential. Now, as well as in the future, the process and equipment choices will have to be made before all aspects are definitely known. Depending on the specific application, it may make more sense to use an older, proven technology instead of a newer one because the application does not require a cutting-edge solution.

A prescriptive approach to safety and environmental management of gathering system and surface facilities relies on a selection of specific regulatory and technical requirements developed for all of the phases from design to decommissioning. The owners must follow existing standards, practices, guidelines, and procedures. The owner of installations must be completely knowledgeable in the regulations of each country in which it operates and is responsible only for complying with individual regulatory requirements. Complex working situations occur on offshore installations, possibly creating dangerous conditions that must be examined carefully and organized in the best way possible.

This Chapter provides basic information regarding the gathering and processing systems and potential environmental impacts of these systems. To avoid or minimize environmental impact of the gathering systems and surface facilities it is very important to implement appropriate activities from designing phase, construction, operational and decommissioning phase which will be also described in this Chapter.

BACKGROUND

Created in 1972, the United Nations Environment Programme (UNEP) addresses environmental issues at the global and regional level. UNEP provides leadership and encourages partnership in caring for the environment by inspiring, informing, and enabling nations and peoples to improve their quality of life without compromising the quality of lives for the future generations.

The E&P Forum is the international association of oil companies and petroleum industry organizations formed in 1974. It was established to represent its members' interests at the specialist agencies of the United Nations, governmental and other international bodies concerned with regulating the exploration and production of oil and gas. While maintaining this activity, the Forum now concerns itself with all aspects of E&P operations, with particular emphasis on safety of personnel and protection of the environment, and seeks to establish industry positions with regard to such matters. At present the Forum has almost 60 members worldwide, the majority being oil and gas companies operating in 60 different countries, but with a number of national oil industry associations/institutes.

Environmental Management in Oil and Gas Exploration and Production, developed by Oil Industry International Exploration and Production Forum and UNEP, provide an overview of environmental issues in the oil and gas exploration and production industry, and of the best approaches to achieving high environmental performance in all parts of the world. Their Technical Report Environmental management in oil and gas exploration and production provides an overview of the environmental issues and the technical and management approaches to achieving high environmental performance in the activities necessary for oil and gas exploration and production in the world.

This Chapter provides guidance on the oil and gas gathering system and surface facilities and their potential environmental impacts. Because of the possible long-term habitat changes, as it will be described in this section, it is very important to implement appropriate activities from designing phase, construction, operational and decommissioning phase. Also, this Chapter proposes practical elements and operational techniques which should be applied on site to minimize the enviromental impact of the gathering and processing system.

MAIN CHARASTERISTICS OF GATHERING SYSTEM AND SURFACE FACILITIES

It is very difficult to classify oil and gas surface production facilities by type, because they differ due to production rates, fluid properties, sale and disposal requirements, location, and operator preference (www.cdphe.state.co.us/ap/onglinks.asp, 2013). Regardless, many parts of the process are quite similar in principle. We find the wellheads and they feed into production and test manifolds called the gathering system. While there are oil or gas only installations, more often the well-stream will consist of a full range of hydrocarbons. With this well flow we will also get a variety of unwanted components such as water, carbon dioxide, salts, sulfur and sand. The purpose of the surface facilities is to process the well flow into clean marketable products: oil, natural gas or condensates. Plants also include a number of utility systems, not part of the actual process, but providing energy, water, air or some other utility to the plant. This section provides an overview of gathering system and main surface facilities.

Manifolds and Flow Lines

The gathering system begins at the wellhead, which should include at least one choke, unless the well is on an artificial lift. On offshore facilities and on shore high-risk facilities, wel-head should contain an automatic shutdown valve. In all cases, block valves are needed so that the maintenance can be performed on the choke in the event of a long flow line. There are two types of the of pipeline networks; radial and trunk system. The radial type brings all the flow lines to a central header, while the trunk-line type uses several remote headers to collect fluid.

Commingled flows from two or more wells in a central facility require an installation of a manifold to allow flow from any well to be produced into any of the bulk or test production systems. A production riser (offshore) or gathering line (onshore) brings the well flow into the manifolds (Figure 1). Check valves allow each well to be routed into one or more of several manifold lines. There will be at least one for each process train plus additional manifolds for test and balancing purposes. The test manifold allows one or more wells to be routed to the test separator. Since there is only one process train, the high pressure and low pressure manifolds allow groups of high pressure and low pressure wells to be taken to the first and second stage separators, respectively. The chokes are set to reduce the wellhead flow and pressure to the desired high pressure and low pressure pressures separators (www.ipt.ntnu.no/~jsg/undervisning, 2013).

Separators

Phase separation of the produced fluids is primary processing operation. A separator is a horizontal, vertical (Figure 2) or spherical vessel in which a mixture of produced fluids is separated. The main principles used to

Figure 1. Manifold (part of author's private collection of INA Plc's facility)

Figure 2. Vertical separator (part of author's private collection of INA Plc's facility)

achieve physical separation of gas and liquids are gravity settling and coalescing. A well stream separator must perform the following (Stewart&Arnold, 1998a):

- Cause a primary phase separation of the liquid hydrocarbon from those that are gas;
- Refine the primary separation by removing most of the entrained liquid mist from the gas;
- Further refine the separation by removing the entrained gas from the liquid;
- Discharge the separated gas and liquid from the vessel.

Because of the multi component nature of the produced fluid, the higher the pressure at which the initial separation occurs, the more liquid will be obtained in the separator. This liquid contains some light components that vaporize in the stock tank downstream of the separator. If the pressure for initial separation is too high, too many light components will stay in the liquid phase at the separator and will be lost to the gas phase at the tank. If the pressure is too low, not as many of these light components will be stabilized into liquid at the separator and they will be lost to the gas phase. The tendency of any component in the process stream to flash to the vapor phase depends on its partial pressure. Thus, if the pressure in the vessel is high, the partial pressure for the component will be relatively high and the molecules of that component will tend toward the liquid phase. As the separator pressure is increased, the liquid flow rate out of the separator increases. The problem with this is that many of these molecules are the lighter hydrocarbons (methane, ethane, and propane), which have a strong tendency to flash to the gas state at stock tank conditions (atmospheric pressure). In the stock tank, the presence of these large numbers of molecules creates a low partial pressure for the intermediate range hydrocarbons (butanes, pentane, and heptanes) whose flashing tendency at stock tank conditions is very susceptible to small changes in partial pressure. Thus, by keeping the lighter molecules in the feed to the stock tank capture of a small amount of them as liquids is enabled, but many more of the intermediate range molecules are lost to the gas phase.

3-Phase Separation

Wells with a high content of gas have a high pressure. This pressure is reduced in high pressure separators, and mid or low pressure separators. This process is a continuous one, where at the front of the separator the liquid/gas is constantly added. The gas is released at the top and the oil-water mixture is separated theough the use of gravity and chemicals. The oil-water mixture contains emulsion, which causes problems during the process. However, the problem of the emulsion is solved by the utilization of chemicals. As the oil is lighter than the water, it rises. The water proceeds downwards at the same time. Once the liquid reaches the end of the separator, the liquids remaining- the oil and water- are discharged separately. Although theoretically, this process is percieved as simple, it remains complex in reality. Addition of a high number of chemicals, as well as the occasional heat utilization to speed up the separation make this process complex.

2-Phase Separation for Liquid

A 2-phase separation operates identically as the 3-phase separation, only lacking in gas which has to be removed. Either the 2-phase separation is a continuous process as described above, or it is done batch-wise. In a batch-wise separation the crude is brought in very big tanks where it is stored for a defined amount of time. During this time gravity separates the oil, the emulsion and the water. The water- being the heaviest- sits at the bottom while the oil- being the lightest- is at the top. Emulsion is found in the middle as it is a mixture of oil and water, and its weight therefore is between that of water and oil.

After some time, normally hours, the crude has separated and is pumped to further processes. The storage time in the tank is known by experience, because the crude will not separate any further, even when stored for a longer time. The issues at this stage are bigger than they initially seem. The quality of the delivered oil might fluctuate, depending on the situation at the wells. The water content is sometime higher or lower resulting in the addition of chemicals in order to speed up the process as well as to reach a better separation with purer components in a shorter time.

Elektrostatic Separation

Electrostatic separation is used as well in order to separate crude with lower percentage of water content. Inside the separator is a mesh of wires where an electrostatic field is introduced to the liquid. This high voltage electrical field speeds up the separation and produces water and oil of purer quality. Yet, there is a drawback with the machines. The electrical field has to be adjusted to the quality and the composure of the incoming crude. If the crude is changing fast, impurities will settle at the mesh and the electrical field will collapse. This leads to a shutdown of the process, as it requires cleaning before restarting the process. High amount of energy required to sustain the electrical field as well as a lot of chemicals are needed in the electrostatic separator.

Well Testing

It is necessary to keep track of the gas, oil, and water production from each well to be able to manage the reserves properly, evaluate where further reserve potential may be found, and diagnose well problems as quickly as possible. Proper allocation of income also requires knowledge of daily production rates as the royalty or working interest ownership may be different for each well. In simple facilities that contain only a few wells, it is attractive to route each well to its own separator and/or treater and measure its gas oil, and water production on a continuous basis. In facilities that handle production from many wells, it is sometimes more convenient to enable each well to flow through the manifold to one or more test separators.

Test separators are used to separate the well flow from one or more wells for analysis and detailed flow measurement (Figure 3). In this way, the behavior of each well under different pressure flow conditions can be defined. This

Figure 3. Test separator (part of author's private collection of INA Plc's facility)

normally takes place when the well is taken into production and later at regular intervals- typically one to two months- and it measures the total and component flow rates under different production conditions. Undesirable consequences such as slugging or sand can also be determined. The separated components are analyzed in the laboratory to determine hydrocarbon composition of the gas, oil and condensate. Test separators can also be used to produce fuel gas for power generation when the main process is not running. Alternatively, a three phase flow meter could be used.

Oil Treating

Most oil treating on offshore facilities is done in vertical or horizontal treaters. A gas blanket is provided to assure that there is always enough pressure in the treater so the water will flow to the water treating. At onshore locations the oil may be treated in a big settling tank. All tanks should have a pressure/vacuum valve with flame arrester and gas blanket to keep a positive pressure on the system and exclude oxygen. This helps prevent corrosion, eliminate a potential safety hazard, as well as conserve some of the hydrocarbon vapors. A pressure in the tank lifts a weighted disk or pallet, which allows the gas to escape. If there is a vacuum in the tank because the gas blanket failed to maintain a slight positive pressure, the greater ambient pressure lifts another disk, which allows air to enter. Although we wish to exclude air, it is preferable to allow a small controlled volume into the tank rather than allow the tank to collapse. Flame arrester contains tubes that prevent the vent flame from traveling back into the tank. Flame arresters have a tendency to gather excess paraffin resulting in the blockage and thus must be installed where they can be inspected and maintained. Since they can get blocked, a separate relieving device must always be installed. The oil is skimmed off the surface of the barrel and the water exits from the bottom either through a water leg or an interface controller and a dump valve. Since the volume of the liquid is fixed by the oil outlet, barrels cannot be used as surge tanks. Flow from the treater goes to a surge tank from which it either flows into a barge or truck or is pumped into a pipeline.

After the third stage separator, the oil can go to a coalescer for final water extraction. In this unit, the water content can be reduced to below 0.1%. The coalescer is completely filled with liquid: water at the bottom and oil on top. Internal electrodes form an electric field, breaking the surface bonds between conductive water thus isolating oil in the oil-water emulsion. The coalescer field plates are generally steel, sometimes covered with dielectric material to prevent short-circuits.

If the separated oil contains unacceptable amounts of salts, they can be removed in an electrostatic desalter. The salts, which may be sodium, calcium or magnesium chlorides come from the reservoir water and are also dissolved in the oil. The desalter will be placed after the first or second stage separator depending on gas/oil ratio and water cut.

Custody Transfer Oil Measurement

In large facilities oil is typically sold through a measurement unit, which is designed to meet API Standards and whatever additional measuring and sampling standards are required by the crude purchaser. The value received for the crude will typically depend on its gravity, sediment and water content, and volume. Therefore, the measurement unit must not only measure the volume accurately, but must continuously monitor the sediment and water content and take a sufficiently representative sample so that the gravity, the sediment and water content can be measured. The crude first flows through a strainer/gas eliminator to protect the meter and to assure that there is no gas in the liquid. An automatic sediment and water content probe is mounted in a vertical run. When measured value exceeds the sales contract quality, this probe automatically actuates the diverter valve, which blocks the liquid from going further in the measurement unit and sends it back to the process for further treatment. Sampler in a vertical run, situated downstream of the diverter, takes a calibrated sample that is proportional to the flow and delivers it to a sample container. The sampler receives a signal from the meter to assure that the sample size is always proportional to flow even if the flow varies. The sample container has a mixing pump so that the liquid in the container can be mixed and made homogeneous prior to taking a sample of this fluid. The liquid then flows through a positive displacement meter. Most sales contracts require the meter to be proven at least once a month and a new meter factor calculated. On large installations, a meter prover is included as a permanent part of the measurement unit or is brought to the location when a meter must be proven. The meter prover contains a known volume between two detector switches. A spheroid pig moves back and forth between the detectors as the four-way valve is automatically switched. The volume recorded by the meter during the time the pig moves between detectors for a set number of traverses of the prover is recorded electrically and compared to the known volume of the meter prover. On smaller installations, a master meter that has been calibrated using a prover may be brought to the location to run in series with the meter to be proven. In many onshore locations a truck-mounted meter prover is used. The sales meter must have a proven repeatability.

Pumps are normally needed to move oil through the measurement unit and deliver it at pressure to a pipeline downstream of the unit. Pumps are sometimes used in water treating and disposal processes. In addition, many small pumps may be required for pumping skimmed oil to higher pressure vessels for treating, glycol heat medium and cooling water service and firefighting.

Water Treating

On an installation where the water cut is high, there will be a huge amount of water produced. Often this water contains sand particles bound to the oil/water emulsion. The environmental regulations in most countries are quite strict regarding the limits of oil in water discharged in the environment. It also places limits on other forms of contaminants.

Water from the separators and coalescers first goes to a sand cyclone, which removes most of the sand. The sand is washed further before it is discharged. The water then goes to a hydro cyclone, a centrifugal separator that will remove oil drops. The hydro cyclone creates a standing vortex where oil collects in the middle and water is forced to the side. Finally the water is collected in the water degassing drum. Dispersed gas will slowly rise to the surface and pull remaining oil droplets to the surface by flotation. The surface oil film is drained, and the produced water can be discharged to the sea. Recovered oil in the water treatment system is typically recycled to the third stage separator.

Compressors

Gas enters the first-stage suction scrubber. Any liquids that may have come through the line are separated at this point and the gas flows to the first stage. Compression heats the gas. Due to the heating, there is a cooler after each compression stage. At the higher pressure, more liquids may separate, so the gas enters another scrubber before being compressed and cooled again. The compressor is equipped with an automatic suction shut-in valve on each inlet and a discharge shut-in valve so that when the unit shuts down, or when an abnormal condition is detected, the shut-in valves actuate to isolate the unit from any new sources of gas. Many operators prefer, and in some cases regulations require, an automatic blow down valve installation so that- as well as in the isolating the unit- all the gas contained within the unit is vented safely at a remote location. Several types of compressors are used for gas compression, each with different characteristics such as operating power, speed, pressure and volume (Devold, 2009; Nooijen H. & Clinton, 2012):

- Reciprocating compressors which use a piston and cylinder design with 2-2 cylinders are built up to about 30 MW power, around 500-1800 rpm (lower for higher power) with pressure up to 5 MPa (500 bars). Used for lower capacity gas compression and high reservoir pressure gas injection.
- Screw compressors are manufactured up to several MW, synchronous speed (3000/3600 rpm) and pressure up to about 2.5 MPa (25 bar). Two counter-rotating screws with matching profiles provide positive displacement and a wide operating range. Typical use is natural gas gathering.
- Axial blade and fin type compressors with up to 15 wheels provide high volumes at relatively low pressure differential (discharge pressure 3-5 times inlet pressure), speeds of 5000-8000 rpm, and inlet flows up to 200000 m^3/hour. Applications include air compressors and cooling compression in LNG plants.
- Larger oil and gas installations use centrifugal compressors with 3-10 radial wheels, 6000-20000 rpm (highest for small size), up to 80 MW load at discharge pressure of up to 50bars and inlet volumes of up to 500,000 m^3/hour. Pressure differential up to 10.

Most compressors will not cover the full pressure range efficiently. The lowest pressure is atmospheric, for gas to pipeline, some 3 to 5 MPa (30-50 bar) pressure is used, while reservoir reinjection of gas will typically require 20 MPa (200 bar) and upwards since there is no liquid in the tubing and the full reservoir pressure must be overcome. Therefore compression is divided into several stages to improve maintenance and availability. Due to single unit power limitations, compression is often divided in several parallel trains.

Compressors are driven by gas turbines or electrical motors (for lower power also reciprocating engines, steam turbines are sometimes used if thermal energy is available). Often several stages in the same train are driven by the same motor or turbine.

The main operating parameters for a compressor are the flow and pressure differentials. The product defines the total loading, so there is a ceiling set by the maximum design power. Furthermore, there is a maximum differential pressure and choke flow, the maximum flow that can be achieved. At lower flow, there is a minimum pressure differential and flow before the compressor will "surge" if there is not enough gas to operate. If variations in flow are expected or difference between common shaft compressors will occur, the situation will be handled with recirculation. A high flow, high pressure differential surge control valve will open to let gas from the discharge side back into the suction side. Since this gas is heated, it will also pass through the heat exchanger and it will be scrubber not to become overheated by circulation.

The operating characteristics are defined by the manufacturer. The objective of compressor performance control is to keep the operating point close to the optimal set point without violating the constraints, by means of control outputs, such as the speed setting. However, gas turbine speed control response is relatively slow and even electric motors are not fast enough, since surge response must be in the 100 ms range. Anti surge control will protect the compressor from going into surge by operating the surge control valve. The basic strategy is to use distance between operating point and surge line to control the valve with a slower response time starting at the surge control line. Crossing the surge trip line will cause a fast response opening of the surge valve to protect the compressor. Operation with recirculation wastes energy (which could result in unnecessary emissions) and produces wear and tear, particularly on the surge valve. Each vendor supplies several variants of compressor control and anti surge control to optimize performance, based on various corrective and predictive algorithms.

If the compressor is turbine driven, a dedicated speed governor handles the fuel valves and other controls on the turbine to maintain efficiency and control rotational speed. For electrical motors, this function is handled by a variable speed drive.

The final function around the compressor itself is lube and seal oil handling. Most compressors have wet seals, which serve as traps around shafts where the oil at high pressure prevents gas from leaking out to the atmosphere or other parts of the equipment. Oil is used for lubrication of the high speed bearings. This oil gradually absorbs gas under pressure and may become contaminated, so it needs to be filtered and degassed.

Gas Dehydrators

Natural gas contain water and should be dehydrated in dehydration unit (Figure 4) to a

Figure 4. Gas dehydrators (part of author's private collection of INA Plc's facility)

controlled water content to avoid hydration as well as to minimize the corrosion problems. Unless gas is dehydrated, liquid water may condense in pipelines and accumulate at low points along the line, reducing its flow capacity. Several methods have been developed to dehydrate gases on an industrial scale.

The three major methods of dehydration are direct cooling, adsorption and absorption (Stewart & Arnold, 1998b).

Molecular sieves, silica gel, and bauxite are the desiccants used in adsorption processes. In absorption processes, the most frequently used desiccants are diethylene and triethylene glycols. Usually, the absorption/stripping cycle is used for removing large amounts of water, and adsorption is used for cryogenic systems to reach low moisture contents.

Direct Cooling

The saturated vapor content of natural gas decreases with increased pressure or decreased temperature. Thus, hot gas saturated with water may be partially dehydrated by direct cooling. Gas subjected to compression is normally "after cooled", and this cooling may remove water from the gas. The cooling process must reduce the temperature to the lowest value that the gas will encounter at the prevailing pressure to prevent further condensation of water.

Absorption of Water in Glycols

Absorption dehydration involves the use of a liquid desiccant to remove water vapor from the gas. Although many liquids possess the ability to absorb water from gas, the liquid that is most desirable to use for commercial dehydration purposes are the glycols, particularly ethylene glycol, diethylene glycol, tri ethylene glycol, and tetra ethylene glycol . Water and the glycols show complete mutual solubility in the liquid phase due to hydrogen-oxygen bonds, and their water vapor pressures are very low.

This is the most common type of gas dehydration system, where bubble-cap glycol contact tower is used. Wet gas enters the base of the tower and flows upward through the bubble caps. Dry glycol enters the top of the tower and because of the down-comer wier on the edge of each tray, flows across the tray and down to the next. There are typically six to eight trays in most applications. The bubble caps assure that the upward flowing gas is dispersed into small bubbles to maximize its contact area with the glycol. Before entering the contactor, the dry glycol is cooled by the outlet gas to minimize vapor losses when it enters the tower. The wet glycol leaves from the base of the tower and flows to the reconcentrator (reboiler) by way of heat exchangers, a gas separator, and filters. In the reboiler the glycol is heated to a sufficiently high temperature to drive off the water as steam. The dry glycol is then pumped back to the contact tower. Most glycol dehydrators use tri ethylene glycol, which can be heated from 444 K to 478 K (340 °F to 400 °F) in the reconcentrator and work with gas temperatures up to 322K (120 °F). Tetra ethylene glycol is more expensive but it can handle hotter gas without high losses and can be heated in the reconcentrator from 478 K to 494 K (400 °F to 430 °F).

Adsorption of Water by a Solid

Adsorption dehydration is the process where a solid desiccant is used for the removal of water vapor from a gas stream. The solid desiccants commonly used for gas dehydration are those that can be regenerated and, consequently, used over several adsorption/desorption cycles. The mechanisms of adsorption on a surface exist in two types; physical and chemical. Chemical adsorbents find a very limited application in the gas processing. Adsorbents that allow physical adsorption hold the adsorbate on their surface by the surface forces. For physical adsorbents, solids used in gas dehydration are bauxite ore, consisting primarily of alumina, silica gels and silica-alumina gels, and molecular sieves. Activated carbon, a widely used adsorbent, possesses no capacity for water adsorption and is therefore not used for dehydration purposes, though it may be used for the removal of certain impurities. Bauxite use is also minimal because it contains iron and is thus unsuitable for sour gases.

Scrubbers and Reboilers

The separated gas may contain mist and other liquid droplets. Drops of water and hydrocarbons also form when the gas is cooled in the heat exchanger, and must be removed before it reaches the compressor. If liquid droplets enter the compressor they will erode the fast rotating blades. A scrubber is designed to remove small fractions of liquid from the gas.

There are various types of gas-drying equipment available, but the most common suction (compressor) scrubber is based on dehydration by absorption in tri ethylene glycol. The scrubber consists of many levels of glycol layers.

A large number of gas traps force the gas to bubble up through each glycol layer as it flows from the bottom up each section to the top. Processed glycol is pumped in at the top from the holding tank. It flows from level to level against the gas flow as it spills over the edge of each trap. During this process it absorbs liquids from the gas and comes out as rich glycol at the bottom. The holding tank also functions as a heat exchanger for liquid to and from the reboilers.

The glycol is recycled by removing the absorbed liquid. This is done in the reboiler, which is filled with rich glycol and heated to boil out the liquids at the temperature of about 400 K to 450 K (260-350 °F) for a number of hours. Usually there is a distillation column on the gas vent to further improve separation of glycol and other hydrocarbons. For higher capacity, there are often two reboilers which alternate between heating rich glycol and draining recycled processed glycol.

On a standalone unit, the heat is supplied from a burner that uses covered vaporized hydrocarbons. In other designs, heating will be a combination of hot cooling substances from other parts of the process and electric heaters, recycling the hydrocarbon liquids to the third stage separator.

Heat Exchangers

For the compressor to operate efficiently, gas temperature should be low. The lower the temperature, the less energy will be used to compress the gas for the given final pressure and temperature. However both gas from separators and compressed gas are relatively hot. When gas is compressed, it must remain in thermodynamic balance so this ends up as a temperature increase.

Heat exchangers of various forms are used to cool the gas. Plate heat exchangers consist of a number of plates where the gas and cooling medium pass between alternating plates in opposing directions. Tube and shell exchangers place tubes inside a shell filled with of cooling fluid. The cooling fluid is often pure water with corrosion inhibitors.

When designing the process, it is important to plan the thermal energy balance. Heat should be conserved e.g. by using the cooling fluid from the gas train to reheat oil in the oil train. Excess heat is dispersed e.g. by seawater cooling. However hot seawater is extremely corrosive, so materials with high resistance to corrosion, such as titanium must be used.

Gas Sweetening

Many natural gases contain hydrogen sulfide (H_2S) in concentration ranging from barely detectable quantities to over 30 mole percent. Gases containing H_2S or CO_2 are classified as sour, and gases free from H_2S and CO_2 are called sweet. With increasing demands to natural gas, natural gases containing H_2S are also being tapped for utilization after purification. Natural gas that is transported to the market must meet legal requirements, which specify a maximum H_2S content less than 4ppm in the gas. These requirements are justified, since H_2S is a toxic gas, and its combustion product is sulfur dioxide or trioxide. Besides emitting a bad odor at low concentrations, H_2S is deadly poisonous and it can be fatal in just three to five minutes at concentrations above 600 ppm. Its toxicity is comparable to cyanide. Thus, it cannot be tolerated in the gas being used as a domestic fuel. Further, H_2S is corrosive to all metals normally associated with gas transporting, processing and handling systems, and may lead to premature failure of most such systems.

The removal of H_2S from natural gas is accompanied by the removal of CO_2 and COS if present, since these have similar acid characteristics. Like dehydration processes, desulfurization processes are primarily of two types: adsorption on a solid (dry process), and absorption into a liquid (wet process). Both the adsorption and absorption processes may be of the physical or chemical type. These processes may also be classified into the following categories:

- Non-regenerative, which means that the materials used in treating the gas are not recovered in these processes.
- Regenerative processes with recovery as H_2S. These include the physical absorption processes, the amine processes, molecular sieves, etc.
- Regenerative processes with recovery as elemental sulfur. With growing environmental concerns regarding sulfur emission, these processes have acquired a prominent role in desulfurization operations.

Mercury Removal

Mercury removal is generally based on molecular sieves. A molecular sieve is a substance containing a material with tiny pores to achieve a large surface area, such as activated carbon. The surface of the material allows certain molecules to bind by surface tension. The molecules can later be extracted and the sieve material regenerated by heating, pressure and/or purging with a carrier gas. A molecular sieve is commonly cyclic with one active unit and one (or more) units in regeneration.

Nitrogen Rejection

Excessive nitrogen is removed by cryogenic distillation and higher concentrations are removed by absorption with lean oil or another special solvent if a smaller fraction is detected. (See acid gas removal for both principles). Cryogenic removal also permits production of helium, if present, as a valuable by-product.

NGL Recovery and Treatment

Remaining NGL is recovered from the gas stream in most modern plants by a cryogenic turbo expander based process, followed by a fractionating process. This process leads the cooled NGLs though distillation columns called de-ethanizer, de-propanizer and de-butanizer, to extract ethane, propane and butane respectively and leave a residual stream of pentane and higher hydrocarbons.

The final step is to remove mercaptans (smelly organic gases e.g. CH_3SH) if present, in a sweetening process, based on molecular sieves adsorption orcatalytic oxidization such as Merox or Sulfrex, where the main difference is in the use of a different type of catalyst.

LNG Liquefaction

LNG is a gas transport product. The gas, primarily methane (CH_4), is converted to a liquid form for the ease of storage or transport. It is produced close to the production facilities in a LNG liquefaction plant, stored, transported in cryogenic tanks on an LNG carrier and delivered to an LNG regasification terminal for storage and delivery to a pipeline system.

LNG processes are generally patented by large engineering, oil and gas companies, but are generally based on a two or three stage cooling process. The actual design varies considerably with the different processes. The most critical component is the heat exchanger, also called the cold box, which is designed for optimum cooling efficiency. Designs may use separate cold boxes, or two stages may combine into one complex common heat exchanger.

Most processes use a mixed refrigerant design. The reason is that the gas has a heat load to temperature curve that must be closely matched to improve stability and efficiency. The curve tends to show three distinct regions, matching the pre-cooling, liquefaction and sub-coiling stages. The refrigerant gas composition will vary based on the individual design, as will the power requirement of each stage. It is often a patented, location-specific combination of one or two main components and several smaller ones, together with careful selection of the compressed pressure and expanded pressure of the refrigerant, to match the LNG gas stream.

The pre-cooling stage cools the gas to a temperature of about 243 K to 223 K (-22 °F to -58 °F) in the pre-cooling cold box. The cooling element is generally propane, or a mixture of propane, ethane and small quantities of other gases. The pre-cooling cold box also cools the cooling medium for the liquefaction and subcooling stage. The liquefaction process takes the gas down from 243K (-22 °F) to about 173 K to 148 K(-148 °F to -193 °F) typically based on a mixture of methane, ethane and other gases. It cools the LNG stream as well as the refrigerant for the final stage. Sub-cooling brings the gas to final stable LNG state at around 111K (-260 °F). The refrigerant is usually methane and/or nitrogen.

Custody Transfer Gas Measurement

Partners, authorities and customers all calculate invoices, taxes and payments based on the actual product that is shipped out. Often custody transfer also takes place at this point, which means transfer of responsibility or title from the producer to a customer, shuttle tanker operator or pipeline operator. Although some small installations are still operated with a dipstick and manual records, larger installations have analysis and metering equipment. To make sure readings are accurate, a fixed or movable prover loop for calibration is also installed.

Custody transfer gas measurement is similar to the oil measurement, but analyzers will measure hydrocarbon content and energy value as well as the pressure and temperature. The meters are normally turbine and orifice meters or ultrasonic meters. Orifice plates with a diameter less than the pipe are mounted in cassettes. The pressure differential over the orifice plate as well as the pressure and temperature are used in standard formulas (such as AGA 3 and ISO 5024/5167) to calculate normalized flow. Orifice plates are sensitive to a buildup of residue and affect the edges of the hole. Larger new installations therefore prefer turbine and ultrasonic gas meters. Gas metering is less accurate than the liquid metering, typically $\pm 1.0\%$ of mass.

LNG is often metered with mass flow meters that can operate at the required low temperature. At various points in the movement of oil and gas, similar measurements are taken, usually in a more simplified way.

Storage

On most production sites, the oil and gas is piped directly to a refinery or tanker terminal. Gas is difficult to store locally, but occasionally underground mines, caverns or salt deposits can be used to store gas. On platforms without a pipeline, oil is stored on the board storage tanks to be transported by shuttle tanker. The oil is stored in storage cells around the shafts on concrete platforms, and in tanks on floating units. On some floaters, a separate storage tanker is used. Ballast handling is very important in both cases to balance the buoyancy when the oil volume varies. Onshore, fixed roof tanks are used for crude oil storage and floating roof for condensate. Rock caves are also used for storage.

Special tank gauging systems such as level radars, pressure or float are used to measure the level in storage tanks, cells and caves. The level measurement is converted to volume via tank strapping tables (depending on tank geometry) and compensated for temperature to provide standard volume. Float gauges can also calculate density, therefore the mass can be established.

Storage LNG at the terminals and on LNG carriers is done in cryogenic tanks at atmospheric pressure or slightly above, up to 125 kPa (1.25 bar). The tanks are insulated, but will not keep LNG cold enough to avoid evaporation. Heat leakage will heat and boil off the LNG. Therefore LNG is stored as a boiling cryogen, which means that the liquid is stored at its boiling point for its storage pressure (atmospheric pressure) i.e. about 111K (-260 °F). As the vapor boils off, heat from the vaporization is absorbed, cooling the remaining liquid. The effect is called auto-refrigeration. With efficient insulation, only a relatively small amount of boil-off is necessary to maintain a steady temperature. Boil-off gas from land based LNG storage tanks is compressed and fed to natural gas pipeline networks. On LNG carriers the boil-off gas can be used for fuel. At the receiving terminal, the LNG is stored in local cryogenic tanks. It is re-gasified to ambient temperature on demand, commonly in the sea water heat exchanger, and then injected into the gas pipeline system.

Marine Loading

Loading systems consist of one or more loading arms/jetties, pumps, valves and a metering system. Tanker loading systems are complex, both because of the volume involved, and because several loading arms will normally interact with the tanker's ballast system to control the loading operation. The tanks must be filled in a certain sequence; otherwise the tanker's structure might be damaged due to uneven stresses. It is the responsibility of the tanker's ballast system to signal data to the loading system and to operate the different valves, while monitoring the tanks on board the ship.

Pipeline Terminal

The gas pipeline is fed from the high pressure compressors. Oil pipelines are driven by separate booster pumps. For longer pipelines, intermediate compressor stations or pump stations will be required due to distance or crossing of the mountain ranges. Block valve stations are placed at regular intervals to limit the potential leakage volume in case of a pipe rupture.

The pipeline terminal includes termination systems for the pipeline. A pig launcher and receiver will be included, to allow insertion of a pipeline pigging device used to clean or inspect the pipeline on the inside. This is essentially a large chamber that can be pressurized and purged to insert and remove the pig or scraper without depressurizing the pipeline. The pig is often driven by pipeline flow.

Utility Systems

Utility systems do not handle the hydrocarbon process flow, but provide some service to the main process safety as well residents. Depending on the location of the installation, many such functions may be available from nearby infrastructure (e.g. electricity). But many remote installations must be fully self-sustaining and must generate their own power and water.

Process Control and SCADA

A process control system is used to monitor data and control equipment on the plant. Very small installations may use hydraulic or pneumatic control systems, but larger plants with up to 30,000 signals to and from the process require a dedicated distributed control system. The purpose of this system is to read values from a large number of sensors, run programs to monitor the process and control valves and switches in order control the process.

Supervisory control and data acquisition (SCADA) is normally associated with telemetry and wide area communications, for data gathering and control over large production sites, pipelines, or corporate data from multiple facilities. With telemetry, the band width is often quite low and based on telephone or local radio systems. SCADA systems are often optimized for efficient use of the available band width. Wide area communication operates with wideband services, such as optical fibers and broadband internet.

Emergency Shutdown and Process Shutdown

The process control system should control the process when it is operating within normal constrains such as level, pressure and temperature. The emergency shutdown and process shutdown systems will take action when the process goes into a malfunction or dangerous state. For this purpose the system maintains four sets of limits for a process value. Separate transmitters are provided for safety systems.

System requirements are set by official laws and regulations and industry standards such as IEC 61508/61511 which set certification requirements for process safety systems and set criteria for the safety integrity level of each loop.

Fire and Gas System

The fire and gas system is not generally related to any particular process. Instead it divides into fire areas by geographical location. Each fire area should be designed to be self-contained and it should detect fire and gas by several types of sensors, and control fire protection and fire fighting devices to contain and fight fire within the fire area. In the event of fire, the area will be partially shut-off by closing ventilation fire dampers. A fire area protection datasheet typically shows what detection exists for each fire area and what fire protection action should be taken in case of an incident.

Power Generation, Distribution and Drives

Power can be provided from mains power or from local gas turbines or diesel generator sets. Large facilities have high power demands, from 30 MW and up to several hundred MW. There is a tendency to generate electric power centrally and use electric drives for large equipment rather than multiple gas turbines, as this decreases maintenance and increases uptime.

The power generation system on a large facility usually consists of several gas turbines diving electric generators, 20-40 MW each. If exhaust heat is not needed in the main process, it can be used to drive exhaust steam turbines (dual cycle) for additional efficiency.

Flare and Atmospheric Ventilation

Flare subsystems include flare (Figure 5), atmospheric ventilation and blow down. The purpose of the flare and vent systems is to provide safe discharge and disposal of gases and liquids resulting from:

- Spill-off flaring from the product stabilization system ;
- Production testing;
- Relief of excess pressure caused by worsened conditions and thermal expansion;

Figure 5. Flare (part of author's private collection of INA Plc's facility)

- Depressurization either in response to an emergency situation or as part of a normal procedure;
- Planned depressurization of subsea production flow lines and export pipelines;
- Venting from equipment operating close to the atmospheric pressure.

The systems are typically divided into a high pressure flare and a low pressure flare systems. The LP system is operated a little above atmospheric pressure to prevent atmospheric gases such as oxygen flowing back into the vent and the flare system thus generating a combustible mixture. With low gas flow, inert gas is injected at the flare nozzle to prevent air ingress.

Traditionally, considerable amounts of hydrocarbons have been more or less continuously flared. In these cases, a continuously burning pilot is used to ensure ignition of hydrocarbons in the flare.

Stronger environmental focus has eliminated continuous flaring and the pilot in many areas. Vapors and flare gas are normally recovered, and only in exceptional situations does flaring occur. To avoid the pilot flame, an ignition system is used to ensure safe ignition even when large volumes are discharged. One patented solution is a "ballistic ignition" system which fires burning pellets into the flare gas flow (Devold, 2009).

Instrument Air

A large volume of compressed air is required for the control of pneumatic valves and actuators, tools and purging of cabinets. It is produced by electrically driven screw compressors and further treated to be free of particles, oil and water.

The Heat, Ventilation and Air Conditioning System

The heat, ventilation and air conditioning system feeds conditioned air to the equipment and accommodation rooms etc. Cooling and heating is achieved by water-cooled or water/steam-heated heat exchangers. Heat may also be taken from gas turbine exhaust. In tropic and sub-tropic areas, cooling is achieved by compressor refrigeration units. In tropical areas, gas turbine inlet air must be cooled to achieve sufficient efficiency and performance. The system is usually delivered as one package, and may also include air emissions cleaning. Some subsystems include:

- **Cool:** Cooling medium, refrigeration system, freezing system.
- **Heat:** Heat medium system, hot oil system.

One function is to provide air to equipment rooms that are secured by positive pressure. This prevents potential influx of explosive gases in case of a leak.

Also water system, chemicals system and telecommunication systems play an important role in laying the foundations for a remote operation, diagnostics and maintenance in gathering system functionality.

Main Characteristics of Offshore Gathering Systems

Many methods are currently used to gather, separate, store and transport production from offshore areas, but process units are more or less similar to previously described units used onshore. Production platforms provide facilities for the separation of formation fluids into oil, gas, and water. Depending on the project, the platform may only be used for production as drilling can be conducted from a separate drilling rig brought alongside. Some platforms are only used to bring the hydrocarbons to surface and directly export them for processing, whilst some gas platforms may be unmanned during routine production operations. Typically, multiple wells are drilled from the platform location using directional drilling techniques. In some cases, where field extremities are unreachable by directional drilling from the fixed location or where small reservoirs exist, subsea production units are installed on the seabed following drilling and the produced hydrocarbons are tied into a nearby platform facility by a system of risers. Following development drilling and well completion in readiness for the flow of formation fluids, a "Christmas tree", which allows the control of flow to the surface, is placed onto the well head. The oil and / or gas are produced by separation of the formation fluid mixture into oil, gas and water, or gas and condensates at the platform.

Oil is exported from the platform by either pumping into a sub-sea pipeline to shore, to a floating storage unit offshore, or directly to a tanker. Typically, gas is exported through a pipeline. Most fields produce in a predictable pattern, called a decline curve, where production increases relatively rapidly to a peak, and then follows a long slow decline. Water or gas injection is often used to maintain reservoir pressure and enhance production.

There are many types of offshore platforms, as:

- **Fixed Platforms:** Used in water depths of up to around 500 m and consisting of steel or concrete legs (jacket) secured directly to the seabed by steel piles that support a steel deck (Figure 6). Drilling Equipment, production facilities and accommodation are typically housed on the deck.
- **Compliant Towers:** Used in water depths ranging from around 500 m up to 1,000 m and consisting of a narrow, flexible tower on a piled foundation supporting a conventional deck.
- **Tension Leg Platforms:** Used in water depths of up to about 2,000 m and consists of a floating facility moored to the seabed and fixed in place by anchors. Mini tension leg platforms exist that are used in water depths of between 200 m and 1,000 m.

Figure 6. Fixed platform (part of author's private collection of INA Plc's facility)

- **Jack-Up Platforms:** Used in shallower water up to around 100m and transported to location where the legs are lowered by hydraulic jacks into position to support the deck.
- **Spar Platforms:** Used in water depths between 500 m and 1,700 m and consisting of a cylindrical hull supporting a floating platform.
- **Floating Production Systems:** Ships equipped with processing facilities and moored on location with a series of anchors

Even though there are many types of platforms, the gathering systems are similar. Individual wells flow full stream by submarine flow lines to one of flow stations. Each flow station is equipped with necessary manifolds, production and test separators and pumping equipment. Gas is separated from the crude oil at the flow station. The gas is flared and the crude oil is pumped through submarine gathering lines to the central gathering platform, where the crude oil is metered and pumped to onshore facilities through a transmission pipeline. Onshore, final separation is completed and the oil is stored in a large tank farm. The crude oil producing and handling system is highly automated and remotely controlled from shore. Crude oil, after metering can be pumped through a pipeline and hose to a permanently anchored storage hulk. From storage hulk crude oil can be pumped through a submarine line to an export tanker anchored seaward of the hulk.

When operating near land, the crude oil is transported full stream to shore separation facilities instead of being separated offshore. Production from several individual wells and/or small multi well platforms can be flowed full stream through gathering lines to a large platform. From that point the crude will flow full stream to onshore separation and storage facilities by pipeline. Also, the production from multi well platforms can be flowed to onshore separation facilities through a separate pipeline.

Permanent drilling-production platforms are used to drill wells and handle production. The crude oil, after separation on the platform, is produced into minimum storage facilities and piped to shore, towards the treating and metering facilities. The gas that is not used is either piped to the shore after treatment and compression, or flared offshore.

Old offshore fields produce more water than oil. The rising water cut increases the force needed to lift a production stream holding ever less oil, as the pressure in the reservoir declines. Ultimately the natural force isn't enough to lift the water and oil from deepwater fields, which results in the termination of production if nothing is done. There are two ways to do that: Increase the lift with pumps, or reduce the load by removing the water on the seafloor, disposing it there. Oil companies are turning to subsea processing for financially challenged projects, looking for ways to make them profitable (Rassenfoss, 2011).

GATHERING SYSTEMS AND PROCESSING FACILITIES RISK ANALYSIS

The main principles of gathering and the most often used processing units are previously described. The significance of their impact depends upon factors such as production quantities, the number and size of units, types of processing facilities, the amount of land disturbed by activities and occupied by the facilities, etc. One of the important factors is impact duration. Impacts on the environment can lead to changes in existing conditions; the impacts can be direct, indirect or cumulative. These changes can be found at different ecological levels, and can vary over space and time as they can be either positive or negative. Direct impacts refer to changes in environmental components that result from direct cause-effect consequences of interactions between the environment and gathering and processing activities. Indirect impacts result from cause-effect consequences of interactions between the environment and direct impacts. Cumulative impacts refer to the accumulation of changes to the environment caused by the activities. Environmental impacts that could occur during the gathering and processing would mostly occur from construction and installation activities, daily activities, maintenance and replacement of facility components and also decommissioning and abandonment activities. Potential impacts from these onshore and offshore activities are presented below, by the environmental impact sources or type of affected resource.

Air Quality and Climate

Atmospheric and climate issues are attracting increasing interest from both owners and government authorities worldwide. In order to examine the potential impacts, it is important to understand the sources and nature of the emissions and their relative contribution to atmospheric impacts- both local and those related to global issues such as stratospheric ozone depletion and climate change. The primary sources of atmospheric emissions from oil and gas gathering and processing arise from:

- Flaring, venting and purging gases;
- Combustion processes such as diesel engines and gas turbines;
- Fugitive gases from loading operations and tankage and losses from process equipment;
- Airborne particulates from soil disturbance during construction and from vehicle traffic;
- Particulates from other burning sources, such as well testing.

On offshore facilities, support and supply vessels as well as helicopters are included.

The principal emission gases include carbon dioxide, carbon monoxide, methane, volatile organic carbons and nitrogen oxides. Emissions of sulphur dioxides and hydrogen sulphide can occur and depend upon the sulphur content of the hydrocarbon and diesel fuel, particularly when used as a power source. In some cases sulphur content can lead to odour near the facility. Ozone depleting substances are used in some fire protection systems, principally halon, and as refrigerants. Following substantial efforts by technology, unplanned emissions have been significantly reduced and alternative agents for existing and new developments have been engineered. The volumes of atmospheric emissions and their potential impact depend upon the nature of the process under consideration. Flaring of produced gas is the most significant source of air emissions, particularly where there is no infrastructure or market available for the gas. However, where viable, gas is processed and distributed as an important commodity. Thus, through integrated development and providing of the markets for all products, the need for flaring will be greatly reduced. Flaring may also occur on occasions as a safety measure, during start-up, maintenance or upset in the normal processing operation.

Flaring, venting and combustion are the primary sources of carbon dioxide emissions from production operations, but other gases should also be considered. Air pollution may cause health effects and reduce visibility.

Terrain and Soils

Potential impacts to soil arise from three basic sources:

- Physical disturbance as a result of construction;
- Contamination resulting from spillage and leakage;
- Solid waste disposal;
- Indirect impact arising from opening access and social change.

Potential impacts that may result from a poor design and construction include soil erosion due to soil structure, slope or rainfall. Left undisturbed and vegetated, soils will maintain their integrity, but, once vegetation is removed and soil is exposed, soil erosion may result. Alterations to soil conditions may result in widespread secondary impacts such as changes in surface hydrology and drainage patterns, increased siltation and habitat damage, reducing the capacity of the environment to support vegetation and wildlife. In addition to causing soil erosion and altered hydrology, the removal of vegetation may also lead to secondary ecological problems, particularly in situations where many of the nutrients in an area are held in vegetation; or where the few trees present are vital for wildlife browsing; or in areas where natural recovery is very slow. Clearing by operators may stimulate further removal of vegetation by the local population surrounding a development.

Due to its simplicity, burial or land-filling of wastes in pits at drilling and production sites has been a popular means of waste disposal in the past. Historically, pits have been used for burial of inert, non-recyclable materials and drilling solids; evaporation and storage of produced water, work over/completion fluids; emergency containment of produced fluids; and the disposal of stabilized wastes. However, the risks associated with pollutant migration pathways can damage soils and usable water resources (both surface and groundwater), if seepage and leaching are not contained.

Land farming and land spreading have also been extensively practiced in the past for the treatment of oily petroleum wastes, and water-based mud and cuttings. However, there are potential impacts where toxic concentrations of constituents may contaminate the soil or water resources, if an exposure pathway is present. In the case of mud and cuttings, the most important consideration is the potential for the waste to have a high salt content.

Land farming and land spreading, however, remain viable treatment options provided a proper assessment is made, and correct procedures are followed. Considerations include the site topography and hydrology, the physical and chemical composition of the waste and resultant waste/soil mixture. With proper assessment, engineering, design, operation and monitoring, land farming provides a cost effective and viable technique for waste disposal. Soil contamination may arise from spills and leakage of chemicals and oil, causing possible impact to both flora and fauna. Simple preventative techniques such as segregated and contained drainage systems for process areas incorporating sumps and oil traps, leak minimization and drip pans, should be incorporated into facility design and maintenance procedures. Such techniques will effectively remove any potential impact arising from small spills and leakage on site.

Routine impacts to soil during the production phase are also soil erosion impacts caused by vehicular traffic.

Impact of Produced Water and Water Resources

Produced water is the largest volume aqueous waste arising from processing operations, and some typical constituents may include in varying amounts inorganic salts, heavy metals, solids, production chemicals, hydrocarbons, benzene, PAHs, and on occasions naturally occurring radioactive material. The environmental impact of produced waters disposed to other receiving waters other than ocean is highly dependent on the quantity, the components, the receiving environment and its dispersion characteristics. The extent of the impact can only be judged through an environmental impact assessment. However, discharge to small streams and enclosed water bodies is likely to require special care. Produced water volumes vary considerably both with the type of gathering and processing and throughout the lifetime of a field. Frequently the water cut is low, early in the production life of a field. As time passes more water is produced from the reservoir and may increase to 80 per cent or more towards the end of field life. Other aqueous waste streams such as leakage and discharge of drainage waters may result in pollution of ground and surface waters. Impacts may be severe particularly where ground and surface waters are utilized for household purposes or where fisheries and ecologically important areas are affected.

Indirect or secondary effects on local drainage patterns and surface hydrology may result from poor construction practice in the development of roads and process sites.

Most produced water is unfit for domestic or agricultural. If it is disposed of by release to the surface without treatment, it can cause soil and surface water contamination. The majority of produced water is disposed of via injection in disposal wells or enhanced recovery wells. Groundwater could be impacted during the injection of produced water if the injection well casing integrity is compromised. Regulations govern the disposal of produced water and maintenance of injection wells. If the injection well is not properly completed or is poorly maintained over time, aquifers can be impacted by the injected produced water.

Other potential impacts to water availability and quality during the processing would include possible minor degradation of water quality resulting from vehicular traffic and machinery operations during maintenance or, if improperly conducted, application of chlorides in dust suppressants or herbicides for vegetation management. A spill or blowout could potentially cause extensive contamination of surface waters or a shallow aquifer.

Hazardous Materials and Waste Management

Industrial wastes are generated during routine operations. These wastes are typically placed in containers, characterized and labeled, possibly stored briefly, and transported by a licensed hauler to an appropriate permitted off-site disposal facility as a standard practice. Impacts could result if these wastes were not properly handled and were released to the environment. Environmental contamination could occur from accidental spills of herbicides or, more significantly, oil. Chemicals in open pits used to store wastes may pose a threat to wildlife and livestock.

Sand separated from produced water must be disposed of properly, and it is often contaminated with oil, trace amounts of metals, or other naturally occurring constituents.

During the processing, large volumes of scale and sludge wastes can accumulate inside pipelines and storage vessels. These wastes consist of precipitated sulfate and carbonate scales, produced sand, heavy hydrocarbons, tight emulsions, rust particles, other metals and chemicals, and salts. They must be removed periodically from the equipment for disposal. These wastes may be transported to offsite disposal facilities. In some instances, they may be disposed of via land spreading, a practice that entails spreading the wastes over the surface of the disposal area and mixing it with the top few inches of soil. This practice often is done specifically to treat the hydrocarbons present in these wastes through biodegradation. In some instances, the land spread area is periodically watered, worked over to remix the soils and increase aeration, and treated with bio nutrients to further enhance hydrocarbon degradation.

Produced water can become a significant waste stream. Regulations govern the disposal of this waste stream; the majority of it is disposed of by underground injection either in disposal wells or, in mature producting fields, in enhanced oil recovery wells (i.e., wells through which produced water and other materials are injected into a producing formation in order to increase formation pressure and production).

In some locations, produced water may carry naturally occurring radioactive materials to the surface. Proper management of naturally occurring radioactive materials bearing in produced water and solid wastes is critical to prevent both occupational and public human health risks and environmental contamination.

Typical non-hazardous and hazardous wastes routinely generated at offshore facilities include general office and packaging wastes, waste oils, oil contaminated rags, hydraulic fluids, used batteries, empty paint cans, waste chemicals and used chemical containers, used filters, fluorescent tubes, scrap metals, and medical waste, among others.

Noise

The main sources of noise during the gathering and processing would include compressor and pumping stations including occasional flaring and vehicle traffic. Compressor stations produce noise levels between 64 and 86 dBA at the station to between 58 and 75 dBA at a distance of about 1 mile (1.6 kilometers) from the station. Use of remote telemetry equipment would reduce daily traffic and associated noise levels within the oil and gas field area. The primary impacts from noise would be localized disturbance to wildlife, recreationists, and residents. Noise associated with cavitations is a major concern for landowners, livestock and wildlife.

Roads and Land Use

Lease roads are constructed and used to support various operations. The environmental impact of the construction of a roadway can have long lasting effects well beyond the limits of the right-of-way. Existing roads should be utilized, where feasible, to limit the extent of new road construction, when they meet regulatory standards, transportation and development needs, and safety and environmental objectives. When it is necessary to build new roadways, they should be developed in an environmentally acceptable manner consistent with landowner recommendations. Impacts to transportation daily traffic levels, particularly heavy truck traffic, would be expected to be lower during the production phase compared to the drilling/development phase. For the most part, heavy truck traffic would be limited to periodic visits to a well site for work over, formation treatment, and collection of produced product or water. The use of pipelines to convey product or produced water to centralized collection facilities or plants would reduce the volume of traffic during the production phase.

Land use impacts during the gathering and processing phase would be an extension of those that occurred during the drilling and development phase. Land use can revert to its original uses after decommissioning and abandonment activities.

Ecological Resources

Adverse impacts to ecological resources could occur from:

- Disturbance of wildlife from noise, extraneous light, changes in vegetation cover and human activity;
- Exposure of biota to contaminants;
- Mortality of biota from colliding with aboveground facilities or vehicles;
- Variations in water, air and soil/sediment quality;
- Ecological resources may continue to be affected by the reduction in habitat quality associated with habitat fragmentation due to the presence of facilities and access roads.

The presence of oil or gas gathering and processing could also interfere with migratory and other behaviors of some wildlife. Discharge of produced water inappropriately onto soil or into surface water bodies can result in salinity levels too high to sustain plant growth. One notable impact to wildlife is the potential for contact with petroleum-based products and other contaminants in reserve pits and water management facilities. Wildlife can become entrapped in the oil and drown, ingest toxic quantities of oil by preening (birds) or licking their fur (mammals); or succumb to cold stress if the oil damages the insulation provided by feathers or fur. Hyper saline water in evaporation ponds can also cause sodium toxicity. If controls are not managed effectively, ecological impacts may also arise from other direct anthropogenic influence such as fires, increased hunting and fishing and possibly poaching. In addition to changing animal habitat, it is important to consider how changes in the biological environment also affect local people and indigenous populations.

Visual Resources

Once production facilities are installed, portions of well pads, access roads, and pipeline rights-of-way that are not needed for production would be reclaimed; however, much of the disturbed area would continue to contrast with the natural form, line, color, and texture of the surrounding landscape. This would impact undisturbed vistas and areas of solitude. The aboveground portions of an oil or gas gathering and processing would be highly visible in rural or natural landscapes, many of which may have few other comparable structures. The artificial appearance of facilities may have visually incongruous "industrial" associations for some, particularly in a predominantly natural landscape. Any nighttime lighting would be visible from long distances.

Health and Safety

Possible impacts to health and safety during production include accidental injury or death to workers and, to a lesser extent, the public. Health impacts could result from water contamination, dust and other air emissions, noise, soil contamination, and stress associated with living close to gathering and processing. Potential fires and explosions would cause safety hazards. Cavitations could ignite grass fires. Increased or reckless driving by oil or gas workers would also create safety hazards. In addition, health and safety issues include working in potential weather extremes and possible contact with natural hazards, such as uneven terrain and dangerous plants, animals, or insects.

In locations where naturally occurring radioactive materials bearing in produced water and solid wastes are generated, occupational and public health risks may occur if the wastes are not properly managed. In particular, workers employed in the cutting and reaming of oilfield pipe, removing solids from tanks and pits, and refurbishing gas processing equipment may be exposed to particles containing alpha-emitting radio nuclides that could pose a health risk if inhaled or ingested.

Methane seeps into drinking water wells and under homes has caused health hazards.

Plans for all gathering and processing operations should incorporate measures to deal with potential emergencies that threaten people, the environment or property. However, even with proper planning, design and the implementation of correct procedures and personnel training, incidents can occur such as:

- Spillage of fuel, oil, gas, chemicals and hazardous materials;
- Oil or gas well blowout;
- Explosions;
- Fires (facility and surrounds);
- Unplanned plant upset and shutdown events;
- Natural disasters and their implications on operations, for example flood, earthquake, lightning;
- War and sabotage.

Planning for emergency events should properly examine risk, size, nature and potential consequences of a variety of scenarios, including combination incidents. A variety of documents is available to describe risk and hazard assessment, contingency planning and effects of emergency events.

Socioeconomic Impact and Cultural Resources

Direct socioeconomic impacts would include the creation of new jobs and the associated royalties and taxes paid. Indirect impacts are those impacts that would occur as a result of the new economic development and would include new jobs at businesses that support the expanded workforce or that provide project materials, and associated taxes. Potential impacts on the value of residential properties located adjacent to an oil or gas field would continue during this phase. A well located in close proximity to a residence may have a small adverse effect on property values. This effect could increase as the number of wells increases. Landowners who own the mineral rights within an oil and gas field would receive royalties on any product removed from their property that could offset potential impacts to property values.

During the production phase, impacts to cultural resources could primarily occur from unauthorized collection of artifacts and from visual impacts. In the later case, the presence of the aboveground structures could impact cultural resources with an associated landscape component that contributes to their significance, such as a sacred landscape or historic trail. Damage to localities caused through off-highway vehicle use could also occur. The potential for indirect impacts (e.g., vandalism and unauthorized collecting) would be greater during the production phase compared to the drilling/development phase, due to the longer duration.

CONCLUSION

Although the oil and gas process is indeed complex, the main principles of gathering and the most often used processing units are described in this Chapter. Like many other manufacturing processes oil and gas gathering and processing affect the environment but there are some specifics in the impact on the environment as described in this chapter. During the gathering and processing there is gas emission which includes emission of carbon dioxide, carbon monoxide, methane, volatile organic carbons and nitrogen oxides.

Emissions of sulphur dioxides and hydrogen sulphide can occur and depend upon the sulphur content of the hydrocarbon. Sometimes air pollution may cause health effects and reduce visibility. Also gathering and processing have impact on terrain and soil, especially in a case of contamination resulting from spillage and leakage. The environmental impact of produced waters disposed to other receiving waters is highly dependent on the quantity, the components, the receiving environment and its dispersion characteristics. Industrial wastes are generated during routine operations. These wastes are typically placed in containers, characterized and labeled, possibly stored briefly, and transported by a licensed hauler to an appropriate permitted off-site disposal facility as a standard practice. Environmental contamination could occur from accidental spills of herbicides or, more significantly, oil. During the gathering and processing the noise from compressor and pumping stations is present. Also, the noise is caused by occasional flaring and vehicle traffic. Lease roads are constructed and used to support various operations. Once production facilities are installed, portions of well pads, access roads, and pipeline would continue to contrast with the natural form, line, color, and texture of the surrounding landscape. Possible impacts to health and safety during production include accidental injury or death to workers and, to a lesser extent, the public. Health impacts could result from water contamination, dust and other air emissions, noise, soil contamination, and stress associated with living close to gathering and processing. Potential fires and explosions would cause safety hazards. Cavitations could ignite grass fires. Increased or reckless driving by oil or gas workers would also create safety hazards. In addition, health and safety issues include working in potential weather extremes and possible contact with natural hazards, such as uneven terrain and dangerous plants, animals, or insects.

REFERENCES

Benmebarek, S., & Gray, T. (1996). *Risk assessment of new added facility at an offshore gathering/ processing plant.* Paper presented at the SPE Health, Safety and Environment in Oil and Gas Exploration and Production Conference. New Orleans, LA.

Bull, A. P. I. 75L. (2007). Guidance document for the development of a safety and environmental management system for onshore oil and natural gas production operations and associated activities. Washington, DC: API.

Definition of Facility. (n.d.). Retrieved from www.cdphe.state.co.us/ap/onglinks.asp

Devold, H. (2009). *Oil and gas production handbook.* Oslo, Norway: ABB Oil and Gas.

JPT. (1999). *Surface production facilities: Separating and treating produced oil and gas.* Retrieved from www.spe.org/JPT1999-06-surface-prod

Nooijen, H., & Clinton, P. (2012). *Design of compressor suction scrubbers.* Paper presented at the TEKNA-Conference. Oslo, Norway.

Rassenfoss, S. (2011, August). Growing offshore water production pushes search for subsea solutions. *JPT.*

Stewart, M., & Arnold, K. (1998a). Surface production operations, volume 1, design of oil-handling systems and facilities. New York: Elsevier Science.

Stewart, M., & Arnold, K. (1998b). Surface production operations, volume 2, design of gas-handling systems and facilities. New York: Elsevier Science.

Transmission and Distribution. (n.d.). Retrieved from www.ipt.ntnu.no/~jsg/undervisning/.../Chap8.TRANSMISSIONAND-DISTRIBUTION.pdf

ADDITIONAL READING

A Regional Case Study, An Assessment of the. Environmental Implications of Oil and Gas Production. A Regional Case Study, September 2008, working draft from www.epa.gov/sectors/pdf/oil-gas-report.pdf.

ANSI/API Spec. 7B-11C (R2008) (1994). Specification for Internal-Combustion Reciprocating Engines for Oil-Field Service, API, Washington D.C., USA.

API E5 (1997) *Environmental Guidance Document: Onshore Solid Waste Management in Exploration and Production Operations*, API, Washington D.C., USA.

API RP 12R1 (R2008) (1997). *Recommended Practice for Setting, Maintenance, Inspection, Operation, and Repair of Tanks in Production Service*, API, Washington D.C., USA.

API RP 2350 (2005). *Overfill Protection for Storage Tanks in Petroleum Facilities*, API, Washington D.C., USA.

API RP 55 (1995), *Recommended Practices for Oil and Gas Producing and Gas Processing Plant Operations Involving Hydrogen Sulfide*, API, Washington D.C., USA.

Background document and technical annexes for biological effects monitoring (2012), OSPAR, from www.ospar.org/documents

Bull, A. P. I. 75L (2007). Guidance Document for the Development of a Safety and Environmental Management System for Onshore Oil and Natural Gas Production Operations and Associated Activities, API, Washington D.C., USA.

Euphemio, M., Oliveira, R., Nunes, G., Capela, C., & Ferreira, L. (2007) *Subsea Oil/Water Separation of Heavy Oil: Overview of the Main Challenges for the Marlim Field—Campos Basin,* Petrobras Offshore Technology Conference, Houston, Texas *Exemption of Oil and Gas Exploration and Production Wastes from Federal Hazardous Waste Regulations*, EPA, United States Protection Agency from www.epa.gov/osw/nonhaz/industrial

Fernando, B., & the Eastern Research Group, Inc. (2007). *Emissions from Oil and Gas MISSIONS FROM OIL AND GAS Production Facilities*, Final Report from www.tceq.texas.gov/.../reports/.

Global Gas Flaring and Venting Reduction Voluntary Standard (2004), Report No4., GGFR

Gruehagen, H., & Lim, D. (2009). *Subsea Separation and Boosting—An Overview of Ongoing Projects.* Jakarta, Indonesia: Asia Pacific Oil and Gas Conference & Exhibition.

Manning, F. S., & Thompson, R. E. (1995). *Oil field Processing of Petroleum: Crude oil.* Tulsa, Oklahoma: Penn Well Publishing Company.

Nooijen, H., & Clinton, P. (2012). *Design of Compressor Suction Scrubbers*, TEKNA-Conference

Rahmawati, S. D. (2012) *Integrated Field Modeling and Optimization*, Thesis for the degree of philosophiae doctor, Trondheim, Norwegian University of Science and Technology from www.iocenter.no/lib/exe/fetch.php?media=news:phds:finalthesis.

Stewart, M., & Arnold, K. (1998). Surface Production Operations, Volume 1, Design of Oil-Handling Systems and Facilities, Elsevier Science, USA.

Stewart, M., & Arnold, K. (1998). Surface Production Operations, Volume 2, Design of Gas-Handling Systems and Facilities, Elsevier Science, USA.

Surface Production Facilities. *Separating and Treating Produced Oil and Gas* (1999). JPT, from www.spe.org/jpt/.../JPT1999_06_surface_prodJP.

KEY TERMS AND DEFINITIONS

Axial Line System: Is system of flow lines where several wells produce into a common flow line.

Due Diligence: Is a term used for a number of concepts involving either an investigation of a business or person prior to signing a contract, or an act with a certain standard of care. It can be a legal obligation, but the term will more commonly apply to voluntary investigations.

Gathering System: Is flow line network and process facilities that transport and control the flow of oil or gas from the wells to a main storage facility, processing plant or shipping point. There are two types of gathering systems, radial and trunk line system.

Radial Line System: Is system of flow lines emanating from several different wellheads converge to a central point where facilities are located.

Surface Facility: Is any grouping of equipment where hydrocarbon liquids are processed, upgraded or stored prior to the point of custody transfer; or where natural gas is processed, upgraded, or stored prior to entering the natural gas transmission and storage source category.

Chapter 12
Activities in Oil and Gas Processing for Avoiding or Minimizing Environmental Impacts

Svijetlana Dubovski
INA Plc, Croatia

ABSTRACT

Formal risk assessments are necessary at various phases of the asset life cycle as they help personnel identify, evaluate, and control hazards that could result in loss of life, injury, pollution, property damage, or business disruption. Hazard evaluations of production development concepts or facility design are well-defined processes, for which much literature is available as guidance. Such evaluations are mandated in some jurisdictions for project regulatory approval. This chapter provides guidance on activities implementation from designing phase, construction phase, operational phase, and decommissioning phase of gathering and processing systems. For each phase, activities that can be implemented to avoid or minimize environmental impact are described, emphasizing the implementation purpose.

INTRODUCTION

The operators are dedicated to responsible development of oil and natural gas resources. Responsible development includes good relationships with our neighbors and a commitment to environmental protection and compliance with all applicable federal, state, and local regulations.

Management leadership and commitment has to be converted into action by the provision of adequate financial and personal resources to ensure that environmental protection measures are incorporated in all operations. Each

DOI: 10.4018/978-1-4666-4777-0.ch012

operating site may need to develop its own specific objectives, and relevant operational targets in line with the company's broader strategic objectives (Rahmawati, 2012). This should be initiated by the site manager, and achieved through a formal communication and consultation process that involves staff, contractors and local stakeholders. The roles, responsibilities, authorities, accountabilities and relationships necessary to implement environmental management must be clearly defined for that site. Line staff in all aspects of operational activity should be assigned specific environmental responsibility and authority within their spheres of control, and must be competent to perform their duties effectively.

Activities during the Design Phase

Designing phase starts by creating a few development solutions and selecting the best one. That process requires integration of engineering, environmental and economic parameters. Environmental forces such as wind, waves, currents, ice, water depth and soil condition have great impact on construction. With platforms and other off shore surface facilities, state of the experience level is important.

The overall basis for designing should be to minimize adverse effects on the environment, consistent with providing an economical means of unit production using safety by design.

Safety by design is a mechanism for moving the discussion of safety hazards and harmful risks into the design and programming phases of construction projects. This involves incorporating hazard analysis and risk assessments during the design phases of construction or renovation of a facility. This also involves taking the necessary steps to mitigate the risk of incidents that can cause damage to the facility or injuries to workers. Risks are mitigated to an acceptable level and various other aspects such as health and environmental concerns are considered in these early phases of development.

Although it can be easier to identify hazards and environmental impact in an existing facility after construction is complete, it's generally safer and more cost effective to implement appropriate measures before facilities or processes are built. By using this proactive approach to safety and environmental protection, engineering controls are put in place in the early stages of construction, resulting in the mitigation of potentially dangerous or fatal accidents. In the early phases of conceptual design development, several strategies can be employed to recognize potential risks:

- Plans and designs can be compared to similar, existing systems or facilities, taking into account what measures were successful, then parlaying them into the safety program;
- People who have experience with similar, existing systems can be interviewed to determine the successes and challenges they have realized in their safety program;
- Plans can be reviewed by a qualified person, that has an engineering and safety background and is capable of creating designs that ensure compliance with industry safety regulations, standards and owner's requirements;
- Organizations can provide standards or guidelines to consultants and equipment manufacturers prior to building design.

Careful implementation of these strategies in the conceptual design phase helps ensure successful hazard identification that will serve as the foundation of a safer work environment.

In gathering and system lines design, appropriate industry codes should be followed and design should consider the following: life of lines; lines environment; nature and quantity of products, impacts on existing facilities, consequences of possible failure; corrosion inhibition measures; optimum location for blow downs, valves, etc.

In planning gathering and system lines, including electrical distribution systems, it is important to consider the impact that construction operations and maintenance activities will have on people, animals, plants, and the land itself, both surface and shallow subsurface. The impact on current use, as well as possible future uses, should be evaluated along with potential future facilities expansion. Because pipelines can be buried, and the surface reclaimed, long-term surface disturbance associated with pipelines can be avoided. The placement of pipelines should avoid steep hillsides and watercourses where feasible. Also, where feasible, pipeline routes should take advantage of road corridors to minimize surface disturbance. Also, when clearing is necessary, the width disturbed should be kept to a minimum and topsoil material should be stockpiled to the side of the routes where cuts and fills or other disturbances occur during pipeline construction. Retaining topsoil for replacement during reclamation can significantly accelerate successful revegetation.

Initial planning for surface facilities within a field should consider future development potential in order to minimize surface disturbance. When it is practical and economic, central field locations should be considered to avoid the use of multiple facilities. Facility sizing should consider future throughput increases to minimize the need for additional tankage and treating vessels.

Surface facilities should be planned to utilize the smallest practical surface area consistent with safe, prudent, and economic operations. Existing roads and rights-of-way should be utilized to the maximum extent possible. The land owner and/or surface tenant should be consulted to consider present and future uses of affected and adjacent land. A site should be selected that minimizes the amount of surface terrain alteration to reduce environmental and aesthetic damages. Cuts and fills which pose possible landslide or slump problems should be avoided.

It is good to follow proposed consideration in designing production facilities.

- Production-related equipment should be sized and designed to provide appropriate safety and utility. Future development and exploration plans should be considered when sizing equipment. Where appropriate, the facilities should be sized to handle current and future production to minimize retrofitting and improper use of equipment. Equipment should be designed with appropriate spill control devices, such as high-/low-level indicators or high-/low-pressure indicators, to improve safety and protection of the environment.
- The anticipated time the equipment is expected to remain active should be considered. Proper design and installation can minimize future equipment failures and downtime.
- Equipment and foundations should be designed and installed giving consideration to adverse natural conditions common to the area, such as floods, excessive snow and rain, earthquakes, tornadoes, hurricanes, and dust storms.

- Equipment installations should comply with industry standards. Air pollution control facilities should be installed whenever practical, economical, and technically feasible. Flaring and venting should be evaluated based on gas volume and composition, safety, economics, and local environmental impact. Pressure requirements for vessels, lines, and other equipment should be considered. Any variance from the manufacturer's recommended rates or pressures should be evaluated.
- Equipment and facilities should be located and designed to minimize the wastes generated by operations and maintenance activities.
- Recyclable products should be used, where possible. Bulk storage, recyclable, and reusable containers should be considered to minimize waste. Appropriate methods of collecting and recycling or disposing of waste generated during construction, operation, and maintenance of the facility should be considered.
- Operators should develop waste management plans.

Activities during the Construction and Installation

During the gathering and system lines construction and installation phase it is very important to follow next principles:

- Line routes and applicable rights-of-way should utilize the smallest practical surface area, consistent with prudent operations.
- Unnecessary damage to trees and other vegetation adjoining lease line routes should be avoided.
- If contractors are used to install lines, the operator should verify that the contractor has implemented a safety program that includes a written hazard communication program.
- The contractor should supply Material Safety Data Sheets for all hazardous materials brought on site.
- Appropriate inspections should be performed during construction to ensure design specifications are met.

Upon completion, lines should be inspected and pressure tested for possible leaks in accordance with state and local codes. Pressure test fluids should be collected and disposed. After installation of a new line, all line routes and rights-of-way should be cleaned up and restored to conditions compatible with existing land use, unless other arrangements have been made with the landowner. Disposal of all waste should be in accordance with the regulations. Line routes and burial depth should be adequately documented to aid in preventing ruptures and/or accidental leaks during future excavation activities. Crossings should be marked.

Before initiating facility construction the following site preparation steps should be taken:

- Soil characteristics should be checked to determine the appropriate foundation design for the site;
- Size and type of equipment to be used during construction should be considered to allow sufficient room to work in a safe manner;
- Adequate culverts and drainage ditches should be provided as required by the local environment;
- The open end of lines under construction should be temporarily capped at the end of each workday if a line could be accessible to wildlife.

During construction, qualified personnel should perform appropriate inspections to ensure that design specifications are met. Upon completion, equipment and facilities should be inspected for possible leaks. If necessary, equipment should be pressure tested in accordance with applicable codes. If fluids are used to pressure test, collect and dispose of the fluids should be in accordance with the regulations. Also, X-raying of welds should be considered in critical areas where extreme pressure or corrosiveness is anticipated or where potential risk to the local environment is of great concern.

The qualifications of personnel working on the construction site should be evaluated to aid in ensuring the work will be properly performed. Also, consideration should be given to requiring contractors to have performance bonds should be considered when facilities are to be constructed in environmentally sensitive areas.

All equipment should be installed in accordance with the original design of the equipment. Any variations from the original specifications should be evaluated thoroughly to ensure safety of the operations.

Upon completion of facilities, the original drawings or schematics should be updated, as required. Changes or modifications from the original design or drawings should be noted for future reference.

Unused and excess construction materials should be properly stored or removed from the site upon completion. During construction, the site should be kept as clean and free of debris as possible.

Interim reclamation consists of minimizing the footprint of disturbance by reclaiming to the extent possible all portions of the site not required for production operations. The portions of the cleared site not needed for operational and safety purposes are recontoured to a final or intermediate contour that blends with the surrounding topography as much as possible. Sufficient level area remains for the set-up of work over or production stimulation and to park necessary equipment. Where practical, the operator should spread topsoil over the entire location and revegetate as closely as possible to the production facilities, unless an all-weather, surfaced access route or turnaround is needed to inspect or operate the well or to complete workover or stimulation operations. It may be necessary to drive, park and operate on restored, interim vegetation within the previously disturbed area. This is acceptable provided damage is repaired and reclaimed following use. To reduce final reclamation costs and effort, to maintain healthy, biologically active topsoil, and to minimize habitat, visual resource, and forage loss during the life of a well, salvaged topsoil should be spread over the areas of interim reclamation rather than stockpiled.

Where the topography is flat and it is, therefore, unnecessary to recontour the well location at the time of final reclamation, the operator should set aside sufficient topsoil for reclamation of the small unreclaimed area around the wellhead. Any topsoil pile set aside should be revegetated to prevent it from eroding and to help maintain its biological viability. On sloped ground, during final reclamation the topsoil and interim vegetation must be restriped from portions of the site that are not at the original contour, the well pad recontoured, and the topsoil respread over the entire disturbed site to ensure successful revegetation.

Activities during the Operational Phase

After design, construction and installation phase starts operational phase. This is the most lasting phase of gathering and processing during which is implemented a lot of activities to minimize or eliminate environmental impact in the same time. These activities are:

- Using operating procedures;
- Asset optimization and maintenance support;
- Information management systems;
- Training simulators;
- Risk assessment;
- Compliance assessment review;
- Initial due diligence site visit.

Using Operating Procedures

Development of a standard operating procedure manual applicable to each major facility should be considered. The manual should contain information as to the equipment located at the facility, safe-operating practices for the equipment, start-up and shutdown procedures, and emergency procedures. Also, consideration should be given to the analysis of failures or malfunctions so that corrective action can be taken to minimize future environmental incidents.

All applicable personnel (both company and contractor) should receive training to provide operation and maintenance in accordance with manual. This training should include start-up and shutdown procedures, normal operating procedures, and emergency response procedures, in the event of a leak or spill of a hazardous substance. Line routes and facilities should be inspected at intervals dictated by evaluation of exposures and/or failures. Appropriate steps should be taken to prevent surface and environmental damage from the use of hot oil, chemicals, and other treatments that are used to maintain lease gathering and system lines. Proper maintenance practices should be exercised with respect to crossing markers, blow down tanks, venting equipment, and corrosion protection equipment. Blow down fluids should be collected and placed in the production system to recover hydrocarbons. Waste materials should be recycled, reclaimed, or disposed. Pressure tests, profile surveys, and other means should be considered to meet operating safety requirements. Operating procedures, also, should provide for early identification of developing corrosion problems, failure-prone equipment, and malfunctions so that corrective action can be taken before environmental or safety consequences occur. Frequency of failure analysis should be considered to aid in scheduling line replacements. Appropriate industry codes should be followed with respect to maintenance of records, repairs, reporting of leaks, etc. Whenever modifications are made to existing lines and facilities, the changes should be considered for evaluation pursuant to management of change principles. Where appropriate, facility drawings should be updated to show modifications and the superseded drawings should be destroyed. Emergency phone numbers should be posted at the entrance to the facility, if located near a populated area.

Waste and residual management practices for production operations should be conducted consistent with lease and landowner obligations. This should include solid wastes and residuals, such as tank bottoms, drilling fluids and cuttings, liquid wastes and residuals, such as produced water and used oil, and gaseous wastes, such as hydrocarbons and carbon dioxide. A sound waste management plan is important to protect human health and the environment and minimize long-term liabilities to the operator.

The final option for management of a waste, after source reduction, recycling, and treatment options have been considered and incorporated, is disposal. The operator should take into consideration the long-term fate of the waste and its constituents before disposal.

Accidental spills (including oil and saltwater) can, besides potentially damaging the environment, create difficult operational, legal, and public relations problems. It is very important to conduct operations in a manner that minimizes the potential for unauthorized spills. Spill prevention, response, and cleanup procedures should be defined and in place before storing any oil or chemicals on site or conducting activities that have a potential for a spill.

The best way to avoid adverse effects of spills is to prevent their occurrence. The key factors in spill incident prevention are adequately trained supervisors and field operating personnel. In the event a spill occurs, it is extremely important for all responsible operating personnel to know how to respond quickly and effectively to control, contain, and clean up the spill. To ensure this capacity exists, a contingency plan should be prepared for inland areas as well as for areas near water. The plans should provide utilization of capabilities of oil spill cooperatives, whenever advantageous.

Spill plans should address the needs to advise the public about significant releases. The plan should include procedures to advise government officials and provide appropriate information and access to the press. In the event a spill occurs, the source of the spill should be stopped, or reduced as much as possible, in a safe manner. The spread of the spilled substance should be controlled or contained in the smallest possible area to minimize the adverse effects. Operators should evaluate the potential for spills and damages and use this information to determine the type and size of primary and secondary containment necessary.

The type and footage of containment boom installed or stored for deployment will vary with the type, size, and location of the facility and spill potential. This information should be developed for each main area or facility and be stated in the facility contingency plan. In addition, the contingency plan should list where emergency equipment is located. The contingency plan should state the type of chemicals that can be used effectively and list sources and procedures for applying these chemicals.

Cleanup procedures should be developed and included in the facility contingency plan. Up-to-date lists of effective cleanup materials and equipment and a list of potential contractors who can supply needed assistance should also be included and maintained in the contingency plan.

Depending on the spill potential at each area, a stock of appropriate cleanup materials sufficient to handle small spills should be maintained on hand at all times. The amount of cleanup material will depend on the time required to obtain more material if the size of the spill should increase.

Sometimes training simulators are used to provide operator training in a realistic plant training environment before ore during operational phase. They use the actual control and safety applications of the plant, running on operator stations. Plant models simulate the feedback from the plant in real time or in fast or slow motion. Training simulator applications include functions for backup and reload including recreation of historical information and snapshots. Offsite training facilities are often connected (read only) to the live plant to give information from the real operating situation.

Asset Optimization and Maintenance Program

During the operational phase the best approach to maintenance is to have asset optimization and maintenance program. The mail goal of that program is to reduce costly production disruptions end environmental impact by enabling predictive maintenance. Recording the maintenance history of an asset and identifying potential problems to avert unscheduled shutdowns, maximize up-time and operate closer to plant production prognoses. These activities can be run manually or using some computerized maintenance management system. Asset optimization and maintenance program is based on continues process condition and structural monitoring (Holiday, 1991).

Process condition monitoring includes both structural monitoring and condition monitoring for process equipment such as valves and rotating machinery. For structural monitoring the devices as corrosion meters (essentially plates that corrode, for which that corrosion may be metered), tension force meters and free swinging strings should be used. Condition monitoring of machinery should be generally used for large rotating apparatus, such as turbines, compressors, generators and large pumps. Input devices are vibration meters, temperature meters as well as the number of start/stops, running time, lubrication intervals measurement. For other process equipment such as valves, the closing times, flow and torque can be registered. A valve which exhibits a negative trend in closing time or torque has to be diagnosed. Asset optimization and maintenance program support functionality of maintenance, based on input from condition monitoring systems and a periodic maintenance plan. This will allow to schedule personnel and activities, for such tasks as lubrication or cleaning, and plan larger tasks such as turbine and compressor periodic maintenance.

Information Management System

Some Information management system can be used to provide information about the operation of gathering and production of the surface facilities. This can be a separate system, or an integral part of the control system or SCADA system. Main functions of such a system are:

- Production monitoring and reporting
- Safety management
- Maintenance
- Operator support

Reports provided by system may be: production overview and prognosis, well tests, chemical injection, chemical consumption, laboratory analysis results, alarm and incidents statistic and overview, etc.

Gathering and Processing Risk Assessment

Risk assessment process consists of both identifying risk and estimating the probability and magnitude of potential losses. The likelihood and consequences to people, company assets and the environment are the main reasons that prompt the development of a formal management process for risk assessment toward loss prevention and risk management. Risk is a term that describes the probability that a specified hazardous event will occur and the severity of the consequences of that event. Also, hazard represents a potential risk to personnel, property or productivity and need to be measured so that appropriate action can be taken to minimize the size. Risk management in itself addresses the value judgments involved in establishing acceptable level of risk and methods of handling identified risk.

Risk assessment is classified here into two parts, risk estimation and risk evaluation. The former is the estimation and assessment of the measures of severity and probabilities including their uncertainties. While the later is the extra efforts required to decide on the importance of a particular value of risk evaluated which support the value judgment.

Gathering and processing risk assessment in the workplace has become practice which has resulted in a major drop in numbers of fires, spills etc. incidents recorded per year. In other to attain a minimal or zero incidents, a number of risk assessment program were developed. Programs are the base for the development of the information gathered during a risk assessment, which has several applications and implications. This ranges from replacing certain equipment, modification of existing process, addition of protective devices, developing and implementing procedure for maintaining, monitoring etc. of process in the facility and sometimes used as a feedback to the design engineers for a new facility or upgrade.

Compliance Assessment Review

During the operating phase, a good practice, in order to recognize the influence of gathering and processing on the environment, is to do compliance assessment review. Compliance assessment review is an internal environmental auditing. The auditing process involves management, audit team members, supervisors and workers, and provides them with a environmental training and monitoring of the effectiveness of the environmental program. These auditing programs were aimed at assuring legal compliance at the field level. Many of the field regulatory problems resulted from lack of regulatory knowledge and training of both the field and engineering staffs. A successful Environmental Compliance Assessment Review must include:

- Senior Management support and active participation;
- An audit team having thorough understanding of the environmental regulations;
- A formalized process including a written report of findings and corrective actions;
- Training of office and field personnel to protect them from penalties;
- Informing Management of the findings and corrective action taken.

Initial Due Diligence Site Visit

If one is not involved in designing and construction phase of gathering and processing it is necessary to visit the facility to assess environmental, social, health, and safety status. This involves walking the site, visiting all of the production areas, and reviewing the documents. For complex facilities, technical environmental staff may be needed.

For understanding the production processes, existing pollution controls, and potential for improvement and cost savings it is necessary to check:

- Housekeeping, (cleanness of the facility, are materials stored, etc.) Housekeeping is often an important indicator of the capacity and drive for environmental compliance;
- Age and type of equipment;
- Pollution control equipment and the maintenance schedule;
- Types of waste produced, where and how disposed;
- Types of liquid effluents, where and how disposed;

- Develop an understanding of the facility's health and safety regime;
- Availability and use of personnel protective equipment;
- Content and frequency of employee health and safety training;
- History of accidents;
- Frequency of health and safety audits.

Also, developing an understanding of the history of the site, the facility, and the surrounding properties is important in order to understand whether there might be historic contamination, permitting issues, or community issues.

Decommissioning and Abandonment

Before removal all equipment such as tankage, separation vessels, meter runs, flow lines, and pumps should be purged and flushed, as appropriate. Whenever possible, materials recovered should be recycled, reclaimed, or disposed. Tanks, separation vessels, meter runs, surface lines, pumps, and any other exposed surface equipment should be removed. Removal of the associated equipment foundations should be considered. Exposed piping segments from surface or subsurface equipment connecting to buried lines should be removed to a depth consistent with subsequent land use or, preferably, to the depth of buried lines. Where feasible or where desired to limit potential future liabilities, consideration should be given to removing buried lines. Where appropriate, each outlet of any abandoned lines should be permanently sealed. Operators should consider removing all crossing markers and other line markers.

Where appropriate, the location of abandoned lines should be identified on facility maps. All pits and surface impoundments should be properly closed after they are dry and free of waste; then they should be backfilled and graded to conform to the surrounding terrain. Closure must also be in accordance with any local and/or state regulations. The location of closed pits should be documented. Materials removed from pits should be reclaimed, recycled or disposed. Upon completion of abandonment activities, all disturbed surface areas should be cleaned up and restored to conditions similar to the adjacent land or to landowner requirements. Timely completion of final reclamation is as important as the initial planning. Incomplete or improperly executed final reclamation can result in the complete loss of a low-impact project opportunity. Reclamation becomes significantly more difficult, more expensive, and less effective if sufficient topsoil is not salvaged, interim reclamation is nhot completed, and if proper care is not taken to construct pads and roads in locations that minimize reclamation costs. Restoration of the original landform is a key element in ensuring that the effects of oil and gas development are not permanent.

Infrastructure associated with formerly producing leases, including water impoundments, power lines, metering buildings, compression facilities and tank batteries must be removed and the footprints or lands disturbed by these facilities and associated foundations reclaimed unless the surface owner requests that items such as impoundments or water wells be kept.

Salvaged topsoil should be respread evenly over the surfaces to be revegetated. The top soiled site should be prepared to provide a seedbed for reestablishment of desirable vegetation. Site preparation may include gouging, scarifying, dozer track-walking, mulching, fertilizing, seeding and planting. In reclamation of sites that are not cultivated for agriculture or grazing, seeding and planting should use plant species indigenous to the area.

Water breaks and terracing should only be installed when absolutely necessary to prevent erosion of fill material and should be removed when the site is successfully revegetated and stabilized.

The decommissioning of offshore facilities occurs when the reservoir is depleted or the production of hydrocarbons from that reservoir becomes unprofitable. Parts of the offshore facility, such as platforms, are treated to remove contaminants and are usually removed, while other production components are rendered safe and left in place. Wells are plugged and abandoned to prevent fluid migration within the wellbore, which could contaminate the surface environment. Internationally-recognized guidelines and standards issued by the International Maritime Organization and decisions issued by OSPAR Convention or similar convention should be followed for the decommissioning of offshore facilities (Benmebarek & Gray, 1996).

All surface lines should be removed. Lines should be purged before removal. Surface and subsurface equipment connected to buried lines should be removed to a depth consistent with subsequent land use or, preferably, to the depth of the buried lines. Harmful or hazardous materials should be displaced from any lines abandoned in place. Where appropriate, each outlet of abandoned lines should be permanently sealed. All crossing markers and other line markers should be removed. The location of abandoned lines should be identified on facility maps. Upon completion of abandonment activities, all disturbed surface areas should be cleaned up and restored to conditions similar to the adjacent lands.

Additional Measures to Avoid, Minimize, or Mitigate Offshore Environmental Risk

Measures for Minimizing Air Emissions

All reasonable attempts should be made to maximize energy efficiency and design facilities for lowest energy use. The overall objective should be to reduce air emissions and evaluate cost-effective options for reducing emissions that are technically feasible. Exhaust gas emissions produced by gas or liquid fuels for power and heat generation, or for water injection or oil and gas export, can be the most significant source of air emissions from offshore facilities. During equipment selection, air emission specifications should be considered. There are different requirements for small sources and sources of more than 50 MW. Measures consistent with the Global Gas Flaring and Venting Reduction Voluntary Standard should be adopted when considering venting and flaring options for offshore activities. The standard provides guidance on how to eliminate or achieve reductions in the flaring and venting of natural gas. Continuous venting of associated gas is not considered current good practice and should be avoided. The associated gas stream should be routed to an efficient flare system, although continuous flaring of gas should be avoided if alternatives are available. Before flaring is adopted, feasible alternatives for the use of the gas should be evaluated to the maximum extent possible and integrated into production design. Alternative options may include gas utilization for on-site energy needs, gas injection for reservoir pressure maintenance, enhanced recovery using gas lift, gas for instrumentation, or export of the gas to a neighboring facility or to market. An assessment of alternatives should be adequately documented and recorded (Harmony, 1998). If none of the options are feasible for the use of associated gas, measures to minimize flare volumes should be evaluated and flaring should be considered as an interim solution, with the elimination of continuous production-associated gas flaring as the preferred goal. Volumes of hydrocarbons flared should be recorded. Methods for controlling and reducing fugitive emissions should be considered and implemented in the design, operation, and maintenance of offshore facilities. The selection of appropriate valves, flanges, fittings, seals, and packing should consider safety and suitability requirements as well as their capacity to reduce gas leaks and fugitive emissions. Additionally, leak detection and repair programs should be implemented.

Measures for Wastewater Discharges

Feasible alternatives for the management and disposal of produced water should be evaluated and integrated into production design. These alternatives may include injection along with seawater for reservoir pressure maintenance, injection into a suitable offshore disposal well, or export to shore with produced hydrocarbons for treatment and disposal. If none of these alternatives are technically or financially feasible, produced water should be treated before disposal into the marine environment. Good practice, for minimizing the volume of hydro test water offshore, is to test equipment at an onshore site before the equipment is loaded onto the offshore facilities or sending offshore pipeline hydro test water to shore facilities for treatment and disposal. Also the same water can be used for multiple tests. Hydro test water disposal into shallow coastal waters should be avoided.

The cooling water discharge depth should be selected to maximize mixing and cooling of the thermal plume to ensure that the temperature is within 3 degrees Celsius of ambient seawater temperature at the edge of the defined mixing zone or within 100 meters of the discharge point. Operators should consider mixing desalination brine from the potable water system with the cooling water or sewage water discharge (Eskijian et al., 2002).

Gray and black water from showers, toilets, and kitchen facilities should be treated in an appropriate onsite marine sanitary treatment unit in compliance with MARPOL 73/78 requirements. Organic waste from the kitchen should, at a minimum, be macerated to acceptable levels and discharged to sea, in compliance with MARPOL 73/78 requirements((MARPOL 73/78; 2009). Water pumped into and out of storage during loading and off-loading operations should be contained and treated before discharge. Bilge waters from machinery spaces in offshore facilities and support vessels should be routed to the facility closed drainage system, or contained and treated before discharge. If treatment to standard is not possible, these waters should be contained and shipped to shore for disposal. Drainage water generated from precipitation, sea spray, or routine operations, such as deck and equipment cleaning and fire drills, should be routed to separate drainage systems on offshore facilities. This includes drainage water from process areas that could be contaminated with oil and drainage water from non-process areas.

Measures for Solid and Liquid Waste Management

Waste materials should be segregated offshore into non-hazardous and hazardous wastes at a minimum, and shipped to shore for re-use, recycling, or disposal. A waste management plan for the offshore facility should be developed that contains a clear waste tracking mechanism to track waste consignments from the originating location offshore to the final waste treatment and disposal location onshore. Efforts should be made to eliminate, reduce, or recycle wastes at all times.

Measures for Hazardous Materials Management

Offshore chemicals should be with least hazard and lowest potential environmental impact, and lowest potential health impact, should be selected, whenever possible. Use of chemicals suspected to cause taint or known endocrine disruptors and ozone depleting substances should be avoided. Also chemicals known to contain heavy metals, other than in trace quantities, should be avoided.

Measures for Minimizing Noise Generation

At first it is necessary to identify areas sensitive for marine life such as feeding, breeding, calving, and spawning areas. Also offshore construction activities should be planned to avoid sensitive times of the year. When marine mammals are observed congregating close to the area of planned activities should begin at least 500 meters away. If marine mammals are sighted within 500 meters of the proposed activities should be postponed until they have moved away, allowing adequate time after the last sighting (www.ifc.org/ifcext/enviro.nsf/Content, 2007).

Soft-start procedures, also called ramp-up or slow buildup, should be used in areas of known marine mammal activity. This involves a gradual increase in sound pressure to full operational levels.

Additional Spill Prevention and Control Measures

Additional spill prevention and control measures specific to offshore oil and gas facilities include conducting a spill risk assessment for offshore facilities and support vessels. Valves, including subsea shutdown valves, leak detection system and Emergency Shutdown System should be installed to allow early shutdown or isolation in the event of an emergency. For export pipelines, maintenance programs should include regular pigging to clean the pipeline, and intelligent pigging should also be considered as required. Personnel also should be adequate trained in oil spill prevention, containment and response.

Occupational Health and Safety

As described for onshore, occupational health and safety issues should be considered as part of a comprehensive hazard or risk assessment. The results should be used for health and safety management planning, in the design of the facility and safe working systems, and in the preparation of safe working procedures. Health and safety management planning should include environmental conditions at the offshore location, adequate living accommodation, protected area at the facility for use by personnel in the event of an emergency, a sufficient number of escape routes, etc. Planning should demonstrate that a systematic and structured approach to managing offshore health and safety will be adopted and that controls are in place to reduce risks to as low as reasonably practical.

Offshore facilities should be equipped, at a minimum, with specialized first aid providers and the means to provide short-term remote patient care. Depending on the number of personnel present and complexity of the facility, provision of an on-site medical unit and doctor should be considered. In specific cases, telemedicine facilities may be an alternative option. An alarm system should be installed which can to be heard throughout the offshore facility. Alarms for fire, gas leak and man overboard should be provided. The formation of a health and safety committee for the facility is recommended. Health and safety inductions should be provided to the entire workforce prior to mobilization to the offshore facilities.

FUTURE RESEARCH DIRECTIONS

Environmental impact of existing gathering systems and surface facilities, as it is described in this Chapter, is mostly recognized, analyzed and activities for minimizing or avoiding impact are proposed. Implementation of new technologies in gathering and processing systems can result in positive and negative direction on theenvironment. There are some examples of new approaches in gathering and processing with very positive environmental impacts.

The Scott field experience with improvements in gas sweetening, in conjunction with advanced sulfur recovery technologies, make it possible to practically eliminate noxious emissions and recover nearly all the acid gas stream's elemental sulfur for late sale or disposal.

Successful tests in New Mexico, a new freeze-thaw/evaporation process purifies produced waters from oil and gas production operations by separating out dissolved solids, metals, and chemicals. These typically brackish waters can be made suitable for beneficial use, significantly lowering environmental risks and furthering resource management.

In Louisiana, better dehydration systems management, including optimization of glycol circulation rates and installation of flash tank separator condensers enables producers to capture up to 90 percent of methane and other emissions. These processes reduce greenhouse gas emissions, improve air quality, and recover substantial gas for on-site use or pipeline sale (www.fe.doe.gov, 2005).

Significant amounts of methane are emitted to the atmosphere during production, transmission, processing, storage, and distribution. Unocal Spill Prevention Task Group uses new technology facilitates accurate, efficient leak detection and measurement by ensuring equipment and pipeline integrity and timely maintenance and repair. These controls capture saleable natural gas, create safer work places, and protect our environment.

Replacing or retrofitting devices, or improving maintenance, can reduce gas emissions substantially, reducing greenhouse gas emissions and potentially saving the industry millions of dollars in lost methane. Chevron installed a low-bleed retrofit valve kit on liquid level and pressure controllers on two platforms. 19 devices were tested on one platform and 30 devices on another. The retrofits yielded average reductions in bleed rates of more than 90 percent.

Offshore operations represent over one quarter of the Nation's oil and natural gas production. Since the early 1990s, Federal regulators and industry have successfully cooperated in the development and implementation of recommended practices for voluntary safety and environmental management programs for Outer Continental Shelf operations. Using that approach, industry is responsible for voluntarily identifying potential hazards in the design, construction, and operation of offshore platforms and for implementing specific processes to improve safety and environmental protection. These measures are designed to reduce the risk and occurrence of accidents, injuries, and oil spills.

Vapor recovery units succeed in the Austin Chalk field, can significantly reduce the fugitive hydrocarbon emissions vaporizing from crude oil storage tanks, particularly tanks associated with high-pressure reservoirs, high vapor releases, and larger operations. These emissions are typically made up of 40 to 60 percent methane, along with other volatile organic compounds and hazardous air pollutants.

CONCLUSION

Once production facilities are installed, portions of well pads, access roads, and pipeline would continue to contrast with the natural form, line, color, and texture of the surrounding landscape. Possible impacts to health and safety during production include accidental injury or death to workers and, to a lesser extent, the public.

Previously listed environmental impacts depend upon factors such as production quantities, the number and size of units, sort of processing facilities, the amount of land disturbed by activities and occupied by facilities, etc. Sometimes it is not possible to environmental impact eliminate, but can be reduced to a minimum by implementing adequate activities.

This Chapter proposes implementation of some activities from the designing phase to abandonment phase with a goal of eliminatiom or minimization of the environmental impact.

As one of the most important factors in environmental impact is implemented technology, this Chapter gives future research directions for implementing new technologies. These technologies will help in minimizing of the environmental impact.

REFERENCES

Benmebarek, S., & Gray, T. (1996). *Risk assessment of new added facility at an offshore gathering/processing plant.* Paper presented at the SPE Health, Safety and Environment in Oil and Gas Exploration and Production Conference. New Orleans, LA.

Environmental, Health, and Safety Guidelines for Onshore Oil and Gas Development. (2007). Retrieved from www.ifc.org/ifcext/enviro.nsf/Content

Eskijian, M., Heffron, R., & Dahlgren, T. (2002). *Engineering standards for marine oil terminal design and maintenance (MOTEMS).* Retrieved from www.slc.ca.gov/.../Presentation www.slc.ca.gov/.../Presentation

Harmony, K. (1998). *Facility risk assessment and risk management in offshore oil & gas industry.* Paper presented at the SPE Asia Pacific Conference on Integrated Modelling for Asset Management. Kuala Lumpur, Malaysia.

Holliday, G. H. (1991). *Environmental compliance assessment review.* Paper presented at the International Arctic Technology Conference. Juno, AK.

MARPOL. 73/78. (2009). Regulations for the prevention of pollution by oil. International Maritime Organization.

Rahmawati, S. D. (2012). *Integrated field modeling and optimization.* (Doctoral Thesis). Norwegian University of Science and Technology, Trondheim, Norway. Retrieved from www.iocenter.no/lib/exe/fetch.php?media=news:phds:finalthesis

U.S. Department of Energy, Office of Fossil Energy. (2005). *Environmental benefits of advanced oil and gas.* Retrieved from www.fe.doe.gov

ADDITIONAL READING

API RP 51R (2009). *Environmental Protection for Onshore Oil and Gas Production Operations and Leases*, API

Belke, D. U. (1995). *Sequential Decision Analysis and Risk Analysis for Antarctic Offshore Petroleum Development*, SPE Annual Technical Conference and Exhibition, Dallas, Texas

Hopkins, J. (2001). *The Practical Application of Process Safety Principles To Determine and Monitor Asset Integrity of Oil and Gas Facilities.* Aberdeen, UK: Europe Oil &Gas Conference and Exibition.

Horn, T., Bakke, W., & Eriksen, G. (2003). *Experience in Operating World's First Subsea Separation and Water Injection Station at Troll Oil Field in the North Sea*, Offshore Technology Conference, Houston, Texas

KEY TERMS AND DEFINITIONS

Environment: Is the sum of all surroundings of a living organism, including natural forces and other living things, which provide conditions for development and growth as well as of danger and damage.

Environmental Impact: Is a change in the make-up, working, or appearance of the environment. These changes may be planned or accidental.

Environmental Risk: Is actual or potential threat of adverse effects on living organisms and environment by effluents, emissions, wastes, resource depletion, etc., arising out of an organization's activities.

Habitat Change: Is change in the local environmental conditions in which a particular organism lives. It can occur naturally through droughts, disease, fire, hurricanes, mudslides, volcanoes, earthquakes, slight increases or decreases in seasonal temperature or precipitation, etc. However, it is generally induced changes by human activities such as land use change and physical modification of rivers or water withdrawal from rivers.

Occupational Safety and Health: Is a cross-disciplinary area concerned with protecting the safety, health and welfare of people engaged in work or employment. It may also protect co-workers, family members, employers, customers, and many others who might be affected by the workplace environment.

Chapter 13
Transportation Risk Analysis

Dragan Crnčević
INA Plc, Croatia

ABSTRACT

Petroleum is transported across the water in barges and tankers, and on land, using pipelines, trucks, and trains. Natural gas is moved, mainly, by pipelines. The most common causes of tanker accidents are: fire/explosions, loading/offloading, structural damage, collision, and grounding. Pipeline accidents are due to: corrosion, third parties activities, mechanical damage, natural events, and operational error. Some of the most commonly applied preventive activities that reduce spills in waterborne transportation are: double-hulled tanker, navigation safety and radio communications equipment, tanker exclusion zone, etc. The pipeline condition can be recorded by using various nondestructive measurement techniques or by chemical analysis of fluid flows. Different types of sensors can be used to locate and determine the size of an anomaly in the pipeline geometry. Mayor methods for detecting leaks are measuring the hydrodynamic parameters or registering abnormal conditions in the fluid flow and detecting phenomena in the immediate vicinity of the pipeline.

INTRODUCTION

Advances in exploration and production have helped to locate and recover a supply of oil and natural gas from major reserves across the globe. At the same time, demand for petroleum-based products has grown in every corner of the world. But supply and demand are rarely concentrated in the same place. Transportation therefore is vital to ensuring the reliable and affordable flow of petroleum we all count on to fuel our cars, heat our homes and improve the quality of

DOI: 10.4018/978-1-4666-4777-0.ch013

our lives. Pipelines, marine vessels, tank trucks, rail tank cars and so forth are used to transport crude oils, compressed and liquefied hydrocarbon gases, liquid petroleum products and other chemicals from their point of origin to pipeline terminals, refineries, distributors and consumers.

Crude oils and liquid petroleum products are transported, handled and stored in their natural liquid state. Hydrocarbon gases can be handled, transported and stored in both the gaseous and liquid states and must be completely confined in pipelines, tanks, cylinders or other containers prior to use. The most important characteristic of liquefied hydrocarbon gases (LHGs) is that they are stored, handled and shipped as liquids, taking up a relatively small amount of space and then expanding into a gas when used. For example, liquefied natural gas (LNG) is stored at -162°C, and when it is released the difference in storage and atmospheric temperatures causes the liquid to expand and gasify. In liquefied form, the volume of LNG is 600 times less than the same amount of natural gas at normal temperature and pressure.

Because of the large volumes of products which are transported by pipelines on a continuous basis, there is opportunity for environmental damage from releases. Depending on company and regulatory safety requirements and the pipeline's construction, location, weather, accessibility and operation, a considerable amount of product may be released should a break in the line or leak occur.

In addition to the usual maritime working hazards, transporting crude oil and flammable liquids by marine vessel creates a number of special health, safety and fire prevention situations. These include surging and expansion of liquid cargo, flammable vapor hazards during transport and when loading and unloading. The economics of operating modern tankers requires them to be at sea for extended periods of time with only short intervals in port to load or unload cargo. This, together with the fact that tankers are highly automated, creates unique mental and physical demands on the few crew members used to operate the vessels.

BACKGROUND

Much of the oil and gas fields are located in countries far away from industrialized areas where the need for energy growing each day. This is why large quantities of hydrocarbons have been transported all over world by land and sea. Increased transport increases risks. However, it can be said that the last decade has been a reduction in accidents and incidents. What is the cause?

Whether transported by sea or by land safety and security of operations is key task. Pipelines and tankers are constructed and operated under strict regulations. During operations are applied national and international laws, conventions and regulations. New technologies that enhance transportation security are also implemented. Despite the safety problems that may occur, maritime oil shipping accidents have fallen steadily. The number of major oil spills (i.e. over 700 tons) decreased eightfold between the '70s and '00s. These encouraging results were made possible by the combined efforts of states, oil companies and ship owners. In fact, almost all the oil transported by sea arrives safely without a hitch (Dhillon, 2007).

In recent decades, oil transportation by pipelines has increased significantly. However, incidents of oil leakage have a decreasing trend in volume and in the number of leaks.

Testing and failures simulation is also one way to find out more about the effects on people and the environment. Therefore, efforts to study the heat impact of LNG thermal on the people intensified the last few years. The first thought that a man comes to mind when he saw the tanker of liquefied natural gas (LNG) from its huge, mostly spherical tanks is: what a bomb! The gas is normally associated in the minds of the explosion. Almost every day in the newspaper can be read by the title of "gas bottle exploded" (and killed people, destroyed several apartment or house into ruins). If only 10 kilograms of liquid gas can cause the kind of damage that would happen to "blow up" the typical gas tanker with 125 000 cubic meters, or 50 million pounds of liquid gas?

With this apocalyptic picture in front of, U.S. Congressmen have asked the state Office of Government Accountability Office (GAO) to investigate and report to them on what is known in the United States and assumed about the possible consequences of a possible terrorist attack on an LNG tanker.

When some hypothetical terrorists decided LNG tanker used as a bomb, you first have to liquefied petroleum gas spilled out of the tank. Spreads along the ground or water, he quickly begins to evaporate as mist rises from the ground. Mixed with air and disperses. Anyone caught up in the mist would suffocate, because they would not have enough air to breathe. And just to be able to ignite the vapor, but even then not always, because natural gas is flammable if it is in the air between five and 15 percent in a concentration lower than five percent is not enough gas in more than 15 - not enough oxygen . But vaporized gas would not be difficult to burn, and then it produces tremendous heat, much more than the burning of crude oil. Spilled gas would quickly burn out and then moved to the fire source, the crack or reservoirs holes. And the biggest disputes arise about how much LNG is dangerous for the people in his vicinity. Experts polled by the GAO were not unique, and thought that exposure to heat from just 30 seconds caused severe burns to people at a distance "from 500 to more than 2,000 meters."

U.S. Sandia National Laboratory (Luketa at al., 2008) in his study concluded that the most probable maximum distance for severe burns about 1,600 meters. Of course, the details depend on the environment, wind direction and speed, the amount of gas escaping, lasting fire. But the U.S. Government Accountability Office has not yet met those findings. Concerned about them is that different experts and studies, such a different estimate of the fire danger zone on an LNG tanker. About this data depends on how the governments of all countries importing liquefied natural gas seek to protect cities, harbors, beaches and waterways from a terrorist attack tanker, but not only intentional, but also against accidental failures. GAO also believes that experiments should be carried out with much more spilled liquid gas than before.

PIPELINES TRANSPORTATION

Main Characteristics of Pipelines

Transmission pipelines are the main arteries' of the oil and gas business; working 24 hours per day, seven days a week, continuously supplying our energy needs. They are critically important to most countries' economies. They have a long history; pipelines have been used to transport liquids and gases for thousands of years: the Chinese used bamboo pipe to transmit natural gas to light their capital, Peking, as early as 400 BC.

The oil and gas are transported in these large transmission pipelines to refineries, power stations, etc., and converted into energy forms such as gasoline for our automobiles, and electricity for our homes.

Without pipelines we would not be able to satisfy the huge oil and gas needs of our planet. Pipelines are also very safe forms of transporting energy (Hopkins, 2007).

- Pipelines are 40 times safer than rail tanks, and 100 times safer than road tanks for transporting energy;
- Oil pipeline spills amount to about 1 gallon per million barrel-miles, according to the USA Association of Oil Pipelines. One barrel, transported one mile (1609 m), equals one barrel-mile, and there are 42 gallons (159 liters) in a barrel. In household terms, this is less than one teaspoon of oil spilled per thousand barrel-miles.

This high pressure, large capacity pipelines carry hazardous products, and consequently, they are designed, constructed and operated using recognized standards that all have a focus on safety. Additionally, these pipelines have to satisfy safety regulations in most countries. These high standards and regulations ensure safe and secure pipelines.

The total length of high pressure transmission pipelines around the world has been estimated at 3,500,000 km. The 'split' is (Hopkins, 2007):

- ~64% carry natural gas;
- ~19% carry petroleum products;
- ~17% carry crude oil.

These systems can be huge; for example, if you laid the Canadian pipeline system, end to end, it would extend 17 times around the world!

We now have many types of pipelines in the world. The types of oil and gas pipelines can be summarized as gathering pipelines, transportation pipelines and distribution pipelines.

Gathering pipelines is a group of smaller interconnected pipelines forming complex network with the purpose of bringing crude oil or natural gas from several nearby wells to a treatment plant or processing facility. In this group, pipelines are usually short- a couple of hundred meters- and with small diameters. Also sub-sea pipelines for collecting product from deep water production platforms are considered gathering systems.

Transportation pipelines are mainly long pipes with large diameters, moving products (oil, gas, refined products) between cities, countries and even continents. These transportation networks include several compressor stations in gas lines or pump stations for crude and multiproduct pipelines.

Distribution pipelines are composed of several interconnected pipelines with small diameters, used to take the products to the final consumer. Those are feeder lines with purpose of distributing gas to homes and businesses downstream. Pipelines at terminals for distributing products to tanks and storage facilities are included in this group

Pipelines are made from high quality steel ('line pipe' steel), constructed using well-established methods, and operated using procedures developed over many decades.

Pigging has been used for many years to clean larger diameter crude oil pipelines. These operations also include inspecting of the pipeline. This is accomplished by inserting the pig into pipeline on launcher/launching station (a funnel shaped Y section). The launcher/launching station is then closed and the pressure (driven flow of the product in the pipeline) is used to push it along down the pipe until it reaches the receiver/receiving station. Figure 1 shows construction of launcher/receiver station at crude oil pipeline.

Aboveground, underwater and underground pipelines, varying in size from several centimeters to a meter or more in diameter (the Russian system has diameters up to 1,422 mm), move vast amounts of crude oil, natural gas, LHGs and liquid petroleum products. Pipelines run throughout the world, from the frozen tundra of Alaska and Siberia to the

Figure 1. Construction of launcher/receiver station at 20" crude oil pipeline (Part of author's private collection of INA Plc's facility)

hot deserts of the Middle East, across rivers, lakes, seas, swamps and forests, over and through mountains and under cities and towns.

The longest pipeline in the world is the Druzhba pipeline (5,327 km), which passes through eight countries - Russia, Belarus, Ukraine, Poland, Germany, the Czech Republic and Hungary.

There are geopolitical and economic issues involved in the building of new oil pipelines.

Generally, the oil industry prefers to ship oil by sea because this mode of transport is more flexible. Unlike a pipeline, a ship does not have to follow the same route, which can therefore be adapted to meet demand. However, sometimes oil has to travel over land, for example to landlocked countries. In this case, it can be easier and less costly to use a pipeline than to transport the oil by road or train. In countries with huge land mass such as Russia, pipelines are useful to deliver oil to ports for exporting by ship.

Western Europe has pipeline networks that transport crude from ports to refineries located further inland. These are also used to send the finished product from refineries to large consumer centers (fuel and raw materials for the petrochemicals industry).

Gas pipelines are the preferred mode of transport for natural gas.

As their name indicates, gas pipelines are similar to oil pipelines except that they transport natural gas. There are two types of gas pipeline:

- Some are underwater, such as those linking Norwegian gas fields to European terminals or North Africa to Sicily.
- Others are overland, such as those that make up the Russian network (the longest in the world, at almost 160,000 km long). For safety and security reasons, these networks are buried underground.

Contrary to oil, natural gas is mostly transported by pipeline. This is because for the same amount of energy, the volume of natural gas is 600 times higher than oil, which is liquid. This means that it is far too costly to transport natural gas by ship.

Causes of Possible Accidents

Despite all the measures taken to preserve the integrity of the pipeline leakage may occur, and thus more or less environmental pollution. Thereby causing leakage affects the size of leaks. Not the same, whether the pipeline leak caused by corrosion millimeter perforated or a relatively large diameter hole in which case there is a sudden release of large the amount of oil into the environment. The causes of failure are one of the important parameters to be analyzed in leakage databases. Location of the leak is also an important element in the assessment of the impact of leakage environment. In case of failure in the area of impervious soil will result in a lower impact on the environment then in the case of leakage in the aquifer. The cost to repair leaks in the area waterways is several times higher than the repair of leaks in the soil. The effective clean spillages of oil, it is very important to know how the spread oil and dissolved constituents, volatilization, biodegradation, dilution, the behavior of oil in heterogeneous soil and effects of oil in the soil and groundwater.

In considering how best to avoid pipeline spillages, it is very important to investigate what are the main causes of failure to be guarded against. The causes of spillage can be divided into few main categories (Dziubinski et al., 2006; CONCAWE,2010):

- Mechanical failure;
- Operational failure;
- Corrosion;
- Natural events;
- Third party activity.

Mechanical Failure

Mechanical failures are ruptures and fissures that occur when stresses in the system exceed the allowable stress. They can be caused by poor material quality or faulty construction. Manufacturing defects can occur in the pipe or fittings, e.g. in the pipe wall or in the longitudinal weld. Poor construction techniques can generate high residual stress levels in the pipeline prior to commissioning. For example, the forming of pipe bends, welding techniques and the handling of materials can all lead to unacceptable construction practices if not carried out according to the specification of the work.

Operational Failure

Operational failures can be due to overpressure or malfunction of systems such as pressure relief or control devices. They are also caused by human error such as failing to observe the correct operating instructions.

Corrosion

In the past, corrosion has been the most common cause of spillage although the quantities involved are usually small. Pipelines are subject to two types of corrosion - internal and external. Crude oils and oil products can give rise to internal corrosion when corrosive products are present usually in combination with water. Corrosion can also occur when pipelines are not in use. External corrosion occurs either because the pipeline coating is found to be inadequate and/or the cathodic protection is inefficient.

Natural Hazard

Natural hazards are phenomena such as landslides, flooding, ground subsidence and earthquakes.

Third Party Activity

The majority of these spillages are caused by accidental damage inflicted after construction of the pipeline by third party excavation in the vicinity of the pipeline. More often than not the right-of-ways are not clearly marked and lines are sometimes broken by bulldozers or similar plant machinery possibly with fatal results. Intentional damage unfortunately is on the increase and pipelines carrying flammable or high value products make ideal targets. Surveillance and inspection procedures are designed to minimize the amount of damage caused by third parties.

As can be seen from Table 1, which present the reported causes of spillages from West European pipelines and USA pipeline, third party damage was responsible for the largest number of spillage. Also, great number of spillages was due to corrosion and mechanical failure.

Accident Risk Analysis

The large volumes of products which are transported by pipelines on a continuous basis, is opportunity for environmental damage from releases. Risk identification involves careful consideration of the possible events that may negatively affect the most commonly observed some activity. These may be risks associated with safety and integrity of physical objects, the risks arising from the environment in which the observed activity takes place and the financial risks associated with the diversity of decision-making. Hazard is defined as a potential cause of injury or damage, but the risk is a combination of several effects and the likelihood that an event will happen. Even the simplest methods of assessment, such as simple verification of the system provide visible results that lead to increased safety. Procedure risk identification is usually carried out in a number of the following steps: identifying or risk identification, determining who may be harmed or that may be affected, the assessment (evaluation) risk and decisions about precautionary measures to reduce the probability of occurrence, record on his findings, and the incorporation of new knowledge into the existing system and to review and repeat the procedure if necessary.

For a good understanding of the concept of risk management are three important questions (Muhlbauer, 2004):

- What can happen?
- How likely is it?
- What are the consequences?

The risk assessment is the basis for decision-making in different situations, as well as to predict the consequences of such decisions. Industrial risk assessment is a careful examination of all the potential sources of danger could affect the operation or the system in a simple or complex form. Recommendations various agencies and international standards state institutions are similar. They are mainly based on statistical data of accidents cases

Table 1. Number of oil spillages by cause

Main causes	CONCAWE statistics for 1971-2006, %	DOT statistics for 1971-1996, %
Third parties	33	34
Corrosion	30	33
Mechanical failure	25	18
Operational failure	7	2,5
Natural events	4	4,5
Others	1	8

by region. They saw many countries have benefited from the introduction of technical risk assessment methodology of leaks as integrated parts of the program analysis risk in the oil companies.

Risk management has been defined by the USA Office of Pipeline Safety as "a comprehensive management decision support process, implemented as a program, integrated through defined roles and responsibilities into the day-to-day operations, maintenance, engineering management, and regulatory decisions of the operator"

Prevention Activities

There are a number of safety precautions and procedures in place to minimize the risk of accidents. Prevention activities can be predicted even in the preparation of initial design studies where defined pipeline route, mechanical design and protection. Pipeline design is regulated by strict national and international standards and specifications which are subject to continuous review and updating (API Std. 1160, 2001). In fact, the transportation of oil and natural gas is one of the safest ways of transporting energy, mostly due to the fact that the infrastructure is fixed, and buried underground. During the operation of the pipeline requires maintenance, control and monitoring.

Pipeline Route - The choice of pipeline route is an important factor in ensuring the safety of the public, protection of the environment and integrity of the pipeline itself. Soil studies, topographical surveys and geological information are assessed to obtain technical solutions for crossing all obstacles, such as difficult terrain, roads, rivers, railways, etc. Special attention is given to the design of crossing important rivers and stretches of water. Underground routes are usually preferred to above ground routes where possible. Hazard assessments and environmental studies which assess the effects of, for example, noise, pollution and accidents on the local population, are reviewed in more detail as the system design develops.

Mechanical Design: The operating pressure can be defined as the pressure required to maintain a given flow rate through the pipeline taking into account the pipeline profile and the residual pressure at the end of the pipeline. The design pressure must be equal to or higher than the maximum operating pressure under any flow condition, including static pressure under no flow conditions (pumps running against closed valves).The minimum wall thickness is calculated (including a safety factor) from a knowledge of the pressures, type of steel pipe and consideration of other factors. These include the ease of handling without damaging pipes, the resistance to stresses imposed during pipeline construction, and the resistance to deformation under external loads. The thermal stresses are also considered, together with hydraulic, mechanical and fatigue stresses, in order to ensure that the maximum allowable stress values are not exceeded for the selected wall thickness for the pipeline.

Mechanical Protection: Isolation valves are installed along the pipeline route so that the pipeline can be quickly divided into sections in the event of an emergency to minimize the available contents that could spill. Various Codes of Practice make recommendations on the spacing of valves, particularly in areas of dense population. In certain circumstances non-return valves are installed in the place of isolation valves. These valves only permit flow in one direction and prevent the contents of the pipeline draining back to a rupture point in an emergency, thereby reducing any spillage. Overpressure protection devices, such as pressure relief valves may be installed at appropriate locations to limit the maximum pressure of the pipeline under transient or incorrect operating conditions.

Corrosion protection: Corrosion is still an important cause of failure. It can occur either on external or internal surfaces and different prevention methods are used in each case (NACE RP0502-2002, 2002).

External corrosion: Corrosion is electrochemical in form. When an unprotected pipe is buried in moist soil conditions, an electric current flows from the metal (the anode) to the soil (acting as cathode). The metal from which the current flows will dissolve in the surrounding moisture. This is apparent as corrosion. There are two ways of preventing this corrosion, either by isolating the pipeline from the soil with an electrically insulating protective coating, or by cathodic protection which ensures that any current flow will be from the soil to the metal and not vice versa. Though it is possible to use the two methods independently, they are commonly used to complement one another. If for example the coating has a fault, then the cathodic protection will protect the line.

Methods of checking the condition of pipe insulation can be indirectly to obtain data on corrosion attack and changes in the thickness of the outer wall piping. These are Pearson methods, and so-called Current Pipeline Mapper (PCM).

Measuring external protecting coating: Pearson detection of the insulation pipe caused a 40-ies of the last century, but still is used. Measurement of exterior insulating coating of pipelines using Pearson device achieved good results. The basic physical principle of his work lies in the fact that the quality of transmission of electromagnetic fields leads to a function with the quality pipe insulation. Transmitter, connected to the pipe, the signals are sent to the receiver placed at a certain distance. Where the insulation is damaged, reducing the resistance between the pipe and the ground, it is evidenced by the receiver in a larger spades or stronger signal. Otherwise, the alignment between the transmitter and receiver are going operators in the space of 6 m, which have put the electrodes which take signals from the ground and transmit them to the receiver cables.

Contrary the Pearson method in which measurements are made using direct current, the Pipeline Current Mapper used by low-frequency alternating current. Since this method is developed many years later, and that in the instrument for the detection of insulation defects found, Global Positioning System - GPS, this method is more commonly used in practice. At the time the pipeline was excavated condition of the insulation can be checked by direct methods such as the determination of bonding insulation.

Catodic protection: Cathodic protection (CP) is a proven method of controlling and reducing corrosion of pipelines, vessels, tanks and other structures, located both above and below ground. The added expense of installing CP systems is usually justified based on pipe or vessel repair/replacement costs. Corroding structures have both anodic and cathodic areas. Electrons naturally flow from the anodic areas through an electrolyte (for example soil) to the cathodic areas and back to the pipe, thereby completing a circuit. Corrosion occurs where the current discharges from the metal into the electrolyte at anodic areas. By applying cathodic protection to a structure, the entire exposed surface can become cathodic to the environment, thereby mitigating the corrosion. Cathodic protection has also been shown to be effective in reducing corrosion underneath biomass colonies, by impeding their attachment to the steel surface.

In a corrosion cell, current flows from the anode through the electrolyte to the cathode. Cathodic protection methods attempt to control the direction of the corrosion current. One such method is the use of a sacrificial anode. A sacrificial anode is an anode electrically connected to the system to be protected. The sacrificial anode will corrode and therefore, the cathodic system will be protected. The basic idea is to force certain parts to be cathodic or anodic with respect to the other in order to control the current direction. If a sacrificial anode is chosen that is anodic to every metal in the pipeline system, the anode will corrode while the pipeline is protected. A prime example of a sacrificial system is the use of magnesium anodes in connection with the pipeline system. The magnesium is buried in the ground within the same electrolyte as the pipeline. The pipeline and the anode are electrically connected, and as the magnesium corrodes, the pipeline is protected. The magnesium should be alloyed to prevent passive layering. A passive layer is an oxide film that forms naturally on the surface of most metals. This passive layer becomes a protective coating on an anode and prevents further corrosion. If the sacrificial anode does not corrode, the protection current is not produced. If this happens, the pipeline will begin to corrode.

Another form of cathodic protection is an impressed current system. In this system, an external source, not the galvanic reaction, forces the direction of the current. The positive terminal from a power is connected to the anode and the negative terminal is connected to the cathode. The pipeline is connected to the negative terminal and consequently becomes cathodic. It is very important that the positive terminal be connected to the ground bed anodes, the material to be corroded. If the positive terminal is connected to the pipeline, the pipeline will become anodic and corrode. Several different current sources are available for use in impressed current systems. Batteries, DC engine-generators, and thermo-electric generators all provide direct current. Alternating current sources such as a power line may be used if a rectifier is used to convert the current to direct current. Alternating current will not provide any protection since the pipeline would be anodic or cathodic depending on the direction of the current. The biggest advantage in cathodic protection systems is its effectiveness. However, in order to provide effective corrosion protection, the system must be carefully maintained.

Preventive maintenance works consist of the following: visual inspection of all measurement places (including appliances, boxes and polarization measurement cell), measuring the potential pipeline - soil, checking for stray current and the impact of power facilities, tuning and optimization of the protection system and, by making field measurements annual reports with measurements of the quality of cathodic protection. Maintained cathodic protection system is an important indicator of the integrity of the pipeline.

Internal corrosion: There are several methods for checking the internal pipe-wall condition of a pipeline and these include corrosion coupons, application inhibitors, internal coatings, iron counts, intelligence pigs, and frequent use of cleaning pigs to displace any corrosive products including water. Various monitoring and inspection tools are used to assess the actual corrosion growth rate. Internal corrosion growth can be assessed using corrosion coupons or probes, and/or external measurement devices like mechanized ultrasonic and field signature monitoring. A corrosion coupon is a strip of metal inserted into a pipeline which is then periodically monitored for corrosion either by electrical methods or by periodical removal and weighing. This can be performed with the pipeline in service. Unfortunately, the corrosion growth is underestimated when these measurement devices are not located at the worst corrosion spots.

Gas and Oil Sampling: Routine sampling of the gas and oil in pipelines ensures its quality, and may also indicate corrosion of the interior of the pipeline, or the influx of contaminants. A common method to detect internal corrosion is the monitoring of the quantity of dissolved iron in the small quantities of entrained water. The accuracy of the method is dependent on the water pH and the level of iron in the liquid. This method is particularly useful in indicating the level of effectiveness of the inhibitors commonly added in oil pipelines. It is not used in isolation, but in combination with other methods of corrosion monitoring, as each method complements the other and provides supporting data on the condition of the pipeline. Another option to assess corrosion growth is by comparison of two intelligence pig surveys. Conventional cleaning pigs are used to scrape away any deposits that have accumulated on the interior wall of the pipeline to reduce pressure loss.

Intelligence pigs: In order to ensure the efficient and safe operation of oil and natural gas pipelines, pipeline companies routinely inspect their pipelines for corrosion and defects. This is done through the use of sophisticated pieces of equipment known as ‘smart pigs.’ Intelligent pigs are robotic devices that are propelled down pipelines to evaluate the interior of the pipe. Intelligence pigs can test pipe thickness, and roundness, check for signs of corrosion, detect minute leaks, and any other defect along the interior of the pipeline that may either impede the flow of gas, or pose a potential safety risk to the operation of the pipeline. Sending a pig down a pipeline is fittingly known as ‘pigging’ the pipeline.

Intelligence pigs have a wider application in the monitoring of internal pipe walls. Different types of pigs can be used to locate and determine the size of the anomalies in the geometry pipeline, such as dents or bulges, and locate and determine the size of the reduction in wall thickness caused by internal or external corrosion, erosion or mechanical damage. Measurement principles of these devices are based on ultrasound ("ultrasonic intelligence pigs") (Figure 2) or the loss of magnetic flux ("magnetic intelligence pigs") (Figure 3).

Figure 2. Ultrasonic intelligence pig (part of author's private collection of INA Plc's facility)

In magnetic devices, an essential part is a pair of permanent magnets in the form of steel wire overlapping the inner wall of the pipeline, and between their poles are located sensitive sensors. Bolting pole permanent magnet on the pipe wall induces a magnetic field in the wall of the pipeline. If it encounters a damaged section of pipe wall at this point, partly interrupted magnetic flux, which is registered by the sensors. Other parts of this device are an electronic device for recording and storing data and a wheeled odometer to measure the distance traveled.

Intelligence pig that operates on the principle of ultrasound allows the measurement of corrosion damage on outside and inside walls of the pipeline. This device works on the principle of rejecting an ultrasonic pulse from the inner and outer walls of the pipeline. Device gets a straight wall thickness, and gaps between the sensor and the inner wall. If the inner wall is damaged of the pipeline, then on the basis of reflection of ultrasonic waves reflected its reduced thickness, while

Figure 3. Magnetic intelligence pig (Adapted from www.tdwilliamson.com)

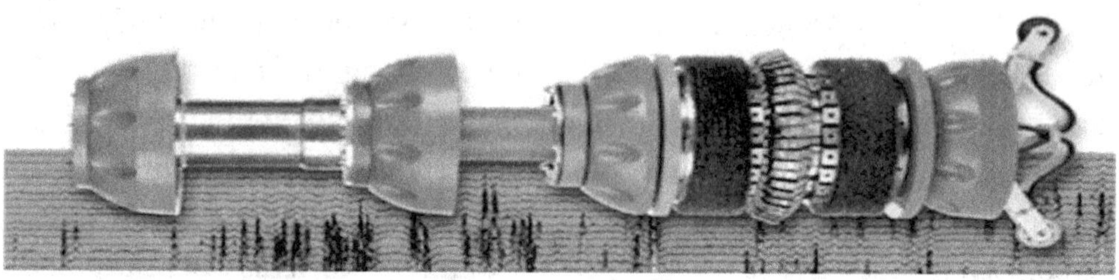

increasing the distance from the sensor to the inner wall. If the outter wall is damaged (as evidenced by the loss of metal), then the reflection of ultrasonic waves also observed reduction in wall thickness, which suggests a short distance from the sensor to the outer wall of the pipeline

Operation: An operating pipeline requires control, maintenance and surveillance. The control centre monitors the variables through the instrumentation on the pipeline in order to ensure safe and efficient operation. The control room is the nerve centre of a pipeline. The essential instrumentation on the pipeline is linked to the control room by cable and/or telemetry. The increasing power of computer systems available to pipeline operators enables ever more sophisticated pipeline supervision and control. Current technology enables pigs and product parcels to be tracked; leak detection on pumping pipelines by mass/volume balance or dynamic modeling is becoming increasingly reliable and accurate. Monitoring of pressure (and temperature) also ensures a leak detection capability for a pipeline that is shut in. The instrumentation monitors variables such as pressures, flow and temperatures in the pipeline and external conditions such as ground movement and air temperature. Maintenance is conducted on a planned formal basis. Routine inspection and maintenance is carried out to ensure that the pipeline is in mechanically sound condition and operating at optimum efficiency. Regular surveys are undertaken to check for corrosion and remedial work is carried out if necessary. All instruments and cathodic protection systems are checked for correct operation.

Deposits are removed from the internal pipe wall with scraper pigs to reduce pressure losses in the pipeline. This involves the testing of valves and the removal of surface impediments to pipeline inspection.

Surveillance: Carried out regularly to check the general condition of the pipeline route and to ensure that no unauthorized work is being started near the pipeline and to check that the pipeline is not endangered in any other way. Such surveillance can be carried out either from the air or on the ground or both.

Aerial Patrols: Planes are used to ensure no construction activities are taking place too close to the route of the pipeline, particularly in residential areas. Unauthorized construction and digging is the primary threat to pipeline safety.

Pipeline Markers: Signs on the surface above pipelines indicate the presence of underground pipelines to the public, to reduce the chance of any interference with the pipeline.

Leak Detection Methods

Although spillages are in fact very rare, leak detection systems are installed to detect and locate a leak as soon and as accurately as possible (Zhang, 1993; API 1130, 2002). Whatever the failure causes, the consequences are almost always the same. These are primarily:

- Loss of life and property;
- Environmental cleanup costs;
- Payment of fines and damages;
- Bad image company:
- Possible fines and legal suits

Today in the world there are a number of methods and techniques based on different principles. The market offers a large number of leak detection devices and monitoring systems. Many are very similar and differ only in the technical performance of different vendors. They differ in commercial names, physical principles, and applications. From over 300 methods available on the market today is summarized view only those that are applicable to detect pipeline leakage. References are grouped and systematized in several ways depending on the source (Martin, 1998; Alaska Dept. of Environmental Conservation, 1999). One of the displays is appropriate to distinguish between three main groups:

- Monitoring of characteristic signals generating by a leak;
- Monitoring hydrodynamic parameters of fluid;
- Surrounding environmental monitoring.

Each of the groups incorporates a number of methods of monitoring which observe and follow the related physical phenomena.

Monitoring of Characteristic Signals Generated by a Leak

Methods of this group observe and record the accompanying effects caused by the leakage and can be divided in:

- Negative pressure wave;
- Wave alert;
- Acoustic reflectometric;
- Ultrasonic leak detection pig.

Leakage or fluid flow through the hole in the pipeline due to expansion and changes in flow direction causes a variety of changes in the steady flow. With leaks will expand the sound waves at a rate that depends on the physical properties of fluids in pipes (http://www.wavealert.com, 2013). Acoustic detectors detect these waves, or failure and the time of their travelling to the sensor and can also locate leaks. Leak detection in pipelines using acoustic emissions technology is based on the principle that escaping liquid creates an acoustic signal as it passes through a perforation in the pipe. Acoustic sensors affixed to the outside of the pipe monitor internal pipeline noise levels and locations. These data are used to create a baseline "acoustic map" of the line. When a leak occurs, the resulting low frequency acoustic signal is detected and analyzed by system processors. Deviations from the baseline acoustic profile would signal an alarm. The received signal is stronger near the leak site thus enabling leak location. Acoustic sensing can be applied externally to buried pipelines by using steel rods driven into the ground to conduct the sound to a sensor mounted on the rod. The rods are inserted at intervals along the pipeline. Furthermore, the pipeline leaks in this group of methods can be detected by using devices equipped with acoustic detector ("intelligence pig"). Liquid escaping under pressure through a defect in the pipeline wall generates ultrasonic noise. This noise can be measured and recorded by a pig propelled through the pipeline by the normal flow of the product. Even small leaks can be detected and located with a good level of accuracy. This method will not alert the operator immediately the leak occurs nor will it indicate the size of it. The technique is used instead for locating and assessing suspected leaks, or conversely, to confirm the integrity of the line.

Monitoring of Hydrodynamic Parameters

Methods of the second group are based on the monitoring and measurement of hydrodynamic parameters of the process or registering abnormal condition in the process fluid flow. All parameters in the process flow at both ends of the pipeline in specific known the bulk, mass, hydrodynamic and thermodynamic equilibrium (http://www.vistaleakdetection.com, 2013). Disruption caused by the failure of the process distorts the equilibrium states and indicates the leakage. The advantage of these methods is that they are measuring devices commonly existing standard processes. These are mainly pressure transducers and temperature, flow meters (for more complex systems and viscometers and densitometers) that are otherwise necessary for the process to take place. Using devices Remote Terminal Units (RTUs) collects data from the measuring device. The whole system is controlled by SCADA, which collects and processes data from the sites, and subjects them to a specific mathematical or statistical analysis. The result is a programmable alarm and procedures (shut-down). Communication between these devices can be accomplished radio, satellite and often dial-up connection. For this group it is necessary to a perfect knowledge of all the conditions of the process, and in particular stationary flow conditions. In the case of unsteady flow conditions required are very sophisticated programs and equipment as more costly equipment and makes it susceptible to the application. Monitoring hydrodynamic parameters of fluid can run the following methods:

- Volume balance;
- Automatic volume balance;
- Modified volume balance;
- Mass balance;
- Real time transient model;
- Statistical analyses;
- Pressure/flow monitoring;
- Hydraulic test;
- Temperature compensated volumetric test.

Surrounding Environmental Monitoring

The third group method is the most numerous and widely spread in the application. It is based on detection of the appearance of the fluid in the immediate vicinity of the pipe. The basic method principle is oversees the ground (below, above and around the pipe), groundwater, vadose zone, corridors and indoor areas at risk of leaks. A very large range of parameters and measurement techniques are implemented in this group. There can be found the simplest examples, visual screening pipeline route or the use of canine olfaction odorant fluids. There are, however, in this group of very sophisticated methods using thermography, electro sensitive cables, optical cables and osmotic diffusion tubes. Thus, the diffusion tubes are also used for this purpose. Principle of this method is to use tubes that selectively permeable vapor and hydrocarbon gases. The pipes are laid in the immediate environment of controlled pipelines or facilities. In case of failure, as soon as the diffusion tube encircled saturated hydrocarbon, the interior is in that zone filled with steam or gaseous phase. Extruding content diffusion tubes, through continuous analyzer receives the data on which the hydrocarbon as well as the location for leaks. A similar method is application of fiber optic sensing cables. Leak detection sensor application of fiber optic cable used to the notion that chemical or physical reaction bothers transmission of light through optical fibers. Fiber optic sensors typically consist of a core (glass) and trim (polymers). In contact with contaminants there are changes in properties lining, which reflects the change in the intensity of the light signal through an optical fiber. This resulted in the identification of pollutants in space monitoring. Some of the methods are in the experimental stage of the application for the detection of leakage. Most often applied to very different purposes, e.g., radars (ground penetrating radar). They reveal the soil pipe, mines, geological structure etc. The use of radar for detecting leakage showed high precision but with difficulty when it is wet, clay layers (http://www.tracerresearch.com, 2013).

Most methods of this group are suitable for remote alarm or automatic locking process since all measured or observations to have electrically output. The advantage of this group over other monitoring methods described above, the higher the sensitivity to small contamination in the immediate vicinity of pipeline. Some of these methods allow for the timely detection and exact location of the leak. The main disadvantage of most methods of this group is that they are the chosen sensors buried piping must pass during the construction of pipelines. On old pipelines, subsequently burying is an additional cost. Environmental monitoring which surrounds the pipeline includes (Novačić, M., Crnčević, D., 2001):

- Visual observations;
- Tracers;
- Electro-chemical sensing cables;
- Fiber optic cables;
- Diffusion hoses;
- Capacitance probe;
- Metal oxide semiconductor (MOS);
- Light detector and ranking (LIDAR);
- Biosensors;
- SCAPS LIF detectors;
- Hydrocarbon detection sensors;
- Product solubility sensors;
- Radio frequency attenuation;
- Product permeability detectors;
- Refractive index of liquid detectors;
- Geochemical survey;
- Gas chromatography;
- Ground penetrating radar;
- Temperature sensing cable;
- Sniffer tube.

TRANSPORTATION BY TANKERS

Another key to maintaining the flow is the global fleet of oil tankers, now numbering around 3,500. Though these vessels are of widely varying size and quality, the one consistent trend throughout the business 100-year history has been toward larger and larger vessels. Today, more than 400 "very large crude carrier" (VLCC) class tankers, carrying over one million gallons of oil each, handle about one-third of the world's tanker volume. As size has increased, the cost of shipping has fallen to roughly 5-10% of the price per barrel. The world's first oil tankers appeared in the late 19th century and carried kerosene for lighting, but the invention of the motor car fuelled demand for oil. During the Second World War, the standard oil tanker was the T2, 16,400 tones deadweight, but tankers grew rapidly in size from the 1950s onwards.

The first 100,000-tone crude oil tanker was delivered in 1959 to cover the route from the Middle East to Europe round the Cape of Good Hope (thereby avoiding the Suez Canal which had been temporarily closed following political conflicts in 1956). Shippers saw economies of scale in larger tankers and by the mid-1960s, tankers of 200,000 tones deadweight- the Very large crude carrier had been ordered (http://gisis.imo.org/Public, 2013).

Tankers Route

Tankers often follow the same routes between producing and consuming countries. Refined petroleum products (gasoline, diesel, and kerosene) are mainly transported within Europe, between Europe and Asia, Europe and the US. The most frequented routes for crude oil start in the Middle East. They pass through the Bab el-Mandeb Strait or the Strait of Hormuz, the world's main oil shipping lane. After this, they travel to:

- The US and the rest of North and South America via the Cape of Good Hope;
- Asia via the Malacca Strait between Sumatra and Malaysia (this route leads to Japan and China, where oil demand has grown significantly since 2000);
- Europe via the Suez Canal. Larger tankers cannot take this route because the Suez canal is too narrow. These tankers pass the Cape of Good Hope to reach European ports. Thus, large oil tanker takes 2 weeks to one month to reach Western Europe from OPEC countries (Organization of the Petroleum Exporting Countries) in Africa, South America, the Arabian Peninsula.

Moreover, oil freight transiting through the Panama canal has increased considerably due to growing demand in Asia.

Main Types of Tankers

In addition to the two main types of oil tankers, crude and product tankers, there are some subcategories of oil tankers in use today. Tankers are divided by type of cargo transported.

Oil tanker: It is tanker for the carriage of crude-oil.

Product tanker: Tanker for the carriage of refined products derived from crude oil (gasoline, Diesel oil, jet fuel) from the refineries.

Shuttle tanker: It is tanker ship for the carriage of crude-oil directly from the offshore oil fields to terminals or refineries.

Tankers for the carriage of condensed gases: Tanker which carriages the gases kept in the liquid state due to high pressures or to very low temperatures, in tanks with highly efficient insulation.

More relevant cargo types:

LNG (liquefied natural gas) tanker: It is tanker which carriages mainly methane with temperatures that can reach –163 ºC.

LPG (liquefied petroleum gas) tanker: Tanker which carriages the gases with a low boiling point (-44 to 0°C) and high vapor pressure. They are carried at environment temperature in pressurized tanks independent from the ship's hull, or at reduced temperatures, at atmospheric pressure. Examples: propane, butane, propylene, butylenes.

Typical Sizes Oil Tankers

While no standardized system for the classification of oil tankers exists; the fleet is typically divided into several major categories based on carrying capacity. These categories are Ultra Large Crude Carrier *(*ULCCs) and Very Large Crude Carrier (VLCCs), Suezmax, Aframax, Panamax and Handysize tankers. To benefit from economies of scale charterers typically charter the largest possible vessel that can be accommodated in a particular voyages arrival and discharge ports. The main categories of vessels are:

Ultra Large Crude Carrier (ULCCs) and Very Large Crude Carrier (VLCCs)

These are the largest vessels in the world tanker fleet: Cape of Good Hope and Asia. The large carrying capacity of ULCCs and VLCCs make them attractive to traders, however, this large size limits their access into some of the worlds ports. The term ULCC describes tankers which range from 300/500,000 dwt. They are mainly used for long haul operations between The Gulf and the Far East, Europe and North America, discharging their cargo at terminals especially constructed to handle such large vessels. VLCC's are vessels ranging from between 150/299,000 dwt and are employed on similar routes to ULCC's but their relatively smaller size allows for greater flexibility. They can discharge at many terminals within the Mediterranean, North West Europe, West Africa, etc. They also have the advantage of being able to transit the Suez Canal in ballast condition.

Suezmax

Suezmax are midsized tankers with displacement between 120,000 and 200,000 dwt. Suezmax tankers offer the relative economies of scale that can be achieved with VLCCs; however, their slightly smaller size offers increased versatility and access to a majority of the world's ports. Suezmax tankers primarily operate in the Atlantic Basin delivering cargoes from West Africa, the North Sea, and the former FSU.

Aframax

Aframax vessels are mid-size tankers with displacement between 80,000 and 120,000 dwt. Widely considered to be the work horses of the fleet, their size makes them ideally suited to operate in areas of lower crude production or where draft and size restrictions prevent the use of larger vessels.

Panamax and Handysize

These tankers are primarily used for both the transportation of crude oil and petroleum products. Panamax tankers have displacement between 50,000 and 80,000 dwt and trade in short haul. Handysize tankers have displacement between 50,000 and 10,000 dwt. They primarily carry finished petroleum products as their smaller size makes them less economic for the transport of crude.

Shuttle Tankers

Generally they are conventional tankers equipped to be moored to a discharge buoy at deep sea. They load cargo directly from the oil field, where it is stored in reservoirs, where generally the sulphur is removed. It is required a great maneuvering capability in comparison with the traditional tankers, and generally are equipped with dynamic positioning systems. Tanker can be easily identified by the raised compartment at the bow and ramp to handle the cargo hose. Typically, a shuttle tanker as 120,000 dwt and a service speed of about 16 knots, higher that conventional tankers.

Product Tankers

Carry refined oil products that can be classified into 2 main groups:

- White products (diesel oil, gasoline, jet fuels, kerosene, lube oils, etc.);
- Black products (fuel oils, residual products).

Product tankers can also be sub-divided into two main tonnage groups; the larger carrying products in quantities between 26,000 and 50,000 tones, whilst the smaller, so-called 'Handy Size' loads between 12,000 and 25,000 tones.

Clean cargo tanks characterized by:

- No structural elements inside the cargo space (double skin);
- Corrugated bulkheads (transverse and longitudinal);
- Cargo tanks painted or internally coated.

Each cargo tank is provided with a segregated cargo system including a submersible pump and its own cargo line to the manifold, on deck. The number of possible cargo segregations on board is equal to the number of cargo tanks. Typically these tankers are smaller than the crude oil tankers, with dwt less than 70,000 t.

Oil Tankers

Oil tankers are tanker ship for transport of liquid cargo which all cargo space divided longitudinal and transverse watertight bulkheads to tight compartments, called tanks. Loading and unloading liquid cargo tanks of oil tankers have some special system of pipelines and pumping equipment. Oil tankers generally have from 8 to 12 tanks. Each tank is split into two or three independent compartments by fore-and-aft bulkheads. The tanks are numbered with tank one being the forward most.

A cofferdam is a small space left open between two bulkheads, to give protection from heat, fire, or collision. Tankers generally have cofferdams forward and aft of the cargo tanks, and sometimes between individual tanks. A pump room houses all the pumps connected to a tanker's cargo lines. Some larger tankers have two pump rooms. A pump room generally spans the total breadth of the ship. A major component of tanker architecture is the design of the hull or outer structure. A tanker with a single outer shell between the product and the ocean is said to be single-hulled. Most of new tankers are double-hulled, with an extra space between the hull and the storage tanks. There are hybrid designs such as double-bottom and double-sided combine aspects of single and double-hull designs. All single-hulled tankers around the world will be phased out by 2026, in accordance with the International Convention.

Tanker Oil Spills

Main cause of oil spills are:

- Fire/explosion;
- Loading/offloading;
- Structural damage;
- Allisions/collision;
- Grounding.

The International Tanker Owners Pollution Federation (ITOPF) database contains information on approximately 10,000 oil spills from tankers, in the period 1970-2011, combined carriers and barges, some 81% of which were less than seven tones (Figure 4).

The average number of large oil spills (more then 700 tones) during the 2000s was just an eighth of that during the 1970s. This dramatic reduction has been due to the combined efforts of the oil/shipping industry and governments (largely through the IMO) to

Figure 4. Numbers of large spills (over 700 tones) 1970-2011 (Adapted from www.itopf.com)

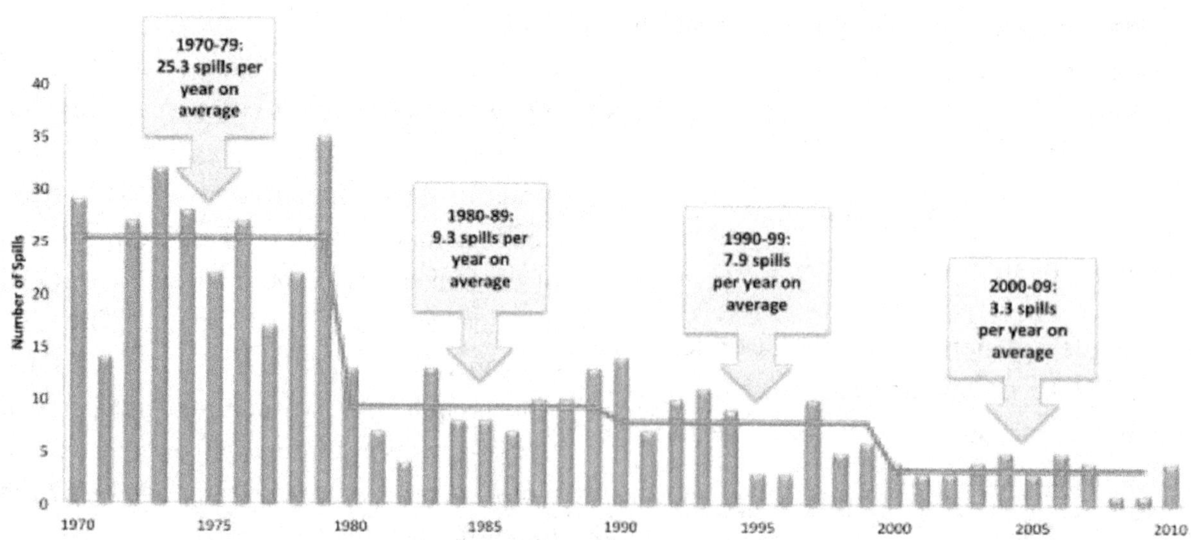

improve safety and pollution prevention. The total amount of oil spilled each year varies considerably, with a few very large spills being responsible for a high percentage of the total annual quantity.

During the period 1970-2011, 50% of larger spills (more than 700 tones) occurred whilst the vessel was underway in open water and 18% whilst underway in inland or restricted waters. The main causes of larger spills were allisions/collisions (29%) and groundings (33%). Other significant causes include hull failures and fire/explosion (from www.itopf.com, 2013).

Transport Safety

Whether oil travels by sea or by land, the safety and security of operations is key task. Every oil company sets its own quality, safety and security criteria for maritime shipping. These criteria include:

- The tanker's construction conditions and features;
- Tanker condition and maintenance;
- Crew recruitment, training, and management;
- Planning routes and the journey itself.

Ship owners are responsible for ensuring that their ships meet these criteria. Independent certification organizations inspect ships on a regular basis and issue sailing permits. The grounding of the Exxon Valdez was a watershed event, triggering fundamental changes in the way oil tankers are managed, operated and maintained. The regulatory bodies recognized a need for continuous improvement in environmental performance, and many new regulations have been introduced over the last two decades. Some of the more significant regulatory actions include: (1) double hull requirements for cargo tanks and subsequently for fuel tanks, (2) cargo tank subdivision requirements intended to mitigate outflow in the event of a collision or grounding, (3) requirements for the coating of ballast tanks and the tops/bottoms of cargo tanks to minimize corrosion, (4) new structural design rules common across classification societies, and (5) requirements for enhanced surveys to maintain structural integrity.

The rules and regulations applied for the design and construction of oil tankers are primarily developed by the International Maritime Organization (IMO) and the classification societies (IMO, 2002; IMO, 2006).

IMO is an agency of the United Nations, charged with developing a regulatory framework that promotes maritime safety and environmental protection. Most of IMO's regulations pertaining to tanker design and construction are contained in the International Convention for the Safety of Life at Sea (SOLAS) and the International Convention for the Prevention of Pollution from Ships (MARPOL 73/78, 2009 and subsequent amendments). IMO regulations focus on the arrangements and outfitting of the vessel as it pertains to safety and environmental performance. For example, these regulations cover load line (freeboard) requirements, intact and damage stability, firefighting and lifesaving equipment. In 2006 IMO adopted regulations on the coating of ballast tanks, and requirements for coating cargo tanks on new buildings will become effective in 2013.

Classification societies develop technical rules and guides for the design and construction of ships, perform plan review to confirm that the design is in conformance with their rules, and survey vessels during construction and trials (commissioning). The classification societies also perform periodic surveys during the life of the vessel to verify continued compliance with the rules and regulations. From time to time, oil companies also appoint specialized inspectors to make sure ship owners comply with regulations. Oil companies share the information gathered by inspectors in a computer database such as SIRE (Ship Inspection Report Exchange), set up in 1993. Before choosing a tanker for shipping crude, the oil company checks all the information available on it. If It may also exclude ships that are too old and that have received negative inspection reports, or turn down a crew that it believes has not received sufficient training.

Some of the most important measures of preventing accidental pollution are: double hull, inert gas systems, equipment duplication and protective location of segregated ballast tanks.

Double Hull

The United States and various international bodies have established deadlines for implementation of the double-hull design. In 2007, the European Union (EU) made it mandatory for ships carrying heavy oils to be double-hulled to use ports in EU nations or to drop anchor in their waters. The United States passed a law in 1990 to phase out single-hull tankers by 2015. The deadline was later moved up to 2010. The International Maritime Organization (IMO), a United Nations agency that regulates global shipping, also has called for the eradication of single hulls by 2010.

A mandatory design feature on newly built oil tankers; double-hull construction means the ship has two hulls, one inside the other. This offers an extra layer of protection against damage that might otherwise result in catastrophic oil spills.

Inert Gas Systems

It is not just fire which is dangerous, in certain circumstances a single spark can cause a disaster, for even tanks which contain no oil are filled with flammable gas which can explode unless proper procedures are followed. The normal method is to fill these tanks with inert (non-explosive) gas from the ship's boiler flue: it is cleaned and then pumped into the empty tanks, or into the spaces left above the oil in loaded tanks. An inert gas system is required on all new tankers and most existing tankers of 20,000 dwt and above.

Equipment Duplication

IMO has introduced several measures over the years which are designed to ensure that, in the event of mechanical failure, the ship can still be controlled. SOLAS makes it necessary for essential parts of the steering gear of tankers to be duplicated. As with other ships, much of the navigational equipment of tankers must also be duplicated

Protective Location of Segregated Ballast Tanks

The 1978 MARPOL Protocol introduced the concept known as protective location of segregated ballast tanks. This means that the ballast tanks (which are empty on the cargo-carrying leg of the voyage and only loaded with water ballast for the return leg) are positioned where the impact of a collision or grounding is likely to be greatest. In this way the amount of cargo spilled after such an accident will be greatly reduced. The 1983 MARPOL amendments ban the carriage of oil in the forepeak tank - the ship's most vulnerable point in the event of a collision.

TRANSPORTATION BY GAS CARRIERS

Shipping LNG by sea began in the 1950s as a result of significant pressure to make use of newly discovered gas reserves and to discontinue flaring of what might otherwise be a useful product.

The first commercial transport of LNG took place in February 1959. This was the transport on the LNG carrier "Methane Pioneer" (capacity of 5,000 m^3) from Lake Charles in the USA to Conveys Island in Great Britain. The "Methane Pioneer" was followed by the carriers "Methane Princess" and "Methane Progress" (each of a capacity of 27,400m^3) which had their first passage in February 1962.

"Methane Princess" "retired" in 1997 after 27 years of use, not due to the state of the carrier but due to the fact that using such an old carrier is not cost-effective.

The first commercial contract for the import of LNG into Japan was signed in 1967 and after that, Japan became the leading power in LNG import. Until this day, Japan remains the biggest importer with a total import of one third of the world's LNG.

The basic design options were standardized by the 1970s, and these have continued with few major changes other than to increase capacity. The size of the LNG carriers was standardized to 125,000 m^3 to 138,000 m^3. Until 2000, the number of LNG carriers was relatively constant, around 120 ships. After that, there was a great boom and expansion in LNG carriers and today, over 300 LNG carriers are sailing in the world. The size and the capacity have also increased and today, the so-called Q-max (260,000 m^3) and Q-flex (215,000m^3) are in use. There are more than 300 LNG carriers currently in operation worldwide.

Gas Carrier Types

Gas carriers can be grouped into six different categories according to the cargo carried and the carriage condition, i.e.:

1. Fully pressurized ships;
2. Semi-refrigerated/semi-pressurized ships;
3. Semi-pressurized/fully refrigerated ships;
4. Fully refrigerated LPG ships;
5. Ethylene ships;
6. LNG carriers

Ship types a), b) and c) are most suitable for the shipment of smaller size cargoes of LPG and chemical gases on short-sea and near-sea routes whereas ship type (d) is used extensively for the carriage of large-size cargoes of LPG and ammonia on the deep-sea routes.

Fully Pressurized Carriers

These ships are the simplest of all gas carriers in terms of containment systems and cargo-handling equipment and carry their cargoes at ambient temperature. Type C tanks - pressure vessels fabricated in carbon steel with a typical design pressure of 17.5 barg, corresponding to the vapor pressure of propane at 45oC, must be used. Ships with higher design pressures are in service: 18 barg is quite common - a few ships can accept up to 20 barg. No thermal insulation or reliquefaction plant is necessary and cargo can be discharged using either pumps or compressors. Because of their design pressure the tanks are extremely heavy. As a result, fully pressurized ships tend to be small with maximum cargo capacities of about 4,000 m3 and they are used to carry primarily LPG and ammonia.

Semi-Refrigerated Carriers

These ships are similar to fully pressurized ships in that they incorporate Type C tanks - in this case pressure vessels designed typically for a maximum working pressure of 5-7 barg. The ships range in size up to 7,500 m3 and are primarily used to carry LPG. Compared to fully pressurized ships, a reduction in tank thickness is possible due to the reduced pressure, but at the cost of the addition of refrigeration plant and tank insulation. Tanks on these ships are constructed of steels capable of withstanding temperatures as low as –10°C. They can be cylindrical, conical or spherical in shape.

Semi-Pressurized/Fully Refrigerated Carriers

Constructed in the size range 1,500 to 30,000 m^3, this type of gas carrier has evolved as the optimum means of transporting the wide variety of gases, from LPG and VCM to propylene and butadiene, found in the busy coastal gas trades around the Mediterranean and Northern Europe. Like the previous two types of ship, SP/FR gas tankers use Type C pressure vessel tanks and therefore do not require a secondary barrier. The tanks are made either from low temperature steels to provide for carriage temperatures of -48°C which is suitable for most LPG and chemical gas cargoes or from special alloyed steels or aluminum to allow the carriage of ethylene at –104° C. The SP/FR ship's flexible cargo handling system is designed to be able to load from, or discharge to, both pressurized and refrigerated storage facilities.

Fully Refrigerated LPG Carriers

Fully refrigerated (FR) ships carry their cargoes at approximately atmospheric pressure and are generally designed to transport large quantities of LPG and ammonia. Four different cargo containment systems have been used in FR ships: independent tanks with double hull, independent tanks with single side shell but double bottom and hopper tanks, integral tanks and semi-membrane tanks, both these latter having a double hull. The most widely used arrangement is the independent tank with single side shell with the tank itself a Type A prismatic free-standing unit capable of withstanding a maximum working pressure of 0.7 barg. The tanks are constructed of low-temperature steels to permit carriage temperatures as low as –48°C. FR ships range in size from 10,000 to 100,000 m^3.

A typical fully refrigerated LPG carrier would have up to six cargo tanks, each tank fitted with transverse wash plates, and a centre line longitudinal bulkhead to improve stability. The tanks are usually supported on wooden chocks and are keyed to the hull to allow expansion and contraction as well as prevent tank movement under static and dynamic loads. The tanks are also provided with anti-flotation chocks. Because of the low temperature carriage conditions, thermal insulation and reliquefaction plant must be fitted. The FR gas carrier is limited with respect to operational flexibility.

However, cargo heaters and booster pumps are often used to allow discharge into pressurized storage facilities. Where Type A tanks are fitted, a complete secondary barrier is required. The hold spaces must be inerted when carrying flammable cargoes. Ballast is carried in double bottoms and in top-side tanks or, when fitted, side ballast tanks.

Ethylene Carriers

Ethylene ships tend to be built for specific trades and have capacities ranging from 1,000 to 30,000 m^3. This gas is normally carried fully refrigerated at its atmospheric pressure boiling point of –104°C. If Type C pressure vessel tanks are used, no secondary barrier is required; Type B tanks require a partial secondary barrier; Type A tanks require a full secondary barrier and because of the cargo carriage temperature of – 104ºC the hull cannot be used as a secondary barrier, so in this case a separate secondary barrier must be fitted. Thermal insulation and a high capacity reliquefaction plant are fitted on this type of vessel. As mentioned, many ethylene carriers can also carry LPG cargoes thus increasing their versatility. Ballast is carried in the double bottom and wing ballast tanks and a complete double hull are required for all cargoes carried below –55°C whether the tanks are of Type A, B or C.

LNG Carriers

LNG carriers are specialized vessels built to transport large volumes of LNG at its atmospheric pressure boiling point of –163°C. LNG carriers are fitted with independent cargo tanks or with membrane tanks.

All LNG ships have double hulls throughout their cargo length, which provides adequate space for ballast; the membranes have a full secondary barrier, the spheres a drip-pan type protection. Another characteristic common to all is that they burn the cargo boil-off as fuel (permitted with methane cargo, being lighter than air at ambient temperature - but not propane or butane which are heavier than air).

During the development of technology transportation of liquefied natural gas, there have been many experimental structural container, but industry was soon excluded from the application of complex and demanding projects so they can maintain two types of containers: an independent, self-supporting and membrane tanks.

Cargo Tank Types

Cargo tank types are categorized as following

- Independent tank (Type A, B and C)
- Membrane tank (Tank type GT No96, Technigaz Mark III tanks, and GTT (CSI) Tanks)

Independent Tanks (Self-Supporting)

Self-supporting, independent tanks are heavy, solid structure, designed to withstand pressures in the transport of liquefied natural gas. They can be shaped prism or sphere (see Figure 5). The hull is designed to support the massive tanks. Self-supporting means that the container walls over in full loads that occur inside the tank due to pressure, and independent means that are built independently of the hull and the hull is only for downloading their weight. Their maximum design pressure is 0.7 barg in the vapor space. In practice, however, such operating pressure usually is maintained below 0.25 barg. They are divided into three types: type A, B and type C.

Type A tanks require a full secondary barrier because the stress analysis is not precise in this design. Tanks of this type are constructed primarily of flat surfaces and were built into ships Methane Princess and Methane Progress, and made them, as well as ships, the company Conch (http://shipbuildinghistory.com/today/highvalueships/lngactivefleet.htm, 2013). Both ships had problems with

Figure 5. Spherical tank construction (Adapted from www.bmtmarinerisk.com)

leakage of liquefied gas during testing. Before mentioned problems plus another total failure in constructing a third larger ship are terminated developments of this design.

Type ‘B’ tanks are divided into prismatic and spherical (Moss tanks). These types of tanks are subjected to more rigorous stress tests. Therefore, a type B tank requires only a partial secondary barrier in the form of a drip tray. This type of containment system is the subject of much more detailed stress analysis compared to Type ‘A’ systems. These controls must include an investigation of fatigue life and a crack propagation analysis. The most common arrangement of Type ‘B’ tank is a spherical tank. This tank is of the Kvaerner Moss design. Because of the enhanced design factors, a Type ‘B’ tank requires only a partial secondary barrier in the form of a drip tray.

The hold space in this design is normally filled with dry inert gas. However, when adopting modern practice, it may be filled with dry air provided that inerting of the space can be achieved if the vapour detection system shows cargo leakage. A protective steel dome covers the primary barrier above deck level and insulation is applied to the outside of the tank. The Type ‘B’ spherical tank is almost exclusively applied to LNG tankers; seldom featuring in the LPG trade.

Type ‘C’ tanks are normally spherical or cylindrical pressure vessels having design pressures higher than 4 barg. The cylindrical vessels may be vertically or horizontally mounted. This type of containment system is always used for semi-pressurized and fully pressurized gas carriers. In the case of the semi-pressurized tankers it can also be used for fully refrigerated carriage, provided appropriate low temperature steels are used in tank construction. Type ‘C’ tanks are designed and built to conventional pressure vessel codes and, as a result, can be subjected to accurate stress analysis. Furthermore, design stresses are kept low. Accordingly, no secondary barrier is required for Type ‘C’ tanks and the hold space can be filled with either inert gas or dry air and for fully pressurized tankers normal air may be allowed.

Membrane Tanks

The concept of the membrane containment system is based on a very thin primary barrier (membrane - 0.7 to 1.5 mm thick) which is supported through the insulation. Such tanks are not self-supporting like the independent tanks. Membrane containment systems must always be provided with a secondary barrier to ensure the integrity of the total system in the event of primary barrier leakage. The membrane is designed in such a way that thermal expansion or contraction is compensated without over-stressing the membrane itself. There are two principal types of membrane system Tank type GT No96 and Technigaz Mark III Tanks, in common use both named after the companies who developed them and both designed primarily for the carriage of LNG. In both membrane systems, boil-off of 0, 15% is guaranteed although in reality its quantity is much smaller. Until 2000, there were only a few membrane ships but due to their significant flexibility and changes in the

LNG spot market, in the past few years they have "surpassed" spherical ships in number. This type of ship is mostly produced in South Korea, but China has "produced" three membrane LNG carriers in the past three years (Huang et al., 2007). After the union between Gaz Transport and Technigaz in 1994, a third type of membrane-type carriers, *GTT (CSI) Tanks* emerged. The two basic concepts of GTT technology are the Mark III and No. 96 System, and the third one is a combination of these two and is appropriately named CSI (Combine System One). The membrane-type ships were developed during the 1960s, and both concepts use a thin flexible metal "membrane" which is in contact with the cargo. The system has the characteristics of a sandwich where the cargo presses on the membrane; the insulation material presses on the membrane and in the end, everything leans on the ship's inner hull.

Risk of Gas Carriers Operation

An LNG carrier is a tank designed for transporting liquefied natural gas. First, it is the ship with all the risks of shipping: grounding, collision, natural events, etc. Secondly, it is a tank containing the cryogenic gas with all its hazardous properties.

The LNG shipping industry has an excellent safety record. Since the first cargoes of LNG were shipped on a regular commercial basis in 1964, over 56,000 shipments have been made without a single incident of LNG being lost through a breach or failure of the ship's tanks.

There have been three major grounding incidents, but none resulted in loss of cargo. The robust design of the ships and cargo tanks and the LNG industry's extraordinary attention to safety details have collectively served to prevent the release of cargo and to facilitate this noteworthy safety record.

Liquefied natural gas (LNG) is composed of mostly methane and is a cryogenic liquid at approximately -162ºC. When vaporized, its flammability range is between approximately 5% and 15% by volume, i.e. a mixture with air within this range of concentrations is flammable. Thus, in addition to possible damages due to its cryogenic temperatures, LNG spills are associated with hazards such as pool fires and ignition of drifting vapour clouds. In its liquid state, LNG is not explosive, and LNG vapour will explode only if ignited in a mixture with air within the flammability range and within an enclosed or semi-enclosed space. Natural gas may also present an asphyxiation hazard. LNG is not toxic and will not be persistent if spilled in a marine environment. LNG weighs less than water, thus LNG spilled on water will float. When handling LNG risks coming from its three characteristics: dispersion, flammable and extremely low temperatures.

LNG Spill

A spill of LNG on water evaporates about five times faster than on land because of the higher heat transfer rates associated with the water and a tendency for the water not to completely freeze. The high heat capacity and the circulation of the water at the surface usually inhibits significant ice formation. Depending upon the quantity spilled and the conditions, LNG has a tendency to spread and form a pool on the water surface because it is insoluble. This pool of LNG will evaporate and create a vapour cloud which expands, begins to dilute, and moves with the ambient wind conditions. The actual size, rate of expansion, movement of a vapour cloud depends upon incident-specific conditions. A first approximation is that the size of the LNG pool will increase until the vapour generation rate equals the LNG release rate. If ignition sources are not present when the part of the vapour cloud that is within flammable limits (5-15% natural gas in air), then no fire will occur (www.GIIGNL.org, 2013).

Some LNG spills on water may have a Rapid Phase Transition (RPT). This is essentially a flameless overpressure caused by the very high transient rates of heat transfer from the water to the LNG. This causes the LNG to change from the liquid to the gas phase so quickly that a rapidly expanding vapour cloud is generated. The cloud can expand so quickly that a sonic boom and localized overpressure is created.

The RPT "explosion" phenomenon for LNG on water has been observed in a number of situations and has been studied extensively in both laboratory and large-scale tests. While this phenomenon is spectacular to observe at large-scales tests, the actual energy release is modest. An RPT is a very unpredictable phenomenon and the exact circumstances of its formation remain unclear. The temperature of the water and the actual composition of the LNG are important factors in predicting whether or not an RPT will take place. Work has also been performed to examine the impact of an RPT on the LNG ship and pier structure. Measured overpressures are insufficient to cause more than minor damage either to the ship or pier.

Small spills of LNG have occurred onto the decks of LNG carriers during the loading and unloading process. The cold of the LNG has caused the carbon steel deck plates to crack in some cases. Fire hoses and deck water sprays are used to warm the spill area and increase the evaporation rate of the LNG.

Unlike deck plates, LNG tanks within a ship are made of metals which do not crack or become brittle in the presence of cryogenic temperatures. All LNG carriers have two systems to prevent leakage from the LNG tanks. If one tank fails for whatever reason, the LNG is held safely by the secondary protection system. LNG tank types are designed to leak before failure, thereby allowing the crew to empty the affected tanks before significant damage can take place (Woodward & Pitblado, 2010).

The LNG tanks and insulation spaces are kept free of air to prevent any leak from creating a flammable mixture. The tanks themselves always contain liquid LNG and some LNG vapor above the liquid.

The insulation spaces on membrane ships are filled with nitrogen (an inert gas) and monitored continuously for the presence of methane (flammable vapor). If an LNG spill does not ignite, it poses an asphyxiation or freezer – burn risk to personnel very near the spill, for example, on unloading platforms or to emergency response personnel. Direct contact with metal at cryogenic temperatures can damage skin tissue more rapidly than exposure to cold vapor. For asphyxiation, the LNG vapors must dilute the oxygen concentration in the breathing zone of people below 15% oxygen for impaired behavior, below 10% for nausea and vomiting or below 6% oxygen for death. Appropriate protective clothing mitigates the exposure risk to cryogenic burns, but this does not mitigate against a cryogenic vapor inhalation hazard. Recommendations are available for personal protective equipment for particular tasks, including breathing air apparatus.

Boiling Liquid Expanding Vapor Explosion

The acronym BLEVE means Boiling Liquid Expanding Vapour Explosion, which is the sudden release of a large mass of pressurized flammable liquid to the atmosphere. A primary cause is if an external flame impinges on the shell of a pressurized tank above the liquid level, weakening the shell and resulting in sudden rupture. A fuller definition explains the mechanism (Mannan, S., 2005): When a vessel containing liquid under pressure is exposed to fire, the liquid heats up and the vapor pressure rises, increasing the pressure in the vessel. When this pressure reaches the set pressure of the relief valve, the valve operates. The liquid level falls as the vapor is released to the atmosphere. The liquid is effective in cooling that part of the vessel wall which is in contact with it, but the vapor is not. The temperature in the proportion of the vessel wall which has the benefit of liquid cooling falls as the liquid vaporizes. After a time, metal which is not cooled by liquid becomes exposed to the fi re; the metal becomes hot and weakens and may then rupture. This can happen even if the relief valve is operating correctly. A pressure vessel is designed to withstand the relief valve set pressure, but only at the design temperature conditions. If the metal has its temperature raised, it may lose it strength sufficiently to rupture.

Sandia Reports

The Sandia National Laboratories report from 2004 (Hightower et al., 2004) concluded that the most significant impacts to public safety exist within 500 m of a spill, with much lower impacts at distances beyond 1,600 m even for very large spills. Because of the increasing size and capacity of many new LNG carriers, the Department of Energy requested that Sandia assess the general scale of possible hazards for a breach and spill from newer LNG carriers with capacities ranging up to 265,000 m3. According to the research and analyses presented in Sandia's 2004 LNG report, was reassessed emerging accidental and intentional threats and then conducted detailed three-dimensional breach analyses for several new large capacity (up to 265,000 m3) LNG carrier designs. Based on the estimated breach sizes, breach locations, and LNG carrier configurations, was estimated LNG spill rates and volumes and conducted thermal hazard and vapor dispersion analyses. The results include analysis of the hazards of potential LNG spills at both near-shore and offshore facilities and operations, which should help improve the understanding of the range of hazards for different marine LNG import options.

RAILROAD TANK CARS TRANSPORTATION

Railroad tank cars are constructed of carbon steel or aluminum and may be pressurized or unpressurized. Modern tank cars can hold up to 171,000 l of compressed gas at pressures up to 40 bar. Non-pressure tank cars may be individual units with one or multiple compartments or a string of interconnected tank cars, called a tank train. Tank cars are loaded individually, and entire tank trains can be loaded and unloaded from a single point. Both pressure and non-pressure tank cars may be heated, cooled, insulated and thermally protected against fire, depending on their service and the products transported.

All railroad tank cars have top- or bottom-liquid or vapour valves for loading and unloading and hatch entries for cleaning. They are also equipped with devices intended to prevent the increase of internal pressure when exposed to abnormal conditions. These devices include safety relief valves held in place by a spring which can open to relieve pressure and then close; safety vents with rupture discs that burst open to relieve pressure but cannot reclose; or a combination of the two devices. A vacuum relief valve is provided for non-pressure tank cars to prevent vacuum formation when unloading from the bottom. Both pressure and non-pressure tank cars have protective housings on top surrounding the loading connections, sample lines, thermometer wells and gauging devices. LNG is shipped as a cryogenic gas in insulated tank truck and rail pressure tank cars. Pressure tank trucks and rail tank cars for LNG transport have a stainless steel inner reservoir suspended in an outer reservoir of carbon steel. The annular space is a vacuum filled with insulation to maintain low temperatures during shipment. To prevent gas from igniting back to the tanks, they are equipped with two independent, remotely controlled fail-safe emergency shut-off valves on the filling and discharge lines and have gauges on both the inside and outside reservoirs.

LPG is transported on land in specially designed rail tank cars (up to 130 capacity) or tank trucks (up to 40 capacity). Tank trucks and rail tank cars for LPG transport are typically uninsulated steel cylinders with spherical bottoms, equipped with gauges, thermometers, two safety relief valves, a gas level meter and maximum fill indicator and baffles.

Rail tank cars transporting LNG or LPG should not be overloaded, since they may sit on a siding for some period of time and be exposed to high ambient temperatures, which could cause overpressure and venting. About 700 LNG tank trucks are in service worldwide. BLEVE can only occur in a road tanker under specific circumstances:

- A fire is ignited beneath or impinging on the road tanker;
- The insulation around the vessel fails to prevent heat reaching the LNG;
- The fire burns sufficiently long that the LNG begins to boil;

- The LNG boils so vigorously that valves installed to relieve the increasing pressure are unable to cope, and pressure within the tank rapidly rises;
- The upper part of the shell not in contact with LNG becomes weakened by overheating;
- The tanker may then rupture with an intense, localized overpressure and fire.

TANK TRUCKS

Petroleum products and crude oil tank trucks are typically constructed of carbon steel, aluminum or a plasticized fibreglass material, and vary in size from 1,900-l tank wagons to jumbo 53,200-l tankers. The capacity of tank trucks is governed by regulatory agencies, and usually is dependent upon highway and bridge capacity limitations and the allowable weight per axle or total amount of product allowed.

There are pressurized and non-pressurized tank trucks, which may be non-insulated or insulated depending on their service and the products transported. Pressurized tank trucks are usually single compartment, and non-pressurized tank trucks may have single or multiple compartments. Regardless of the number of compartments on a tank truck, each compartment must be treated individually, with its own loading, unloading and safety-relief devices. Compartments may be separated by single or double walls. Regulations may require that incompatible products and flammable and combustible liquids carried in different compartments on the same vehicle be separated by double walls. When pressure testing are compartmented, the space between the walls should also be tested for liquid or vapor.

Tank trucks have either hatches which open for top loading, valves for closed top- or bottom-loading and unloading, or both. All compartments have hatch entries for cleaning and are equipped with safety relief devices to mitigate internal pressure when exposed to abnormal conditions. These devices include safety relief valves held in place by a spring which can open to relieve pressure and then close, hatches on non-pressure tanks which pop open if the relief valves fail and rupture discs on pressurized tank trucks. A vacuum relief valve is provided for each non-pressurized tank truck compartment to prevent vacuum when unloading from the bottom. Non-pressurized tank trucks have railings on top to protect the hatches, relief valves and vapour recovery system in case of a rollover. Tank trucks are usually equipped with breakaway, self-closing devices installed on compartment bottom loading and unloading pipes and fittings to prevent spills in case of damage in a rollover or collision.

Transportation of petroleum products by motor vehicle or railroad tank car is regulated by government agencies throughout most of the world. Agencies such as the US DOT have established regulations governing the design, construction, safety devices, testing, preventive maintenance, inspection and operation of tank trucks and tank cars. Regulations governing railroad tank car and tank truck operations typically include tank pressure and pressure relief device testing and certification before being placed into initial service and at regular intervals thereafter.

FUTURE RESEARCH DIRECTIONS

Due to the growing needs of the world's oil and gas production is increasing every year. Almost 2/3 of the known conventional oil and gas reserves are located in the Russian confederation and the Middle East region where the long distance between the reservoirs and final consumption markets is the major obstacle in efficient exploitation of the resources. Hence the need to keep increasing the network of oil and gas pipelines as well as a fleet of tankers. Increasing transport capacity increased potential risks and incidents. It is improving and preventive measures activities, making the task of national and international organizations as well as the owner of the pipeline and tanker fleet. Application of new technologies for these purposes is certainly one of the main drivers to improve preventive measures and activities. Do now applied technology, such as SCADA systems, leak detection systems, intelligence pig technology increased safety of transport of oil and gas pipelines. It is imperative that technology be applied to all pipelines and are continuously improved. One of the serious problems of pipeline infrastructure is ageing. A very large part of the oil and gas pipeline system in the USA and Russia is over 40 years old or at the end of its design life. It's a big engineering challenge. How will solve the problem of rehabilitation: applying some of the newer technologies, inserting flexible composite pipe in the old existing pipelines, or simply decommissioning?

The LNG technology is necessary to complete the experiments SANDIA and Government Accountability Office (GAO) studies. Based on these studies will be provided making more quality regulations to protect the environment and people. About 71% of Earth's surface is covered by water. To a large space that will be the development of technology research certainly increasingly exploited.

There are transportation challenges which should be fulfilled especially for remote and offshore reserves. While the option for transporting oil is limited to pipeline and tanker transportation, there are options to utilize offshore and remote natural gas production and transport it to markets or to another location for further processing or utilization. Volume reduction options as an alternative to pipeline then have to be considered next, followed by true conversion alternatives where the "methane" molecule is chemically converted into another liquid products. These technologies are called gas-to-liquids or GTL. Although GTL is used more for conversion of gas to middle distillates such as naphtha, kerosene, diesel and gasoline, it is alternatively used for other gas conversion technologies such as methanol and dimethyl ether (DME) production.

CONCLUSION

Every day millions of barrels of oil must be transported from their origin in oilfields around the world, moved to refineries in many countries, and distributed to consumer. Tankers by the thousands ply the seas, shipping the majority of the world's internationally traded crude oil from producing to consuming nations. Pipelines and trains handle much of the rest, moving oil and gas overland and sometimes underground from oil fields to ports, often across international borders. Liquefied gas is transported by specially constructed vessels, offshore or on land.

In this Chapter the causes of accidents during transportation are presented. Consequences of accidents are leakage, exposure and fire. Pollution vary from small spills to catastrophic accidents as in the case of some famous tanker accidents. No matter what kind of scale for accidents they can have a big environmental impact.

In case leakage of oil in soil it is possible that it comes in contact with groundwater. One liter can contaminate a huge amount of ground water and the recovery of water is impossible. What does oil in the sea, clearly illustrated by the following data: only 8 g of oil is enough to contaminate 1 m^3 sea. One m^3 discharged oil depletes oxygen from 400.000 m^3 sea!

To avoid or minimize environmental impact of oil and gas transportation it is very important to implement appropriate activities from designing phase, construction and operational phase. The greatest number of activities required by national and international organizations: API, ANSI, DOT, CONCAWE, IMO. There have been the conventions of which the MARPOL one of the most important conventions of sea protection. For instance, the International Convention of Safety of Life at Sea (SOLAS) includes special requirements for tankers. Fire safety provisions, for example, are much more stringent for tankers than for ordinary dry cargo vessels, since the danger of fire on board ships carrying oil and refined products is greater. In addition to a variety of regulations that prescribe preventive action to prevent or reduce the risk of accident during shipping, significant role had the development of certain types of technology such as, SCADA system, communication technology, smart pig, new leak detection methods, etc. Some of new technologies are applied on ships, especially on oil and LNG tankers. These ships are simultaneously processing units. For this it requires special training of specialized personnel on these ships. For all before mentioned reasons it is necessary to increased transportation security and this chapter contribute to it.

REFERENCES

Alaska Department of Environmental Conservation. (1999). *Technical review of leak detection technologies*. Washington, DC: EPA.

API 1130. (2002). *Computational pipeline monitoring for liquid pipelines*. Washington, DC: American Petroleum Institute.

CONCAWE's Oil Pipeline Management Group's Special Task Force on Oil Pipeline Spillages. (2010). *Performance of European cross-country oil pipelines, statistical summary of reported spillages in 2008 and since 1971*. Brussels, Belgium: EU.

Dhillon, B. S. (2007). *Human reliability and error in transportation systems*. London: Springer.

Dziubinski, M., Fratczak, M., & Markowski, A. S. (2006). Aspects of risk analysis associated with major failures of fuel pipelines. *Journal of Loss Prevention in the Process Industries*, *19*, 399–408. doi:10.1016/j.jlp.2005.10.007

Fuel Pipeline Leak Detection. (n.d.). Retrieved from http://www.vistaleakdetection.com

Hightower, M. et al. (2004). *Guidance on risk analysis and safety implications o a large liquefied natural gas (LNG) spill over water*. Albuquerque, NM: Sandia National Laboratories.

Hopkins, P. (2007). Oil and gas pipelines: Yesterday and today. London: Pipeline Systems Division (PSD), Pens pen Ltd.

Huang, S., Chiu, C., & Elliot, D. (2007). *LNG: Basics of liquefied gas*. Austin, TX: The University of Texas at Austin.

IMO. (2002). *Guidelines for formal safety assessment (FSA) for use in the IMO rule-making process. MSC/Circ.1023*. International Maritime Organization.

IMO. (2006). *Amendments to the guidelines for formal safety assessment (FSA) for use in the IMO rule making process. MSC–MEPC.2/Circ 5 (MSC/Circ.1023 – MEPC/Circ.392)*. International Maritime Organization.

International Maritime Organisation (IMO). (2013). *Global integrated shipping information system*. Retrieved from http://gisis.imo.org/Public

International Tanker Owners Pollution Federation Limited (ITOPF). (2013). *Handbook 2012/13*. Author.

Leak Detection Services. (n.d.). Retrieved from http://www.tracerresearch.com

LNG Carriers in Service or Under Construction. (n.d.). Retrieved from http://shipbuildinghistory.com/today/highvalueships/lngactivefleet.htm

Luketa, A. et al. (2008). *Breach and safety analysis of spills over water from large liquefied natural gas*. Albuquerque, NM: Sandia National Laboratories.

MARPOL. 73/78. (2009). Regulations for the prevention of pollution by oil. International Maritime Organization.

Martin, D.E. (1998). *Methods of prevention, detection and control of spillages in European oil pipelines*. Brussels: EU.

Mazzucchelli, L., Pizzorni, D., Comini, A., & Scanarotti, N. (2002). *Rap project: An innovative approach to the risk assessment of pipeline*. Paper presented at the SPE International Conference on Health, Safety and Environment in Oil and Gas Exploration and Production. Kuala Lumpur, Malaysia.

Muhlbauer, W. K. (2004). *Pipeline risk management manual* (3rd ed.). Oxford, UK: Elsevier Inc.

NACE RP0502-2002. (2002). *Standard recommended practice: Pipeline external corrosion direct assessment methodology*. Author.

Novačić, M, Crnčević, D. (2001.) *Gas pipeline monitoring after rehabilitation*, Proceedings of the XVII International Meeting of Gas Experts, Opatija, Croatia

49. Part, C. F. R. 195. (2002). Transportation of hazardous liquids by pipeline: Code for federal regulations title 49. Washington, DC: US Printing Office.

Pipeline Leak Detection. (n.d.). Retrieved from http://www.wavealert.com

Std, A. P. I. 1160. (2001). Managing system integrity for hazardous liquid pipelines. Washington, DC: American Petroleum Institute.

Woodward, J. L., & Pitblado, R. M. (2010). *LNG risk based safety: Modeling and consequence analysis*. Hoboken, NJ: John Wiley & Sons, Inc. doi:10.1002/9780470590232

Zhang, X. J. (1993). Statistical leak detection in gas and liquid pipelines. *Pipes & Pipelines International*, 26-29.

ADDITIONAL READING

Alaska Department of Environmental Conservation. (1999). *Technical Review of Leak Detection Technologies* (Vol. 1). USA: EPA.

49. CFR. Part 193 (2005). Liquefied natural gas facilities, Federal safety standards, USA.

Clifton, A. H. (2005). Hazard analysis techniques for system safety, Wiley-Interscience, A John Wiley & Sons Inc. Publication, New Jersey, USA.

CONCAWE's Oil Pipeline Management Group by D.E.Martin,(1998). *Methods of prevention, detection and control of spillages in europian oil pipelines*, Concawe, EU.

Dumitru I., Negut N. (2005). *The Structural Integrity Approach Using the S.I.N.T.A.P. Programme*, Buletinul Stiintific Al Universitat Politehnica din Timisoara, Seria Mecanica, Tom 54 (60)

Geiger, G., Werner, T., & Matko, D. (2003). *Leak Detection and Locating – A Survey*, PSIG Annual Meeting, 15-17 October 2003, Bern, Switzerland.

Government Accountability Office (GAO), (2007).Public Safety Consequences of a Terrorist Attack on a Tanker Carrying

International Maritime Organisation. Maritime Security from http://www.imo.org/Newsroom/mainframe.asp?topic_id=551

International Tanker owners Pollution Federation Limited (ITOPF) (2013), *Handbook 2012/13*

ISGINTT. (2010). *International Safety Quide for Inland Navigation Tank-barges and Terminals Central Commission for the Navigation of the Rhine*, Strasbourg And Oil Companies International Marine Forum

Liou, C. P. (1993). *Pipeline Leak Detection Based on Mass Balance*, Proceedings of the 1993 International Conference on Pipeline Infrastructure II, ASCE

Mannan, S. (Ed.). (2005). *Lees loss prevention in the process industries*. Elsevier Publishers.

MARPOL. 73/78 (2009). Regulations for the Prevention of Pollution by Oil, Annex I of MARPOL 73/78, International Maritime Organization

Mazzucchelli, L., Pizzorni, D., Comini, A., & Scanarotti, N. (2002). *Rap Project: An Innovative Approach To The Risk Assessment Of Pipeline*, SPE International Conference on Health, Safety and Environment in Oil and Gas Exploration and Production, Kuala Lumpur, Malasya

Simic, Z., Mikulicic, V., & Vukovic, I. (2007). *Use of Probabilistic Safety Assessment for Infrastructure Risk Modeling, Computational Models of Risks to Infrastructure*. Amsterdam: IOS Press.

Society of International Gas Tanker and Terminal Operators (SIGTTO)from website-www.sigtto.org

Stewart, M., & Melchers, R. (1997). *Probabilistic Risk Assessment of Engineering Systems*. Australia: Chapman & Hall.

TRFL (2003). *Technishe Regeln fur Fernleitungen*

www.GIIGNL.org

www.sigtto.org

KEY TERMS AND DEFINITIONS

Boiling Liquid Expanding Vapor Explosion (BLEVE): Is an explosion caused by the rupture of a vessel containing a pressurized liquid above its boiling point.

Boiling Point: Is the temperature at which the vapor pressure of a liquid equals that of the atmosphere above its surface; this temperature varies with pressure.

Crude Oil: Basic raw material pumped from the earth. There many different grades of vrude, each containing various vapors, liquids and solids. This crude is changed at a refinery into products.

Cryogenic Liquids: Liquefied gases that are kept in their liquid state at very low temperatures and have a normal boiling point below -238 degrees Fahrenheit (-150 degrees Celsius) normally stored at low pressures. All cryogenic liquids are gases at normal temperature and pressure. These include methane, oxygen, nitrogen, helium and hydrogen.

Environmental Impact: Is a change in the make-up, working, or appearance of the environment. These changes may be planned or accidental.

Inert Gas: Is a gas (e.g. nitrogen) or mixture of gases, containing insufficient oxygen to support combustion makes it possible to remotely monitor and control pipeline facilities.

Primary Barrier: Is the inner element designed to contain the cargo when the cargo containment system includes two boundaries.

Remote Terminal Unit (RTU): A SCADA system component, typically installed at a field site, that gathers process data from instruments for transfer to the MTU.

SCADA: An acronym for Supervisory Control and Data Acquisition, the technology that makes it possible to remotely monitor and control pipeline facilities.

Secondary Barrier: Is the liquid-resisting outer element of a cargo containment system designed to afford temporary containment of any envisaged leakage of liquid cargo through the primary barrier and to prevent the lowering of the temperature of the ship's structure to an unsafe level. Types of secondary barriers are more fully defined in the 1MO Codes.

Chapter 14
Oil and Gas Storage Tank Risk Analysis

Katarina Simon
University of Zagreb, Croatia

ABSTRACT

Storage tanks are widely used in the oil refinery and petrochemical industry in storing a multitude of different products ranging from gases, liquids, solids, and mixtures. Design and safety concerns have become a priority due to tank failures causing environment pollution as well as fires and explosions, which can result in injuries and fatalities. The chapter illustrates different types of crude oil and oil product storage tanks as well as the risks regarding the storage itself. Considering that the natural gas, in its gaseous state, is stored in underground storages like oil and gas depleted reservoirs, aquifers or salt caverns, and there are numerous publications and books covering the subject in detail, this chapter only illustrates the storage of liquefied natural gas and the risks posed by its storage.

INTRODUCTION

In oil and gas industry the movement of crude and refined oil products from the places of origin to the various markets would not be possible without the existence of economic and safe storage facilities. Storage tanks are located at the ends of feeder lines and gathering lines, along truck pipelines, at marine loading and unloading facilities and in refineries, terminals and bulk plants.

Oil produced from the early wells was stored in whisky barrels. Barrels, although

DOI: 10.4018/978-1-4666-4777-0.ch014

unsuitable for storage use because of leaking, remained in use for a long time due to their ease of transportation and were eventually replaced by steel ones. The measurement of the oil volume is often expressed in barrels (1 bbl=0,159 m^3) precisely because of such practices. With the ever growing need for oil, the storage tank sizes increased with time. Today, crude oil is commonly stored in large aboveground atmospheric vertical cylindrical storage tanks at oil fields, terminals and refineries where storage tanks are typically installed with similar or identical vessels in a group.

At the oilfield, when the oil is brought to the surface under high pressure conditions it is passed through either a two (where the associated gas is removed and any oil and water remain together) or a three phase separator (where the associated gas is removed and the oil and water are also separated). The remaining low pressure oil is then directed to a storage vessel where it is stored for a period of time before being transported off-site. The remaining hydrocarbons in the oil are released from the oil as vapours in the storage vessels. At the oil fields produced crude oil is usually stored in fixed roof storage tanks while stabilised oil and oil products at terminals and refineries are stored in floating roofs tanks. Within the European Union (EU) the specification for the design of such tanks is covered by the British Standard BS EN 14015:2004 (2005) which is generally equivalent to American Petroleum Institute (API) standards, API 650 (Welded steel tanks for oil storage, 2007) and API 620 (The design and construction of large, welded, low-pressure storage tanks, 2002).

Terminals are storage facilities which generally receive crude oil and petroleum products by trunk pipeline or marine vessel. Crude oil and petroleum product terminals are designed to receive and dispatch crude oil and petroleum products to refineries, other terminals, bulk plants and consumers by pipelines, railroad tank cars and tank trucks. Tanks should ideally be located alone, but storage tanks are often grouped in tank farms. Within tank farms, individual tanks or groups of two or more tanks are usually surrounded by a catchment area in the form of a retaining wall, known as a bund or dike. Bunds would normally be constructed from earth or preferably concrete and should be largely impervious to liquid and capable of withstanding hydrostatic and hydrodynamic pressures to which they could be subjected. Whenever tanks share a common bund, intermediate walls up to half the height of the main bund walls and no more than 0,5 metres high should be provided to control small spillages from one tank affecting another. Bund floors should drain to a single location complete with sump for the regular removal of water from rainfall or firewater testing. Drains should normally be kept closed, with the drain isolation valve situated outside the bund. Bunds are normally designed to hold 110% of that of the largest tank, the excess height notionally to prevent stored liquid surging over the top of the bund in the event of a catastrophic failure of the primary containment. This allowance has proved to be inadequate as, even with storage volume excess, large quantities of liquid can still overtop bunds as discussed by Thyer and Jagger (2002) and Atherton (2005).

Natural gas can be stored in gaseous or liquefied state. The natural gas in gaseous state will be stored in underground storages (depleted oil or gas reservoirs, aquifers or salt caverns). By converting natural gas to liquefied natural gas (LNG) it can be economically transported over the oceans and great distances from the places where it is produced to those where it is in demand. Currently, the liquefied natural gas trade, accounting for the bringing to markets of some 30% of the world's natural gas (BP Statistical Review, 2013), would not be able without the development of large scale cryogenic storage units at both export and import terminals. Liquefaction of natural gas also provides the opportunity to store the fuel for use during high consumption periods close to demand centres.

BACKGROUND

The storage of crude oil and petroleum product creates the potential for leaks or accidentally releases from tanks, pipes, hoses, and pumps during loading and unloading of products. The storage and transfer of these materials also poses a risk of fire and explosion due to the flammable and combustible nature of the materials stored. Basic requirements for tank design and construction are mandated by various industrial codes and standards but the risk of an accident, although minimal, is present. There are many examples of storage tank accidents. One of the often mentioned is storage tank explosion at the Hertfordshire Oil Storage Limited part of the Buncefield (2005). The accident was caused by operational error because two level gauges were inoperable at the moment: a gauge that enabled the employees to monitor the filling operation; and an independent high-level switch which was meant to close down operations automatically if the tank was overfilled. So there was no means to alert the control room staff that the tank was filling to dangerous levels and as a consequence over 300 tons of petrol overflowed from the top of the tank. As a result, a vapour cloud formed which ignited causing a massive explosion and a fire that lasted three days (Herbert, 2010). Some bizarre incidents are also possible like a case in Texas when a 24-year-old man and a 24-year-old woman were critically injured in an explosion and fire after smoking on a catwalk over oil storage tanks along state highway. Three of the six storage tanks caught fire and exploded thus inducing a loud explosion. Luckily, it seems that the tanks had very little crude oil in them at the time of the explosion (http://www.heavy.com/news/2013/01/massive-explosion-sparked-by-2-smokers-on-top-of-oil-tank/). From the above mentioned examples, the presence of a risk is evident. Thus it is crucial to clearly define the causes and consequences of the tank damages. Possible storage tanks accident can be: spills, gas releases, fire and explosions. The most common failure causes for fixed and floating roof tanks according to Argyropoulos et al. (2012) are operational errors, equipment or instrument failure and maintaince errors while lightning and static electricity can causing fluid or vapour ignition if flammable vapours are present.

In the article prepared by Chang & Lin (2006) the analysis of 242 tank accidents occurred in industrial facilities in last 40 years was presented. The analysis has shown that accidents occurred most frequently at petroleum refineries (116 cases or 47.9%). The second most frequently involved place were terminals and pumping stations (64 cases, 26.4%) and finally only 25.7% of accidents occurred in petrochemical plants (12.8%), oil fields (2.5%), and other types of industrial facilities (10.3%). Crude oil, gasoline and oil products such as fuel oil, diesel, etc. were major contents. The atmospheric external floating roof tank was the most frequently damaged storage tank type while the atmospheric cone top tank was the second most frequent type. Both types were used extensively for the storage of crude oil, gasoline, and diesel oil.

Fire and explosion together accounted for 85% of total cases. Oil spill and toxic gas/liquid release were the third and the fourth most frequent consequence while the tank body distortion and the worker's falling only occurred a few times.

Renni et al., (2010) state that one of most frequent cause of accidents in chemical facilities caused by natural events is lightning strikes. Author's analysis has shown that in 91% recorded cases, lightning caused accident in oil and gas facilities (mainly oil refineries) and the petrochemical sector, including storage sites and tank farms. In 485 recorded accidents, storage tanks were most frequently affected by lightning impact (59.6% or 289 cases). The atmospheric tanks, and in particular floating-roof tanks, are the most vulnerable equipment while only 3 out of 289 accidents involved pressurised storage tanks.

STORAGE TANKS

Oil Storage Tanks

Atmospheric oil storage tanks (Figure 1) come in a range of sizes, from small to truly gigantic and can be constructed as aboveground, in ground and belowground. The aboveground tank is usually easier to construct, costs less, and can be built in far larger capacities than underground storage tanks. In shape they are most usually of vertical cylindrical form, but also come in horizontal cylindrical, spherical and rectangular forms. Vertical cylindrical storage tank can be constructed as the open top tank, fixed roof tank, external and internal floating roof tank. According their construction tanks can be those with single walls, double walls and insulated tanks. Temperatures range from high temperature heated storage tanks through to -163°C for the storage of liquefied natural gas. According the internal pressure the tanks can be atmospheric or low-pressure tanks. By far the most common type of tank is the atmospheric tank. Although called atmospheric, these tanks are usually operated at internal pressures slightly above atmospheric pressure while low pressure in the context of tanks means tanks designed for a higher pressure than atmospheric tanks. These tanks are designed to operate from atmospheric pressure up to 1.0342 bar (15 psig) (Myers 1997; Long & Garner, 1998).

Figure 1 illustrates various types of storage tanks commonly used for crude oil and oil products storage.

Figure 1. Tank types

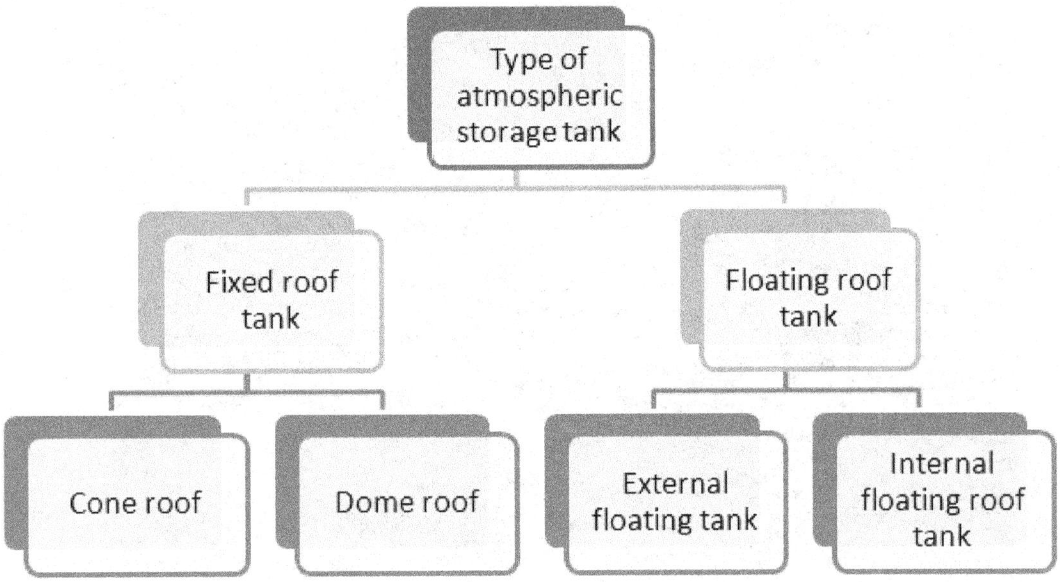

Generally, for crude oil and oil products storage only fixed and floating roof tanks are used. Open top tanks shall not be used for oil or oil product storage but may be used for fire water/ cooling water.

Fixed Roof Tanks

Fixed roof storage tanks (Figure 2) are typically cylindrical in shape and can be either horizontal or vertical in orientation. In general, tank has vertical cylinder shell made of a painted steel and fixed cone-shaped roof that is welded to each other. Fixed roof tanks can be divided into cone roof and dome roof types. They can be selfsupported or rafter/ trusses supported depending on the size. Main disadvantage of the fixed roof tank is the loss of product vapour into the atmosphere because of usual tank operation (filling and empting) as well as ambiental temperature changes. At the production site, where fixed roof tanks are often used, flashing losses occur when crude oils or condensates flow into a storage vessel from a processing vessel operated at a higher pressure. Typically, the larger the pressure drop, the more flash emissions will occur in the storage stage. Temperature of the liquid may also influence the amount of flash emissions. The volume of gas vapour emitted from a storage vessel depends on many factors. Lighter crude oils flash more hydrocarbons than heavier crude oils. In storage vessels where the oil is frequently cycled and the overall throughput is high, working losses are higher. Additionally, the operating temperature and pressure of oil in the separator dumping into the storage vessel will affect the volume of flashed gases coming out of the oil. The composition of the vapours from storage

Figure 2. Fixed roof storage tank

vessels varies, and the largest component is methane, but also includes ethane, butane, propane and other flammable components.

To minimize emissions of volatile organic compounds into the atmosphere fixed roof tanks may be fitted with internal floating roofs. This can be done when tank is built or it may be retro-fitted at later date.

Floating Roof Tanks

Floating roof tanks are used for volatile liquid hydrocarbons such as crude oil and "white" light products (jet, diesel and gasoline) storage and during normal tank operation the majority of the vapour stays under the roof.

In floating roof tanks the roof floats directly on the top of the product. Floating roof tanks may have external and internal floating roofs. The former has no fixed roof while he latter has both a fixed and a floating roof. In construction of external floating roof there is a rim seal that covers the space between the floating roof and the tank shell. Floating roofs are provided with legs which may be set in high or low positions depending on the type of operation. Legs are normally maintained in the low position so that the greatest possible amount of product can be withdrawn from the tank without creating a vapour space between the top of the product and the bottom of the floating roof. In both type constructions, the floating roof consists of a deck, fittings, and rim seals, and typically involves a pontoon and double-deck system. The roof rises and falls with the liquid level in the tank to minimize volatile organic compounds emissions. An internal floating roof storage tank is a combination of the above two types of tanks, as the tank consists of a conical roof with the addition of the internal floating roof or pan that floats directly on the fuel surface.

The type and nature of the stored product is most important criteria in type storage tank selection.

In the process of the tank design following factors must be considered:

- Operation needs - influencing tank capacity
- Type and properties of the stored fluid
- Construction materials (corrosion)
- Foundation
- Tank loadings (influence of internal and external factors)
- Tank venting

During the tank design process factors like actual site-specific conditions, local regulations, required operating life, space available, potential for fires and explosion and finally cost considerations, should also be evaluated and considered.

Oil Storage Tanks Risk Analysis

Oil Tank Spills

Leaks and spills from storage tanks have had great impact on the way tanks are regulated and will be regulated as well as upon the design, maintance and operation of tanks. There are numerous causes of tank leaks and spills. Some of them can be caused by:

- **Tank overfilling:** Can be caused by operational or human error. Operational error can be result of failed equipment (for example: level indicators, wrong setting or failure of valves) while human error can resulted from lack of attention or lack of training. According Marsh Ltd. (2011) in the USA overfilling of atmospheric storage tanks occurs once in every 3300 filling operations.
- **Tank rupture:** Can be caused by corrosion, foundation problems, shell distortion or buckling, sabotage, extreme weather conditions, earthquake.
- **Piping rupture or crack:** Very often cause of leaks and spills. According Myers perhaps the most significant leaks have resulted from pressurized underground piping.

Basic input data in every risk analysis is frequency of the specific event. The frequency describes the number of times that certain accidents will happen in a particular type of unit per year. It can be estimated by performing historical analyses or using event trees. Storage tank accident frequency data is most often estimated on the historical data of tank failures and depends on the used database. It is illustrated by the frequency of catastrophic failure of the tanks used for the storage of hazardous liquid. According to Thyer et al. (2002) this frequency is 5,0 x 10^{-6} per tank year while frequency determined in publication of International Association of oil and gas producers (2010) is 3.0 x 10^{-6} per tank year. In both cases, probability of catastrophic failure of a crude oil storage tank appears to be quite remote. The probability of the minor failure (hole and leak) which may result in minor oil spill will be relatively more. That is illustrated by the estimated frequencies of leak for atmospheric storage tanks shown in Table 1.

Table 1. Atmospheric storage tank leak frequencies (International Association of oil and gas producers, 2010)

Type of tank	Type of release	Leak frequency (per tank year)
Floating roof	Liquid spill on roof	$1.6x10^{-3}$
	Sunken roof	$1.1x10^{-3}$
Fixed/floating roof	Liquid spill outside tank	$2.8x10^{-3}$
	Tank rupture (Catastrophic failure)	$3.0x10^{-6}$

Fire and Explosion

Whenever hydrocarbons are present in closed containers such as storage tanks, the potential exists for release of liquids and vapours. These vapours could mix with air in the flammable range and, if subjected to a source of ignition, cause an explosion or fire. Fire and explosion hazards may result from the presence of combustible gases and liquids, oxygen, and ignition sources during loading and unloading activities, and/or leaks and spills of flammable products. Possible ignition sources include sparks associated with the build up of static electricity, lightning and open flames. Lightning is a major accident initiator. A survey of industrial accident databases showed that up to 5% of industrial accidents were triggered by natural events which impact industrial facilities that process or store hazardous substances. Analysis of the accidents at storage and processing activities initiated by natural events conducted by Renni et al. (2010) is shown that in 61% of recorded cases accident were triggered by lightning strikes. Lightning was also found to be the most frequent cause of failure in the set of storage tank accidents analysed in the study of Chang and Lin (2006). Lightning can interact with tank structure by direct strike, indirect lightning currents that induce secondary sparks, or by the disruption of control systems and electrical circuitry. Direct lightning strikes can severely damage structures due to thermal heating. If flammable vapours are present, immediate ignition can occur. There are many other sources of ignition as welding/cutting operations, circuit shortcuts, use of non explosion-proof motor and tools, transformer sparks, adjacent open flames or remains of smoking, escalation from another unit (domino effect), previous accident caused by energy/fuel transportation lines and sabotage. Potential fire scenarios that can be developed in a tank accident are rim seal fire, spill on roof fire, full surface fire, bund fire, pontoon explosion and boilover. The most severe are the full surface fire and boilover. If open-top floating roof storage tanks are out of round or if the seals are worn or not tight against the tank shells, vapours can escape and mix with air, forming flammable mixtures. In such situations, when lightning strikes, fires may occur at the point where the roof seals meet the shell of the tank. Such type of fire named rim seal fire is considered as the most frequent fire cause for floating roof tanks containing petrol, crude oil and kerosene (Argyropoulos et al., 2012). Existence of an effective grounding system is indispensable for every tank and for tank farm must be combined with lightning conductors positioned at sufficient height and distance from the farm. However, several accidents suggest that this type of protection often is not sufficient. Data about large atmospheric storage tanks fires can be finding in LASTFIRE bases – a consortium of international oil companies reviewing the risks associated with fires in storage tanks and developing the best industry practice to mitigate the risks. The base is updated in 2010 and gives a valuable data about floating storage tank fire accidents and its consequences. Data from the LASTFIRE base are often used in estimation of the frequencies of different types of fire/explosion. Atmospheric Storage tank fire frequencies are shown in Table 2.

The data presented confirms that the rim seal fire mostly occurs around storage tanks with floating roof, while the occurrence of fires in other storage tanks is significantly less. Because of detrimental impact of tank spills on the environment and the negative perception of such incidents by public, storage tank design, control and maintance requirements are becoming more rigorous. The existing laws regarding such practices determine the risks and attempt to minimize those same risks.

Liquified Natural Gas Storage Tanks

Liquefied natural gas (LNG) is natural gas that is cooled below 160 °C. It is a mixture of hydrocarbons, ranging from methane to butane, with approximately 85 to 95% methane. The properties of LNG vary with composition, which depends on the location of the gas reservoir. The produced natural gas may contain many other materials including water vapour, carbon dioxide, nitrogen, and helium, some of which must be removed for liquefaction.

In its liquefied state, natural gas takes up 1/600th of the space, making it much easier to ship and store when pipeline transport is not feasible. In liquid state natural gas is not combustible. LNG weighs slightly less than half as much as water, so it floats on fresh or sea water. However, when LNG comes in contact with any warmer surface such as water or air, it evaporates very rapidly ("boil"),

Table 2. Atmospheric storage tank fire frequencies (International Association of oil and gas producers, 2010)

Type of fire	Floating roof tank (per tank year)	Fixed roof tank (per tank year)
Rim seal fire	$1.6x10^{-3}$	
Full surface fire on roof	$1.2x10^{-4}$	
Internal explosion&full surface fire		$9.0x10^{-5}$
Internal explosion without fire		$2.5x10^{-5}$
Vent fire		$9.0x10^{-5}$
Small bund fire	$9.0x10^{-5}$	$9.0x10^{-5}$
Large bund fire (full bund area)	$6.0x10^{-5}$	$6.0x10^{-5}$

returning to its original, gaseous volume. As the LNG vaporizes, a vapour cloud resembling ground fog will form under relatively calm atmospheric conditions. The vapour cloud is initially heavier than air since it is so cold, but as it absorbs more heat, it becomes lighter than air, rises, and can be carried away by the wind and when concentration of gas is between 5% and 15% it can ignite. At defined concentration, an explosion can occur if gas is ignited in confined space. If not confined, the gas will burn until there is no longer sufficient fuel and/or air to maintain needed ratio.

Liquefied natural gas has a positive safety records. There were a few storage and plant accidents to date and only 15 marine accidents (8 with and 7 without spills). None of the marine accidents resulted in cargo fires, explosions or fatalities.

After liquefaction process, liquefied natural gas is pumped into a cryogenic storage tanks at atmospheric pressure until a transport vessel loading. After loading in vessel containment system, LNG is transported to the receiving terminal where is again pumped into the cryogenic tank. There are several different designs of refrigerated storage tanks. Three most common used types of LNG storage tanks are (Huang et al. 2007; Long& Garner, 1998):

- Single containment tank (Figure 3)
- Double containment tank (Figure 4)
- Full containment tank (Figure 5).

The cryogenic tank is a tank within a tank, with insulation between the walls of the tanks. The inner tank, in contact with the LNG, has to be made of material suitable for cryogenic temperatures and is usually made of 9% nickel steel. The hydrostatic head of the LNG impose a load on the tank and determines the thickness of the inner tank side walls. All tanks must be designed in accordance with international LNG codes (EMMUA 147, EN 1473).

Single containment tanks. These are a single primary container and an outer shell divided by isolation layer. The primary con-

Figure 3. Single containment tank

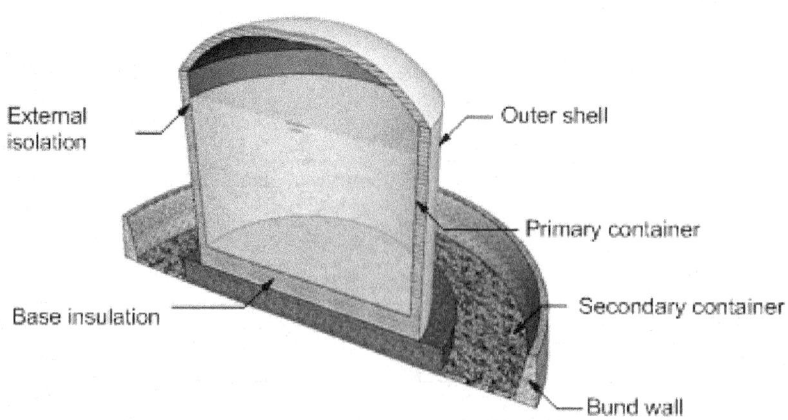

Figure 4. Double containment tank

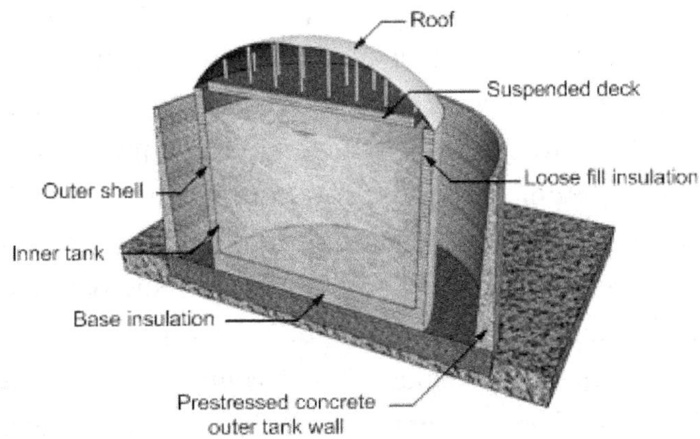

Figure 5. Full containment tank

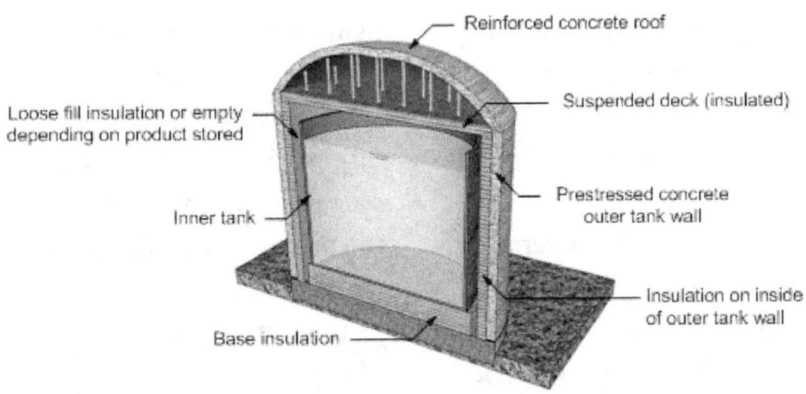

tainer made of 9% nickel – steel is required to meet the low temperature ductility requirements for storage of the product while the outer tank generally made of carbon steel, serves primarily to retain insulation and vapour and is not designed to contain LNG due to leakage from the primary tank. Tank has a flat metallic bottom and a cylindrical metal wall both built of materials suitable for cryogenic temperatures. Single containment tanks require larger land areas for LNG storage facilities because of the larger potential spill area of the dike impoundment. Single containment tank shall be surrounded by a dyke to contain any leakage in the unlikely event of tank failure. The dikes are designed to contain 100 percent to 110 percent of tank volume and are usually low structures made of earth, reinforced earth or reinforced concrete. Impoundments not only limit the spread of an LNG spill, they reduce the surface area of the liquid pool, thereby decreasing and controlling the size of the vapour cloud. The construction of this tank type remains most economical.

The safety records of the onshore LNG facilities around the world demonstrate that the primary containment of the LNG tanks is safe, because secondary spill containment systems installed around all of the tanks, have never been required to hold liquid. However, due to safety reasons, the construction of storage capacity for LGN consists of constructing double or full containment tanks.

Double containment tanks. These tanks are designed and constructed so that both the inner self supporting primary container and the secondary container are capable of independently containing the refrigerated liquid stored. The primary container contains the refrigerated liquid under normal operating conditions. The secondary container or wall is intended to contain any LNG leakage from the inner container but is not intended to contain any vapour resulting from this leakage. To minimise the pool of escaping liquid, the secondary container should be located at a distance not exceeding 6 m from the primary container.

Full containment tanks. In this tank type the primary (inner) container contains the LNG under standard operating conditions but both self supporting primary and the secondary container are capable of independently containing the stored LNG. The secondary container composed of concrete is 1 m to 2 m distance from the primary container and supports the outer roof. The secondary containment shall be capable both of containing the refrigerated liquid and of controlled venting of the vapour resulting from product leakage after a credible event.

The majority of LNG storage tanks built recently around the world are designed as double or full containment tanks.

Membrane tank. This tank type is utilized for the storage of liquefied natural gas during the transportation with ships. However, there are similarly constructed tanks used for the storage of the liquefied natural gas on land and on the gravity based storage systems. These tanks are designed and constructed so that the primary container, constituted by a stainless steel corrugated membrane, is capable of containing both the liquefied gas and its vapour under normal operating conditions. Such tanks located on land have the concrete secondary container, which supports the primary container and should be capable of containing all the liquefied gas stored in the primary container and of controlled venting of the vapour resulting from product leakage of the inner tank. The vapour of the primary container is contained by a steel liner which forms with the membrane an integral gastight containment. The action of the liquefied gas acting on the primary container (the metal membrane) is transferred directly to the prestressed concrete secondary container through the load bearing insulation.

Modern storage tanks have no side or bottom penetrations. All penetrations, including those for LNG send out, are through the roof.

This design substantially reduces the amount of LNG spilled in the unlikely event of a rupture or leakage in the send out piping.

LNG Storage Risk

Sources of LNG hazards connected to storage tanks are spills and rollover. The most serious hazard with LNG is the unplanned release from storage tank and contact with an ignition source. Release can occur by failure of the container or valves. Experts agree that the risks posed by shipping, loading and offloading of the LNG are greater than risks posed by the land-based storage facilities. As the containment opportunity for water spills is limited, the spilled LNG disperses faster in the ocean than on land. In addition, due to the ocean's heat, the LNG vaporizes rapidly on water.

Rollover occurs under certain conditions as stratified LNG comes to equilibrium. Stratification occurs when liquid of a different composition than that which is already in the tank is introduced into the tank. If rollover occurs, pressure inside the storage tank may rise to excessive levels. The tanks are equipped with safety vent valves that are designed to keep the pressures from rising to levels that could cause structural damage. However, when these valves operate, LNG is vented to the atmosphere at an uncontrolled rate, which is an additional safety concern. In both cases, the real danger is the potential ignition of vapour cloud. The detection of the development of "rollover" conditions in modern LNG storage tanks is possible through the use of instruments that monitor the pressure, temperature, and density of the LNG along the entire height of the liquid column.

LNG's cryogenic temperatures, dispersion characteristics and flammability characteristics are the causes of hazards.

The LNG, due to its coldness, can directly cause injury or damage although momentary contact on the skin can be harmless. However, the prolonged contact can result in severe freeze burns. On contact with certain metals LNG can cause immediate cracking. Although not poisonous, exposure to the center of a vapour cloud could cause asphyxiation due to the absence of oxygen. LNG vapour clouds can ignite within the portion of the cloud where the concentration of natural gas is between 5 and 15 percent (by volume) mixture with air. Without ignition source LNG vapour cloud will simply dissipate into the atmosphere. An ignited LNG vapour cloud is very dangerous, because of its tremendous radiant heat output. According to Woodword & Pitblado (2010), there are four types of fires in general and of LNG fires in particular:

1. **Pool Fire and Trench Fire:** A fire burning over a pool of liquid fuel.
2. **Jet Fire:** A narrow, long turbulent jet, like a blowtorch.
3. **Flash Fire:** A flame front that moves through a vapour cloud.
4. **Fireball:** A nearly spherical, rising, burning mass of vapour or flashing liquid.

Pool fire: A consequence of LNG spill and a vapour cloud forming and its ignition is the most dangerous type of hazard. The damage to life and property is extensive and takes place at a distance far from the actual fire itself. The evaporating natural gas above the pool burns quickly as the flame burns back towards the evaporating pool until all of the fuel is consumed.

An LNG jet fire: occuring when the liquid is released under pressure through an orifice or a crack and is ignited - has a limited reach that is less than 50 m. As such, it is confined to the location of the LNG processing plant. Disabling the buildings and the process equipment, jet fires generate high heat fluxes and burn through most objects given enough exposure time.

A jet fire requires a pressurized release that can occur in process plants but is not typically a threat to the public. An LNG spill on land or on water results in a wind driven vapour cloud caused by a rapidly evaporating pool. Contact with an ignition source results on an ignition, resulting in a *flash fire*. Burning downwind, the flash fire burns faster along the premixed (diluted by air) edges. A flash fire is inherently transient, and exposure normally lasts no more than a few tens of seconds. While fatal to people inside the fire, the total radiation reaching an object near a flash fire is substantially lower than that from a longer - lasting pool or jet fire the same distance away. A flash fire likely does not produce secondary ignition or burns to people outside of the flaming region. Appearing in spherical shapes, *fire balls* rise while burning. Often they appear as a result from the rupture of a bursting pressure vessel or from the ignition of a large accumulation of flammable vapours at atmospheric pressure.

Due to a small number of accidents regarding the leakage of the storage tanks and the vapour cloud ignition, numerous research projects intend to clarify and create a greater understanding of the causes and the consequences of the fires.

CONCLUSION

The data presented suggests that the accidents of greater proportions resulting in serious fatalities and damage related to the storage of oil and, especially the LNG, occur rarely. However, when they do occur, the consequences are usually severe in a form of a loss of human lives. The establishment of the accident risks connected to storing is therefore important for the procedure definitions, defining of the control time intervals and the maintenance of the storage capacity. Regardless of the storing location - on the oil field, terminal or refinery - safety procedures and safe work practices should be established and implemented, and personnel training or education provided. These should include, as a minimum, information concerning the basics of hydrocarbon fire ignition, control and extinguishment, hazards and protection from exposures to toxic substances, emergency response actions, and normal physical and climatic hazards associated with this activity. Storage tank should be regularly inspected for corrosion, erosion, settling or other visible damage to prevent loss or degradation of product. Tank pressure/vacuum valves, seals and shields, vents, foam chambers, roof drains, water draw-off valves and overfill detection devices should be inspected, tested and maintained on a regular schedule, including removal of ice in the winter. Where flame arrestors are installed on tank vents or in vapour recovery lines, they have to be inspected and cleaned regularly and kept free of frost in the winter to ensure proper operation. Valves on tank outlets which close automatically in case of fire or drop in pressure should be checked for operability. Dike surfaces should drain or slope away from tanks, pumps and piping to remove any spilled or released product to a safe area. Dike walls should be maintained in good condition, with drain valves kept closed except when draining water and dike areas excavated as needed to maintain design capacity. Stairways, ramps, ladders, platforms and railings to loading racks, dikes and tanks should be maintained in a safe condition, free of ice, snow and oil. Leaking tanks and piping should be repaired as soon as possible. The use of victaulic or similar couplings on piping within diked areas which could be exposed to heat should be discouraged to prevent lines from opening during fires.

The physical and chemical properties of LNG are such that risks and hazards are well understood and incorporated into technology and operations. LNG storage safety increases with the construction of a great number of double and full containment tanks as the standards, while codes and regulations that apply to the LNG industry further ensure safety.

REFERENCES

API 653. (2001). *Tank inspection, repair, alteration, and reconstruction* (3rd ed.). Washington, DC: American Petroleum Institute.

API Standard 620. (2002). *Design and construction of large, welded, low-pressure storage tanks* (3rd ed.). Washington, DC: American Petroleum Institute.

API Standard 650. (2007). *Welded steel tanks for oil storage* (11th ed.). Washington, DC: American Petroleum Institute.

Argyropoulos, C. D., Christolis, M. N., Nivolianitou, Z., & Markatos, N. C. (2012). A hazards assessment methodology for large liquid hydrocarbon fuel tanks. *Journal of Loss Prevention in the Process Industries*, *25*, 329–335. doi:10.1016/j.jlp.2011.12.003

Atherton, W. (2005). *An experimental investigation of bund wall overtopping and dynamic pressure on the bund wall following catastrophic failure of a storage vessel*. Liverpool, UK: John Moores University.

Bernechea, J. E., & Viger, A. J. (2013). Design optimization of hazardous substance storage facilities to minimize project risk. *Safety Science*, *51*, 49–62. doi:10.1016/j.ssci.2012.06.007

BS EN 14015. (2005). *Specification for the design and manufacture of site built, vertical, cylindrical, flat-bottomed, above ground, welded, steel tanks for the storage of liquids at ambient temperature and above*. Author.

Chang, I. J., & Lin, C.-C. (2006). A study of storage tank accidents. *Elsevier. Journal of Loss Prevention in the Process Industries*, *19*, 51–59. doi:10.1016/j.jlp.2005.05.015

Herbert, I. (2010). The UK buncefield incident - The view from a UK risk assessment engineer. *Journal of Loss Prevention in the Process Industries*, *23*, 913–920. doi:10.1016/j.jlp.2010.09.001

Huang, S., Chiu, C., & Elliot, D. (2007). *LNG: Basics of liquefied gas*. Austin, TX: The University of Texas at Austin.

International Asociation of Oil and Gas Producers. (2010). *Storage incident frequencies* (Report No. 434 – 3 March 2010). Author.

Krausman, E. (2010). *Analysis of Natech risk reduction in EU member states using a questionnaire survey*. Retrieved from http://publications.jrc.ec.europa.eu/repository/bitstream/111111111/16071/1/jrc_eu_natech.pdf

Long, B., & Garner, B. (1998). *Guide to storage tanks&equipment*. London: Professional Engineering Publishing.

Massive Explosion Sparked by 2 SMOKERS on Top of Oil Tank Risk Engineering Position Paper. (n.d.). *Atmospheric storage tanks*. Retrieved from http://usa.marsh.com/NewsInsights/ThoughtLeadership/Articles/I

Myers, E. P. (1997). *Aboveground storage tanks*. New York: McGraw-Hill Companies, Inc.

Natech Accident Database. (2010). Retrieved from http://enatech.jrc.ec.europa.eu

Renni, E., Krausmann, E., & Cozzani, V. (2010). Industrial accidents triggered by lightning. *Journal of Hazardous Materials*, *184*, 42–48. doi:10.1016/j.jhazmat.2010.07.118 PMID:20817399

Statistical Review of World Energy. (2013). Retrieved from www.bp.com/en/global/corporate/about-bp/statistical-review-of-world-energy-2013.html

Thyer, A. M., Hirst, I. L., & Jagger, S. F. (2002). Bund overtopping – The consequence of catastrophic tank failure. *Journal of Loss Prevention in the Process Industries*, *15*, 357–363. doi:10.1016/S0950-4230(02)00023-2

Thyer, A. M., Jagger, S. F., Atherton, W., & Ash, J. W. (2009). A review of catastrophic failures of bulk liquid storage tanks. *Loss Prevention Bulletin*, 205.

Woodward, J. L., & Pitblado, R. M. (2010). *LNG risk based safety: Modeling and consequence analysis*. Hoboken, NJ: John Wiley & Sons, Inc. doi:10.1002/9780470590232

ADDITIONAL READING

Bates, S., & Morrison, S. D. (1997). Modelling the behaviour of stratified liquid natural gas in storage tanks: a study of the rollover phenomenon. *International Journal of Heat and Mass Transfer*, *40*(8), 1875–1884. doi:10.1016/S0017-9310(96)00218-9

Chang, M. K., Chang, R. R., Shua, C. M., & Lin, K. N. (2005). Application of risk based inspection in refinery and processing piping. *Journal of Loss Prevention in the Process Industries*, *18*, 397–402. doi:10.1016/j.jlp.2005.06.036

Cozzani, V., Bandini, R., Basta, C., & Christou, M. (2006). Application of land-use planning criteria for the control of major accident hazards: a case study. *Journal of Hazardous Materials*, *A136*, 170–180. doi:10.1016/j.jhazmat.2005.12.031 PMID:16487657

Cozzani, V., Gubinelli, G., Antonioni, G., Spadoni, G., & Zanelli, S. (2005). The assessment of risk caused by domino effect in quantitative area risk analysis. *Journal of Hazardous Materials*, *A127*, 14–30. doi:10.1016/j.jhazmat.2005.07.003 PMID:16102897

Crippa, C., Fiorentini, L., Rossini, V., Stefanelli, R., Tafaro, S., & Marchi, M. (2009). Fire risk management system for safe operation of large atmospheric storage tanks. *Journal of Loss Prevention in the Process Industries*, *22*, 574–581. doi:10.1016/j.jlp.2009.05.003

Cruz, A. M., Steinberg, L. J., Vetere-Arellano, A. L., Nordvik, J. P., & Pisano, F. (2004). State of the Art in Natech (Natural Hazard Triggering Technological Disasters) Risk Assessment in Europe. Report EUR 21292 EN, DG Joint Research Centre, European Commission and United Nations International Strategy for Disaster Reduction, Ispra, Italy.

DiGrado, D. B. Thorp. A.G. (2004). The Aboveground Steel Storage Tank Handbook, Hoboken, New Jersey: John Wiley&Sons, Inc.

Drogaris, K. G. (1991). *Major Accidents in Oil and Gas Industries.* Paper SPE 23216 presented at the First International Conference on Health, Safety and Environment held in The Hague, The Netherlands, 10-14 November 1991.

Duguid, I. M. (2008). *Analysis of past accidents in the process industries*, Paper 87 and handouts, *Hazards XX: Process Safety and Environmental Protection: Harnessing Knowledge—Challenging Complacency,* Proceedings of a Conference held in Manchester on 14 – 17 April *2008,*Symposium Series No. 154, Institution of Chemical Engineers, Rugby, UK, 2008, p. 1070.

Fabbrocino, G., Iervolino, I., Orlando, F., & Salzano, E. (2005). *Quantitative risk analysis of oil storage facilities in seismic areas*, Elsevier. *Journal of Hazardous Materials, A123*, 61–69. doi:10.1016/j.jhazmat.2005.04.015

Fay, J. A. (2003). Model of Spills and Fires from LNG and Oil Tankers. *Journal of Hazardous Materials, B96*, 171–188. doi:10.1016/S0304-3894(02)00197-8 PMID:12493207

Griffin, L. M. (2000). Protecting atmospheric storage tanks against vacuum collapse. *Journal of Loss Prevention in the Process Industries, 13*, 83–89. doi:10.1016/S0950-4230(99)00074-1

Kletz, T. A. (2009). Accident reports may not tell us everything we need to know. *Journal of Loss Prevention in the Process Industries, 22*, 753. doi:10.1016/j.jlp.2009.08.017

Koseki, H., & Mulholland, G. W. (1991). Effect of diameter on the burning of crude oil pool fires. *Fire Technology, 27*(1), 54–65. Available at fire.nist.gov/bfrlpubs/fire91/art021.html doi:10.1007/BF01039527

Koseki, H., Natsuma, Y., Iwata, Y., Takahashi, T., & Hirano, T. (2003). A study on largescale boilover using crude oil containing emulsified water. *Fire Safety Journal, 39*, 143–155. doi:10.1016/j.firesaf.2003.09.003

Krausmann, E., Cozzani, V., Salzano, E., & Renni, E. (2011). Industrial accidents triggered by natural hazards: an emerging risk issue. *Natural Hazards and Earth System Sciences, 11*, 921–929. doi:10.5194/nhess-11-921-2011

Medina, H., Arnaldos, J., & Casal, J. (2009). Process design optimization and risk analysis. *Journal of Loss Prevention in the Process Industries, 22*, 566–573. doi:10.1016/j.jlp.2009.04.007

Pitblado, R. (2010). Global process industry initiatives to reduce major accident Hazards. *Journal of Loss Prevention in the Process Industries, 24*, 57–62. doi:10.1016/j.jlp.2010.08.006

Raj, K. P., & Lemoff, T. (2009). Risk analysis based LNG facility siting standard in NFPA 59A. *Journal of Loss Prevention in the Process Industries, 22*, 820–829. doi:10.1016/j.jlp.2009.08.019

Ronza, A., Muńoz, M., Carol, S., & Casal, J. (2006). Consequences of mayor accidents:assessing the number of injured people. *Journal of Hazardous Materials, A133*, 46–52. doi:10.1016/j.jhazmat.2005.10.011

Shaluf, I. M., & Abdullah, S. A. (2011). Floating roof storage tank boilover. *Journal of Loss Prevention in the Process Industries*, *24*, 1–7. doi:10.1016/j.jlp.2010.06.007

Shebeko, N. Y., Bolodian, A. I., Molchanov, P. V., Deshevih, I. Yu., Gordienko, M. D., Smolin, M. I., & Kirillov, S. D. (2007). Fire and explosion risk assessment for large-scale oil export terminal. *Journal of Loss Prevention in the Process Industries*, *20*, 651–658. doi:10.1016/j.jlp.2007.04.008

Shuai, J., Han, K., & Xu, X. (2012). Risk-based inspection for large-scale crude oil tanks. *Journal of Loss Prevention in the Process Industries*, *25*, 166–175. doi:10.1016/j.jlp.2011.08.004

KEY TERMS AND DEFINITIONS

Bund, Dike: Tank enclosures with primary function to contain, direct and divert rain water, physically separate tanks to prevent the spread of fire in one area to another, and to contain a spill, release, leak or overflow from a tank, pump or pipe within the area.

Leak: Any loss of liquid because of a defect in a storage tank system.

Liqufied Natural Gas: Odorless, colorless, non-corrosive and non-toxic product of natural gas consisting primarily of methane (CH_4 that is stored and transported in liquid form at the atmospheric pressure and temperature of – 161 °C.

LNG Storage Tank: Specially constructed vessels to contain LNG. The tanks are generally constructed of nickel steel (steel containing 9% nickel) to withstand cryogenic temperatures, and are insulated to maintain the LNG at -161 °C.

Storage Tank: A container designed to store liquids or gases.

Tank Farm: Crude oil and petroleum product terminals designed to receive and dispatch bulk shipments of crude oil, gasoline, middle distillates and specialty products from pipelines, ships, railcars, and trucks. Crude oil and petroleum product terminals are often located at the sea coast but may also be situated inland.

Vapours: Vapours released from a tank stored liquids that occupied space above liquid. These releases occur when tanks are emptied and filled and as a result of an increase in temperature.

Chapter 15
CO_2 Underground Storage and Wellbore Integrity

Nediljka Gaurina-Medjimurec
University of Zagreb, Croatia

Borivoje Pasic
University of Zagreb, Croatia

ABSTRACT

Geologic storage is the component of Carbon Capture and Storage (CCS) in which the carbon dioxide (CO_2) is disposed in the appropriate underground formation. To successfully inject CO_2 into the subsurface to mitigate greenhouse gases in the atmosphere, the CO_2 must to be trapped in the subsurface and must not be allowed to leak to the surface or to potable water sources above the injection zone. For the purposes of risk assessment, a priority is to evaluate what would happen if CO_2 migrated unexpectedly through the confining unit(s), potentially resulting in undesirable impacts on a variety of potential receptors. One of the main risks identified in geological CO_2 storage is the potential for CO_2 leakage through or along wells. To avoid leakage from the injection wells, the integrity of the wells must be maintained during the injection period and for as long as free CO_2 exists in the injection zone.

INTRODUCTION

Carbon capture and storage (CCS) is becoming a critically important part of global warming mitigation efforts, and this trend is expected to continue, with more and more wells being drilled for this purpose. Geologic storage is also known as geologic sequestration or geosequestration. The CO_2 for the geologic storage is not removed from the atmosphere but comes from industrial facilities that emit large amounts of CO_2 such as coal power stations, petroleum refineries, oil and gas production facilities, iron and steel mills, cement plants

DOI: 10.4018/978-1-4666-4777-0.ch015

and various chemical plants. The large-scale underground storage of CO_2 has the potential to play a key role in reducing global greenhouse gas emissions. Typical underground storage reservoirs would lie at depths of 1 km or more and contain 10 or even $100x10^6$ tones of CO_2. To date, the technology as a whole has only been deployed at a few pilot sites around the world: Sleipner field in Norway, Weyburn field in Canada, In Salah field in Algeria (Gallo et al., 2002; Jimenez & Chalaturnyk, 2002). These projects have demonstrated successful operational experience with the injection of CO_2 into the subsurface at rates of approximately 1 x 10^6 metric tons of CO_2 per year. According to the International Energy Agency (IEA), by the year 2050, $5x10^9$ tons of CO_2 per year could be avoided through CO_2 capture and storage, representing a 16% contribution to the reduction of global emissions needed to limit climate disruption.

An important component of a CCS project is risk assessment. A comprehensive risk assessment is needed early in the project, but risks should be continually assessed, and updated throughout the project. The importance of assessing the potential for environmental impacts from CO_2 storage is recognized by certain regulations, such as the *European council directive on storage* (Directive 2009/31/EC, 2009) and the *OSPAR Guidelines for Risk Assessment and Management of Storage of CO2 Streams in Geological Formations* (OSPAR, 2007), and other sources of guidance such as the United States Environmental Protection Agency (US EPA) *Vulnerability Evaluation Framework* (United States Environmental Protection Agency, 2008) and the *CO2QUALSTORE Guideline for Selection and Quantification of Sites and Projects for Geological Storage of CO_2* (Det Norske Veritas, 2009).

They all recognize that the primary issue to be assessed and demonstrated for CCS sites is long-term containment. However, they all also recognize the importance of assessing the potential for environmental impacts should there be any leaks.

One key component of a risk assessment is identifying potential leakage pathways (e.g., faults, wells, fractures). This identification is then integrated with the Measurement, Monitoring, and Verification (MMV) program and includes, if injected CO_2 migrates toward an identified pathway, approaches to mitigate the risk of negative impacts upon the surface, a Underground Sources of Drinking Water (USDW), or outside the project footprint or to implement remediation measure. One method for ensuring this needed integration is to link the models used for risk assessment and the subsurface models that are developed and informed by MMV (Forbes et al., 2008). A lot of literature is published with the aim of assessing the risk of CO_2 storage (Benson et al., 2002; Vendrig et al., 2003; White et al., 2003)

Various methodologies are currently being applied to risk assessment of geological CO_2 storage and only some of them will be mentioned here (Le Guen et al., 2008; Forbes et al., 2008; Le Guen et al., 2009).

STORAGE SITE SELECTION AND CHARACTERISATION

The basic concept of CO_2 storage is to prevent CO_2 emissions being released into the atmosphere by injecting it into deep porous geological formations so that it remains permanently trapped. For a carbon storage method to be effective, CO_2 must be stored for several hundred or thousands of years. The majority of locations that are being considered for CO_2 injection and sequestration are typically found in areas that have a history of oil, natural gas, and/or coalbed methane production. This is due to value-added opportunities such as Enhanced Oil Recovery (EOR), enhanced gas recovery (EGR), and Enhanced Coal Bed Methane (ECBM) recovery (Bellarby, 2009). There also exists a greater knowledge base for saline formations that lie either above or below oil and gas reservoirs due to well logging and exploration activities. As a result of human activity, these formations are typically punctured by a significant number of wells from both exploration and production. No matter how impermeable an overlying cap rock is, the sealing integrity may be compromised by the presence of wells. Diffusion through the cap rock is normally less of a concern than fracturing or fault-related leakage (Jimenez & Chalaturnyk, 2002). There are a number of different types of CO_2 injection scheme (Ennis-King & Paterson, 2002; Gallo et al., 2002; Nezhad et al., 2006; Bellarby, 2009):

- Injection of CO_2 into a producing reservoir either to promote oil production (Enhanced Oil Recovery – EOR) or to sequester the CO_2;
- Water-alternating CO_2 injection for production enhancement;
- Injection of CO_2 into a previously producing reservoir (depleted oil and gas reservoirs);
- Sequestering CO_2 into a deep saline aquifers that is independent of any oil or gas development;
- Sequestering CO_2 into deep unmineable coal beds (often thin and therefore uneconomic);
- Storage in salt caverns, etc.

Storage sites need to be chosen carefully on the basis of their geological characteristics. For safe storage the storage site must have: (1) a good reservoir at a depth of greater than about 800 meters (where CO_2 will be in the dense or 'supercritical' phase under typical subsurface conditions of temperature and pressure); (2) a geological trap; (3) a good seal above the reservoir (typically mudrock or salt, usually hundreds of meters thick); and (4) a geologically stable area. In the case of an aquifer, the reservoir salinity must be sufficiently high as to be unusable as a future source of drinking water.

Site screening in many cases can be executed with existing well control, seismic data, and/or literature published in the course of earlier oil and gas exploration and development. The various criteria are considered important in the evaluation potential CO_2 storage site such as technical, economic, commercial, organizational and political because of their potential to add short and long-term risk to the storage project

In the case of insufficient data (especially data provided by wells) the operator may elect to drill additional wells in order to find suitable storage sites that can accommodate planned volumes of supercritical CO_2. The success of these wells will be determined by the permeability (k) and height (h) product within suitable formation for supercritical CO_2 storage. The number of wells required depends upon the site specifics environment, project risks and uncertainties that need better constraint, and the presence of previous well and seismic data to control and calibrate the geological predictions (Jenvey et al., 2008).

Choosing the best storage site is the best way to manage the uncertainties and minimize the risks associated with CO_2 storage. There is a range of methods, based around risk assessment, for screening geological basins and comparing different sites. The first step is collecting the available data, rank the sites, and select candidates for further characterization. Some considerations include capacity (Will the site hold as much CO_2 as needed?), injectivity (Can we fill the store at the required rate?) and containment (Will the stored CO_2 remain safely in place?). Once selected, the site has to be characterized in detail using proprietary (risk-based) detailed evaluation techniques, tailored to the special needs of CO_2 storage. The geological structure of the site may be imaged from the surface using seismic data or other technologies. Wells can be drilled and rock properties investigated using logging tools. Rock and fluid samples can be taken for laboratory measurements. Injectivity can be assessed using fluid flow testing. The results are integrated to produce models of storage performance, together with plans for monitoring and risk mitigation. These models examine geochemical and geomechanical processes to simulate and test a range of scenarios covering injection rates, fluid displacement, CO_2 trapping, and containment performance.

A geological formation shall only be selected as a storage site, if under the proposed conditions of use there is no significant risk of leakage, and if no significant environmental or health risks exist.

When assessing the performance of a storage system it is useful to consider relevant Features, Events and Processes (FEPs), which are affecting the evolution of the storage system (Quintessa, 2010). A generic FEP database has been produced and can be freely accessed at www.quintessa.org/CO2fepdb/. The features include a list of parameters, such as cap rock porosity, a number of wells and reservoir permeability. Events are processes like seismic events and well blowouts. The processes are physical and chemical processes such as geomechanical or geochemical processes and multi-phase flow behavior (Wildenborg, 2013). A methodological framework for screening and ranking potential geological sites based on their posed leakage risk has been developed (van Egmond, 2006). This framework has been applied to several Dutch onshore aquifer traps and nearly depleted gas fields. The application showed that an evaluation and ranking of prospective sites with very limited site specific characterization data can be accomplished based on the potential CO_2 leakage risk.

The injection of CO_2 in deep geological formations involves many of the same technologies that have been developed in the oil and gas exploration and production industry. Well-drilling technology, injection technology, computer simulation of storage reservoir dynamics and monitoring methods from existing applications are being developed further for design and operation of geological storage (Intergovernmental Panel on Climate Change, 2005). Once injected into the storage formation, the fraction retained depends on a combination of physical and geochemical trapping mechanisms.

Storage Mechanisms and Storage Security

Once underground, a variety of mechanisms keep the supercritical CO_2 securely stored. Four different storage mechanisms have been distinguished: (1) structural/stratigraphic (or physical) trapping, (2) solubility trapping, (3) residual trapping, and (4) mineral trapping (Figure 1). Structural/stratigraphic trapping is the most important CO_2 storage mechanism during an injection process of several ten of years. The other three mechanisms are likely to be of a possible quantitive contribution on larger time scales.

Figure 1. Trapping mechanisms of CO_2 (Intergovernmental Panel on Climate Change, 2005)

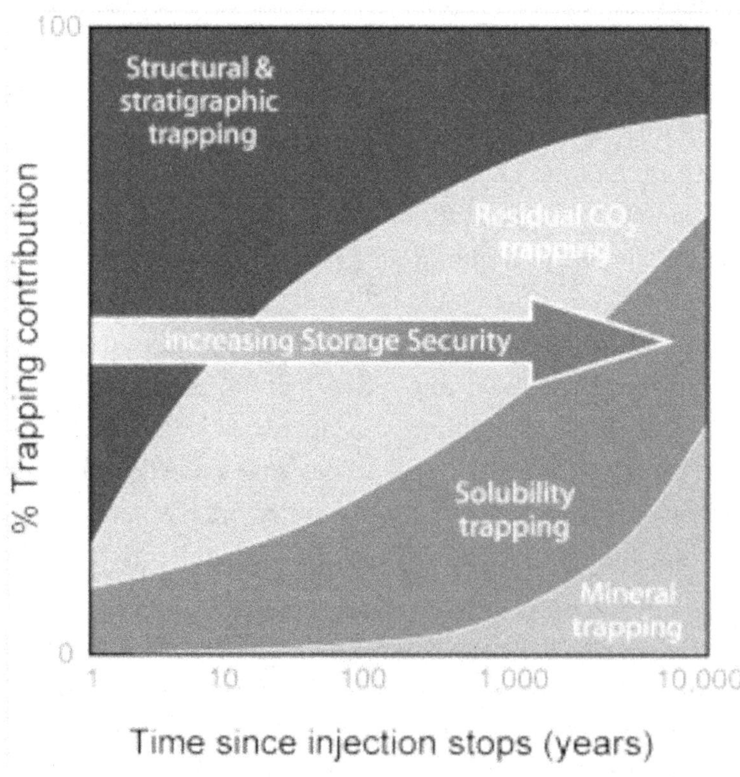

Structural traps are formed by a folded or fractured rock, whereas stratigraphic traps are formed by changes in a rock type caused by variations in the deposited setting. Fluid migration is happening very slowly and over long distances. Carbon dioxide remains under the sealing formation and on the longer term dissolves in the formation water. Physical trapping to block the upward migration of CO_2 is provided by a layer of shale and clay rock above the storage formation. This impermeable layer or sealing formation is known as the "cap rock". Additional physical trapping can be provided by capillary forces that retain CO_2 in the pore spaces of the formation but one or more sides of the formation remain open, allowing for lateral migration of CO_2 beneath the cap rock.

Solubility trapping is in progress when CO_2 dissolves in the formation water and once CO_2 is dissolved it no longer exists as a separate phase and buoyant forces are eliminated.

Residual CO_2 trapping takes place when CO_2 dissolves and becomes trapped in the water in the pore space of the rock minerals (residual water) after the displacement of formation brine by CO_2. Besides, small droplets of supercritical CO_2 will become trapped within the pore space by the surface tension between the formation water and the CO_2.

Mineral trapping is the result of chemical reactions between the CO_2 and the rock minerals to form a solid mineral structure. It is the most permanent form of geological storage, but it is relatively slow, and could take a thousand of years or longer.

These trapping processes take place over many years, and lead to immobilization of injected CO_2 at different rates from days to years to thousands of years, but in general, geologically stored CO_2 becomes more securely trapped with time (Figure 1). The best estimates of leakage rates by geologists are well below levels that would cause any significant increase in atmospheric CO_2 or risk to public safety. The studies of geologic storage test sites suggest leakage rates of less than 1 percent over thousands of years (Imbus et al., 2006).

Carbon Dioxide and its Behavior under Reservoir Conditions

Carbon dioxide is colourless and nearly odourless, but toxic. At concentrations of about 5% by volume, CO_2 in the air causes dizziness, confusion and breathing and hearing difficulties. Above around 8%, CO_2 leads to a loss of consciousness after exposures of between 5 to 10 min (Bellarby, 2009). CO_2, being denser than air, will also follow terrain downhill. Carbon dioxide has 30 to 40% less density and 5 to 10 times less viscosity than formation brine under typical subsurface conditions (Ennis-King & Paterson, 2002).

Under most reservoir conditions, CO_2 does not behave like a gas, but more like a low-viscosity liquid. This will affect the injection performance (in the tubing and near wellbore area). In geologic storage, CO_2 is injected at sufficiently high pressures (>7.3 MPa) and temperatures (> 31 °C) and it becomes a *supercritical fluid* below 800 m (Onan, 1984). Supercritical fluids take up much less space and diffuse better than either gases or ordinary liquids through the tiny pore spaces in storage rocks (Price & Smith, 2008). The supercritical CO_2 compresses further as the depth increases, increasing the amount that can be stored in the same volume of rock. High pressure at sufficient depths (i.e., greater than 800 meters) maintains the supercritical fluid state.

Except at very high pressures, CO_2 is less dense than most subsurface fluids, but it is denser than hydrocarbon gases such as methane. It will therefore naturally migrate to the top of oil or water bearing structures due to buoyancy and spread laterally beneath the reservoir cap rock. This is important as this CO_2 will then potentially interact with wells and completions at the top of the reservoir. Over time, carbon dioxide will tend to gradually dissolve in formation water, and then slowly sink, being denser than unsaturated water by around 1% (Ennis-King & Paterson, 2002). In both of these processes CO_2 can come into contact with existing wells and change the physical and chemical properties of cements under deep-well conditions (e.g., elevated pressure, temperature and salinity).

Carbon dioxide also has the potential to be corrosive when mixed with water. The well needs to maintain its integrity for the life of the project since the goal of geological sequestration (GS) is the long term storage of carbon dioxide. An improperly constructed well can lead to loss of well integrity that could lead to carbon dioxide or formation fluid leakage from the wellbore and into USDWs (Gasda et al., 2004; Gasda et al., 2005).

CARBON DIOXIDE INJECTION WELL DESIGN

The technologies for drilling and equipping CO_2 injection wells are well developed. American Petroleum Institute published a number of Specification and Recommended Practices for Casing and Tubing, and Well Cements such as: *API Specification 5CT – Specification for Casing and Tubing, API RP 5C1 – Recommended Practices for Care and Use of Casing and Tubing, API RP 10B-2 – Recommended Practice for Testing Well Cements, API Specification 10A – Specification on Cements and Materials for Well Cementing, API RP 10D-2 – Recommended Practice for Centralizer Placement and Stop Collar Testing, and API Specification 11D1 – Packers and Bridge Plugs.*

Most aspects of drilling and completing such wells are similar or identical to that of drilling and completing a conventional gas (or other) injection well or a gas storage well, with the exception that much of the downhole equipment (e.g. casing and tubing, safety valves, cements, blowout preventers) must be upgraded for high pressure and corrosion resistance. The well is completed at the surface by installing a wellhead and "Christmas Tree" that sits on top of the wellhead and is an assembly of valves, pressure gauges and chokes. Devices are connected to the "Christmas Tree" that allow the monitoring of pressure, temperature, and injection rates (Figure 2). The combined wellhead has casing annulus valves to access all annular spaces to measure the pressure between the casing strings and between the casing and production tubular.

Figure 2. Typical CO_2 injection wellhead

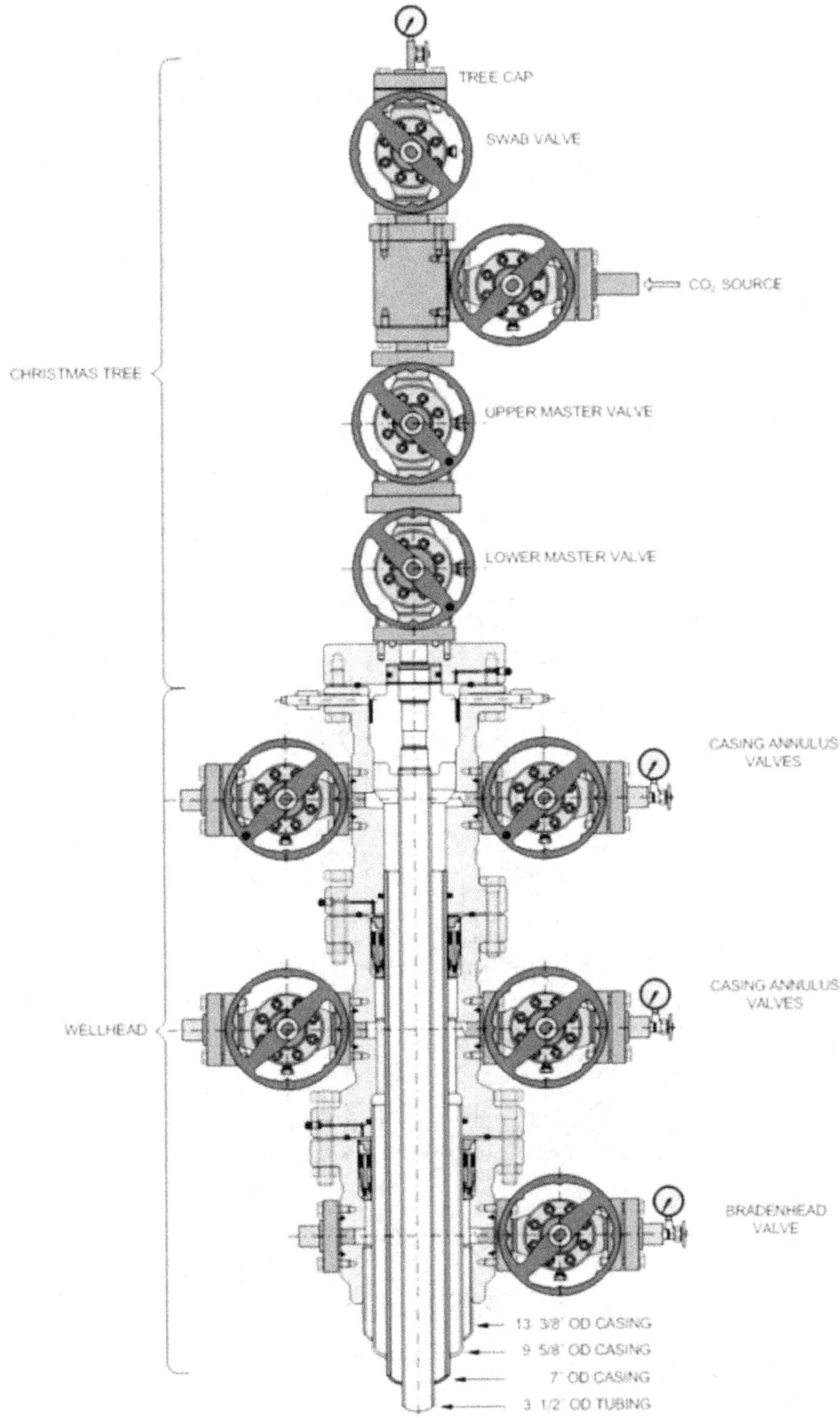

Above the Christmas tree a CO_2 injection valve is mounted and an access valve for running wirelines from the top.

The typical components of an injection well that are relevant to maintaining mechanical integrity and to ensuring that fluids do not migrate from the injection zone into USDWs are the casing, tubing, cement, and packer (Figure 3). The well components should be designed to withstand the maximum anticipated stress in each direction – axial direction (tensile, compressive) or radial (collapse, burst), and include a safety factor. The loading in each of the stress directions should be compared to the strength of the material in that direction. The loadings correspond to the burst, collapse, and tensile strengths of the material.

Figure 3. Schematic of a CO_2 injection well

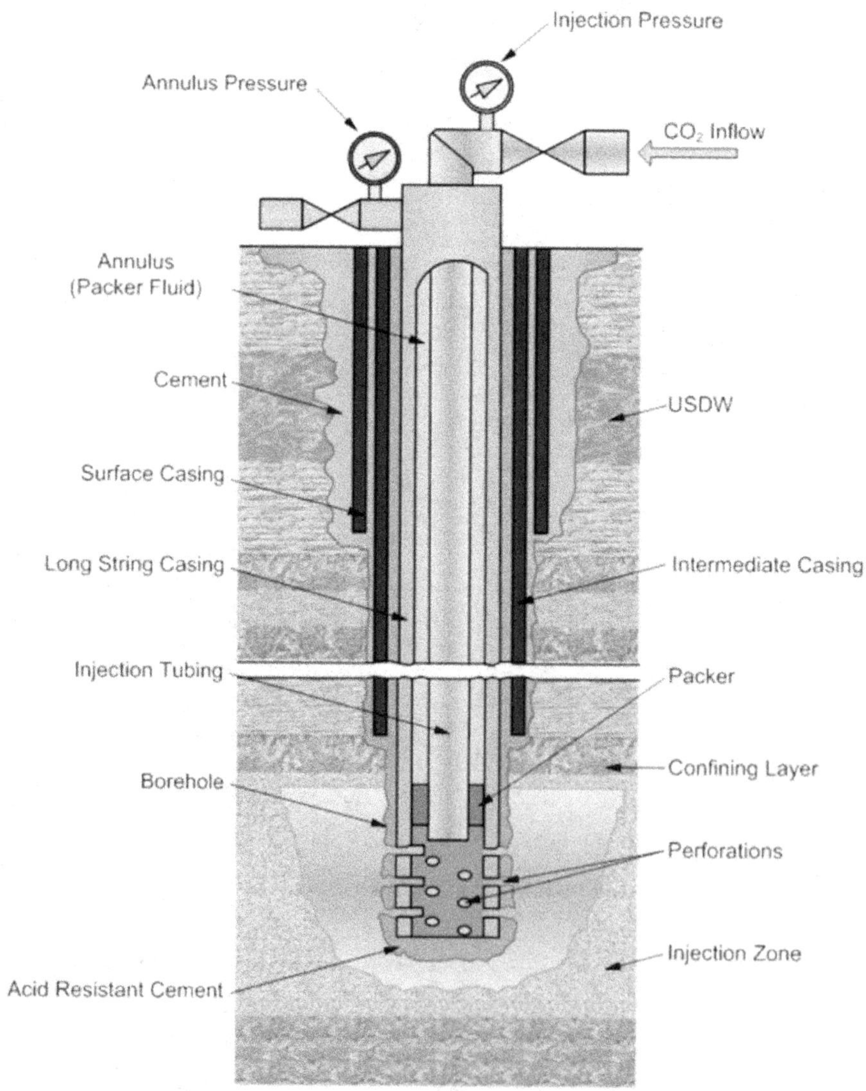

Casing

An injection well typically consists of one or more casings. Leaks in the casing can allow a fluid to escape into unintended zones or allow fluid movements between zones. The construction materials selected for the casing and the casing design must be appropriate for the fluids and stresses encountered at the site-specific down-hole environment. Carbon dioxide in combination with water forms carbonic acid, which is corrosive to many materials. Native fluids can also contain corrosive elements such as brines and hydrogen sulfide. In CO_2 injection wells, the spaces between the long string casing and the intermediate casing, and between the intermediate casing and the surface casing as well as between the casings and the geologic formation are required to be filled with cement, along all casings.

Tubing

The tubing is placed inside the long string casing from the ground surface down to the injection zone. The injected fluid moves down the tubing, out through the perforations in the long string casing, and into the injection zone. The tubing ends at a point just below the packer. The space between the long string casing and tubing must be filled with a noncorrosive packer fluid. The tubing forms another barrier between the injected fluid and the long string casing. It must be designed to withstand the stresses and fluids with which it will come into contact. The tubing and long string casing act together to form two levels of protection between the carbon dioxide stream and the geologic formations above the injection zone. A safety valve/profile nipple can be used to isolate the wellbore from the formation to allow the tubing string to be replaced. The injection will be conducted through the perforated casing. In the base case there is no stimulation method used, but hydro fracturing may be an option. Using acids to improve injectivity is not recommended because of the possible damage to the cement sheath and casing.

Packer

A packer is a sealing device which keeps fluid from migrating from the injection zone into the annulus between the long string casing and tubing. The tubing and retrievable packer are set on above the injection zone to ease the changing of the tubing if pitting is identified during regular inspections. A packer must also be made of materials that are compatible with fluids which it will come into contact.

Construction Requirements

All new CO_2 injection wells have to be cased and cemented to prevent the migration of fluids into or between underground sources of drinking water. The casing and cement used in the construction of each newly drilled well has to be designed for the life expectancy of the well. In determining and specifying casing and cementing requirements, the following factors have to be considered: (1) depth of the injection zone; (2) injection pressure, external pressure, internal pressure, axial loading, etc.; (3) hole size; (4) size and grade of all casing strings (wall thickness, diameter, nominal weight, length, joint specification, and construction material); (5) corrosiveness of injected fluids and formation fluids; (6) lithology of injection and confining zones; and (7) type and grade of cement. The following information concerning the injection zone has to be determined or calculated for new wells: (1) fluid pressure; (2) fracture pressure; and (3) physical and chemical characteristics of the formation fluids. Appropriate logs and other tests have to be conducted during the drilling and construction of new wells. Mandatory technical requirements for CO_2 injection well are presented in Table 1.

Cement

Cement is important for providing structural support of the casing, preventing contact of the casing with corrosive formation fluids, and preventing vertical movement of carbon dioxide. Some of the most current research indicates that a good cement job is one of the key factors in effective zonal isolation. The proper placement of the cement is critical, as errors can be difficult to fix later on. Failing to cement the entire length of casing, failure of the cement to bond with the casing or formation, not centralizing the casing during cementing, cracking, and alteration of the cement can all allow migration of fluids along the wellbore. If carbon dioxide escapes the injection zone through the wellbore because of a failed cement job, the injection process must be interrupted to perform costly remedial cementing treatments. In a worst case scenario, failure of the cement sheath can result in the total loss of a well. During the

Table 1. Mandatory technical requirements for CO_2 injection well (National Energy Technology Laboratory, 2009)

Technical Requirements for CO_2 Injection Well (Class VI)	
Siting	Extensive site characterization needed, including well logs, maps, cross-sections, USDW locations, determine injection zone porosity, identify any faults, and assess seismic history of the area.
Fluid Movement	No fluid movement to a USDW.
Area of Review (AoR)	Determined by computational model and reevaluated during project duration.
Construction	Two layers of corrosion-resistant casing required and set through lowermost USDW. Cement compatible with subsurface geology.
Operation	Injection pressures may not initiate or propagate fractures into the confining zone or cause fluid movement into USDWs. Quarterly reporting on injection, injected fluids and monitoring of USDWs within the AoR. Must report changes to facility, progress on compliance schedule, lose of mechanical integrity, or noncompliance with permit conditions. Permit valid for 10 years.
Mechanical Integrity Test (MIT)	Continuous internal integrity monitoring and annual external integrity testing.
Monitoring	Analyze injectant. Continuous temperature and pressure monitoring in the target formation. Plume tracking required.
Closure	50 day notice and flush well. Must be plugged to prevent injectant from contaminating USDWs.
Proof of Containment and Post-Closure Care	Post-closure site care for 50 years or until proof of non endangerment to USDWs demonstrated. (No-migration petition demonstration; fluids remain in injection zone for 10 000 years).
Financial Responsibility	Periodically update the cost estimate for well plugging, post injection site care and site closure, and remediation to account for any amendments to the area of review and corrective action plan. Environmental Protection Agency (EPA) is also proposing that the owner or operator submit an adjusted cost estimate if the original demonstration is no longer adequate to cover the cost of the injection well plugging, post-injection site care, and site closure.

injection phase, cement will only encounter dry CO_2. However after the injection phase and all the free CO_2 around the wellbore is dissolved in the brine, the wellbore will be attacked by carbonic acid (H_2CO_3). The carbonic acid will only attack the reservoir portion of the production (long string) casing, therefore special consideration of CO_2 cement needs only to be considered for the reservoir, the primary seal and a safety zone above the reservoir. Regular cement should be placed over the CO_2-resistant cement. However since two different cement slurries will be used, a CO_2-resistant cement that is compatible with regular Portland cement has to be used to prevent flash setting. The cement must be able to maintain a low permeability over lengthy exposure to reservoir conditions in a CO_2 injection and storage scenario. Long-term carbon sequestration conditions include a contact of set cement with supercritical CO_2 (>31 °C at 7,3 MPa) and brine solutions at increased pressure and temperature and decreased pH (Kutchko et al., 2007).

Degradation of Wellbore Cement Due to CO_2 Injection

Portland cement systems are used conventionally for zonal isolation in oil or gas production wells. It is thus crucial to study how such cement behaves at depth in CO_2-rich fluids and understand the chemical interactions between injected CO_2 and existing cements that could potentially lead to leakage. The properties of Portland cement are determined by the mineralogical composition of the Portland cement clinker (Table 2). The total content of minor compounds such as calcium oxide (CaO), magnesium oxide (MgO), potassium oxide (K_2O), sodium oxide (Na_2O), titanium dioxide (TiO_2), Manganese (III) oxide (Mn_2O_3) and sulfur trioxide (SO_3) is normally under 5%. When Portland cement is mixed with water, its compounds form hydration products. The main products formed by the cement hydration process are calcium silicate hydrate gel – C-S-H and calcium hydroxide - $Ca(OH)_2$. C-S-H is a semi-amorphous gel-like material that comprises approximately 70 wt % of the hydrated cement and is the primary binding material. $Ca(OH)_2$ is crystalline and comprises roughly 15 to 20 wt % of the hydrated cement (Nelson & Guillot, 2006).

Portland cement is thermodynamically unstable in CO_2-rich environments and can degrade rapidly upon exposure to CO_2 in the presence of water (Onan, 1984). As CO_2-laden water diffuses into the cement matrix, the dissociated acid (H_2CO_3) reacts with the

Table 2. Mineralogical composition of Portland cement clinker

Oxide Composition	Cement Notation	Common Name	Concentration (wt %)
$3CaO{\cdot}SiO_2$	C_3S	Alite	55 - 65
$2CaO{\cdot}SiO_2$	C_2S	Belite	15 - 25
$3CaO{\cdot}Al_2O_3$	C_3A	Aluminate	8 - 14
$4CaO{\cdot}Al_2O_3{\cdot}Fe_2O_3$	C_4AF	Ferrite phase	8 - 12

free calcium hydroxide and the calcium-silicate-hydrate gel. The reaction products are soluble and migrate out of the cement matrix. Eventually, the compressive strength of the set cement decreases and the permeability and porosity increase leading to loss of zonal isolation (Gaurina-Medjimurec, 2010).

There are mainly three different chemical reactions involved in cement-CO_2 interaction: (1) formation of carbonic acid, (2) carbonation of calcium hydroxide and/or cement hydrates, and (3) dissolution of calcium carbonate ($CaCO_3$) which are shown below (Onan,1984; Bellarby, 2009; Santra et al., 2009):

1. Formation of Carbonic Acid

Carbon dioxide diffuses into the capillary pores of the cement which contain, to some extent, water and dissolves in it to form carbonic acid as shown in Equation 1:

$$CO_2 + H_2O \rightarrow H_2CO_3 \quad (1)$$

Depending on temperature, partial pressure of CO_2, and other ions present in water (such as salt, etc), forming of carbonic acid causes lowering in pH value.

2. Carbonation of Calcium Hydroxide and/or Cement Hydrates

When carbonic acid comes in contact with hydrated cements, corrosion can occur. Conventional Portland cement will react with dissolved carbon dioxide and revert to calcium carbonate (the reverse reaction to the manufacturing of cement). Carbonic acid reacts with calcium hydroxide (also named as hydrated lime or portlandite) in the cement causing carbonation of $Ca(OH)_2$ (Eq. 2a) and/or decomposition of calcium silicate hydrate gel, the main binding component in hydrated cement, into calcium carbonate and an amorphous silica (Equation 2b):

$$Ca(OH)_2 + H_2CO_3 \rightarrow CaCO_3 + 2H_2O \quad (2a)$$

$$\text{C-S-H and/or crystalline phases} + H_2CO_3 \rightarrow CaCO_3 + SiO_2\,(gel) + H_2O \quad (2b)$$

The carbonation reactions cause densification, leading to increased hardness and reduced permeability thereby decreasing CO_2 diffusion and up to 6% volume expansion, which can lead to development of micro to macro cracks in extreme cases. The rate at which cement degradation occurs depends primarily on temperature, but also on cement type, cement composition, water/cement ratio, moisture content, CO_2 partial pressure, and porosity/permeability (Kutchko et al., 2007; Santra et al., 2009). Carbonation is extremely fast in the early days but later slows down drastically because of the time dependant reduced porosity/permeability caused by the initial carbonation itself (Santra et al., 2009). Moderate degrees of carbonation can be beneficial to cement porosity, permeability, and strength; extensive carbonation can result in the loss of cement structural integrity (Carey et al, 2007).

3. Dissolution of $CaCO_3$

It is a long-term phenomenon and happens only when the set cement is surrounded by liquid water containing dissolved CO_2 (Equation 3):

$$CaCO_3 + H_2CO_3 \rightarrow Ca(HCO_3)_2 \quad (3)$$

The effects of this reaction are increased porosity/permeability and loss of overall mechanical integrity, leading to inefficient or even potential loss of zonal isolation in extreme cases (Santra et al., 2009). In a long-term scenario where the dissolution of $CaCO_3$ to $Ca(HCO_3)_2$ is unavoidable, Portland cement with higher pozzolanic content wherein the total amount of $CaCO_3$ formed is much lower after carbonation will face less-severe effects of dissolution simply because there is less $CaCO_3$ to dissolve either by CO_2 or H_2CO_3 (Santra et al., 2009).

For over three decades, mostly laboratory based studies have indicated that pure Portland-cement-based formulations could fail to maintain zonal isolation over a long period of time.

Onan (1984) studied the effects of supercritical CO_2 on present cement, as well as the influence of carbonation on the early stages of the cement hydration process. He concluded that after prolonged exposure to CO_2 under supercritical conditions, the hydration products formed in the hydration of Portland cement undergo decomposition into calcium carbonate and a siliceous residue. Cement samples exposed to the lower extremes (temperature and pressure) of a supercritical CO_2 environment exhibited greater reactivity under dynamic conditions as compared to static conditions, while increasing CO_2 pressure increased the degree of reaction regardless of the carbonation conditions used.

The carbonation front is reported by Barlet-Gouedard et al. (2006 and 2009) to progress at 5 to 6 mm over 3 weeks under specific laboratory conditions, with the front slowing over time as diffusion through the carbonated cement reduces the feed of acidic water.

Kutchko et al. (2007) conducted experiments to assess the durability of cements in wells penetrating candidate formations for geological sequestration of CO_2. Cement samples were prepared using Class H Portland cement and a water-to-cement ratio of 0.38. Cement was allowed to cure 28 days under different condition (A: 22 °C, 0.1 MPa; B: 22 °C, 30.3 MPa; C: 50 °C, 0.1 MPa; D: 50 °C, 30.3 MPa) in a 1% NaCl/brine solution in a high-pressure vessel to simulate well cement in contact with a deep saline aquifer. Portland cement cured under each condition was subsequently exposed to a brine solution saturated with CO_2 at 50°C, 30.3 MPa in a high-pressure vessel. At this temperature and pressure, the CO_2 is supercritical. These experiments showed a significant variation in the initial degradation (9 days of exposure) based on the curing conditions. The high temperature (50°C) and high pressure (30.3 MPa) curing environment increased the degree of hydration and caused a change in the microstructure and distribution of the $Ca(OH)_2$ phase within the cement. Cement cured at sequestration conditions proved to be more resistant to carbonic acid attack and it exhibited a shallower depth of degradation (0.22 mm) than cement cured at 22°C and 0.1 MPa (0.59 mm). This is likely due to smaller, more evenly distributed $Ca(OH)_2$ crystals that provide a uniform and effective barrier to CO_2 attack.

Carey et al. (2007) investigated the impact of CO_2 – cement interactions on cement performance by collecting wellbore samples from the world´s second oldest continuous CO_2-flooding operation (since 1972), the SACROC Unit, located in West Texas. The coring operation recovered samples of casing and cement. The casing was in excellent conditions and showed little evidence of corrosion. The SACROC samples consisted of neat Portland cement exposed to CO_2 in the subsurface at 54ºC and 18 MPa for 30 years. The recovered cement had air permeability in the tenth of a μm^2 range and thus retained its capacity to prevent a significant flow of CO_2. There was evidence, however, for CO_2 migration along both the casing-cement and cement-shale interfaces. A 0.1 to 0.3 cm thick carbonate precipitate occurs adjacent to the casing. The CO_2 producing this deposit may have traveled up the casing wall or may have infiltrated through the casing threads or points of corrosion. The cement in contact with the shale (0.1 to 1 cm thick) was heavily carbonated. The CO_2 causing this reaction originated by migration along the cement-shale interface where the presence of a filter cake may have provided a fluid pathway. Santra et al. (2009) studied the role of pozzolanic substitution (silica fume and fly ash) and concluded that, before CO_2 treatment, addition of either silica fume or fly ash, the amount of $Ca(OH)_2$ still present in cement either decreases or completely disappears. After 15 days of CO_2 treatment (at 93.3 ºC and 13.79 MPa) the cement samples with the increasing amount of pozzolanic substitutions (up to 50%) were less affected by the CO_2, and the depth of carbonation during a period of up to 15 days is directly proportional to the amount of pozzolanic supplement. After 90 days of CO_2 treatment (at 93.3 ºC and 13.79 MPa) carbonation slows down drastically because of the time dependant reduced porosity/permeability caused by the initial carbonation itself.

Several approaches have been adopted to help reduce detrimental effects of carbonation (Santra et al., 2009): (a) reduce the amount of Portland cement by incorporating filler, (b) reduce porosity/permeability, (c) add reactive supplementary materials to reduce the $Ca(OH)_2$, as well as changing the C-S-H composition to a more CO_2-resistive one. Certain mismatching between the research lab tests results and actual field measurements implies the necessity of further investigation and standardization of CO_2 resistance-testing devices, conditions and procedures.

WELL INTEGRITY ASSESMENT

One of the main CO_2 injection candidates may be new wells, or old wells that are active, closed or abandoned. In all cases, it is critical to ensure that the long-term integrity of the storage wells is kept; in others words wellbore integrity is one of the key performance criteria in the geological storage of CO_2.

Underground gas storage operations and CO_2 sequestration in aquifers relay on both proper wellbore construction and sealing function of the cap rock. The potential leakage paths are the migration CO_2 along the wellbore due to poor cementation and flow through the cap rock. The permeability and integrity of the cement will determine how effective it is in preventing leakage. The integrity of the cap rock is assured by an adequate fracture gradient and by sufficient cement around the casing across the cap rock and without a micro-annulus.

Well Integrity

Well integrity has been identified as the biggest risk contributing to leakage of CO_2 from underground storage sites. Wellbore represents the most likely route for the leakage of CO_2 from geologic carbon sequestration. Abandoned wells are typically sealed with cement plugs intended to block vertical migration of fluids. In addition, active wells are usually lined with steel casing, with cement filling the outer annulus in order to prevent leakage between the casing and formation rock.

Several potential leakage pathways can occur along active injection well (Figure 4a) and/or abandoned well (Figure 4b). These include leakage: through deterioration (corrosion) of the tubing (1), around packer (2), through deterioration (corrosion) of the casing (3), between the outside of the casing

Figure 4. Possible leakage pathways in an active CO_2 well (a) and abandoned well (b)

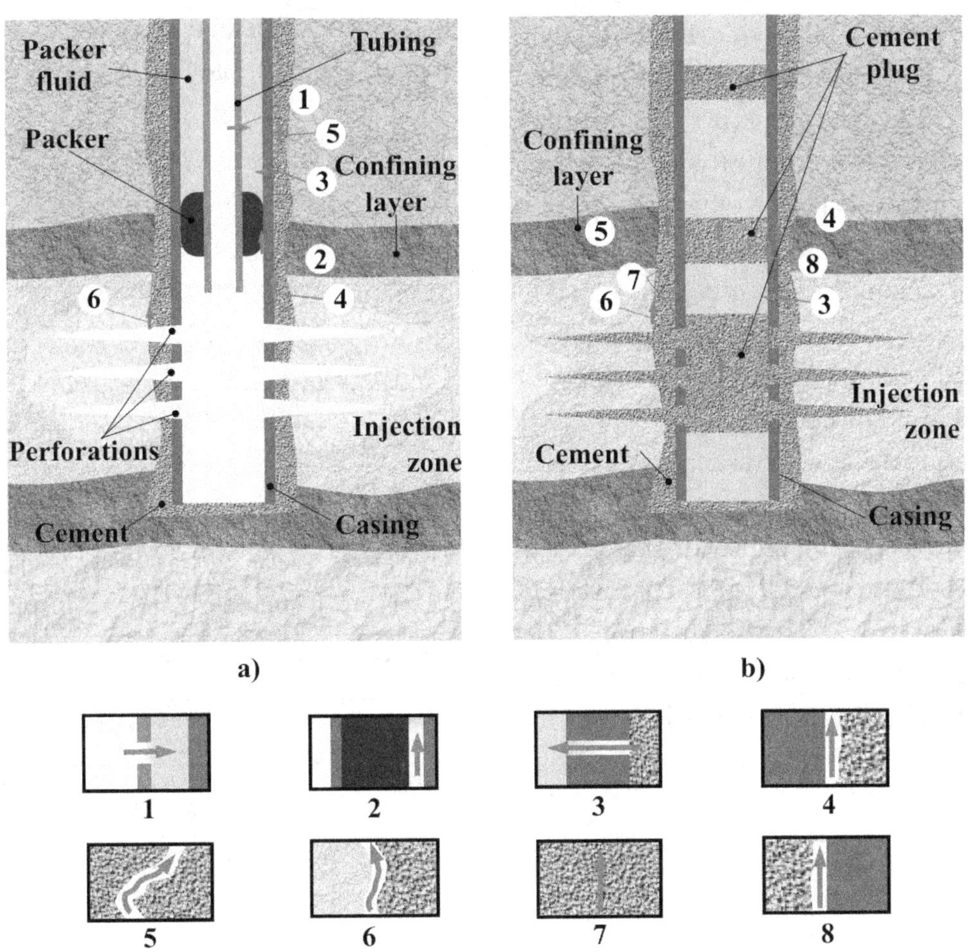

and the cement (4), through deterioration of the cement in the annulus (cement fractures) (5), leakage in the annular region between the cement and the formation (6), through the cement plug (7), and between the cement and the inside of the casing (8).

The permeability and integrity of the cement in the annulus and in the wellbore will determine how effective the cement is in preventing fluid leakage. The cement must be able to maintain a low permeability over lengthy exposure to reservoir conditions in a CO_2 injection and storage scenario. Long-term carbon sequestration conditions include contact with supercritical CO_2 and brine solutions at increased pressure and temperature and decreased pH (Kutchko et al., 2007).

The greatest risk for the escape of CO_2 may come from other wells, typically for oil and gas, which penetrate the storage formation. Such wells need to be properly sealed in order to ensure that they do not provide pathways for the CO_2 to escape into the atmosphere. Planning for geologic storage must take such wells into account. The escaping of CO_2 through water wells is much more unlikely since water wells are usually much shallower than the storage formation.

In injection wells, the presence of CO_2 may lead to corrosion on those parts coming in contact with CO_2 which normally means the production tubing and part of the production casing below the packer. Corrosion may be limited by: the selection of high alloy chromium steels, resistant to corrosion, and by inhibitor injection, if using carbon steel casing.

Mechanical Integrity

Mechanical integrity is a key concept related to the performance of an injection well, and the prevention of injected fluid movement into or between USDWs or other zones. Mechanical integrity of the well is achieved by ensuring that each of the components of the well are constructed with appropriate materials and are functioning together as intended. Typical corrosion resistant materials include 316 stainless steel, fiberglass, or lined carbon steel for casing and tubing. Casing and tubing can be lined with glass reinforced epoxy, plastic, or cement. If lined casing or tubing is used, care is recommended during installation to avoid damaging the lining (Meyer, 2007). Other metal parts such as packers and valves can be nickel plated or made of other high nickel alloys (Table 3).

Table 3. Materials of construction for CO_2 injection wells (Meyer, 2007)

Component	Materials of construction
Upstream Metering & Piping Runs	316 SS, Fiberglass
Christmas Tree (Trim)	316 SS, Nickel, Monel
Valve Packing and Seals	Teflon, Nylon
Wellhead (Trim)	316 SS, Nickel, Monel
Tubing Hanger	316 SS, Incoloy
Tubing	GRE (*Glass Reinforced Epoxy*) lined carbon steel, IPC carbon steel, CRA
Tubing Joint Seals	Seal ring (GRE) Coated threads and collars (IPC)
ON/OFF Tool, Profile Nipple	Nickel plated wetted parts, 316 SS
Packers	Internally coated hardened rubber of 80-90 durometer strenght (Buna-N), Nickel plated wetted parts

The selection of materials has to be made by considering the characteristics of the injected fluid, planned well life, and probability that the pipe comes into contact with reservoir water (D'Alesio et al., 2011).

Maintaining mechanical integrity helps prevent the well and wellbore from becoming conduits for fluid migration out of the injection zone. There are two aspects of mechanical integrity: internal and external.

Internal Mechanical Integrity

Internal mechanical integrity is defined as the absence of significant leaks in the casing, tubing, or packer. These well components act as the main barriers preventing contact between the injectate (the injected carbon dioxide stream) and the surrounding geologic formations through which the well has been drilled and constructed. Ensuring that these components are constructed properly with appropriate materials and that they remain undamaged when subject to stresses or corrosive (and other) operational conditions may prevent carbon dioxide from moving out of the well bore during injection (National Energy Technology Laboratory, 2009). The pressure applied during an internal mechanical integrity test should be limited to prevent casing ballooning that could create cement defects.

The absence of significant leaks in the casing, tubing, or packer is demonstrated through the use of (1) the Standard Annulus Pressure Test (SAPT), (2) the Standard Annulus Monitoring Test (SAMT), and (3) the Radioactive Tracer Survey (RTS).

External Mechanical Integrity

External mechanical integrity is defined as the absence of significant leakage outside of the casing. Maintaining external mechanical integrity helps to ensure that the injected carbon dioxide, which tends to be more buoyant than native formation fluids, does not migrate upwards from the injection zone after it has been injected; therefore helping to ensure zonal isolation of the injected carbon dioxide. The main construction component ensuring external mechanical integrity is the set cement. Properly emplaced cement should both prevent fluid movement by sealing the space between the casing and the formation, and protect the well casing from stress and corrosion.

The absence of significant fluid movement into an USDW through vertical channels adjacent to the injection well bore is demonstrated through the use of (1) the results of a temperature log, (2) noise log, (3) Oxygen Activation Log (OAL), (4) the results of a Radioactive Tracer Survey (RTS) (when the injection zone is separated from the lowermost USDW by a single confining layer), or (5) cementing records (Ultrasonic well logging; Cement Bond Log – CBL) demonstrating the presence of adequate cement to prevent fluid migration into USDWs.

D'Alesio et al. (2011) has been carried out well integrity assessment, as a part of ENI Exploration and Production (E&P) GreenHouse Gases (GHG) project launched in 2004 in order to comply with the Kyoto protocol and the Italian Legislation about the reduction of CO_2 emissions. They have been developed and implemented as operational methodology for three Italian fields with the aim of verifying and, if necessary, reinstating their structural integrity. Well integrity assessment process includes comparison between the Sustained Casing Pressures (SPSs) measured in the annuli and the relevant Maximum Allowable Pressures (MAPs) to estimate how close each drilled well in selected fields is to losing its safety conditions (well integrity assessment) (Figure 5). The criteria for the classification of the annular space criticalities have been established according to Sustained Casing Pressure (SCP) and Maximum Allowable Pressure (MAP) as: low (SCP < 50% MAP), medium (SCP = 50 to 100% MAP), and high (SCP > MAP). According to the NORSK Standard D-010 they use terms "primary barrier" (the production tubing, packer and safety valve) and "secondary barrier" (the cement outside the producing casing, the production casing and the wellhead valves) instead of internal and external mechanical integrity.

Figure 5. Well integrity assessment process (after D'Alesio, 2011)

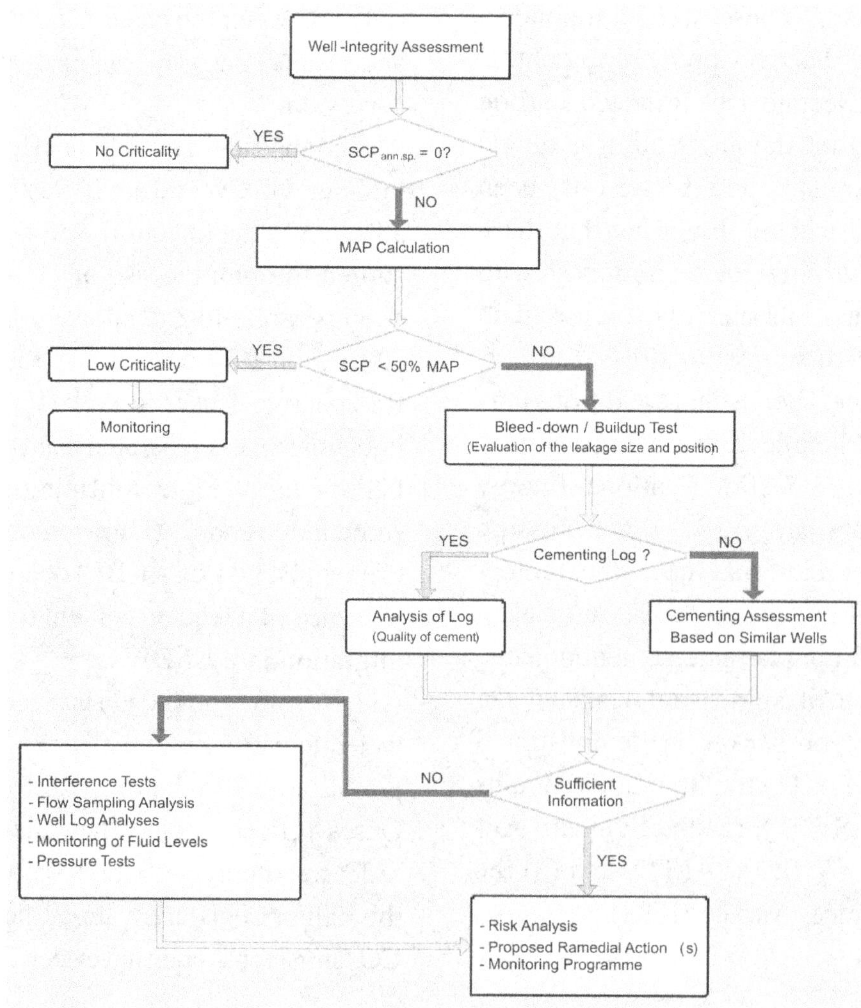

The methodology requires gathering all available data including: (1) list of all wells (active, plugged and abandoned), (2) construction and operational data of existing wells (fracture gradient of formation, drilling and completion fluid characteristics, wells "as built" characteristic and operations, collapse and burst resistance of tubing and casing), (3) construction and abandonment related data of the plugged and abandoned wells, and (4) historical pressure values measured in each annular space.

On the basis of gathered data, the applied methodology consists of five steps: (1) calculation of the MAP of each annular space, (2) preliminary classification of the criticality of each SCP well on the basis of the comparison between the highest historical pressure value measured in its annular spaces and the relevant MAPs, (3) execution of diagnostic tests (including pressure-bleed off and build up performance, interference tests, flow-sampling analysis, well-log analysis, monitoring of fluid level, pressure test) to define size and location of leakage in the critical wells, (4) risk analyses and final classification of the wells criticality levels, and (5) selection of the most appropriate remedial actions to reinstate the integrity of each well.

RISK ASSESSMENT OF CO_2 STORAGE AND ENVIRONMENTAL IMPACT

Facilities for the geological storage of carbon dioxide (CO_2) will be designed to prevent any leakage back to the surface. However, it is important to be able to assess the consequences of any leakage in the unlikely event that it should occur. Risk assessment involves the identification (or qualification) and quantification of hazards, including the probability of features, events and processes that can result in undesirable impacts on human health and the environment.

Risk is generally defined as the product or combination of the probability (likelihood) of occurring a failure event (cause, hazard) and the magnitude of its impact or the severity of the consequences.

The uncertainties of the system can be converted into the term of probability and the quantitative CO_2 leak mass assessed into the term of severity (Le Guen et al., 2009). Any viable site for storing CO_2 will most likely include some number of identified hazards. When appropriately characterized and managed, the hazards for most sites will not present a substantial leakage risk (Intergovernmental Panel on Climate Change, 2005). The risks associated with the hazards identified should also influence measurement, monitoring, and verification (MMV) strategies for a given site.

Risk assessments should (Forbes et al., 2008):

- Examine (encompass) the potential for leakage of injected or displaced fluids via wells, faults, fractures and seismic events, and with a focus on the fluids' potential impacts on the integrity of the confining zone and endangerment to human health and the environment;
- Address the potential for leakage during operations, as well as over the long term;
- Help identify priority locations and approaches for enhanced MMV activities;

- Provide the basis for mitigation/remediation plans for response to unexpected events;
- Inform operational decisions, including setting an appropriate injection pressure that will not compromise the integrity of the confining zone;
- Include site-specific information, such as the terrain, potential receptors, proximity of USDWs, faults and the potential for unidentified borehole location within the project footprint;
- Include non-spatial elements or non-geologic factors (such as population, land use, or critical habitat) that should be considered in evaluating a specific site.

US EPA has developed a Vulnerability Evaluation Framework (VEF) for geologic sequestration of carbon dioxide (United States Environmental Protection Agency, 2008). The conceptual VEF model is shown in Figure 6. The vulnerability assessment incorporated in the VEF was developed to systematically identify those conditions that could increase the potential for adverse impacts from geologic sequestration, regardless of likelihood or broad applicability. It is not a quantitative, probabilistic risk assessment tool. Vulnerability assessment examines conditions that lead to increased or decreased susceptibility to consequences, whereas risk assessment measures the probability and severity of consequences. The conceptual approach to the VEF shown in Figure 5 has the following components (United States Environmental Protection Agency, 2008):

1. First, the GS system has characterized in terms of the injected CO_2 stream, the confining system, the injection zone, and a series of geologic attributes that could influence (i.e., increase or decrease) the vulnerability of the GS system to unanticipated migration, leakage and undesirable pressure changes (first column).
2. An approach is then provided for defining the spatial area that should be evaluated for adverse impacts associated with unanticipated migration, leakage, or undesirable pressure changes (middle column).
3. Potential impact categories and associated key receptors are then identified, including human health and welfare, the atmosphere, ecosystems, groundwater and surface water, and the geosphere (last column).

An extensive program of modeling, monitoring and verification activities was deployed at a set of active storage sites worldwide including Sleipner, In Salah, Ketzin, Weyburn, K12-B and Snřhvit (Wildenborg, et al. 2013). At Sleipner 16 years of CO_2 injection, monitoring and performance assessment indicate that CO_2 has been contained within the storage reservoir with negligible pressure increase. At Snřhvit downhole monitoring and 4D seismic confirm retention of CO_2 in the reservoir albeit with a higher than expected pressure response. The extensive monitoring and verification program at In Salah has shown that the CO_2 is currently contained within the storage complex. At Ketzin a very comprehensive deep- and shallow-focused

Figure 6. Vulnerability Evaluation Framework (VEF) conceptual model (United States Environmental Protection Agency, 2008)

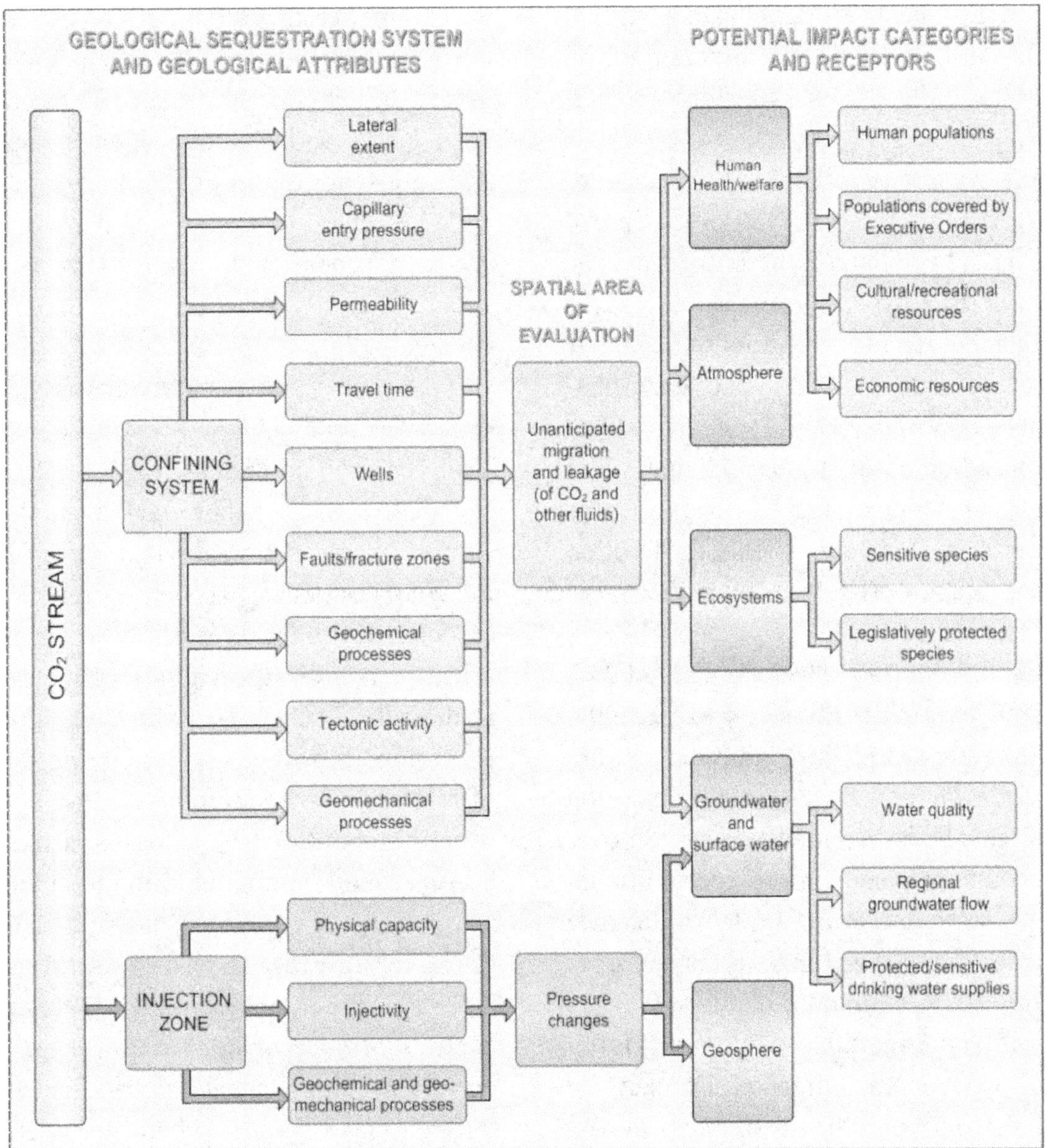

research monitoring program has tested a large number of tools and performance assessment methods. Again, no migration out of the storage complex has been detected. At Weyburn, 4D seismic has proved its worth not just for imaging plume development in the reservoir but also, via accurate measurement of time-shifts at top reservoir level, for providing a robust upper bound on the maximum amounts of CO_2 that could have migrated into the overburden. Results confirm that no CO_2 from the oil reservoir has been detected at surface. All investigated storage sites were well managed and did not have a negative impact on humans or the environment (Wildenborg et al., 2013) Time lapse seismic and pressure monitoring appeared to be the key in verifying the performance of the storage sites.

The risks due to leakage from storage of CO_2 in geological reservoirs fall into two broad categories: (1) global risks and (2) local risks.

Global risks involve the release of CO_2 that may contribute significantly to climate change if some fraction leaks from the storage formation to the atmosphere. With regard to global risks, based on observations and analysis of current CO_2 storage sites, natural systems, engineering systems and models, the fraction retained in appropriately selected and managed reservoirs is very likely („Very likely" is a probability of 90 to 99%) to exceed 99% over 100 years, and is likely to exceed 99% over 1000 years. Similar fractions retained are likely for even longer periods of time, as the risk of leakage is expected to decrease over time as other mechanisms provide additional trapping. In addition, if CO_2 leaks out of a storage formation, local hazards may exist for humans, ecosystems and groundwater.

Local risks include: public/workforce safety, environmental impacts, impacts to water, oil, gas or other resources, assets, ground displacement/induced seismicity. The elevated concentrations of CO_2 could lead to health effects to human and animals and negatively influences the ecosystems. These effects depend on the concentration and duration of exposure time (Benson et al., 2002).

Releasing of CO_2 from storage to atmosphere reduces the benefit of storing it and can pose higher risks with increasing concentrations. Clearly, the more CO_2 that is stored, the greater the potential that leakage from geologic sinks could result in adverse environmental and atmospheric impacts.

With regard to local risks, there are two types of scenarios in which leakage may occur (Intergovernmental Panel on Climate Change, 2005).

In the first scenario, injection well failures or leakage up abandoned wells could create a sudden and rapid release of CO_2. This type of release is likely to be detected quickly and stopped using techniques that are available today for containing well blowouts. Hazards associated with this type of release primarily affect workers in the vicinity of the release at the time it occurs, or those called in to control the blowout. Containing these kinds of releases may take hours to days and the overall amount of CO_2 released is likely to be very small compared to the total amount injected. These types of hazards are managed effectively on a regular basis in the oil and gas industry using engineering and administrative controls. *In the second scenario*, leakage could occur through undetected faults, fractures or through leaking wells where the release to the surface is more gradual and diffuse. In this case, hazards primarily affect drinking-water aquifers and ecosystems where CO_2 accumulates in the zone between the surface and the top of the water table. Groundwater can be affected both by CO_2 leaking directly into an aquifer and by brines that enter the aquifer as a result of being displaced by CO_2 during the injection process. There may also be acidification of soils and displacement of oxygen in soils in this scenario. Additionally, if leakage to the atmosphere were to occur in low-lying areas with little wind, or in sumps and basements overlying these diffuse leaks, humans and animals would be harmed if a leak were to go undetected. Humans would be less affected by leakage from offshore storage locations than from onshore storage locations.

Health and Ecological Effects

Humans, flora and fauna experience virtually no adverse effects when exposed to concentrations of CO_2 below 1% (10,000 ppm). Carbon dioxide can pose adverse health effects for humans when exposure concentrations are elevated. When humans are exposed to atmospheric concentrations of CO_2 between 3% and 5%, they experience respiratory rate increase, and physical discomfort such as headache and dizziness may occur. When the CO_2 concentration is above 5%, physical and mental capacities are impaired, and loss of consciousness can occur. Exposure to more than 10% CO_2 can result in rapid loss of consciousness, possible coma, or death. Generally, a concentration of CO_2 greater than 7 to 10% in air would cause immediate dangers to human life and health. Plants, insects, and soil organisms have a higher tolerance than humans to elevated CO_2 concentrations. In general, only a few microbes, invertebrates, insects and fungi can survive in CO_2 concentrations in excess of 20% (Benson et al., 2002; Vendrig et al., 2003).

Leakage Rate

A maximum leakage rate of approximately 0.001 to 0.01% per year of stored CO_2 per annum has been suggested as being acceptable. The Intergovernmental Panel on Climate Change (IPCC) assessed 0.001% per year as likely for well-designed storage; this leakage rate was the benchmark for the risk assessment. Also, it has been proposed that leakage rates of 0.01% per year be established as a performance requirement for geologically sequestered CO_2 (White et al., 2003). A leakage rate of 0.01% per year would ensure that 90% of the CO_2 would remain underground over a 1000 year time period. Leakage rates of less than 0.1% per year to the atmosphere are needed to ensure effective climate abatement by CCS. Leakage rates of more than 1% per year are unacceptable (Benson & Hepple, 2005).

Leakage Pathways

A key component of a risk assessment is identifying potential leakage pathways. The injected CO_2 could migrate upwards if (1) CO_2 gas pressure exceeds capillary pressure and passes through silstone, (2) free CO_2 leaks from silstone into upper aquifer up fault, (3) CO_2 escapes through "gap" in cap rock into higher aquifer, (4) injected CO_2 migrates up dip, increases reservoir pressure and permeability of fault, (5) CO_2 escapes via poorly plugged new or old abandoned wells, (6) natural flow dissolves CO_2 at CO_2/water interface and transports it out of closure and (7) dissolved CO_2 escape to atmosphere or ocean.

Remedial Measures

Once leakages are detected, some remediation techniques are available to stop or control them. Depending on the type of leakage, these techniques could involve standard well repair techniques or the extraction of CO_2 by intercepting its leak into a shallow groundwater aquifer. Techniques to remove CO_2 from soils and groundwater are also available, but they are likely to be costly. Experience will be needed to demonstrate the effectiveness and ascertain the costs, of these techniques for use in CO_2 storage.

Monitoring and Verification

Monitoring is a very important part of the overall risk management strategy for geological storage projects. It is essential in assessing whether injected CO_2 is behaving as expected, whether any migration or leakage occurs, and whether any identified leakage is damaging the environment or human health. Standard procedures or protocols have not been developed yet but they are expected to evolve as technology improves, depending on local risks and regulations. However, it is expected that some parameters such as injection rate and injection well pressure will be measured routinely. Repeated seismic surveys have been shown to be useful for tracking the underground migration of CO_2. Newer techniques such as gravity and electrical measurements may also be useful. The sampling of groundwater and the soil between the surface and water table may be useful for directly detecting CO_2 leakage. Carbon dioxide sensors with alarms can be located at the injection wells for ensuring worker safety and to detect leakage. Surface-based techniques may also be used for detecting and quantifying surface releases. High-quality baseline data improve the reliability and resolution of all measurements and will be essential for detecting small rates of leakage. Since all of these monitoring techniques have been adapted from other applications, they need to be tested and assessed with regard to reliability, resolution and sensitivity in the context of geological storage. All of the existing industrial-scale projects and pilot projects have programmers to develop and test these and other monitoring techniques. Given the long-term nature of CO_2 storage, site monitoring may be required for very long periods.

Le Guen et al. (2008 & 2009) proposed a quantitive performance and risk assessment methodology associated with well integrity within CO_2 geological storage and applied it to a synthetic case based on an existing well. The methodology is based on a systematic and functional approach. The concept of the function of each well component with respect to the zonal isolation and their failure modes is central to the performance and risk analysis. The methodology gathers different steps to go through for risk quantification and for recommending risk mitigation actions to ensure the well integrity performance (Table 4). The main steps are: (1) data collection and interpretation, (2) static model of the well, (3) dynamic model, (4) consequence grid, (5) risk mapping, and (6) recommendation.

In the first step, all relevant data are collected (source: drilling and cementing reports, cement and corrosion logs, well comple-

Table 4. Risk matrix dedicated to well integrity

Risk=Probability x Severity			Frequency level					
			1	2	3	4	5	6
			Probability					
Severity levels		Loss of injected CO_2 over 1000 years	≤0,00001	≤0,0001	≤0,001	≤0,01	≤0,1	≤0,1
1	Minor	≤0,05%	1	2	2	3	3	4
2	Low	≤0,01%	2	2	3	3	4	4
3	Serious	≤0,25%	2	3	3	4	4	5
4	Major	≤0,5%	3	3	4	4	5	5
5	Critical	≤0,1%	3	4	4	5	5	6
6	Extreme	>1%	4	4	5	5	6	6

tion design documents, production history, workover reports, geological profile etc.), checked and interpreted in order to make the well description (well trajectory, completion details) and the characteristics of its components (metallurgy of tubulars and packers, cement composition and properties). After data collection and interpretation, a functional analysis and a static model of the well are performed. The physical environment of the well (geology) is also included in the description of the overall system because it can interact with the well components and have a negative impact on hydraulic isolation. It encompasses: (1) specific formations located above the CO_2 reservoir and the cap rock, (2) subsurface fluids, (3) shallow subsurface or soils, (4) sea floor and sea water in case of an offshore well, (5) surface in case of an onshore well, and (6) atmosphere. Uncertainties about well components integrity are assessed and interpreted in terms of ranges of values distributed according to probabilistic lows.

After static model of the well has been performed, a dynamic model is built by combining degradations phenomena (cement degradation, casing corrosion due to CO_2) and associated kinetics to the static model. Such kinetics comes from the mechanical and chemical interactions between geological formations, the different fluids (brines, CO_2 etc.) and the well components (cement, casing steel etc.). The lack of knowledge on "exact" in situ conditions and long term effect are considered by studying the impact of these parameters on the CO_2 migration along the wellbore through a sensitivity study. Boundary (initial and limit) conditions (hydrostatic and reservoir pressure, temperature, flow/no flow conditions etc) are based on the geological environment of a well).

A "scenario" approach is then introduced to account for uncertain parameters. To build a scenario, system parameters are set to a given value within defined ranges. Therefore, each scenario corresponds to a possible state of the system. The probability of the scenario depends on the probability of each parameter to be equal to the selected value. Then all identified and relevant scenarios (in this case 1296 possible failure scenarios were studied) are simulated in order to quantify the CO_2 migration along the wellbore over a time period. The key outputs for each scenario are (Le Guen et al., 2009): (1) the CO_2 mass leakage towards a sensitive zone (surface, aquifer(s) etc) vs. time and (2) the associated CO_2 leakage pathways along the wellbore. Carbon dioxide mass that escapes out reservoir along the wellbore is then converted into a severity level by assessing the impact vs. a set of specific stakes relevant to the project (health, safety, performance of the confinement, financial).

This connection is ensured by using a *consequence grid* initially defined with the stakeholders involved in the CO_2 storage project. The consequence grid enables to link the result of simulations to severity levels via the criteria defined by the stakeholders. By crossing both severity and probability of each scenario, a risk level can be estimated. Once all risk levels are quantified, they populate a risk matrix. Severities higher than 5 and criticality values (frequency level + severity level) higher than 8 are not considered acceptable. Black line presents possible the Risk Acceptance Limit.

CONCLUSION

Facilities for the geological storage of carbon dioxide (CO_2) have to be designed to prevent any leakage back to the surface. However, it is important to be able to assess the consequences of any leakage in the unlikely event that it should occur. In order to have the safe underground storage of carbon, the injection wells as well as any well penetrating through the cap rock have to maintain sufficient integrity over a long time period. The expectation is that appropriately selected and managed storage sites will retain the injected CO_2 for thousands of years, and that leaks to the atmosphere are unlikely. For the purposes of risk assessment, a priority is to evaluate what would happen if CO_2 migrated unexpectedly through the confining unit(s), potentially resulting in undesirable impacts on a variety of potential receptors e.g. human populations, ecosystems, groundwater or other resources, and relevant aspects of the wider environment. Among the major risks identified to CO_2 geological storage, wells are often considered to be the weakest spots with respect to CO_2 confinement in the injection zone. Because of that, well integrity considerations should be present during all phases of well life including design phase, drilling, completion, injection, workover (service) and abandonment. Both existing and new wells must be fully evaluated and tested for integrity because there are many different possible leakage pathways. It is necessary to examine the condition of the casing and the cement and identify any annuli or defects that exist within the well. There is no one tool or method capable of looking at all of these features at the same time, so a suite of measurements must be run to analyze the integrity of a well. These measurements can be acquired using wireline tools such as caliper and ultrasonic tools to measure the integrity of the casing, sonic and ultrasonic tools to measure the integrity of the well cement, and tools to sample the casing, cement, formation, and formation fluid. The choice of well equipment and materials must be carefully considered to achieve the desired integrity. CO_2 corrosion may be limited by: the selection of high alloy chromium steels, resistant to corrosion, and by inhibitor injection, if using carbon steel casing. In addition, the use of acid resistant cement is highly recommended. Monitoring is essential in assessing whether injected CO_2 is behaving as expected, whether any migration or leakage occurs, and whether any identified leakage is damaging the environment or human health.

REFERENCES

Barlet-Gouedard, V., Rimmele, G., & Porcherie, O. (2006). *Mitigation strategies for the risk of CO_2 migration through wellbores.* Paper presented at the IADC/SPE Drilling Conference. Miami, FL.

Barlet-Gouedard, V., Rimmele, G., Porcherie, O., Quisel, N., & Desroches, J. (2009). A solution against well cement degradation under CO_2 geological storage environment. *International Journal of Greenhouse Gas Control*, *3*(2), 206–216. doi:10.1016/j.ijggc.2008.07.005

Bellarby, J. (2009). *Well completion design.* Dordrecht, The Netherlands: Elsevier.

Benson, S. M., & Hepple, R. (2005). *Prospects for early detection and options for remediation of leakage from CO_2 sequestration projects.* London: Elsevier Publishing.

Benson, S. M., Hepple, R., Apps, J., Tsang, C. F., & Lippmann, M. (2002). *Lessons learned from natural and industrial analogues for storage of carbon dioxide in deep geological formations.* Berkeley, CA: Lawrence Berkeley National Laboratory. doi:10.2172/805134

Carey, J. W., Wigand, M., Chipera, S. J., WoldeGabriel, G., Pawar, R., & Lichtner, P. C., … Guthrie, G. D. (2007). Analysis and performance of oil well cement with 30 years of CO_2 exposure from the SACROG unit, West Texas, U.S.A. *International Journal of Greenhouse Gas Control*, *1*, 75–85. doi:10.1016/S1750-5836(06)00004-1

D'Alesio, P., Poloni, R., Valente, P., & Magarini, P. A. (2011, May). Well-integrity assessment and assurance: The operational approach for three CO_2-storage fields in Italy. *SPE Production & Operations*, 140-148.

Det Norske Veritas. (2009). *CO2QUALSTORE guideline for selection and quantification of sites and projects for geological storage of CO_2.* Retrieved from http://www.dnv.com/binaries/co2qualstore_guideline_tcm4-412142.pdf

Directive 2009/31/EC of the European Parliament and of the Council of 23 April 2009 on the Geological Storage of Carbon Dioxide. Retrieved from http://eur-lex.europa.eu/en/index.htm

Ennis-King, J., & Paterson, L. (2002). *Engineering aspects of geological sequestration of carbon dioxide.* Paper presented at the SPE Asia Pacific Oil and Gas Conference and Exhibition. Melbourne, Australia.

Forbes, S. M., Verma, P., Curry, T. E., Friedmann, S. J., & Wade, S. M. (2008). *Guidelines for carbon dioxide capture, transport, and storage.* Washington, DC: World Resources Institute.

Gallo, Y. L., Couillens, P., & Manai, T. (2002). *CO_2 sequestration in depleted oil or gas reservoirs.* Paper presented at the SPE International Conference on Healt, Safety and Environment in Oil and Gas Exploration and Production. Kuala Lumpur, Malaysia.

Gasda, S., Celia, M., Nordbotten, J., & Dobossy, M. (2005). $_2$ *storage and the potential for leakage along existing wellbores*. Geological, CO: Intergovernmental Panel on Climate Change - IPCC.

Gasda, S. E., Bachu, S., & Celia, M. A. (2004). The potential for CO_2 leakage from storage sites in geological media: Analysis of well distribution in mature sedimentary basins. *Environmental Geology*, *46*(6-7), 707–720. doi:10.1007/s00254-004-1073-5

Gaurina-Medjimurec, N. (2010). The influence of CO_2 on well cement. *The Mining-Geology-Petroleum Engineering Bulletin*, *22*, 19–25.

Gaurina-Medjimurec, N., & Pasic, B. (2011). Design and mechanical integrity of CO_2 injection wells. *The Mining-Geology-Petroleum Engineering Bulletin*, *23*, 1–8.

Imbus, S., Orr, F. M., Kuuskraa, V. A., Kheshgi, H., Bennaceur, K., & Gupta, N. … Hovorka, S. (2006). *Critical issues in CO_2 capture and storage: Findings of the SPE advanced technology workshop (ATW) on carbon sequestration*. Paper presented at SPE Annual Technical Conference and Exhibition. San Antonio, TX.

Intergovernmental Panel on Climate Change – IPCC. (2005). *IPCC special report on carbon dioxide capture and storage*. Working Group III of the Intergovernmental Panel on Climate Change.

Jenvey, N. J., Moen, R. S., Muller, N., Buehler, J., Costello, M. S., & Schwander, M. M. (2008). *CO_2 geological storage field development: Initial experiences in an emerging industry*. Paper presented at the 2008 SPE North Africa Technical Conference and Exhibition. Marrakech, Morocco.

Jimenez, J. A., & Chalaturnyk, R. J. (2002). *Integrity of bounding seals for geological storage of greenhouse gases*. Paper presented at the SPE/ISRM Rock Mechanics Conference. Irving, TX.

Kutchko, B. G., Strzisar, B. R., Dzombak, D. A., Lowry, G. V., & Thaulow, N. (2007). Degradation of well cement by CO_2 under geologic sequestration conditions. *Environmental Science & Technology*, *41*, 4787–4792. doi:10.1021/es062828c PMID:17695930

Le Guen, Y., Le Gouevec, J., Chammas, R., Gerard, B., Poupard, O., Oxand, S. A., & Van Der Beken, A. (2008). *CO_2 storage – Managing the risk associated with well leakage over long timescales*. Paper presented at the 2008 SPE Asia Pacific Oil and Gas Conference and Exhibition. Perth, Australia.

Le Guen, Y., Meyer, V., Poupard, O., Houdu, E., Chammas, R., & Oxand, S. A. (2009). *A risk-based approach for well integrity management over long term in a CO_2 geological storage project*. Paper presented at the 2009 SPE Asia Pacific Oil and Gas Conference and Exhibition. Jakarta, Indonesia.

Meyer, J. (2007). *Summary of carbon dioxide enhanced oil recovery (CO2EOR) injection well technology*. Washington, DC: American Petroleum Institute.

National Energy Technology Laboratory – NETL. (2009). *Monitoring, verification, and accounting of CO_2 stored in deep geologic formations*. National Energy Technology Laboratory.

Nelson, E. B., & Guillot, D. (2006). *Well cementing*. Sugar Land, TX: Schlumberger.

Nezhad, S. A. T., Mojarad, M. R. R., Paitakhti, S. J., Moghadas, J. S., & Farahmad, D. R. (2006). *Experimental study on applicability of water-alternating-CO_2 injection in the secondary and tertiary recovery.* Paper presented at First International Oil Conference and Exhibition. Cancun, Mexico.

Onan, D. D. (1984). *Effects of supercritical carbon dioxide on well cements*. Paper presented at the 1984 Permian Basin Oil Gas Recovery Conference. Midland, TX.

OSPAR. (2007). *OSPAR guidelines for risk assessment and management of storage of CO_2 streams in geological formations*. OSPAR 07/24/1-E.

Price, J., & Smith, B. (2008). *Geological storage of carbon dioxide, staying safely underground (booklet). The International Energy Agency (IEA)*. Greenhouse Gas R&D.

Quintesa. (2010). *Quintesa database*. Retrieved from http://www.quintessa.org/co2fepdb/

Santra, A., Reddy, B. R., Liang, F., & Fitzgerald, R. (2009). *Reaction of CO_2 with Portland cement at downhole conditions and the role of pozzolanic supplements*. Paper presented at the 2009 SPE International Symposium on Oilfield Chemistry. The Woodlands, TX.

United States Environmental Protection Agency. (2008). *Vulnerability evaluation framework for geologic sequestration of CO_2*. Retrieved from http://www.epa.gov/climatechange/emissions/downloads/VEF-Technical_Document_072408.pdf

van Egmond, B. (2006). *Developing a method to screen and rank geological CO_2 storage site on the risk of leakage*. (Master Thesis). Utrecht University, Utrecht, The Netherlands.

Vendrig, M., Spouge, J., Bird, A., Daycock, J., & Johnsen, O. (2003). *Risk analysis of the geological sequestration of carbon dioxide: No. R DTI/pub*. Department of Trade and Industry's Cleaner Coal Technology Transfer Programme.

White, C. M., Straziser, B. R., Granite, E. J., Hoffman, J. S., & Pennline, H. W. (2003). Separation and capture of CO_2 from large stationary sources and sequestration in geological formations - Coalbeds and deep saline aquifers. *Journal of the Air & Waste Management Association*, *53*, 645–715. doi:10.1080/10473289.2003.10466206 PMID:12828330

Wildenborg, T., Chadwick, A., Deflandre, J-P., Eiken, O., Mathiesone, A., Metcalfef, R., … Wollenwebera, J. (2013). Key messages from active CO_2 storage sites, GHGT-11. *Energy Procedia*, 1-9.

ADDITIONAL READING

Aines, R. D., Leach, M., Weisgraber, T. J., Simpson, M. D., Friedmann, S. J., & Bruton, C. J. (2009). Quantifying the potential exposure hazard due to energetic releases of CO_2 from a failed sequestration well. *Energy Procedia*, *1*, 2421–2429. doi:10.1016/j.egypro.2009.02.003

Arts, R., Eiken, O., Chadwick, R. A., Zweigel, P., van der Meer, L., & Zinszner, B. (2004). Monitoring of CO_2 injected at Sleipner using time-lapse seismic data. *Energy*, *29*(9-10), 1383–1392. doi:10.1016/j.energy.2004.03.072

Bachu, S. (2000). Sequestration of carbon dioxide in geological media: Criteria and approach for site selection. *Energy Conversion and Management*, *41*(9), 953–970. doi:10.1016/S0196-8904(99)00149-1

Bachu, S., & Watson, T. L. (2009). Review of failures of wells used for CO_2 and acid gas injection in Alberta. *Energy Procedia*, *1*, 3531–3537. doi:10.1016/j.egypro.2009.02.146

Baklid, A., Korbřl, R., & Owren, G. (1996, October). *Sleipner Vest CO_2 disposal, CO_2 injection into a shallow underground aquifer.* Paper SPE 36600 presented at the 1996 SPE Annual Technical Conference and Exhibition, Denver, Colorado, U.S.A.

Benson, S. M., & David, D. R. (2008). CO_2 sequestration in deep sedimentary formations. *Elements*, *4*, 325–331. doi:10.2113/gselements.4.5.325

Bérard, T., Jammes, l., Lecampion, B. Vivalda, V., & Desroches, J. (2007, September). *CO_2 storage geomechanics for performance and risk management.* Paper SPE 108528presented at Offshore Europe 2007, Aberdeen, Scotland, U.K.

Blackford, J. C., Chen, B., Widdicombe, S., & Lowe, D. (2010). Marine environmental risks of CO2 sequestration In Developments and innovation in carbon dioxide (CO2) capture and storage technology. Maroto-Valer, M (Ed.), Woodhead.

Bowden, A. R., & Rigg, A. (2004). Assessing risk in CO_2 storage projects. *Austarlian Petroleum Production and Exploration Association Journal*, *44*, 677–702.

Bře, R., Magnus, C., Osmundsen, P. T., & Rindstad, B. I. (2002). CO2 point sources and subsurface storage capacities for CO2 in aquifers in Norway. NGU Report 2002.010, 132 pp.

Chen, B., Nishio, M., Song, Y., & Akai, M. (2009). The fate of CO_2 bubble leaked from seabed. *Energy Procedia*, *1*, 4969–4976. doi:10.1016/j.egypro.2009.02.329

De Lary, L., Loschetter, A., Bouc, O., Rohmer, J., & Oldenburg, C. M. (2012). Assessing health risk of CO_2 leakage from a geological storage site into buildings: role of attenuation in the unsaturated zone and building foundation. *International Journal of Greenhouse Gas Control*, *9*, 322–333. doi:10.1016/j.ijggc.2012.04.011

Deel, D., Mahajan, K., Mahoney, C. R., Mcilvried, H. G., & Srivastava, R. D. (2007). Risk assessment and management for long-term storage of CO_2 in geologic formations. United States DOE R&D, *Greenhouse Gases. Sciences et Techniques (Paris)*, *1*(2), 160–174.

Evans, D. (2008). *An appraisal of underground gas storage technologies and incidents, for the development of risk assessment methodology*. Health and Safety Executive Research Report RR605.

Friedmann, S. J. (2007). Geological carbon dioxide sequestration. *Elements.*, *3*, 179–184. doi:10.2113/gselements.3.3.179

Gale, J., & Davison, J. (2004). Transmission of CO_2 - safety and economic considerations. *Energy*, *29*(9-10), 1319–1328. doi:10.1016/j.energy.2004.03.090

Gasda, S. E., Bachu, S., & Celia, M. A. (2004). Spatial characterization of the location of potentially leaky wells penetrating a deep saline aquifer in a mature sedimentary basin. *Environmental Geology*, *46*, 707–720. doi:10.1007/s00254-004-1073-5

Gerstenberger, M., Nicol, A., Stenhouse, M., Berryman, K., Stirling, M., Webb, T., & Smith, W. (2009). Modularised logic tree risk assessment method for carbon capture and storage projects. *Emergy Procedia*, *1*, 2495–2502. doi:10.1016/j.egypro.2009.02.012

Gough, C. (2008). State of the art in carbon dioxide capture and storage in the UK: An experts' review. *International Journal of Greenhouse Gas Control*, *2*, 155–168. doi:10.1016/S1750-5836(07)00073-4

Hepple, R. P., & Benson, S. M. (2002). Implications of surface seepage on the effectiveness of geological storage of carbon dioxide as a climate change mitigation strategy. In Gale, J., & Kaya, Y. (Eds.), 6th International Conference on Greenhouse Gas Control Technologies, 1, 261-266, Pergamon, Oxford (U.K.).

Iding, M., & Ringrose, P. (2009). Evaluating the impact of fractures on the long-term performance of the In Salah CO_2 storate site. *Energy Procedia*, *1*, 2021–2028. doi:10.1016/j.egypro.2009.01.263

Jordan, P. D., Oldenburg, C. M., & Nicot, J.-P. (2011). Estimating the probability of carbon dioxide plumes encountering faults. *Greenhouse Gases: Science and Technology*, *1*(2), 160–174. doi:10.1002/ghg.17

Lagneau, V., Pipart, A., & Catalette, H. (2005). Reactive transport modeling and long term behavior of CO_2 sequestration in saline aquifers. *Oil & Gas Science and Technology*, *60*, 231–247. doi:10.2516/ogst:2005014

Lewicki, J. L., Birkholzer, J., & Tsang, C.-F. (2007). Natural and industrial analogues for leakage of CO_2 from storage reservoirs: identification of features, events, and processes and lessons learned. *Environmental Geology*, *52*, 457–467. doi:10.1007/s00254-006-0479-7

Lindeberg, E. (2002, October).The quality of a CO_2 repository: What is the sufficient retention time of CO_2 stored underground? Paper presented at the 6th International Conference on Greenhouse Gas Control Technologies, Kyoto, Japan.

Linderberg, E. (1997). Escape of CO_2 from aquifers. *Energy Conversion and Management*, *38*, 235–240. doi:10.1016/S0196-8904(96)00275-0

Loizzo, M., Akemu, O. A. P., Jammes, L., Desroches, J., Lombardi, S., & Annunziatellis, A. (2011). Quantifying the risk of CO_2 leakage through wellbores. *SPE Drilling &Completion*, *9*, 324–331.

Metz, B., Davidson, O., de Coninck, H. C., Loos, M., & Meyer, L. A. (Eds.). Chapter 5 underground geolological storage. Benson, S., & Cook, P. (Eds.). Cambridge University Press, Cambridge, United Kingdom and New York, NY, U.S.A.

Meyer, L., & Benson, S. (2006, September). *Critical issues in CO_2 capture and storage: findings of the SPE advanced technology workshop (ATW) on carbon sequestration.* Paper SPE 102968 presented at the SPE Annual Technical Conference and Exhibition, San Antonio, Texas, U.S.A.

Miles, N., Davis, K., & Wyngaard, J. (2005). Detecting leakage from CO_2 reservoirs using micrometeorological methods. In *Carbon Dioxide Capture for Storage in Deep Geologic Formations – Results from the CO_2 Capture Program. 2: Geologic Storage of Carbon Dioxide with Monitoring and Verification* (pp. 1031–1044). London, UK: Elsevier Science.

Moberg, R., Stewart, D. B., & Stachniak, D. (2003). The IEA Weyburn CO_2 monitoring and storage project. In Gale, J., Kaya, Y. (Eds.). Sixth International Conference on Greenhouse Gas Control Technologies, Kyoto, Japan.

Nelson, C. R., Evans, J. M., Sorensen, J. A., Steadman, E. N., & Harju, J. A. (2005, October). *Factors affecting the potential for CO_2 leakage from geologic sinks.* Plains CO_2 Reduction (PCOR) Partnership. The Energy & Environmental Research Center (EERC).

Oelkers, E. H., Gislason, S. R., & Matter, J. (2008). Mineral carbonation of CO_2. *Elements*, *4*, 333–337. doi:10.2113/gselements.4.5.333

Oldenburg, C. (2007). Migration mechanisms and potential impacts of CO2 leakage and seepage. In Wilson. E. J., & Gerard, D. (Eds.), Carbon Capture and Sequestration: Integrating Technology, Monitoring, Regulation (pp. 127-146). Blackwell Publishing.

Oldenburg, C. M. (2005). *HSE screening and ranking framework for geological CO_2 storage site selection. Paper LBNL-58873.* Lawrence Berkeley National Laboratory. doi:10.2172/885235

Papanikolau, N., Lau, B. M. L., Hobbs, W. A., & Gale, J. J. (2006, June). *Safe storage of CO_2: Experience from the natural gas storage industry*. Paper presented at 8th International Conference on Greenhouse Gas Control Technologies (GHGT-8). Trondheim, Norway.

Pawar, R. J., Watson, T. L., & Gable, C. W. (2009). Numerical simulation of CO_2 leakage through abandoned wells: Model for an abandoned site with observed gas migration in Alberta, Canada. *Energy Procedia*, *1*, 3625–3632. doi:10.1016/j.egypro.2009.02.158

Polak, S., Lundin, E., Zweigel, P., Bře, R., Lindeberg, E., & Olesen, O. (2004). *Storage potential for CO_2 in the Frohavet Basin, Mid-Norway. NGU Report 2004.049, SINTEF report 54.5272.00/02/04*. Trondheim: Norges Geologiske Undersřkelse.

Senior, B., Espie, T., & Christopher, C. (2004). *Developing a framework for managing the long term storage of CO_2 in geological formations*. In Rubin, E.S., Keith, D.W., & Gilboy, C.F. (Eds.). Seventh International Conference on Greenhouse Gas Control Technologies, Cheltenham, U.K.

Sweatman, R. E., Santra, A., Kulakofsky, D. S., & Calvert, D. G. J. (2009, November). *Effective zonal isolation for CO_2 sequestration wells*. Paper SPE 126226 presented at the 2009 SPE International Conference on CO2 Capture, Storage, and Utilization, San Diego, California, U.S.A.

Themann, S., Schmidt, H. M., & Esser, D. (2009, November). *Measurement, monitoring, and verification of CO_2 storage: An integrated approach.* Paper SPE 129127 presented at the 2009 SPE International Conference on CO_2 Capture, Storage, and Utilization, San Diego, California, U.S.A.

United States Environmental Protection Agency - US EPA (2011). *Geologic sequestration of carbon dioxide, draft UIC program class VI well construction guidance for owners and operators*, March 3, from http://water.epa.gov/type/groundwater/uic/class6/gsguidedoc.cfm

United States Environmental Protection Agency - US EPA. (2010). *Federal requirements under the underground injection control program for carbon dioxide geologic sequestration wells*. December 10 (The GS Rule), from http://water.epa.gov/type/groundwater/uic/class6/gsregulations.cfm.

Vignes, B., Enoksen, A. M., & Ovesen, M. (2010, May). CO_2 injection and storage – quantification of well barriers in adjacent wells. Paper OTC 20724 presented at the 2010 Offshore Technology Conference, Houston, Texas, U.S.A.

Walton, F. B., Tait, J. C., Le Neveu, D., & Sheppard, M. I. (2004, September). Geological storage of CO_2: A statistical approach to assessing performance and risk. In Rubin, E.S., Keith, D.W., & Gilboy, C.F. (Eds.), *Proceedings of 7th International Conference on Greenhouse Gas Control Technologies, Vancouver, Canada. Volume 1: Peer-Reviewed Papers and Plenary Presentations*, IEA Greenhouse Gas Programme, Cheltenham, UK.

Wilson, E. J., Friedmann, S. J., & Pollak, M. F. (2007). Research for deployment: Incorporating risk, regulation, and liability for carbon capture and sequestration. *Environmental Science & Technology, 41*, 5945–5952. doi:10.1021/es062272t PMID:17937265

Wilson, M., & Monea, M. (2004, September). *IEA GHG Weyburn CO_2 monitoring and storage project summary report 2000-2004.* Proceedings of the 7th International Conference on Greenhouse Gas Control Technologies, Vancouver, Canada, vol. 3. Petroleum Technology Research Centre, Regina.

Winter, E. M., & Bergman, P. D. (1993). Availability of depleted oil and gas reservoirs for disposal of carbon dioxide in the United States. *Energy Conservation and Management, 9-11*, 1177–1187. doi:10.1016/0196-8904(93)90067-K

Zhou, W., Stenhouse, M. J., Arthur, R., Whittaker, S., Law, D. H. S., Chalaturnyk, R., & Jzrawi, W. (2004). The IEA Weyburn CO_2 monitoring and storage project – Modelling of the long-term migration of CO_2 from Weyburn. *Proceedings of the 7th Int. Conf. on GHG Control Technologies (GHGT-7).*

KEY TERMS AND DEFINITIONS

CO_2 Leakage: Migration of injected CO_2 from injection zone through wells, faults or cap rock.

Geologic CO_2 Storage: Storage of CO_2 in subsurface formations for a long term.

Hazard: A potentially damaging physical event, phenomenon or human activity that may cause the loss of life or injury, property damage, social and economic disruption or environmental degradation.

Injection Well Mechanical Integrity: There is no significant leak in the casing, tubing, or packer (internal MI) and no significant fluid movement into an underground source of drinking water (USDW) through vertical channels adjacent to the wellbore (external MI).

Mitigation: Actions taken to reduce the impact of hazards in order to protect lives, property and the environment, and to reduce economic disruption.

Portland Cement: The most common type of cement made by grinding clinker, used for well cementing.

Receptor: Any component of the environmental system that could be subject to adverse (or positive) impacts as a result of leakage, e.g. human populations, ecosystems, groundwater or other resources, and relevant aspects of the wider environment.

Risk: The product of the likelihood (probability) and the consequence (impact) of a specified hazard (event) being realized; refers to the vulnerability, proximity or exposure to hazards, which affects the likelihood of adverse impact.

Risk Assessment: The overall process of risk identification, risk analysis and risk evaluation.

Storage Site: A defined volume within a geological formation used for the geological storage of CO_2 and associated surface and injection facilities.

Chapter 16
Petroleum Industry Environmental Performance and Risk

Lidia Hrncevic
University of Zagreb, Croatia

ABSTRACT

The petroleum industry holds long- and short-term environmental risks. Besides production fluids, all petroleum industry activities involve either use of fluids, which contain abundant substances, or waste generation, both associated with potential risk to the environment. The principal environmental risk associated with the petroleum industry is the risk of fluid spill/emission to the environment. Although in recent decades the risk analysis methodologies have matured, to date there is still no universally accepted methodology for environmental risk assessment in petroleum industry. In this chapter, the petroleum industry's environmental incident history and statistics are presented. The environmental impact of the petroleum industry's activities, its extent, and trends are analyzed. The overview of pollution sources with associated environmental risk is given along with the analysis of the causes and consequences of incidents in the petroleum industry.

DOI: 10.4018/978-1-4666-4777-0.ch016

INTRODUCTION

All petroleum industry's activities hold the potential for a variety of impacts on the environment's components: soil, water, air and consequently all living species. Since different activities in different ecosystems and conditions may result in significant variations in the extend of a potential impact, the potential for oil and gas operations to cause environmental impact has to be addressed on a case- by-case basis. The assessment of the potential impacts and resulting mitigation measures is commonly carried out through Environmental Impact Assessment Study (EIA). Estimation and quantification of the probability of an unwanted consequence of a particular activity's impact on the environment in a specific time period (the environmental risk) is done by Environmental Risk Assessment (ERA). The environmental risk assessment is based on calculating the probability for an ecosystem to come to contact or to receive a dose of pollutant. Though the first application of risk analysis to petroleum industry was done in 1960 by Caryson (Yanting & Liyun, 2011), environmental risk is a relatively recent concept, which has quickly become an important consideration in environmental assessment of the new projects, facility and process design and overall petroleum sector management. A number of definitions of environmental risk have been proposed. Most commonly environmental risk is defined as the product of the probability (or frequency) and consequence (Weiner & Matthews, 2003). The consequences are adverse effects on different components of the environment. Crichton (1999) defines the risk as the "probability of loss" including the hazard and the concepts of vulnerability and exposure (Olita et. al., 2012) where vulnerability is a measure of the sensitivity of a specific ecosystem to a given hazard.

Petroleum industry poses a long and short-term environmental risk. Besides production fluids, that pose significant environmental risk (if found uncontested in environment), all petroleum industry activities (oil and gas exploration, production, processing, storage and transportation) involve either use of fluids, that contain abundant and diverse toxic chemicals, or waste generation, all associated with a potential risk to the environment. The imperative of contemporary petroleum industry practice is to understand those risks, evaluate it and quantify, in order to use the best available technical and technological solutions to design facilities and/or processes to prevent or mitigate it.

The principal environmental risk associated with petroleum industry is primary seen as the risk of fluid (production, operating or waste) spill, discharge or emission to the environment. An important part of protecting the environment, and thus minimizing the risk, is ensuring that there are as few spills, discharges or emissions as possible. Along with the mentioned best available techniques and technologies applied to control the risk, other, indirect, risk restrictive clauses are the introduction of new and strict regulations, both on national and international level resulting with high fines, stringent operating codes and high costs of cleanup/restoration processes (20 – 200 $/l of spilled oil depending on type of oil and location (Fingas, 2001). Even though petroleum industry's incidents, resulting with large releases of fluids, rarely occur, they attract particular public and media attention resulting with global awareness of the risks and consequences they cause to the environment, so it is important to identify the sources, size and the frequency of these releases.

Although in recent decades the risk analysis methodologies have matured and there is already a range of methods of risk assessment with different focus, advantages and disadvantages, to the date there is still no universally accepted methodology for environmental risk assessment in petroleum industry. If there is an adequate data base of previous identical or similar events, environmental risk assessment of petroleum industry activities is commonly done by the analysis of former reported spill data for a specific area. However, if there are no sufficient or available data to determine the extents of risks, some of analytical methods for accidental risk are used (Table 1). In 1975 the Department of Interior (DOI) of the United States of America (USA) developed the OSRA (Oil Spill Risk Analysis) model. This model is used for the analysis of possible oil spill impact from offshore oil and gas operations (Price et. al. 2003). On basis of historical spill patterns, meteorological (wind intensity) and oceanographic (ocean currents) data the model enables estimations of oil spill occurrence probability and probability of oil slick contact with biological and economic resources in outer continental shelf of the USA. It is not designed for use in "real time". The OSRA model summarizes the results of thousands of spill trajectory simulations. Recently it has become a common practice by several authors to use modeling tools to study oil trajectories for planning or oil spill hazard estimates purposes.

In order to quantify the environmental risk form either a specific facility or process or from a complete oil/ gas field, a number of steps are required. These are (Dumitran & Onutu, 2010, Woods et. al., 1991):

1. Characterization of the main sources of the adverse impacts on the environment (facilities and activities);
2. Identification of cases which may lead to adverse impacts on the environment (pollution causes);
3. Identification of the types of substances that may cause adverse impacts on the environment (pollutants);
4. Estimation of the frequency of the cases which may lead to adverse impacts on the environment;
5. Scenarios for the exposure to the pollutant (receptors submitted to risk and spill consequences);
6. Quantification of the impact on the environment (environmental risk).

In this chapter the petroleum industry's environmental incidents history and statistics will be presented. Also the environmental impacts of petroleum industry's activities, its extend and trends will be analyzed. The overview of pollution sources with associated environmental risk will be given along with the analysis of the causes and consequences of incidents in petroleum industry.

Table 1. Risk analysis methods for accident risk (Andersen & Mostue, 2012)

Method	Type	Description
FMEA/FMECA, PDS	Technical reliability analysis	Quantification of technical reliability, Input to quantitative risk analysis
FTA, ETA, Consequence and escalation models	Consequence analysis	Identification of effects of accident loads by the use of models of physical phenomena
QRA/TRA	Quantitative risk analysis	Identification and analysis of hazards based on quantitative data
HTA	Task analysis	Human factor methods for analysis of human actions and decisions. Basis for human reliability analysis
HRA	Human reliability analysis	Quantification of the probability of human errors in work process. Input to quantitative risk analysis
BORA	Organizational risk analysis	Analysis of the effect of organizational and management factors in relation to quantitative risk analysis
MANAGER, MACHINE, SAM, I- RISK, ORIM OMT, WPAM, PHA	Qualitative risk analysis	Identification and analysis of possible unwanted events; qualitative assessment
SWIFT, HAZOP, HAZID, JSA, FAM, STAMP based risk analysis or hazard analysis	Systemic model based analysis	Analysis of the resilience of socio- technical systems
CRIOP	Verification analysis	Verification and validation of human, technical and organizational conditions

Legend: FMEA- Failure Mode and Effects Analysis
FMECA: Failure Mode, Effects and Criticality Analysis
PDS: Reliability and Availability of Computer based Safety System
FTA: Fault Tree Analysis
ETA: Event Tree Analysis
QRA: Quantitative Risk Analysis
TRA: Total Risk Analysis
HTA: Hierarchical Task Analysis
HRA: Human Reliability Analysis
BORA: Barrier and Operational Risk Analysis
MANAGER: Management Safety Systems Assessment Guidelines in the Evaluation of Risk
MACHINE: Model of Accident Causation using Hierarchical Influence Network
SAM: System- Action- Management
I-RISK: Integrated Risk
ORIM: Organizational Risk Influence Model
OMT: A model for quantifying barrier performance related to major risk hazards
WPAM: Work Process Analysis Model
PHA: Preliminary Hazard Analysis
SWIFT: Structured What If Technique
HAZOP: Hazard and Operability Analysis
HAZID: Hazard Identification
JSA: Job Safety Analysis
FRAM: Function Resonance Analysis Method
STAMP: System Theoretic Accident Model and Process
CRIOP: Crisis Intervention in Offshore Production

OIL SPILL TRENDS

Looking back in the past, recorded environmental accidents that happened during petroleum industry's activities are primarily associated with crude oil or petroleum products inland or, more frequent, offshore spills. An "*oil spill*" as such is defined as discrete event in which oil is discharged though neglect, by accident or with intent over a relatively short time. It does not include an event in which oil leaks slowly over a long period of time, nor does it include operational spillage allowed by international or national regulations or that occur over a relatively long period of time (such as >5 ppm oil discharge in refinery effluents) even if those discharges violate pollution regulations (Schmidt Etkin, 2001). According to The International Tanker Owners Pollution Federation (ITOPF) oil spills are generally divided by size into three categories: spills less than 7 tonnes, 7 – 700 tonnes and spills greater than 700 tonnes (www.itopf.com, 2013).

Even with the best practices applied, with the significant use of oil and petroleum products in everyday life, frequent occurrence of oil/petroleum products spills of different proportion, ranging from few liters to several thousands of tonnes, can be henceforth expected. The ten worst oil spills, in terms of the volume spilled, that happened to the date are shown in Table 2. Most of the spills are relatively small and cause localized impacts, but occasionally large spills occur and cause tremendous environmental and socioeconomic damages. The damage caused by an oil spill is

Table 2. Ten worst oil spills (http://earth.tryse.net//oilspill.html; 2013)

Oil spill	Oil spilled (10^3 t)	Date	Location	Type	Cause
Gulf War oil spill	546 -1 500	21. Jan 1991 –May 1991	Persian Gulf	Oil wells	War
BP Deepwater Horizon	400 – 700	20 Apr 2010 – 15 Jul 2010	Gulf of Mexico	Oil wells	Blowout
Ixtoc I oil well	454 – 480	3 Jun 1979 – 23 Mar 1980	Gulf of Mexico	Exploratory oil well	Drilling error, blowout
Atlantic Empress	276 – 287	19 Jun 1979 – 2 Aug 1979	Trinidad & Tobago	Oil tanker	Ship collision
Fergana Valley oil spill	259 – 285	2 Mar 1992 – May 1992	Uzbekistan	Oil well	Blowout
ABT Summer	260	28 May 1991	700 nautical miles off Angola	Oil tanker	Explosion, fire, sank
Nowruz oil field	250 – 260	4 Feb 1983 – May 1985	Persian Gulf	Oil well	War, tanker – platform collision
Castillo de Bellver	254	5 Aug 1983	Saldanha Bay South Africa	Oil tanker	Fire, broke in two
Amoco Cadiz	223 – 229	16 Mar 1978	Portsall Rocks, Brittany, France	Oil tanker	Ran around in bad weather
Chevron Texaco Amazon oil spill	165 – 171	1972 – June 1992	Ecuadorian Amazon	Drilling operations	Waste product dumping

not necessarily proportional to spill size. The location itself and the type of oil spilled play the crucial role in determining the degree of environmental and socioeconomic damage. Naturally, the bigger the accident, namely the greater amount of oil spilled, the bigger public attention, the greater impact, but also the greater data base of the event.

As already mentioned, most of the spills are relatively small (72% of the spills are 0.003 to 0.03 tonnes or less) with the share of 0.4% in total spillage. The largest spills make 0.1% of the incidents but involve 60% of the total amount spilled (Fingas, 2011). Most of the largest oil spills have been related to oil transportation by tankers (marine environment) or pipelines (inland and marine environment). Tanker oil spills dominated the global oil spill stage until 1990s when large pipeline and facility spills occurred. In period from 1970 to 2012 almost 6 million tonnes (5.75×10^6 t) of oil was spilled in tanker accidents all over the world (www.itopf.com, 2013). The tanker accident that for the first time turned the public eye to the problem of oil spill pollution on an international level was the Torey Canyon tanker accident that happened near the Scilly Isles off the western coast of the United Kingdom in 1967 (119×10^3 t of oil spilled (www.itopf.com, 2013). The Torrey Canyon spill was not the first oil tanker spill, but it was the biggest spill to the date and the environmental damage caused by the spill was even more magnified by the application of inadequate and toxic cleanup methods (inadequate sinking agents and toxic first-generation dispersants). Since 1967 there was a number of significant tanker oil spills- the tanker Sea Star (1972, Oman), the tanker Metula (Chile, 1974), the tanker Jakob Maersk (1975, Portugal), the tanker Urquiola (1976, Spain), the tanker Hawaiian Patriot (1977, Hawaii), the tanker Amoco Cadiz (1978, France), the tanker Atlantic Empress (1979, Trinidad & Tobago), the tanker Independenta (1979, Turkey), the tanker Castillo de Bellver (1983, South Africa), the tanker Odyssey (1988, Canada), the tanker Exxon Valdez (1989, Alaska), the tanker ABT Summer (1999, Angola), the tanker Sea Empress (1996, UK), the tanker Prestige (2002, Spain), the tanker Hebei Spirit (2007, Republic of Korea) etc. Perhaps the most significant, in terms of having the biggest impact on passage of oil spill prevention and liability legislation and international conventions, was the tanker oil spill Exxon Valdez, which happened in Prince William Sound, Alaska on March 24th 1998.

Oil spill occurrences and the total amount of oil spilled have decreased significantly over the last 30 years, especially last decade, even though there was increase in both marine (tanker) and pipeline transportation of oil and petroleum products. The average number of worldwide large oil spills (>700 tonnes) records considerable decrease from 1970s, with average of 25 spills per year, to just 3.7 spills per year nowadays (Psarros et. al., 2011). Considering the global oil spillage rates the decrease was from about 635×10^3 t/yr in 1960s and 1970s to about 300×10^3 t/yr nowadays (Fingas, 2011). Consistent with the reduction in the number and volume of worldwide oil spills, the number and volume of tanker oil spills also show a remarkable reduction. The

Figure 1. Number of large tanker oil spills in period 1970 - 2012 (www.itopf.com, 2013)

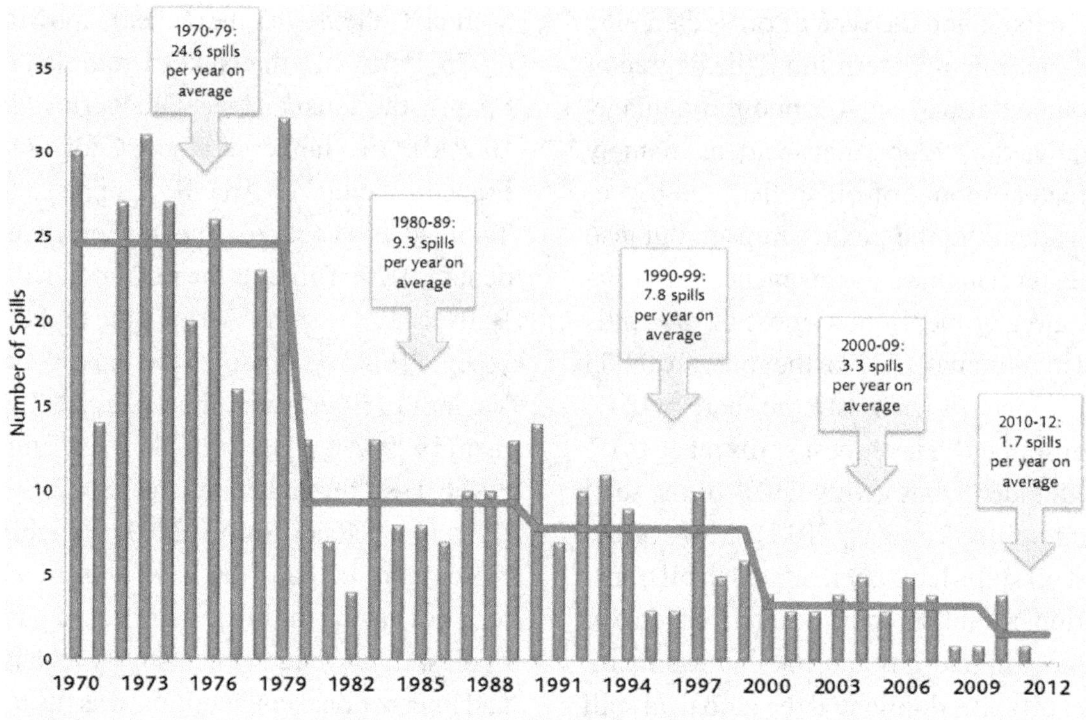

biggest reduction was observed from 1970s to the 1980s. The number of average annual large tanker oil spills (>700 t) decreased from 24.6 spills per year in the 1970s to only 3.3 spills per year in the 2000s (Figure 1) and the quantities of oil spilled in tanker accidents decreased from about 320×10^3 t/yr in 1970s to only about 21×10^3 t/yr in 2000s (www.itopf.com, 2013).

There are several reasons that influenced the decrease of the global oil spill trends. The adoption of The International Convention for the Prevention of Pollution from Ships, 1973, modified by the Protocol of 1978 (MARPOL 73/78), as response to a spate of tanker accidents in 1976 – 1977, being the most important one. MARPOL (short from "*Marine Pollution*") is the main international convention covering the pollution of the marine environment by ships from operational and accidental causes (www.imo.org, 2013). It currently contains six technical Annexes. Annex I covers the regulations for the prevention of pollution by oil. The 1992 amendments to Annex I made it mandatory for all new oil tankers to have double hulls and brought in a phase- in schedule for existing tankers to fit double hulls (www.imo.org, 2013). MARPOL contains a set of prescriptive technical and operational regulations, but there is also an opening for acceptance of alternative design of tankers (regulation 13F(5) of Annex I) (Vanem et. al., 2008). Some of other reasons of the decrease of the global oil spill trends are the increase of cleanup costs and environmental damage liability.

Spill statistics are collected by a number of organizations and agencies both on national and international scale around the world. The data collected on an international level have to be taken with certain precaution due to different reporting practices, particularly in minimum spill amounts/volumes that must be reported, between different jurisdictions and organizations. The accuracy of it is particularly questionable for spills from particular remote regions, small size spills and spills from non- vessel sources. Also, comparing oil spill statistics from different databases can be misleading because different data collecting methods are used. In most of the cases of oil spill the most difficult fact to determine or estimate is the volume of oil spilled, unless it is a case of a vessel or tank storage oil spill in which the exact volume of oil is know even before the accident.

The spill history data are used to generate statistics to estimate the average spill size in specific conditions for a specific area. In order to discern trends and devise spill scenarios, with the final goal to identify and rank high-risk areas, besides historical spill data the data on potential spill sources have to be complied.

SOURCES AND CAUSES OF POLLUTION IN PETROLEUM INDUSTRY

All petroleum industry's activities (exploration, production, processing, storage, transportation and refining) with its related facilities and processes, due to the fact that they primarily involve dealing with highly environmentally damageable substance- oil, gas or petroleum products, depending upon the type of a facility or a stage in petroleum's "life cycle", pose a potential source of pollution. Their role as a potential pollution source is even more amplified if all working fluids and solids used, as well as associated wastes generated, are considered. Along with the environmental changes, these activities also induce social, cultural and economic changes.

The principal environmental risk associated with petroleum industry is primarily seen as the risk of pollutant spill, discharge or emission to the environment. Beside danger of accidental spills and leaks, oil and gas exploration and production also hold a potential of blowouts. A substance is considered a pollutant if it has perceived to have an adverse impact on the components of the environment. The pollutants occur in form of liquid, solid, gaseous, aerosol and noise discharges and emissions.

Sources of Pollution in Petroleum Industry

During oil and gas exploration and production there are several pollutants that pose a real threat to the environment, either through accidental, intentional or permitted release. These are (The E&P Forum, UNEP, 1997; Rana, 2008):

1. Drilling fluids (muds), cuttings and well treatment chemicals;
2. Produced water;
3. Process, cooling, wash and drainage water;
4. Emissions of adverse gases;
5. Produced hydrocarbons (oil and gas);
6. Sewerage, sanitary and domestic wastes.

The type and the volume of the pollutant depend on the stage of the exploration and production process. In the early phases of exploration (aerial surveys and seismic surveys) all the activities are temporary and short-term in the environment. Principal pollutants during seismic operations are noise and waste, which are small in volume and mainly associated with camp activities.

Drilling operations usually last from few weeks (exploratory drilling) to one to several months depending on situation. The main pollutants during drilling operations are drilling fluids (muds) and borehole cuttings. Besides air and water, and air/water, air/polymer combinations, there are basically three types of drilling fluids used in drilling operations: water-based drilling fluid, oil-based drilling fluid and synthetic-based drilling fluid. Among the three, water- based drilling fluids are most frequently used due to their inexpensiveness and simple composition. Its main components are clay (usually bentonite) and water, which make this kind of drilling fluid, compared with other two, also the most environmental friendly one. Oil-based drilling fluids are primarily composed of diesel oil while the base fluid of synthetic-based mud is synthetic oil. All drilling fluids, beside their main components, contain certain additives (weighting agents as barium sulfate (barite), calcium carbonate (chalk) or hematite, polymers, oxygen, scavengers, various thickeners e.g. xanthan gum, guar gum, glycol, carboxymethylcellulose, polyanionic cellulose (PAC), then, deflocculants, lubricants, shale inhibitors, fluid loss additives, biocides, surfactants, emulsifying agents, etc.) which provide various specific functional characteristics. In view to their potential environmental impact due to their composition the water-based drilling fluids are least toxic, with toxicity rising from synthetic-based fluids to oil-based fluids.

On their way back from the well, after their usage, drilling fluids contain certain amount of borehole cuttings. The volume of this drilling waste ranges from 1,000 – 5,000 m^3 per well (Rana, 2008). After separating cuttings from drilling fluid, the fluid is used again, and the cuttings, as solid drilling waste, are treated in appropriate way. After the treatment, on land they are usually deposited by burial in pits or by land spreading, and during offshore activities cuttings are usually discharged to the water body. Drilling cuttings have a complex composition depending, among other factors, mainly on type of rocks drilled, drilling regime, composition of drilling fluid used and drilling fluid-cuttings separation technology.

Drilling muds may enter the environment by being directly discharged into a body of water (usually water-based muds in case of offshore drilling), seeping into groundwater or surface water from land-based impoundments, accidental release from onshore or offshore holding facilities (Rana, 2008). If found uncontained in the environment, drilling fluids will primarily have impact on water and terrestrial ecosystems. Ocean discharges of water-based muds and drilling cuttings affect species diversity (primarily benthic organisms) up to 100 m, and oil- based muds and cuttings up to 800 m from the discharge point (The E&P Forum, UNEP, 1997).

Produced water (brine) is the main type of waste resulting from production operations.

Its main constituents are inorganic salts (mineralization of produced water may reach up to 300 g/l (Rana, 2008), heavy metals, solids, production chemicals (corrosion inhibitors, descalers, biocides, dispersants, emulsion breakers, acids etc.), hydrocarbons (oil, gas, benzene, polycyclic aromatic hydrocarbons-PAHs) and occasionally radionuclides (naturally occurring radioactive material, NORM). The volume of produced water depends upon the type of production (oil or gas production), type of production formation and type of production methods applied (primary, secondary or tertiary production method) and it varies throughout a lifetime of a production fields (usually low at the beginning of the production with the increase in volume towards the end of field life). Industry practice is to treat produced water and then dispose it by reinjection back into the reservoir. In some cases, after the treatment, produced water is directly released to surface water body. In case of spill, leakage or blowout untreated produced waters, due to their hydrocarbon and salt content, pose a threat to terrestrial and aquatic ecosystems (principally soil and surface and underground water). The primary environmental impact of other aqueous wastes, such as process, cooling, wash and drainage water, is also on soil and surface and underground water.

Emissions of adverse gases occur in different ranges at all stages of all petroleum industry activities. The main types of emissions that take place in petroleum industry are (Hrncevic, 2008):

1. Combustion emissions which include combustion of fuels in energetic units (internal combustion engines, turbines, generators, pumps, heaters etc.) and constant or periodical flaring (burning of associated gas and excessive amounts of hydrocarbons during start-up, production, maintains, well testing or upset in normal processing operation, as well as burning of gas from storage tanks and pressure controlling equipment). In spite the fact that some countries prohibit flaring of oil associated gases it remains the main source of emissions in the upstream activities of petroleum industry. The combustion emissions comprise mainly of carbon dioxide (CO_2) with certain amounts of nitrous oxide (N_2O), carbon monoxide (CO), volatile organic compounds (VOC), polycyclic organic hydrocarbons (PAHs) and sulfur compounds (carbon disulfide (CS_2) and carbonyl sulfide (COS).
2. Venting emissions, which include releases of gas through vents due to operational or security reasons. These are principally methane (CH_4) and carbon dioxide (CO_2) emissions with small quantities of non-methane volatile organic compounds (NMVOC), sulfur oxides (SO_x) and nitrogen oxides (NO_x). The main sources of venting emissions are storage tanks, pumps and pressure controlling equipment (if gas is not flared) i.e. all pressurized processing equipment, but also some processes in

which this kind of emissions is a direct result of chemical transformations (gas dehydration process, sour gas treatment processes etc.)

3. Fugitive emissions include emissions, mainly of methane (CH_4) and other volatile organic compounds, from pressurized equipment due to leaks through valves, pipe connections, mechanical seals, or related equipment. Fugitive emissions also occur at evaporative sources such as waste water treatment ponds and storage tanks.

The primary sources of petroleum industry activities impact on air (atmosphere) are (The E&P Forum, UNEP, 1997):

1. Flaring, venting and purging gases;
2. Combustion processes;
3. Fugitive gases;
4. Airborne particulates (from burning sources, vehicles and soil disturbance during construction operations).

The volumes of atmospheric emissions and their potential impact depend upon the nature of the process under consideration. The primary environmental impact of the emissions is on air (atmosphere) with a potential through precipitates, after interactions with atmospheric components and solar radiation, of secondary impact on soil and underground and surface water.

Causes of Pollution in Petroleum Industry

All petroleum industry activities are, apart from emissions, also characterized by the high potential of fluid spill or leak. Even though spills and leaks of all working fluids are possible, the majority of environmental accidents in petroleum industry are caused by spills and leaks (or, in a case of exploration and production, also blowouts) of hydrocarbons (crude oil, gas and petroleum products). Considering specific type of activity, most of environmental accidents are related to accidental spills, leaks and blowouts during production and transportation.

The most common causes of oil spills are equipment failures, operational errors, material failure, corrosion, weather-related events (storms or floods), earthquakes, human errors and recently terrorism. Pipe leaks are the most common cause of spills representing 22% of total causes with more than 65% of them being in petroleum sector (Rana, 2008).

Drilling and production spills, as already mentioned, if it is not a case of, previously mentioned working fluids leaks or spills, are usually associated with unexpected blowouts of liquid and gaseous hydrocarbons from the well. On average 7 out of every 1,000 exploratory wells will result in blowout (Rana, 2008). In case of a well blowout environmental impact is expected on all components of the environment.

In storage facilities the risk of crude oil/petroleum products spill or leak exists especially during loading/offloading operations or severe weather conditions. Since most of the oil storage tanks contain containment areas around the tank, it is expected that most of the spill will be retained. Principal environmental impact of crude oil/petroleum products storage tanks both, during normal activities or in a case of a spill (if there is a containment around the tank), is on air (during normal activities there are venting emissions and during the spill there are emissions of spilled fluid due to its volatility). If there is no containment around the storage tank, during crude oil/ petroleum product spill or leak from the tank, environmental impact on organisms, soil, surface and underground water and air is expected.

Oil is transported by a variety of means within the production field, from the production field to refineries and further to final consumers. Transport of oil from the production site to the consumers can involve up to 10 – 15 transfers between different modes of transportation such as pipelines, tank vessels (tank ships, tank barges), tank trucks and railcars, each potentially a source of oil spill or leak. Even though there was much of progress in technology of pipeline construction and utilization, pipeline failures are still the most common cause of oil spill/ leak in petroleum transportation sector. Depending on the cause and nature of pipeline damage (crack, rupture or other), pipeline is either a source of small and long-term leakage or an abrupt blowout. The most common causes of pipeline spills are respectively: 81% corrosion, 7% third party involvement, 6% mechanical failure, 4% natural hazards and 2% operational causes (www.concawe.eu, 2013). The potential spill volumes form pipelines are highly variable, depending on the pipeline diameter, flow rate, distance between shut- off valves and type of control and leak detection system applied. Principal environmental impact of oil pipeline failures, regardless whether the pipeline is under - or above ground, is on organisms, soil and surface and underground water, while, in case of gas pipeline failure, most significant impact is on air (atmosphere). Gas pipeline failures, compared to oil pipelines, pose an additional risk to the environment because of the hazardous nature of methane (CH_4), which carries the risk of fire and explosion. Marine oil pipeline hold environmental impact on ocean ecosystems.

There is general misconception that oil spills from tankers are the primary source of oil pollution in the marine environment, however these spills make up less than about 5% of all oil entering marine environment (Fingas, 2001). Half of the oil pollution in marine environment comes from land-based sources usually from waste water. Considering the volumes involved and frequencies of worldwide tanker transportation, tanker accidents occur relatively rarely with variations in number of such events and volumes of oil spilled differing from year to year. The main causes of big tanker accidents (> 700t) in period 1970 - 2012 are respectively: 33% grounding, 29% collision, 13% structural failure (hull failure), 12% fire/explosion, 4% equipment failure and 9% of causes is unknown (www.itopf.com, 2013). There are several hazards identified that hold the potential risk of possible tanker accident. These are (Michel et. al., 1994; Donovan & Owen, 1976):

1. **Areas of high currents or wave conditions:** Narrow channels with strong tidal currents, river-entrance bars, river crossing and harbor entrances where tides, waves and river currents can

create conditions hazardous for tanker operations.

2. **Congested traffic routes:** Ports entrances, narrow straits or bridges pose increased risk of collision.
3. **Draft constraints:** The areas of decreased maneuverability for deep-draft vessels and shoaling zones indicate increased risk of spill due to groundings or collision.
4. **Bottom substrate:** Groundings are more likely to result in a discharge in areas with rocky substrates than in softer substrates such as sand or mud.
5. **Lightering:** Some tankers, due to their size or draft limitations, sometimes offload their cargo at sea or in anchorage.
6. **Seasonal considerations:** Time of the year dictates weather conditions, water surface conditions, sensitivity of a particular area, and sometimes the type of cargo carried. Three parameters have been identified as being critical for tanker transportation. These are wind, fog and sea state, each of them occurring at different levels of intensity and in different combinations.

Compared to marine transportation, the potential spill volumes from road and railway oil transportation is relatively small. The data on road and railway oil transportation are rather difficult to obtain because of a large number of transport companies, many possible traffic routes and poor information on types, frequencies and volumes of oil transported in this was in general. In case of the spill, spills from tank trucks often end up on pavements and don't directly impact soil and water bodies.

Spills from refining facilities are usually restricted to legal discharges of effluents. The effluents consist of wastewater containing low concentrations of diluted oil. The effluents are generally discharges in rivers or coastal areas, depending on the location of the refinery. The exact environmental impact of the effluents is difficult to define, since refineries are usually located in industrial zones characterized with more than one environmental impact source, but, due to high legislative restrictions, they are considered to be low and localized. The total amount and toxicity of aqueous effluent discharges from oil refineries have decreased over the last 40 years due to implementation of improved technological solutions (air cooling, recirculation of cooling water, improved wastewater treatment systems etc.) (Fingas, 2011).

Since there is known volume of oil involved in spills from storage and transport facilities, the best way to determine potential spill risk from these facilities is to determine the worst-case discharge scenario. The worst- case discharge scenario presumes that all of the volume of fluid contained in storage facility or transport mean is released into the environment under adverse weather conditions.

Oil spills involve many types of oil ranging from different types of crude oil to variety of petroleum products. Beside its impact on the environment, the composition of oil also determines its physical and chemical properties (density, viscosity, surface tension, pour point, flash point, solubility), which then influence the behavior of oil in the environment and consequently the selection of remediation methods. In order to classify crude oils petroleum industry often characterizes crude oils according to their geographical origin (Saudi Light Crude, West Texas Intermediate, Venezuela crude etc.). Even though this is a common practice and a good way to distinguish different kinds of crude oil, because different oils from different geographical locations will naturally have different and unique properties, this kind of crude oil characterization does not give the complete insight into fate and impacts if oil in case of an oil spill. That is why crude oils are commonly divided into five general types (Michel et. al. 1994, http://oilprice.com, 2013 Fingas, 2011):

1. **Very light crude oils/very light oil products:** In this category are gasoline, light virgin naphtha, heavy virgin naphtha, petroleum ether, petroleum spirit and petroleum naphtha. These oils tend to be highly volatile and highly flammable (flash point near 40 °C) with high evaporation rates (complete removal by evaporation can be expected within few days), low viscosity (spread rapidly to a thin sheen) and high acute toxicity to biota (due to high-water solubility and ability to penetrate cellular membranes). Their specific gravity is less than 0.8 (float on water). They don't tend to adhere to surfaces. In case of onshore spill this kind of oil will penetrate porous surfaces and cause subsurface pollution.
2. **Light crude oils/ medium (diesel- like) oil products:** In this category are kerosene, jet fuels, marine diesel, West Texas crude and Alberta crude. These oils tend to be moderately volatile (flash point from 40 to 65 °C), less evaporative (up to 2/3 of the spill volume will evaporate), with low to moderate viscosity (spread rapidly into thin slicks) and moderate to high acute toxicity to biota (depending upon type and concentration of aromatic compounds in the water- soluble fraction). Their specific gravity is 0.8 to 0.85 (slick will float on water surface except under turbulent mixing conditions). In case of onshore spill this kind of oil will coat and penetrate substrate and causing some subsurface pollution.
3. **Medium oils/ intermediate oil products:** In this category are North Slope crude, South Louisiana crude, intermediate fuel oils, lube oils. These oils tend to be moderately volatile (flash point > 52 °C), with low evaporation rates (1/3 of the spill volume will evaporate), moderate to high viscosity and variable acute toxicity to biota (dependent upon amount of light fractions). Their specific gravity is 0.85 – 0.9 (slick will float on water surface except under turbulent mixing conditions). These types of oils can form stable emulsions. In case of onshore spill this kind of oil will coat and penetrate porous surfaces and causing heavy subsurface pollution.
4. **Heavy Crude Oils/residual oil products:** In this category are Venezuela crude, San Joaquin Valley crude, heavy marine fuels and Bunker C. These oils tend to be slightly volatile (flash point > 65 °C), with very low evaporation rates, high viscosity (very viscous to semi- solid) and low acute toxicity to

biota (due to low- water solubility). Their specific gravity is 0.95 to 1.0. These types of oils form stable emulsions. In case of onshore spill this kind of oil will slightly penetrate substrate.

5. **Very heavy residual products:** Similar properties as heavy crude oils/ residual oil products expect their specific gravity is >1.0 (will not flow on water surface).

There are two types of potential oil spill sources that, although they are generally outside the realm of petroleum industry itself, have to be considered when analyzing oil spills. These are natural oil seepages and seepages from sunken shipwrecks. Oil seeps are natural springs from which liquid and gaseous hydrocarbons leak out of sea bass or ground into the environment. Some of the authors consider natural oil seeps to be single most important source of oil entering the ocean. The estimated volumes of oil from natural seeps are 0.2 to 6.0×10^6 t (Fingas, 2011).

Sunken ships also act like natural seeps. Both of the sources release oil randomly in relatively small amounts over a longer period of time with occasional releases of larger quantities that can have the same environmental impacts as tanker oil spills, with the exception that most of the oil contained in sunken shipwreck's reservoirs is heavy oil which would most likely form tar balls rather than larger slicks. Another difference between natural oil seeps and these "artificial seeps" is that in case of a shipwreck oil will not be necessarily discharged (depending on vessel integrity and condition, age, depth, water temperatures and type of soil). According to American Petroleum Institute's (API) study from 2005 ("*Potentially Polluting Wrecks in Marine Waters*" commissioned in association with IMO, US Coast Guard, USEPA, IPIECA, Minerals Management Service and NOAA), which covered tankers of at least 150 gross registered tons (GRTS) carrying oil and non-tank vessels of at least 400 GRTS that carried oil as fuel/bunkers, there are at least 8,569 recorded vessel sinkings worldwide of which 1,583 were tankers and 6,986 were non-tank vessels. It is estimated that these shipwrecks contain 2.5 to 20.4×10^6 t of oil (Fingas, 2011).

Identification of potential sources of pollution, their placement on maps with combination of maps of the past spills is a good practice to indentify and rank high- risk areas.

ENVIRONMENTAL IMPACTS OF PETROLEUM INDUSTRY AND CONSEQUENCES OF OIL SPILL

All the activities conducted in the environment tend to have some impact on the environment i.e. its components (air, soil, water and biodiversity). Environmental impact of petroleum industry activities ranges from very low to very high primarily depending on type of activity, pollution prevention and control techniques applied and the sensitivity of the environment in which these activities are conducted. The most widespread and dangerous consequence of petroleum industry activities impacts is pollution. Pollution is, as already mentioned previously, due to handling with adverse substances, potentially associated with all the activities throughout all stages of oil and gas production, form exploration to refining.

Environmental Impacts of Petroleum Industry

During seismic operations the main sources of potential environmental impacts are seismic equipment, base camps and vessel activities. Principal impact of seismic equipment is noise, which temporarily affects humans and wildlife on the location. Base camps also pose a short-term impact by noise and light on local environment (humans, biosphere and atmosphere). Base camps through their activities leave a certain ecological footprint (vegetation is cleared leading to possible erosion and changes in surface hydrology, immigration of labour, waste creation and discharges, etc.) which in different ranges affects all components of the environment. Potential long-term impact from the camps is through construction of new access routes. Offshore seismic activities, beside noise from seismic equipment, which causes short-term disturbances to marine organisms, also temporary impact the environment through emissions (from vessel engines), discharges (bilges, sewage, etc) and interference with other resource users (e.g. fishery).

The main sources of environmental impacts from drilling activities are access roads, site preparation operations (drilling site preparation and camp construction), drilling and decommissioning. The environmental impact of using access roads is mainly short-term affecting all the components of the environment (humans, terrestrial, aquatic and atmospheric systems and biosphere), however this impact can become permanent if new access roads are constructed. The impact is manifested as vegetation ruination (causing possible erosion and changes in surface hydrology), emissions, vibrations and noise. Site preparation has significant footprint on the environment, which affects all environments' components (humans, terrestrial, aquatic and atmospheric systems and biosphere) through removal of vegetation and surface soil, causing possible erosion and changes in surface hydrology and drainage patterns, then loss of habitat, noise, vibrations and emissions from used equipment, aesthetic visual intrusion and possible land use conflict. These impacts are usually restricted to abidance of site preparation activities.

Main potential impacts during drilling operations are discharges, emissions and waste production, which affect all environments' components (humans, terrestrial, aquatic and atmospheric systems and biosphere) through noise, vibrations and emissions from transport, equipment used on drilling rig well test operations and flaring, liquid discharges (muds, wash and drainage water), disposal of solid waste (drilling cuttings), spillages and leakages, sanitary waste disposal and water supply requirement. As in case of site preparation activities, all these impact are restricted to drilling period. Depending on location, drilling operations usually also have significant socio-economic and cultural impact due to interactions with local population (immigration, impact on local infrastructure- employment, education, roads, services, hunting, fishing etc).

Offshore drilling operations, due to discharges to ocean (muds, cuttings, wash water, drainage and kitchen waste) and emissions, cause changes in sediment, water and air quality. Primary impact of these activities is on benthic and pelagic organisms. These operations also cause interference with other resource users (e.g. fishery).

After decommissioning, if it is conducted in a proper way, all the stated impact should be eliminated. Improper decommissioning can result in soil and water pollution, erosion and changes in surface hydrology, impacts on biodiversity, changes in land use and loss of habitat.

Principal sources of production activities' environmental impacts are access roads, site preparation, production operations and decommissioning. The impacts resulting from access roads, site preparation and decommissioning are the same as in case of drilling with the exception in duration of the impacts, namely all of the impacts in this case are long-term.

Potential impacts of oil and gas production - discharges, waste generation and emissions potentially have long- term impact on all the components of the environment (humans, terrestrial, aquatic and atmospheric systems and biosphere). These impacts have potential effects on biota, wildlife, habitats, biodiversity, water, soil and air quality. They also pose increased risk of soil, water and air pollution from spills and leakages.

Consequences of Oil Spill

Each oil spill is a unique event in terms of impacts and damages because each spill occurs under different circumstances of oil type and volume, weather and site conditions and in different ecosystems. Spilled oil varies widely in its composition from crude oils to refine products. Once released into the environment oil is subjected to numerous chances in composition, thus also changes of physical and chemical properties. This is known as "*weathering*". These changes include the following processes: evaporation, emulsification, dispersion, dissolution, oxidation and biodegradation. Most of these processes play important role in reducing the concentration of spilled oil in the environment. Considering previously mentioned, determining the impacts of oil spill is very complex. Spilled oil impact on the environment is often demonstrated as the dose- toxicity- exposure ratio (Fingas, 2011).

$$\text{Oil impact} = \text{Dose} \times \text{Toxicity} \times \text{Exposure} \quad (1)$$

The absence of one or more factors in Equation1 can amplify or negate another factor. The above equation holds all the necessary information that have to be considered during oil spill analysis and cleanup. Dose (volume of oil spilled) and exposure (what can be done to contain/ divert/ collect spilled volume) can be relatively easily determined. The problem lies in toxicity. Namely, toxicity is defined as "*the degree to which a substance is able to damage an exposed organism*" (Fingas, 2011). Determining oil toxicity is rather difficult due to the fact that oil is not a single substance but a complex mixture of many substances. It is more difficult to determine the toxicity of chemical mixture than a pure chemical, because each component displays its own toxicity, and components may interact to produce enhanced or diminished effects. Hydrocarbons that are most often associated with toxic effects are aromatic hydrocarbons. In this group there are hydrocarbons that consist of single and multiple benzene rings. Monocyclic aromatic hydrocarbons (MAHs) include so called BETEX compounds (benzene, toluene, ethyl benzene and xylenes), which are highly volatile and relatively water- soluble. BETEX compounds are mostly found in light crude oils and light petroleum products. Hydrocarbons that consist of multiple benzene rings are known as polycyclic aromatic hydrocarbons (PAHs). These compounds have been identified as carcinogenic, mutagenic, and teratogenic.

Toxicity is manifested through acute and chronic effects mutually differing in time span of toxic effect manifestation after organisms' exposure to the toxic substance. Namely, acute toxic effects manifest within a short period of time in relation to the organisms' life time, while it will take much longer for chronic effects to be observable. Toxic substances are usually identified when associated adverse effects are noticed. In most cases the first information that a particular substance is toxic is its' association with an unusual number of deaths. Standard measures of acute toxicity are LD_{50} (mg of toxic substance per kg of body weight) or LC_{50} (ppm, mg/m^3, mg/l), which stand for "*lethal dose*" or "*lethal concentration*". Lethal dose/concentration is a dose/concentration of the toxic substance which will cause death of 50% of tested population within a specific period of exposure time.

In assessing potential effects of a substance and risk that the substance poses to the environment, besides LC_{50}, there are several other expressions that characterize concentration of a substance used, like predicted environmental concentration (PEC), predicted no-effect concentration (PNEC), no-observed effect concentration (NOEC) and lowest observed effect concentration (LOEC). If the PEC exceeds the PNEC there is a risk of environmental damage in proportion to the PEC/PNEC ratio, which stands as a measure of probability that harm will occur (UNEP/IPCS, 1999).

Individual responses of organisms to a particular toxic substance differ widely. Every organism, depending on its sex, age and general physical state, is characterized with its own threshold dose value. A threshold dose is the lowest dose at which there is an observable effect. Threshold value in a population usually follows a Gaussian distribution. The effect of a dose of toxic substance depends on the exposure pathway. The exposure pathway determines the biochemistry of the substance in the organisms, thus the same substance may have different effects depending on whether it is ingested, inhaled or absorbed through the skin (Weiner & Matthews, 2003).

The relationship between the dose of a toxic substance and the organisms' response can be expressed by the dose-response curve (Figure 2).

Figure 2. Different types of dose- response curves (Weiner & Matthews, 2003)

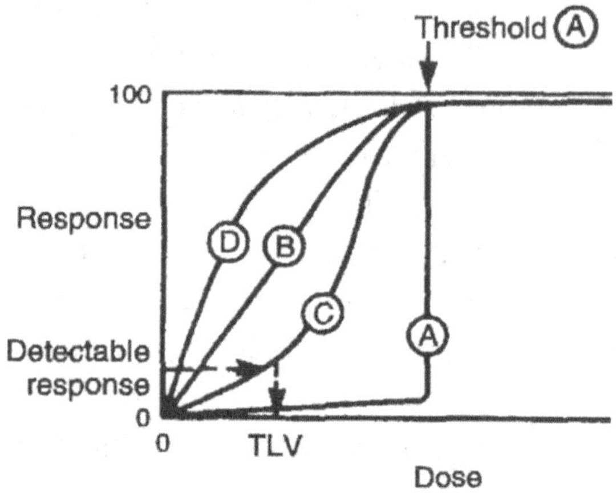

A- *illustrates a threshold response: there is no observed effect until a particular concentration is reached.*
B- *linear response with no threshold, i.e. the intensity of the effect is directly proportional to the toxic substance dose.*
C- *sublinear or sigmoidal dose- response curve- the lowest dose at which a response can be detected is called "threshold limit value (TLV)".*
D- *displays a supralinear dose- response relationship, which is found when low doses of a toxic substance provoke a disproportionately large response.*

Some characteristic features of the dose-response curve are (Weiner & Matthews, 2003):

1. Threshold
2. Total Body Burden
    a. If an organism is simultaneously exposed to several different sources of a toxic substance the concentration of the substance in organism is then a sum of all the quantities of the substance from all the sources.
3. Physiological Half- Life (PHL)
    a. PHL of a toxic substance is the time needed for an organism to eliminate 50% of internal concentration of the substance through metabolism or other normal physiological function.
4. Bioaccumulation and Bioconcentration
    a. Bioaccumulation occurs when a substance is concentrated in one organ or type of tissue of an organism. Bioconcentration occurs with movement up the food chain.
5. Exposure Time and Time vs. Dosage
    a. Time needed for toxic substance to react.
6. Synergism
    a. Phenomena when two or more toxic substances enhance each others' effects.
7. LC_{50} and LD_{50}
8. Hormesis
    a. Some substances have beneficial effect in small doses but harmful in large ones.

Oil can affect organisms in many ways depending on the exposure pathway (physical exposure, ingestion, absorption, adsorption or through food chain), the exposure location (water surface, shorelines or land), sensitivity of an organism, organisms' recovery potential, it's tenancy to avoid oil, possibility of helped rehabilitation (some animals can be caught and some can't) and age (young specimens are more sensitive than old ones). The sensitivity varies with species, season and weather conditions. The ability of an organisms or ecosystem to return to the state they were before exposure to oil spill is specified by the recovery potential. The time needed for an organism/ ecosystem to recover ranges from couple of days to several years. The most common consequences of organisms' exposure to oil spill are loss of habitat, changes in feeding and reproductive behavior, tainting phenomena (fish and bivalves) and general behavioral changes such failure to take care of their offspring.

Pollution of soil with oil or oil products will result in: changes of some soil properties (e.g. sorption properties or soil reaction), changes of soil microbial diversity, disabled transport of water and nutrients, disturbance of carbon nitrogen ratio and transition from oxidation to reduction conditions resulting with secondary pollution with metal oxides and changes in pH values of ground water.

It is difficult to determine the full impacts and consequences of an oil spill in marine environment because, along with the complexity of weathering processes, many physical and biological processes in the marine and coastal environment are still poorly understood. The factors that affect toxicity of hydrocarbons in aquatic environment are salinity and temperature of water, abudance of food and the general health of species. There is no evidence of aquatic organisms' tendency to bioaccumulate hydrocarbons. Over a certain period of time (up to one year) they tend to "purify" themselves and lose the hydrocarbons they have taken. There are seven general groups of organisms under threat in case of oil spill in aquatic environment:fish, plankton, benthos organisms, marine mammals, marine plants, intertidal fauna and birds.

Concerning fish, the most vulnerable to oil spill are species whose habitats are close to water surface, sea shore or sea floor. The effects of an oil spill on fish are mostly sub-lethal and primarily manifested through toxic effects of aromatic hydrocarbons. Lethal concentrations are rarely found in open seas however such concentrations can occur in confined waters like bays and estuaries, directly under or in the vicinity of the spill.

Plankton (phytoplankton and zooplankton) are very sensitive to hydrocarbons. Relatively low hydrocarbon concentrations will affect plankton, but plankton occurs in enormous numbers and lost individuals are relatively quickly replaced. Plankton is at the bottom of aquatic food chain and hydrocarbons taken by plankton are passed up the food chain affecting higher life forms.

Benthic organisms (organisms that live on the sea floor) are usually divided into immobile (benthic infauna- fauna within bottom sediments and benthic epifauna - fauna on top of bottom sediments) and mobile benthic organisms (starfish, gastopods, sea urchins, amphipoda and isopods) (Fingas, 2001). These organisms are usually killed when large amounts of oil accumulate on the sea floor. This happens in case of oil sedimentation due to increase of oil specific gravity (result of the weathering processes) or as a result of oil adsorption onto solid particles (e.g. sand particles). Benthic organisms are particularly endangered in shallow areas of near- shore environments. Oil spills usually affect benthic organisms on sub-lethal level causing changes in growth (slow growth, deformity of body parts), changes in molting times, occasional anomalies in organ development, smaller brood sizes and changes in feeding practice (Fingas, 2001).

Marine mammals, which come to contact with oil spill, usually suffer from heat lost (their fur loses waterproofing), eye irritation and lost of buoyancy. Especially sensitive to oil spills are marine mammals covered with fur, like polar bears or otters, because they, by grooming themselves, ingest oil resulting in death or severe illness.

Intertidal fauna includes mobile organisms (crabs, snails and shrimp) and immobile barnacles and mussels and sedentary limpets, periwinkles and tube worms (Fingas, 2001). These organisms, besides being the most vulnerable to oil spill, are also very sensitive to shoreline cleanup methods (e.g. washing).

Marine plants include a wide range of plant families and algae. Most affected marine plants by an oil spill are intertidal algae, macroalgae (kelp and fucus) and sea grasses. Intertidal algae, which are an important food source for intertidal fauna, usually are the first biota that recovers after the spill.

Oil spill effects on birds manifest through loss of insulation and buoyancy (oiled feathers), ingestion of oil or oiled pray, which, depending on oil concentration, results with death or rarely gastrointestinal dysfunction, liver problems, pneumonia and behavioral disorders.

In case of an oil spill that occurs under ice, epontic organisms (microscopic plants and animals that live under ice) have similar reactions to hydrocarbons as plankton, but are much more vulnerable because oil remains directly under ice.

Direct consequence of emissions from petroleum industry is contribution to greenhouse effect and overall global climate change. Even though petroleum industry is one of greenhouse gases emission intensive industries (beside direct greenhouse gas emissions (CO_2, CH_4 and N_2O) with oil and gas production, petroleum industry is also characterized with high-energy intensity and final products with high emission potentials) contribution of petroleum industry's emissions (less than 1.5 Gt CO_{2eq}/yr) (Hrncevic, 2008) to total global greenhouse gas emissions, which at the moment amount more than 30 Gt CO_{2eq}/yr, are relatively low.

ENVIRONMENTAL RISK ASSESSMENT OF PETROLEUM INDUSTRYS' ACTIVITIES

Risk is the category that is perceived very differently- e.g. the risks associated with nuclear waste transport are usually perceived as far greater than the risks associated with crude oil or petroleum products transport even though the latter are considerably larger. In general, the risk perception is influenced by the knowledge about the risk, whether the risk is undertaken (or perceived as undertaken) voluntarily and the estimated perceived benefits of the activity the considered risks are associated with. Environmental risk assessment, even though it is done rather in a similar way as, for instance, human health risk assessment, differs from risk assessments in other disciplines. The assessment becomes even more complex if it is case of petroleum industry activities, due to possible exposure pathways and identification of all environmental components at risk with their integral constituents and it is usually difficult to perceive all the consequences of an activity. Since there is still no universally accepted methodology for petroleum industry activities risk assessment, the best one can expect to achieve is a general estimation of the level of risk by means of analogy with similar systems.

As already mentioned, risk is defined as the probability of an adverse effect in a specific period of time. The environmental risk assessment is based on calculating the probability for an ecosystem to come to contact with a dose of a toxic substance. Environmental risk is usually described by the following equation (Dumitran & Onutu, 2010):

$$RISK = Probability \times Severity \quad (2)$$

Probability is often confused with frequency, because frequencies are used to estimate probabilities in risk assessments. Probability is dimensionless number and is always less or equal to 1. A probability of 1 means that the event has a 100% chance of occurrence. Depending on how the frequency is defined, it has dimensions and can have value larger than 1.

The biggest risk related to petroleum industry activities is the risk of an oil spill during transportation. Following the analogy with the Equation (2), this risk of tanker oil spill can be defined as (Edie, 2007):

$$R(x,y,z) = F(x,y,z) \times C(x,y,z) \quad (3)$$

where:

R= Risk of an oil spill, tonnes of oil spilled/ship year
F= Frequency
C= Consequence
x, y= Position coordinates
t= Time component

The consequence in that case can be defined as (Edie, 2007):

$$C = S \times I \quad (4)$$

where:

S= Expected spill size, m^3
I= Impact of 1 tonne of oil

Expected spill size combines probability of an oil spill accident with the expected outflow (Edie, 2007):

$$S = P \times O \quad (5)$$

$$O = \alpha \times \beta \times D_{wt} \quad (6)$$

where:

P= Probability of an oil spill accident
O= Expected outflow, m^3
α= Expected outflow rate given as a percentage of the tank content volume (depended on accident type)
β= Volume of cargo or bunker oil, m^3
D_{wt}= Vessel deadweight tonnage

The spill impact (I) per tonne oil depends on the type of oil and on the affected area's vulnerability to oil spill (Edie, 2007):

$$I = E \times T \tag{7}$$

where:

E= Environmental sensitivity index
T= Oil type significance index

Methods used to estimate the frequency and the size (with associated consequences) of an oil spill include both quantitative and qualitative techniques. Quantitative risk assessment takes into account the dangers and consequences based on so called SMART indicators (SMART- specific, measurable, available, relevant and timely) estimating the size of such consequence and the probability. Qualitative risk assessment is based on research of three factors: "*threat/source*" (equipment and specific pollutants generated from it), "*path of action*" (pollutant's migration pathways to the receptor) and "*target/receptor*" (components of the environment which are affected by harmful effects of the pollutant).

To quantify the probability of oil spill the number of spill is quantitatively related to the type and level of activities which may affect the number of spills. The number of drilling and production spills is likely to be proportional to the number of wells drilled and in production. The number of process spills depends, among other things, on the number of processing units, valves, pumps, flares, operators and the volume of oil processed. The number of crude oil storage tanks is likely to be proportional to the number and size (volume of oil contained) of the tanks. The number of pipeline spills is a function of many factors: pipeline diameter, age, degree of burial, location, facility operations and maintenance practice. The number of tanker spills depends on the number and size (deadweight) of tankers and the frequency and length of marine transport. The number of loading spills is proportional to the number of loading operations and their duration.

Spill frequency is determined by a Poisson process. This means that the probability of a spill in specific time frame is proportional to the amount of exposure in this time frame and that it is not affected by the last occurrence of a spill. The Poisson distribution has two parameters; the contemplated exposure and the intensity. The intensity of the Poisson process is the mean spill rate. This is an uncertain quantity lying between zero and infinity. Before examining the spill data, which are in many cases unavailable, the nature of the intensity variable can be suitably characterized with the gamma distribution with parameters equal to zero. The parameters of gamma distribution are determined by the data concerning numbers of spills observed and observed exposure. The exposure variable is taken to be the volume of oil handled. Given a contemplated exposure of "*t*" units, the distribution of the number of future spills "*n*" can be determined. For each "*n*" the distribution is the probability of "*n*" spills given each possible value of intensity multiplied by that value of intensity summed over all possible intensity values (Equation 8) (Donovan & Owen, 1976).

$$p\left(n / t, \nu, \tau\right) = \frac{\left(n + \nu - 1\right)!\, t^{n} \tau^{\nu}}{n!\left(\nu - 1\right)!\left(t + \tau\right)^{n+\nu}} \tag{8}$$

where:

ν= Number of spills observed
τ= Observed exposure
t= Contemplated exposure
n: Number of spills over facility lifetime

Assessment of the environmental risk from either a specific facility or process or from a complete oil/gas field involves several steps. These are (Dumitran & Onutu, 2010, Woods et. al., 1991):

1. Hazard identification;
2. Hazard assessment;
3. Risk estimation;
4. Risk assessment;
5. Environmental risk management.

Each step includes a structured series of stages in a number of steps that determine the data needed for the final evaluation. Steps and their main stages are presented in Figure 3. Each identified hazard should be classified according to the level of risk to the environment. The relations of frequencies and consequences of the identified hazards are then presented in so called risk assessment matrix (Figure 4, Table 3).

CONCLUSION

Principles of environmental protection were initiated and became embedded in petroleum business for more than 80 years, but it is only few decades ago that environmental protection started to be seen as an integral part of sustainable development approach. There are at least two indicators serving the fact that, over the decades, the environmental performance of petroleum industry has improved. Namely the environmental expenditures of petroleum companies have increased and the world-wide total spill discharges have decreased. Nowadays it could be said that environmental management techniques have successfully integrated into all segments of upstream operational procedures and activities. Since petroleum industry, regarding its activities, which are environmentally intensive, takes up special public attention, great emphasis has been put on quality environmental system based on effective management with systematic approaches and adequate mechanisms which have to ensure compliance with present, as national as the global, environmental policies and regulations. Nevertheless, accidents with adverse environmental impact still happen.

Petroleum industry has a long and short-term environmental risk. Along with production fluids, that pose significant environmental risk (if found uncontested in environment), all petroleum industry activities involve either use of fluids, that contain abundant and diverse toxic chemicals, or waste generation, both associated with a potential risk to the environment. The principal environmental risk associated with petroleum industry is seen primary as the risk of fluid (production, operating or waste) spill/ emission to the environment.

Oil spill occurrences and the total amount of oil spilled have decreased significantly over the last 30 years, especially last decade, even though there was increase in both marine (tanker) and pipeline transportation of oil and petroleum products. Even though incidents, resulting with large releases of fluids in pe-

Figure 3. Environmental risk assessment steps (Dumitran & Onutu, 2010, Woods et. al., 1991)

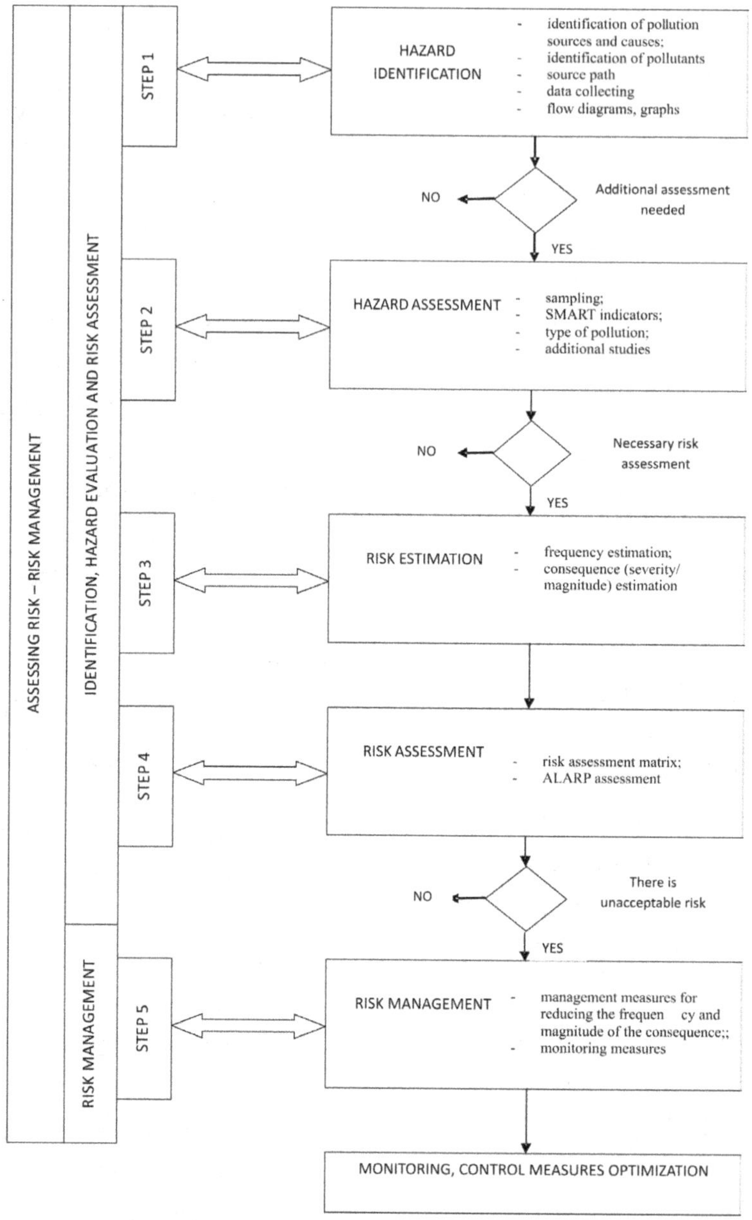

troleum industry, occur rarely, it is important to identify the sources, size and the frequency of these releases. Oil spill can be minor in its impacts but public or political interest and media coverage may elevate perceived impact to much higher level. Despite significant progress in reducing spills/ emission through a variety of technological measures and regulatory restrictions, better industry practices in proper planning and design, incidents in petroleum industry can occur and the risk for significant environmental damage remains, however, the consequences of these incidents can be minimized, mitigated or even avoided

Figure 4. Example of risk assessment matrix (Pinceratto & Casey, 1998, Dumitran & Onutu, 2010, Woods et. al., 1991)

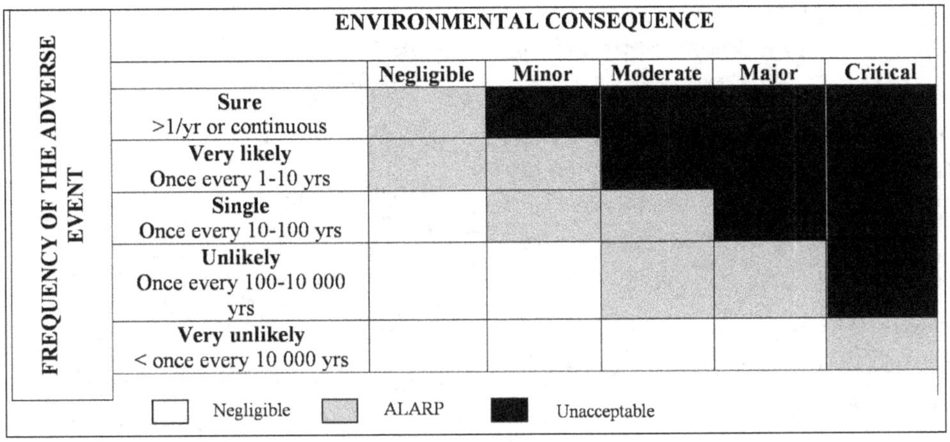

Table 3. Assigning environmental consequences (Pinceratto & Casey, 1998)

Negligible	No measurable or detectable damage to the components of the environment
Minor	Minor damages (0-25% of ecosystem damaged); localized impact; less than 1 year needed for the environment to recover
Moderate	Localized environmental impact (25-50% of ecosystem damaged); less than 5 year needed for the environment to recover
Major	Major environmental damage (50-75% of ecosystem damaged); 5 to 10 years needed for the environment to recover
Critical	Massive environmental damage (75-100% of ecosystem damaged); more than 10 years needed for the environment to recover

if proper risk-based measures and monitoring, including use of modern technologies are implemented.

Environmental impact of petroleum industry activities ranges from very low to very high primarily depending on type of activity, pollution prevention and control techniques applied and the sensitivity of the environment in which these activities are conducted. Petroleum industry has been proactive in the development of technologies, operational practices and management systems in general targeted at minimizing environmental impact. This has resulted with significant reduction of the number of environmental accidents.

To the date there is still no universally accepted methodology for environmental risk assessment in petroleum industry. If there is an adequate data base of previous identical or similar events, environmental risk assessment of petroleum industry activities is commonly done by the analysis of former reported spill data for a specific area. However, if there are no sufficient or available data to determine the extents of risks, the combination of qualitative and quantitative methods is used.

REFERENCES

Andersen, S., & Mostue, B. A. (2012). Risk analysis and risk management approaches applied to the petroleum industry and their applicability to IO concepts. *Safety Science*, *50*, 2010–2019. doi:10.1016/j.ssci.2011.07.016

CONCAWE. (2013). Concawe report no. 3/13, performance of european cross- sounty oil pipelines, statistical summary of reported spillages in 2011 and since 1971. Retrieved june 18, 2013, from https://www.concawe.eu

Donovan, L. J., & Owen, J. J. (1976). *A method to estimate the pollution risk and cost of OCS oil transportation*. Paper presented at the Eighth Annual Offshore Technology Conference. Houston, TX.

Dumitran, C., & Onutu, I. (2010). Environmental risk analysis for crude oil soil pollution. *Carpathian Journal of Earth and Environmental Sciences*, *5*(1), 83–92.

Edie, M. S. (2007). Prevention of oil spill from shipping by modeling of dynamic risk. *Marine Pollution Bulletin*, *54*, 1619–1633. doi:10.1016/j.marpolbul.2007.06.013 PMID:17765267

Editorial Dept. (2009). *A detailed guide on the many different types of crude oil*. Retrieved June 22, 2013, from http://oilprice.com/energy/crude-oil/a-detailed-guide-on-the-many-different-types-of-crude-oil.html

E&P Forum. UNEP. (1997). *Environmental management in oil and gas exploration and production*. Oxford, UK: Words and Publications.

Fingas, M. (2001). *The basics of oil spill cleanup*. Boca Raton, FL: CRC Press LLC.

Fingas, M. (Ed.). (2011). *Oil spill science and technology*. New York: Elsevier Ltd.

Hrncevic, L. (2008). *Analiza utjecaja provedbe Kyoto protokola na naftnu industriju i poslovanje naftne tvrtk*. (Unpublished doctoral dissertation). University of Zagreb, Zagreb, Croatia.

IMO. (2005). *International convention for the prevention of pollution from ships (MARPOL)*. Retrieved June 21, 2013, from http://www.imo.org

ITOPF. (2013). Oil tanker spill statistics 2012. Retrieved June 20, 2013, from http://www.itopf.com/news-and-events/documents/StatsPack.pdf

Michel, J., Christopherson, S., & Whipple, F. (1994). *Mechanical protection guidelines*. Retrieved March 8, 2012, from http://response.restoration.noaa.gov/sites/default/files/mechanical-protection-guidelines.pdf

Olita, A. et al. (2012). Oil spill hazard and risk assessment for the shorelines of a Mediterranean coastal archipelago. *Ocean and Coastal Management*, *57*, 44–52. doi:10.1016/j.ocecoaman.2011.11.006

Pinceratto, E., & Casey, J. (1998). *Environmental risk assessment- Case study of an offshore petroleum development*. Paper presented at the 1998 SPE International Conference on Health, Safety and Environment in Oil and Gas Exploration and Production. Caracas, Venezuela.

Price, J. M. et al. (2003). Overview of the oil spill risk analysis (OSRA) model for environmental impact assessment. *Spill Science & Technology Bulletin*, *8*(5-6), 529–533. doi:10.1016/S1353-2561(03)00003-3

Psarros, G., Skjong, R., & Vanem, E. (2011). Risk acceptance criterion for tanker oil spill risk reduction measures. *Marine Pollution Bulletin*, *62*, 116–127. doi:10.1016/j.marpolbul.2010.09.003 PMID:20888015

Rana, S. (2008). *Facts and data on environmental risks- Oil and gas drilling operations*. Paper presented at the 2008 SPE Asia Pacific Oil & Gas Conference and Exhibition. Perth, Australia.

Schmidt Etkin, D. (2001). Analysis of oil spill trends in the united states and worldwide. In *Proceedings of International Oil Spill Conference* (vol. 2, pp. 1291-1300). Washington, DC: American Petroleum Institute.

UNEP/IPCS. (1999). *Training module No. 3, Section B: Environmental Risk Assessment*. Retrieved October 18, 2012, from http://whqlibdoc.who.int/hq/1999/WHO_PCS_99.2_eng.pdf

Vanem, E., Endresen, Ø., & Skjong, R. (2008). Cost- effectiveness criteria for marine oil spill preventive measures. *Reliability Engineering & System Safety*, *93*, 1354–1368. doi:10.1016/j.ress.2007.07.008

Weiner, R. F., & Matthews, R. A. (2003). *Environmental engineering* (4th ed.). New York: Elsevier Science.

Woods, I. T., Shirley, D., & Bottelberghs, P. H. (1991). *Environmental risk in the offshore industry*. Paper presented at the First International Conference on Health, Safety and Environment. The Hague, The Netherlands.

Yanting, Z., & Liyun, X. (2011). Research on risk management of petroleum operations. *Energy Procedia*, *5*, 2330–2334. doi:10.1016/j.egypro.2011.03.400

ADDITIONAL READING

Al- Majed. A. A., Adebayo, A. R., & Hossaion, M. E. (2012). A Sustainable Approach to Controlling Oil Spills, *Journal of Environmental Management* 113, 213-227. Retrieved October 22, 2012, from http://www.sciencedirect.com

Aven, T., & Vinnem, J. E. (2007). *Risk Management With Applications from the Offshore Petroleum Industry*. London, UK: Springer-Verlag.

Cheremisinoff, N. P. (2002). *Handbook of Air Pollution Prevention and Control*. USA: Elsevier Science.

Cheremisinoff, N. P., & Rosenfeld, P. (2009). *Best Practices in the Petroleum Industry* (Vol. 1). USA: Elsevier Ltd.

Cline, J. T., & Førde, R. (1991, November). *Environmental Program With Operational Cases To Reduce Risk to the Marine Environment Significantly*, Paper presented at the First International Conference on Health, Safety and Environment, The Hague, The Netherlands.

Committee on Oil Spill Risks From Tank Vessel Lightering et al. (1998). *Oil Spill Risks From Tank Vessel Lightering*. Washington, D.C., USA: National Academy Press.

CONCAWE. (2003). Concawe Report No. 3/03, European Oil Industry Guideline for Risk- Based Assessment of Contaminated Sites. Retrieved June 18, 2013, from https://www.concawe.eu.

Conference on Health, Safety and Environment in Oil and Gas Exploration and Production, Kuala Lumpur, Malaysia.

Frost, T. K., Johsen, S., & Hjelsvold, M. (2002, March). *Environmental Risk Management of Discharges from E & P Activities in the North Sea*, Paper presented at the SPE International

Godec, M. L. (2009, March). *Environmental Performance of the Exploration and Production Industry: Past, Present and Future*, Paper presented at the 2009 SPE Americas E & P Environmental and Safety Conference, San Antonio, Texas, USA.

Handbook, I. T. O. P. F. 2010/11, Retrieved October 9, 2012, from http://www.itopf.com

Iakovou, E. T. (2001). An Interactive Multiobjective Model for The Strategic Maritime Transportation Of Petroleum Products: Risk Analysis And Routing [from http://www.sciencedirect.com]. *Safety Science*, *39*, 19–29. Retrieved December 19, 2012 doi:10.1016/S0925-7535(01)00022-4

Mills, M. A. (1999). Method for Quantifying the Fate of Petroleum in the Environment [from http://www.sciencedirect.com]. *Chemosphere*, *39*(14), 2563–2582. Retrieved April 23, 2013 doi:10.1016/S0045-6535(99)00163-0

O'Hara, P. D., & Morandin, L. A. (2010). Effects of Sheens Associated With Offshore Oil and Gas Development on the Feather Microstructure of Pelagic Seabirds [from http://www.sciencedirect.com]. *Marine Pollution Bulletin*, *60*(Issue 5), 672–678. Retrieved January 3, 2013 doi:10.1016/j.marpolbul.2009.12.008 PMID:20060137

Ornitz, B. E., & Champ, M. A. (2002). Oil Spill First Principles: Prevention and Best Response, Oxford, Uk: Elsevier Science Ltd.

Orszulik, S. T. (Ed.). (2008). *Environmental Technology in the Oil Industry* (2nd ed.). Springer Science. doi:10.1007/978-1-4020-5472-3

Rana, S. (2009, August). *Environmental Risks-Oil & Gas Operations Reducing Compliance Cost Using Smarter Technologies*, Paper presented at the 2009 SPE Asia Pacific Health, Safety, Security and Environment Conference and Exhibition, Jakarta, Indonesia.

Reis, J. C. (1996). *Environmental Control in Petroleum Engineering*. Huston, Texas, USA: Gulf Publishing Company.

Samiullah, Y. (1985). Biological Effects of Marine Oil Pollution, *Oil and Petrochemical Pollution*, Volume 2 Issue 4, 235-264. Retrieved January 3, 2013, from http://www.sciencedirect.com

Skogdalen, J. E., & Vinnem, J. E. (2012). Combining Precursor Incidents Investigations and QRA in Oil and Gas Industry [from http://www.sciencedirect.com]. *Reliability Engineering & System Safety*, *101*, 48–58. Retrieved January 3, 2013 doi:10.1016/j.ress.2011.12.009

Suslick, S. B., & Schiozer, D. J. (2004). Risk analysis Applied to Petroleum Exploration and Production: An Overview [from http://www.sciencedirect.com]. *Journal of Petroleum Science Engineering*, *44*, 1–9. Retrieved January 3, 2013 doi:10.1016/j.petrol.2004.02.001

KEY TERMS AND DEFINITIONS

Environmental Risk: Probability of an unwanted consequence of a particular event on the environment in a specific time period.

Oil Spill: Discrete event in which oil is discharged through neglect, by accident or with intent into the environment.

Oil Toxicity: The degree to which oil is able to damage an exposed organism.

Petroleum: Naturally occurring flammable liquid consisting of a complex mixture of hydrocarbons of various molecular weights and other organic and inorganic compounds, that are found in geologic formations beneath the Earth's surface.

Petroleum Industry: Industry which involves exploration, production, processing, storage, transportation and refining of oil and gas.

Pollutant: Substance that has an adverse effect on the components of the environment (terrestrial, aquatic and atmospheric systems and biosphere).

Risk Assessment: Calculation/ estimation of probability of an event with adverse consequences.

Compilation of References

(1971). Well servicing and workover. InBoyd, W. E. (Ed.), *Control of formation pressure, lesson 9*. Austin, TX: Petroleum Extension Service, The University of Texas at Austin.

49. Part, C. F. R. 195. (2002). Transportation of hazardous liquids by pipeline: Code for federal regulations title 49. Washington, DC: US Printing Office.

Aadnøy, B. S., Larsen, K., & Berg, P. C. (1999). *Analysis of stuck pipe in deviated boreholes*. Paper presented at the 1999 SPE Annual Technical Conference and Exhibition. Houston, TX.

Abass, H., Shebatalhamd, A., Khan, M., Al-Shobaili, Y., Ansari, A., Ali, S., & Mehta, S. (2006). *Wellbore instability of shale formation, Zuluf field, Saudi Arabia*. Paper presented at Technical Symposium of Saudi Arabia Section. Dhahran, Saudi Arabia.

Adams, A. J., Parfitt, S. H., Reeves, T. B., & Thorogod, J. L. (1993) *Casing system risk analysis using structural reliability*. Paper presented at the SPE/IADC Drilling Conference. Amsterdam, The Netherlands.

AIChE. (2009). *Guidelines for developing quantitative safety risk criteria: Appendix B survey of worldwide risk criteria application*. New York: American Institute of Chemical Engineers, Inc. Centre for Chemical Process Safety.

Akamine, A. Y., Bratton, T., Onyia, E., & Romanchock, M. (2003). *Apllication of real-time resistivity and annular pressure data in reducing lost-circulation events*. Paper presented at the SPE/IADC Drilling Conference. Amsterdam, The Netherlands.

Al Ubaidan, A., Esmail, W. J., Aborshaid, H., & Al Umair, M. (2000). *Successful isolation of a loss circulation zone of a short radius horizontal in a highly fractured environment*. Paper presented at the 2000 IADC/SPE Asia Pacific Drilling Technology. Kuala Lumpur, Malaysia.

Al-Ashhab, J., Afzal, M., & Emenike, C. O. (2004). *Well integrity management system (WIMS)*. Paper presented at the International Petroleum Exhibition and Conference. Abu Dhabi, UAE.

Alaska Department of Environmental Conservation. (1999). *Technical review of leak detection technologies*. Washington, DC: EPA.

Al-Baiyat, I., & Heinze, L. (2012). *Implementing artificial neural networks and support vector machines in stuck pipe prediction*. Paper presented at the SPE Kuwait International Petroleum Conference and Exhibition. Kuwait City, Kuwait.

Al-Bazali, T. M., Zhang, J., Chenevert, M. E., & Sharma, M. M. (2006). *Factors controlling the membrane efficiency of shales when interacting with water-based and oil-based muds*. Paper presented at SPE International Oil & Gas Conference and Exhibition. Beijing, China.

Aldred, W., Plumb, D., Bradford, I., Cook, J., Gholkar, V., & Cousins, L. et al. (1999). Managing drilling risk. *Oilfield Review*, *11*(2), 2–19.

Algu, D. R., van Oort, E., Landgrave, S., Grant, L., & Hillburn, D. (2007). *A systematic approach to drillling and delivering wells in severely depleted sands at ram powell (GOM)*. Paper presented at the 2007 SPE/IADC Drilling Conference. Amsterdam, The Netherlands.

Andersen, S., & Mostue, B. D. (2012). Risk analysis and risk management approaches applied to the petroleum industry and their application to IO concepts. *Safety Science*, *50*, 2011–2019. doi:10.1016/j.ssci.2011.07.016

Apache. (n.d.). Retrieved from https://semsportal.apachecorp.com/

API 1130. (2002). *Computational pipeline monitoring for liquid pipelines*. Washington, DC: American Petroleum Institute.

API 653. (2001). *Tank inspection, repair, alteration, and reconstruction* (3rd ed.). Washington, DC: American Petroleum Institute.

API RP 14B. (1994). *Design, installation, repair and operation on subsurface safety valve system* (4th ed.). Washington, DC: American Petroleum Institute.

API RP 14C. (2001). *Analysis, design, installation and testing of basic surface safety systems for offshore production platforms*. Washington, DC: American Petroleum Institute.

API RP 53. (1997). *Blowout prevention equipment systems for drilling wells* (3rd ed.). Washington, DC: American Petroleum Institute.

API RP 53. (1999). *Recommended practice for blowout prevention equipment systems for drilling wells* (3rd ed.). Washington, DC: American Petroleum Institute.

API RP 6AR. (1994). *Recommended practice for repair and remanufacture of wellhead and Christmas tree equipment*. Washington, DC: American Petroleum Institute.

API RP 75. (2004). *Development of a safety and environmental management program for offshore opeartions and facilities*. Washington, DC: American Petroleum Insitute.

API RP 7G. (1998). *Recomended practice for drill stem design and operating limits* (16th ed.). Washington, DC: American Petroleum Institute.

API RP 90. (2006). *Annular casing pressure management for offshore wells*. Washington, DC: American Petroleum Institute.

API Standard 620. (2002). *Design and construction of large, welded, low-pressure storage tanks* (3rd ed.). Washington, DC: American Petroleum Institute.

API Standard 650. (2007). *Welded steel tanks for oil storage* (11th ed.). Washington, DC: American Petroleum Institute.

Arcanum Deep Secrets. (n.d.). Retrieved from http://arcanumdeepsecrets.wordpress.com/2010/07/01/bp-oil-disaster-simultaneous-operations-overview-infographic-diagram/

Argyropoulos, C. D., Christolis, M. N., Nivolianitou, Z., & Markatos, N. C. (2012). A hazards assessment methodology for large liquid hydrocarbon fuel tanks. *Journal of Loss Prevention in the Process Industries*, *25*, 329–335. doi:10.1016/j.jlp.2011.12.003

Arthur, J. D., Bohm, B., Coughlin, B. J., & Layne, M. (2009). *Evaluating implications of hydraulic fracturing in shale gas reservoirs*. Paper presented at the SPE Americas Environmental and Safety Conference. San Antonio, TX.

Arthur, J. D., Hochheiser, H. W., & Coughlin, B. J. (2011). *State and federal regulation of hydraulic fracturing: A comparative analysis*. Paper presented at the SPE Hydraulic Fracturing Technology Conference and Exhibition. The Woodlands, TX.

ASME. (2003). *Standard for probabilistic risk assessment for nuclear power plant applications*. ASME RA-Sa-2003. Addenda to ASME RA-S-2002.

Atherton, W. (2005). *An experimental investigation of bund wall overtopping and dynamic pressure on the bund wall following catastrophic failure of a storage vessel*. Liverpool, UK: John Moores University.

Attard, M. (1991). *The occurrence of annulus pressures in the North West Hutton Field: Problems and solutions*. Paper presented at the Offshore Europe Conference. Aberdeen, UK.

Aven, T., & Vinnem, J. E. (2007). *Risk management with applications from the offshore petroleum industry*. London, UK: Springer-Verlag.

Awal, M. R., Khan, M. S., Mohiuddin, M. A., Abdulraheem, A., & Azeemuddin, M. (2001). *A new approach to borehole trajectory optimisation for increased hole stability*. Paper presented at the SPE Middle East Oil Show. Bahrain.

Ayers, R. C., O'Reilly, J. E., & Henry, L. R. (1989). *Offshore operators committee Gulf of Mexico spotting fluid survey*. Paper presented at the 1989 SPE/IADC Drilling Conference. New Orleans, LA.

Ayyub, B. M. (2003). *Risk analysis in engineering and economics*. London: Chapman & Hall. doi:10.1201/9780203497692

Baker Hughes. (2003). *Coiled tubing solutions- Solve downhole problems with reliable, cost-effective technology*. Houston, TX: Author.

Barker, J. W., Feland, K. W., & Tsao, Y-H. (1994, September). Drilling long salt sections along the U. S. Gulf Coast. *SPE Drilling & Completion*, 185-188.

Barlet-Gouedard, V., Rimmele, G., & Porcherie, O. (2006). *Mitigation strategies for the risk of CO_2 migration through wellbores*. Paper presented at the IADC/SPE Drilling Conference. Miami, FL.

Barlet-Gouedard, V., Rimmele, G., Porcherie, O., Quisel, N., & Desroches, J. (2009). A solution against well cement degradation under CO_2 geological storage environment. *International Journal of Greenhouse Gas Control*, *3*(2), 206–216. doi:10.1016/j.ijggc.2008.07.005

Barree, R. D., Fisher, M. K., & Woodrood, R. A. (2002). *A practical guide to hydraulic fracture diagnostics technologies*. Paper presented at the SPE Annual Technical Conference and Exhibition. San Antonio, TX.

Baugh, B. F. (1991). *RU (reusable) profile-second generation application*. ETTCE, ASME 91-Pet-32.

Baugh, B. F., & Rash, P. (1992, August). Frictionless stab plate engages new API 17D profile. *Ocean Industry*.

Baugh, B. F., Vozniak, J., & Schmidt, N. (2011). *Post-Macondo BOP safety upgrades*. Paper presented at the Offshore Technology Conference. Rio de Janeiro, Brasil.

BCO&GC. (2008). *Surface casing vent flow/gas migration testing, reporting and repair requirements*. Vancouver, Canada: British Columbia Oil and Gas Commission.

Bell, R. J., Jr., & Davis, J. M. (1987). *Lost circulation chellenges: Drilling thick carbonate gas reservoir, natuna D-alpha block*. Paper presented at the 1987 SPE/IADC Drilling Conference. New Orleans, LA.

Bellarby, J. (2009). Well completion design. *Developments in Petroleum Science, 56*.

Benmebarek, S., & Gray, T. (1996). *Risk assessment of new added facility at an offshore gathering/ processing plant*. Paper presented at the SPE Health, Safety and Environment in Oil and Gas Exploration and Production Conference. New Orleans, LA.

Benson, S. M., & Hepple, R. (2005). *Prospects for early detection and options for remediation of leakage from CO_2 sequestration projects*. London: Elsevier Publishing.

Benson, S. M., Hepple, R., Apps, J., Tsang, C. F., & Lippmann, M. (2002). *Lessons learned from natural and industrial analogues for storage of carbon dioxide in deep geological formations*. Berkeley, CA: Lawrence Berkeley National Laboratory. doi:10.2172/805134

Bernechea, J. E., & Viger, A. J. (2013). Design optimization of hazardous substance storage facilities to minimize project risk. *Safety Science*, *51*, 49–62. doi:10.1016/j.ssci.2012.06.007

Blauch, M. E. (2010). *Developing effective and environmentally suitable fracturing fluids using hydraulic fracturing flowback waters*. Paper presented at the SPE Unconventional Gas Conference. Pittsburgh, PA.

BOEMRE. (2010). *Report regarding the cause of the April 20, 2010 Macondo well blowout*. Retrieved from http://deepseanews.com/2011/09/boemre-final-report-on-deepwater-horizon/

Bois, A-P., Garnier, A., Galdiolo, G., & Laudet, J-B. (2012, June). Use of a mechanistic model to forecast cement-sheat integrity. *SPE Drilling & Completion,* 303-314.

Boomer, R. J. (1995). *Predicting production using a neural network*. Paper presented at the Petroleum Computer Conference. Houston, TX.

Bourgoyne, A. T., Jr., Scott, S. L., & Manowski, W. (1999). *A review of sustained casing pressure occurring on the OCS*. Retrieved from ftp://www.mms.gov/TubingStudy/

Bourgoyne, A. T., Jr., Scott, S. L., & Regg, J. B. (1999). *Sustained casing pressure in offshore producing wells*. Paper presented at the Offshore Technology Conference. Houston, TX.

Bower-White, G. (2013). *Demonstrating adequate management of risk: The move from quantitative to qualitative risk assessment.* Paper presented at the SPE European HSE Conference and Exhibition. London, UK.

Bowes, C., & Procter, R. (1997). *Drillers stuck pipe handbook, 1997 guidelines & drillers handbook credits.* Ballater, Scotland: Schlumberger.

BP. (2010). *Deepwater horizon accident investigation report.* Retrieved from http://www.bp.com/liveassets/bp_internet/globalbp/globalbp_uk_english/incident_response/STAGING/local_assets/downloads_pdfs/Deepwater_Horizon_Accident_Investigation_Report.pdf

Brandsæter, A. (2002). Risk assessment in the offshore industry. *Safety Science, 40*, 231–269. doi:10.1016/S0925-7535(01)00048-0

Bratton, T. R., Rezmer-Cooper, I. M., Desroches, J., Gillie, Y.-E., Li, Q., & McFayden, M. (2001). *How to diagnose drilling induced fractures in wells drilled with oil-based muds with real-time resistivity and pressure measurements.* Paper presented at the SPE/IADC Drilling Conference. Amsterdam, The Netherlands.

Brehm, A. D. K., Davis, J. B., Ward, C. D., & Bowman, G. R. (2004). *Incorporating geomechanics into the well design process - A Gulf of Mexico case study.* Paper presented at Gulf Rocks 2004, the 6th North America Rock Mechanics Symposium (NARMS), Rock Mechanics Across Borders and Disciplines. Houston, TX.

Brickley, P., Coholan, P., Sharma, N., & Cadwallader, M. (2011). *When mother nature smiles – Multi-year analysis of currents at the Macondo well site.* Paper presented at the Offshore Technology Conference. Houston, TX.

Brill, J. P. (1974). *API activities for improving subsurface safety valve reliability.* Tulsa, OK: The University of Tulsa.

Britt, L. K., & Schoeffler, J. (2009). *The geomechanics of a shale play: What makes a shale prospective!* Paper presented at the SPE Eastern Regional Meeting. Charleston, WV.

Bruton, J. R., Ivan, C. D., & Heinz, T. J. (2001). *Lost circulation control: Evolving techniques and strategies to reduce downhole mud losses.* Paper presented at the SPE/IADC Drilling Conference. Amsterdam, The Netherlands.

BS EN 14015. (2005). *Specification for the design and manufacture of site built, vertical, cylindrical, flat-bottomed, above ground, welded, steel tanks for the storage of liquids at ambient temperature and above.* Author.

Bull, A. P. I. 75L. (2007). Guidance document for the development of a safety and environmental management system for onshore oil and natural gas production operations and associated activities. Washington, DC: API.

Bulletin, A. P. I. 5C2. (1999). Bulletin on performance properties of casing, tubing and drill pipe (21st ed.). Washington, DC: American Petroleum Institute.

Bulletin, A. P. I. 5C3. (1994). Bulletin on formulas and calculations for casing, tubing, drill pipe, and line pipe properties (6th ed.). Washington, DC: American Petroleum Institute.

Busch, J. M., Policky, B. J., & Llewelyn, D. C. G. (1985, October). Subsurface safety valves: Safety asset or safety liability. *Journal of Petroleum Technology*, 1813–1818.

Bushnell-Watson, Y. M., & Panesar, S. S. (1991). *Differential sticking laboratory tasts can improve mud design.* Paper presented at the 66th Annual Technical Conference and Exhibition of the Society of Petroleum Engineers. Dallas, TX.

Caenn, R., Darley, H. C. H., & Gray, G. R. (2011). *Composition and properties of drilling and completion fluids* (6th ed.). London: Elsevier.

Carey, J. W., Wigand, M., Chipera, S. J., WoldeGabriel, G., Pawar, R., & Lichtner, P. C., … Guthrie, G. D. (2007). Analysis and performance of oil well cement with 30 years of CO_2 exposure from the SACROG unit, West Texas, U.S.A. *International Journal of Greenhouse Gas Control, 1*, 75–85. doi:10.1016/S1750-5836(06)00004-1

Castillo, H., Turner, D., & Smalley, E. (2008). *Extreme production logging in long horizontal wells with high flow rates using downhole tractors.* Paper presented at the SPE Annual Technical Conference and Exhibition. Denver, CO.

Caughron, D. E., Renfrow, D. K., Bruton, J. R., Ivan, C. D., Broussard, P. N., Bratton, T. R., & Standifird, W. B. (2002). *Unique cross-linking pill in tandem with fracture prediction model cures circulation losses in deep water of Gulf of Mexico*. Paper presented at the SPE/IADC Drilling Conference. Dallas, TX.

Chamkalani, A., Pordel Shahri, M., & Poordad, S. (2013). *Support vector machine model: A new methodology for stuck pipe prediction*. Paper presented at the SPE Middle East Unconventional Gas Conference and Exhibition. Muscat, Oman.

Chang, I. J., & Lin, C.-C. (2006). A study of storage tank accidents. *Elsevier. Journal of Loss Prevention in the Process Industries*, *19*, 51–59. doi:10.1016/j.jlp.2005.05.015

Chen, M. Y. G., Chenevert, M. E., & Sharma, M. M. (2001). *Chemical and thermal effects on wellbore stability of shale formations*. Paper presented at SPE Annual Technical Conference and Exhibition, New Orleans, LA.

Chenevert, M. E., & Pernot, V. (1998). *Control of shale swelling pressures using inhibitive water-base muds*. Paper presented at SPE Annual Technical Conference and Exhibition. New Orleans, LA.

Chenevert, M. E., & Osisanya, S. O. (1989, September). Shale/mud inhibition defined with rig-site methods. *SPE Drilling Engineering*, 261–268.

Cheng, R., Wang, H., Shi, J., Ge, Y., Sun, Z., & Tian, H. (2013). *Drilling risk management in offshore China: Insights and lessons learned from the deepwater horizont incident*. Paper presented at the International Petroleum Technology Conference. Beijing, China.

Chilingarian, G. V., & Vorabutr, P. (1981). *Drilling and drilling fluids*. New York: Elsevier.

Cikes, M. (2000). Long-term hydraulic-fracture conductivities under extreme conditions. *SPE Production and Facilities*, *15*(4), 255–261.

Cipolla, C. L., Lolon, E. P., Mayerhofer, M. J., & Warpinski, N. R. (2009). *Fracture design considerations in horizontal wells drilled in unconventional gas reservoirs*. Paper presented at the SPE Hydraulic Fracturing Technology Conference. The Woodlands, TX.

Cipolla, C. L. (2009). Modeling production and evaluating fracture performance in unconventional reservoirs, SPE 118536. *Journal of Petroleum Technology*, (9): 84–90.

Coastal Energy. (n.d.). Retrieved from http://www.nu-coastal.co.th/manuals_nct/SIMOP%20Manta%20RM/Manta%20to20Richmond%20SIMOPS%20Risk%20Matrix%20FINAL.pdf

CONCAWE. **(2013).** Concawe report no. 3/13, performance of european cross- sounty oil pipelines, statistical summary of reported spillages in 2011 and since 1971. Retrieved june 18, 2013, from https://www.concawe.eu

CONCAWE's Oil Pipeline Management Group's Special Task Force on Oil Pipeline Spillages. (2010). *Performance of European cross-country oil pipelines, statistical summary of reported spillages in 2008 and since 1971*. Brussels, Belgium: EU.

Corneliussen, K. (2006). *Well safety- risk control in the operational phase of offshore wells*. (Doctoral Thesis). Department of Production and Quality Engineering, The Norwegian University of Science and Technology, Trondheim, Norway.

Corneliussen, K., Sorli, F., Haga, H. B., Tenold, E., Menezes, C., Grimbert, B., & Owren, K. (2007). *Well integrity management system (WIMS)- A systematic way of describing the actual and historic integrity status of operational wells*. Paper presented at the Annual Technical Conference and Exhibition. Anaheim, CA.

Courteille, J. M., & Zurdo, C. (1985). *A new approach to differential sticking*. Paper presented at the SPE Annual Technical Conference and Exhibition. Las Vegas, NV.

Crabtree, A. R. (2008). *CT-failure monitoring: A decade of experience*. Paper presented at the Coiled Tubing and Well Intervention Conference and Exhibition. Dallas, TX.

CTES. (2005). *Coiled tubing manual*. Conroe, TX: National Oilwell Varco-CTES.

Cunha, J. C. (2004). *Risk analysis application for drilling operation*. Paper presented at the Petroleum Society's 5th Canadian International Petroleum Conference (55th Annual Technical Meeting). Calgary, Canada.

Cunha, J. C., Demirdal, B., & Gui, P. (2005). *Use of quantitative risk analysis for uncertainty quantification on drilling operations - Review and lessons learned.* Paper presented at the SPE Latin American and Caribbean Petroleum Engineering Conference. Rio de Janeiro, Brazil.

Darley, H. C. H. (1969, July). A laboratory investigation of borehole stability. *Journal of Petroleum Technology*, 883–892.

Davidson, E., Richardson, L., & Zoller, S. (2000). *Control of lost circulation in fractured limestone reservoirs.* Paper presented at the 2000 IADC/SPE Asia Pacific Drilling Technology. Kuala Lumpur, Malaysia.

Definition of Facility. (n.d.). Retrieved from www.cdphe.state.co.us/ap/onglinks.asp

Deflandre, J. P., & Huguet, F. (2002). Microseismic monitoring on gas storage reservoirs: A ten-year experience. In *Proceedings of Worldwide Gas Production, Monitoring and Storage, Forum* (pp. 441-445). World Petroleum Congress.

Det Norske Veritas. (2009). *CO2QUALSTORE guideline for selection and quantification of sites and projects for geological storage of CO_2*. Retrieved from http://www.dnv.com/binaries/co2qualstore_guideline_tcm4-412142.pdf

Devold, H. (2009). *Oil and gas production handbook.* Oslo, Norway: ABB Oil and Gas.

Dhar, R. (2011). *Performance standards for safety critical elements – Are we doing enough?* Paper presented at the SPE European Health, Safety and Environmental Conference in Oil and Gas Exploration and Production. Vienna, Austria.

Dhillon, B. S. (2007). *Human reliability and error in transportation systems*. London: Springer.

Dick, M. A., Heinz, T. J., Svoboda, C. F., & Aston, M. (2000). *Optimizing the selection of bridging particles for reservoir drilling fluids.* Paper presented at the SPE International Symposium on FormationDamage. Lafayette, LA.

Directive 2009/31 / EC of the European Parliament and of the Council of 23 April 2009 on the Geological Storage of Carbon Dioxide. Retrieved from http://eur-lex.europa.eu/en/index.htm

Donovan, L. J., & Owen, J. J. (1976). *A method to estimate the pollution risk and cost of OCS oil transportation.* Paper presented at the Eighth Annual Offshore Technology Conference. Houston, TX.

Doytchev, D. E., & Szwillus, G. (2009). Combining task analysis and fault tree analysis for accident and incident analysis: A case study from Bulgaria. *Accident; Analysis and Prevention*, *41*(6), 1172–1179. doi:10.1016/j.aap.2008.07.014 PMID:19819365

Drakeley, B. K., Douglas, N. I., Haugen, K. E., & Willmann, E. (2001) *Application of reliability analysis techniques to intelligent wells.* Paper presented at the Offshore Technology Conference. Houston, TX.

Duan, S., & Wojtanowicz, A. K. (2005). *Surface induced removal of sustained casinghead pressure by sealing the well's leaking cement with condensate from flowing gas.* Paper presented at the Petroleum Society's 6th Canadian International Petroleum Conference (56th Annual Technical Meeting). Calgary, Canada.

Dumitran, C., & Onutu, I. (2010). Environmental risk analysis for crude oil soil pollution. *Carpathian Journal of Earth and Environmental Sciences*, *5*(1), 83–92.

Dupriest, F. E., Elks, W. C., Jr., & Ottesen, S. (2010). *Design methodology and operational practices eliminate differential sticking*. Paper presented at the 2010 IADC/SPE Drilling Conference and Exhibition. New Orleans, LA.

Dusseault, M. B., Maury, V., Sanfilippo, F., & Santarelli, F. J. (2004). *Drilling around salt: Risks, stresses and uncertainties*. Paper presented at Gulf Rocks 2004, the 6th North America Rock Mechanics Symposium (NARMS), Rock Mechanics across Borders and Disciplines. Houston, TX.

Dykhno, L. A., Hatton, G. J., & Nisbet, W. J. (2005). *Bullheading optimization for hydrate management in deepwater wells.* Paper presented at the Offshore Technology Conference. Houston, TX.

Dziubinski, M., Fratczak, M., & Markowski, A. S. (2006). Aspects of risk analysis associated with major failures of fuel pipelines. *Journal of Loss Prevention in the Process Industries*, *19*, 399–408. doi:10.1016/j.jlp.2005.10.007

D'Alesio, P., Poloni, R., Valente, P., & Magarini, P. A. (2011, May). Well-integrity assessment and assurance: The operational approach for three CO_2-storage fields in Italy. *SPE Production & Operations*, 140-148.

E&P Forum. UNEP. (1997). *Environmental management in oil and gas exploration and production.* Oxford, UK: Words and Publications.

Economides, M. J., & Nolte, K. G. (2000). *Reservoir stimulation.* West Sussex, UK: John Wiley & Sons.

Economides, M. J., Watters, L. T., & Dunn-Norman, S. (1998). *Petroleum wells construction.* Chichester, UK: John Wiley and Sons Ltd.

Edie, M. S. (2007). Prevention of oil spill from shipping by modeling of dynamic risk. *Marine Pollution Bulletin, 54*, 1619–1633. doi:10.1016/j.marpolbul.2007.06.013 PMID:17765267

Editorial Dept. (2009). *A detailed guide on the many different types of crude oil.* Retrieved June 22, 2013, from http://oilprice.com/energy/crude-oil/a-detailed-guide-on-the-many-different-types-of-crude-oil.html

Edmondson, J. N., & Hide, D. K. (1996). *Risk assessment of hydrocarbon releases during workover and wireline operations on completed wells on offshore platforms.* Paper presented at the European Petroleum Conference. Milan, Italy.

Engel, S. P., & Mackey, P. (2001, March). *Opportunities to improve the success rate of coiled-tubing operations.* Paper presented at the SPE/ICoTA Coiled Tubing Roundtable. Houston, TX.

Engen, G., & Rausand, M. (1982). *Reliability of downhole safety valves used in north sea.* Paper presented at 14th Annual OTC. Houston, TX.

ENI S.p.A., E & P Division. (2005). *General best practice for well site simultaneous operations for onshore and offshore installation.* Author.

Ennis-King, J., & Paterson, L. (2002). *Engineering aspects of geological sequestration of carbon dioxide.* Paper presented at the SPE Asia Pacific Oil and Gas Conference and Exhibition. Melbourne, Australia.

Environmental, Health, and Safety Guidelines for Onshore Oil and Gas Development. (2007). Retrieved from www.ifc.org/ifcext/enviro.nsf/Content

Eskijian, M., Heffron, R., & Dahlgren, T. (2002). *Engineering standards for marine oil terminal design and maintenance (MOTEMS).* Retrieved from www.slc.ca.gov/.../Presentation www.slc.ca.gov/.../Presentation

EU Parliament. (2011). Draft report on facing the challenges of the safety of offshore oil and gas activities. *Committee on Industry, Research and Energy, 18*(8).

Everdij, M. H. C., & Bloom, H. A. P. (2006). *Safety methods database.* Paper presented at the 8th International Conference on Probabilistic Safety Assessment and Management (PSAM8). New Orleans, LA.

Ewy, R. T., & Morton, E. K. (2009, September). Wellbore-stability performance of water-based mud additives. *SPE Drilling & Completion*, 390-397.

Ezine Articles . (n.d.). Retrieved from http://www.EzineArticles.com/?expert=Tina_Wallace

Falk, K., Fossli, B., Lagenberg, C., Handal, A., & Sangesland, S. (2011). *Well control when drilling with a partly-evacuated marine drilling riser.* Paper presented at the IADC/SPE Managed Pressure Drilling and Underbalanced Operations. Denver, CO.

Ferras, M., Galal, M., & Power, D. (2002). *Lost circulation solutions for severe sub-salt thief zones.* Paper presented at the AADE 2002 Technology Conference Drilling & Completion Fluids and Waste Management. Houston, TX.

Fidan, E., Babadagli, T., & Kuru, E. (2004). *Use of cement as lost circulation material – Field case studies.* Paper presented at the IADC/SPE Asia Pacific Drilling Technology Conference and Exhibition. Kuala Lumpur, Malaysia.

Fingas, M. (2001). *The basics of oil spill cleanup.* Boca Raton, FL: CRC Press LLC.

Fingas, M. (Ed.). (2011). *Oil spill science and technology.* New York: Elsevier Ltd.

Fischhoff, B., Lichtenstein, S., Slovic, P., Derby, S. L., & Keeney, R. L. (1981). *Acceptable risk.* New York: Cambridge University Press.

Fisher, K., & Warpinski, N. (2011). *Hydraulic fracture height-growth: Real data*. Paper presented at the SPE Annual Technical Meeting and Exhibition. Denver, CO.

Fontaine, J., Johnson, N., & Schoen, D. (2008). *Design, execution and evaluation of a typical marcellus shale slickwater stimulation: A case history*. Paper presented at the SPE Eastern Regional/AAPG Eastern Section Joint Meeting. Pittsburgh, PA.

Forbes, S. M., Verma, P., Curry, T. E., Friedmann, S. J., & Wade, S. M. (2008). *Guidelines for carbon dioxide capture, transport, and storage*. Washington, DC: World Resources Institute.

Fowler, J. H., & Roche, J. R. (1993). *System safety analysis of well control equipment*. Paper presented at the 25th Annual OTC. Houston, TX.

Fuel Pipeline Leak Detection . (n.d.). Retrieved from http://www.vistaleakdetection.com

Gallo, Y. L., Couillens, P., & Manai, T. (2002). *CO_2 sequestration in depleted oil or gas reservoirs*. Paper presented at the SPE International Conference on Healt, Safety and Environment in Oil and Gas Exploration and Production. Kuala Lumpur, Malaysia.

Garg, T., & Gokavarapu, S. (2012). *Lessons learnt from root cause analysis of Gulf of Mexico oil spill 2010*. Paper presented at the SPE Kuwait International Petroleum Conference. Kuwait City, Kuwait.

Gasda, S. E., Bachu, S., & Celia, M. A. (2004). The potential for CO_2 leakage from storage sites in geological media: Analysis of well distribution in mature sedimentary basins. *Environmental Geology*, *46*(6-7), 707–720. doi:10.1007/s00254-004-1073-5

Gasda, S., Celia, M., Nordbotten, J., & Dobossy, M. (2005). *$_2$ storage and the potential for leakage along existing wellbores*. Geological, CO: Intergovernmental Panel on Climate Change - IPCC.

Gaskari, R., & Mohaghegh, S. D. (2006). *Estimating major and minor natural fracture pattern in gas shales using production data*. Paper presented at the SPE Eastern Regional Meeting. Canton, OH.

Gaurina-Medjimurec, N. (2010). The influence of CO_2 on well cement. *The Mining-Geology-Petroleum Engineering Bulletin*, *22*, 19–25.

Gaurina-Medjimurec, N., & Pasic, B. (2011). Design and mechanical integrity of CO_2 injection wells. *The Mining-Geology-Petroleum Engineering Bulletin*, *23*, 1–8.

Gdanski, R. D., Fulton, D. D., & Johnson, B. J. (2010). *Returns matching reveals new tools for fracture/reservoir evaluations*. Paper presented at the SPE Tight Gas Completions Conference. San Antonio, TX.

Ghalambor, A., Ali, S. A., & Norman, W. D. (Eds.). (2009). *Frac packing handbook*. Richardson, TX: Society of Petroleum Engineers.

Gilbert, D. L. (2001). *New simultaneous operations guidelines enhance safety of offshore operations*. SPE 22802.

Goins, W. C., & Dawson, D. D. (1953). Temperature survey to locate zone of lost circulation. *Oil & Gas Journal*, *52*, 170–171.

Goodvin, K. J., & Crook, R. J. (1992). Cement sheath stress failure. *SPE Drilling Engineering*, *7*(4), 291–296.

Grassick, D. D., Kallos, P. S., Dean, S., & King, S. D. J. (1992, May). Blowout risk analysis of gas-lift completions. *SPE Production Engineering*, 172–180.

Grassick, D. D., Kallos, P. S., Jardine, I. J. A., & Deegan, F. J. (1990, November). Risk analysis of single- and dual-string gas-lift completions. *JPT*, *1374-1376*, 1418–1419.

Gunt, W. G. (1976). *Subsurface safety valves for gas storage wells*. Paper presented at Eastern Regional Meeting of the Society of Petroleum Engineers of AIME. Columbus, OH.

Haimes, Y. Y. (2009). Risk modeling, assessment and management (3rd ed.). Chichester, UK: A John Wiley & Sons Inc. Publication.

Halliburton- Boots & Coots. (2012). *Hydraulic workover services- pressure control*. Online free purchase manual.

Handal, A., & Øie, S. (2013). *Safety barrier analysis and hazard identification of blowout using managed pressure grilling compared with conventional drilling*. Paper presented at the IADC/SPE Managed Pressure Drilling and Underbalanced Operations Conference and Exhibition. San Antonio, TX.

Harmony, K. (1998). *Facility risk assessment and risk management in offshore oil & gas industry*. Paper presented at the SPE Asia Pacific Conference on Integrated Modelling for Asset Management. Kuala Lumpur, Malaysia.

Harms-Ringdahl, L. (2001). *Safety analysis, principles and practice in occupational safety* (2nd ed.). CRC Press. doi:10.1201/9780203302736

Hempkins, W. B., Kingsborough, R. H., Lohec, W. E., & Nini, C. J. (1987, September). Multivariate statistical analysis of stuck drillpipe situations. *SPE Drilling Engineering*, 237–244.

Hendrick, K., & Benner, L. Jr. (1987). *Investigating accidents with STEP*. New York: Marcel Dekker Inc.

Herbert, I. (2010). The UK buncefield incident - The view from a UK risk assessment engineer. *Journal of Loss Prevention in the Process Industries*, *23*, 913–920. doi:10.1016/j.jlp.2010.09.001

Heriot-Watt University. (2000). *Production technology I*. Edinburgh, UK: Author.

Hightower, M. et al. (2004). *Guidance on risk analysis and safety implications o a large liquefied natural gas (LNG) spill over water*. Albuquerque, NM: Sandia National Laboratories.

Hilbert, L. B., Jr., & Kalil, I. A. (1992). *Evaluation of premium threaded connections using finite-element analysis and full.scale testing*. Paper presented at the IADC/SPE Drilling Conference. New Orleans, LA.

Hoeman, K., Klasner, S., & Karcher, A. (2011). *Method for identification and analysis of polysaccharide-based, hydraulic-fracture flowback fluid*. Paper presented at the SPE International Symposium on Oilfield Chemistry. The Woodlands, TX.

Holand, P. (1997). *Offshore blowouts- Causes and control*. Houston, TX: Gulf Publishing Company.

Holand, P. (2006). *Blowout and well release characteristics and frequencies*. Trondheim, Norway: SINTEF Technology and Society, Safety and Reliability.

Holliday, G. H. (1991). *Environmental compliance assessment review*. Paper presented at the International Arctic Technology Conference. Juno, AK.

Hollnagel, E. (2008). Risk + barriers = safety? *Safety Science*, *46*, 221–229. doi:10.1016/j.ssci.2007.06.028

Holmberg, J.-E., & Knochenhauer, M. (2007). Probabilistic safety goals, phase 1 – Status and experiences in Sweden and Finland- Nordic nuclear safety research. ISBN 978-87-7893-216-7

Holton, G. A. (2004). Defining risk. *Financial Analysts Journal*, *60*(6), 19–25. doi:10.2469/faj.v60.n6.2669

Hong, E.-S., Lee, I.-M., Shin, H.-S., Nam, S.-W., & Kong, J.-S. (2009). Quantitative risk evaluation based on event tree analysis technique: Application to the design of shield TBM. *Tunnelling and Underground Space Technology*, *24*(3), 269–277. doi:10.1016/j.tust.2008.09.004

Hopkins, C. J., & Leicksenring, R. A. (1995). *Reducing the risk of stuck pipe in The Netherlands*. Paper presented at the 1995 SPE/IADC Drilling Conference. Amsterdam, The Netherlands.

Hopkins, P. (2007). Oil and gas pipelines: Yesterday and today. London: Pipeline Systems Division (PSD), Pens pen Ltd.

Houtchens, B., Foster, J., & Tessari, R. (2007). *Applying risk analysis to casing while drilling*. Paper presented at the 2007 SPE/IADC Drilling Conference. Amsterdam, The. Netherlands.

Howard, J. A., & Glover, S. B. (1994). *Tracking stuck pipe probability while drilling*. Paper presented at 1994 IADC/SPE Drilling Conference. Dallas, TX.

Howard, G. C., & Scott, P. P. Jr. (1951). An analysis and control of lost circulation. *Journal of Petroleum Technology*, *3*(6). doi:10.2118/951171-G

Hrncevic, L. (2008). *Analiza utjecaja provedbe Kyoto protokola na naftnu industriju i poslovanje naftne tvrtk*. (Unpublished doctoral dissertation). University of Zagreb, Zagreb, Croatia.

HSE. (2006). *Offshore installations*. Retrieved from http://www.hse.gov.uk/offshore/is2-2006.pdf

Huang, S., Chiu, C., & Elliot, D. (2007). *LNG: Basics of liquefied gas*. Austin, TX: The University of Texas at Austin.

Imbus, S., Orr, F. M., Kuuskraa, V. A., Kheshgi, H., Bennaceur, K., & Gupta, N. ... Hovorka, S. (2006). *Critical issues in CO_2 capture and storage: Findings of the SPE advanced technology workshop (ATW) on carbon sequestration*. Paper presented at SPE Annual Technical Conference and Exhibition. San Antonio, TX.

IMO. (2002). *Guidelines for formal safety assessment (FSA) for use in the IMO rule-making process. MSC/Circ.1023*. International Maritime Organization.

IMO. (2005). *International convention for the prevention of pollution from ships (MARPOL)*. Retrieved June 21, 2013, from http://www.imo.org

IMO. (2006). *Amendments to the guidelines for formal safety assessment (FSA) for use in the IMO rule making process. MSC–MEPC.2/Circ 5 (MSC/Circ.1023–MEPC/Circ.392)*. International Maritime Organization.

Intergovernmental Panel on Climate Change – IPCC. (2005). *IPCC special report on carbon dioxide capture and storage*. Working Group III of the Intergovernmental Panel on Climate Change.

International Asociation of Oil and Gas Producers. (2010). *Storage incident frequencies* (Report No. 434 – 3 March 2010). Author.

International Association of Drilling Contractors. (2009). *Health, safety and environment case guidelines for offshore drilling contractors*. Author.

International Marine Contractors Association (IMCA). (2010). *Guidance on simultaneous operations* (SIMOPS). Retrieved from www. Imca-int.com

International Maritime Organisation (IMO). (2013). *Global integrated shipping information system*. Retrieved from http://gisis.imo.org/Public

International Tanker Owners Pollution Federation Limited (ITOPF). (2013). *Handbook 2012/13*. Author.

Isambourg, P., Ottesen, S., Benaissa, S., & Marti, J. (1999). *Down-hole simulation cell for measurement of lubricity and differential pressure sticking*. Paper presented at the 1999 SPE/IADC Drilling Conference. Amsterdam, The Netherlands.

ISO. (2002). *Petroleum and natural gas industry – Offshore production installations – Guidelines on tools and techniques for identification and assessment of hazards*. ISO 17776.

ISO/IEC. (2012). *Guide 51.2, safety aspects – Guidelines for the inclusion in standards* (3rd ed.). Author.

ITOPF. **(2013).** Oil tanker spill statistics 2012. Retrieved June 20, 2013, from http://www.itopf.com/news-and-events/documents/StatsPack.pdf

Ivan, C. D., Bruton, J. R., Thiercelin, M., & Bedel, J. P. (2002). *Making a case for rethinking lost circulation treatments in induced fractures*. Paper presented at the SPE Annual Technical Conference. San Antonio, TX.

Jackson, P.-B., & Murphey, C. E. (1993). *Effect of casing pressure on gas flow through a sheath of set cement*. Paper presented at the SPE/IADC Drilling Conference. Amsterdam, The Netherlands.

Jahanbakhshi, R., Keshavarzi, R., Shoorehdeli, M. A., & Emamzadeh, A. (2012, December). Intelligent prediction of differential pipe sticking by support vector machine compared with conventional artificial neural networks: An example of iranian offshore oil fields. *SPE Drilling & Completion,* 586-595.

Jahn, F., Cook, M., & Graham, M. (2008). *Hydrocarbon exploration and production* (2nd ed.). Amsterdam, The Netherlands: Elsevier.

Jenvey, N. J., Moen, R. S., Muller, N., Buehler, J., Costello, M. S., & Schwander, M. M. (2008). *CO_2 geological storage field development: Initial experiences in an emerging industry*. Paper presented at the 2008 SPE North Africa Technical Conference and Exhibition. Marrakech, Morocco.

Jimenez, J. A., & Chalaturnyk, R. J. (2002). *Integrity of bounding seals for geological storage of greenhouse gases*. Paper presented at the SPE/ISRM Rock Mechanics Conference. Irving, TX.

Joint Industry Project Report (JIP). (2001). *Best practices for prevention and management of sustained casing pressure*. Houston, TX: Stress Engineering Services, Inc.

JPT. (1999). *Surface production facilities: Separating and treating produced oil and gas*. Retrieved from www.spe.org/JPT1999-06-surface-prod

Kavianian, H. R. (2003). *Process safety management: A legal and technical overview*. Paper presented at the ASSE Professional Development Conference and Exposition. Denver, CO.

Kefi, S., Lee, J. C., Shindgikar, N. D., Brunet-Cambus, C., Vidick, B., & Diaz, N. I. (2010). *Optimizting in four steps composite lost-circulation pills without knowing loss zone width*. Paper presented at the IADC/SPE Asia Pacific Drilling Technology Conference and Exhibition. Ho Chi Minh City, Vietnam.

Kercheville, J. D., Hinds, A. A., & Clements, W. R. (1986). *Comparison of environmentally acceptable materials with diesel oil for drilling mud lubricity and spotting fluid formulations*. Paper presented at the 1986 IADC/SPE Drilling Conference. Dallas, TX.

Khan, F. I., & Abbasi, S. A. (1997). Mathematical model for HAZOP study time estimation. *Journal of Loss Prevention in the Process Industries*, *10*(4), 249–251. doi:10.1016/S0950-4230(97)00010-7

Khan, F. I., & Haddara, M. R. (2003). Risk-based maintenance (RBM), a quantitative approach for maintenance/inspection scheduling and planning. *Journal of Loss Prevention in the Process Industries*, *16*(6), 561–573. doi:10.1016/j.jlp.2003.08.011

Khodja, M., Canselier, J. P., Bergaya, F., Fourar, K., Khodja, M., Cohaut, N., & Benmounah, A. (2010). Shale problems and water-based drilling fluid optimisation in the Hassi Messaoud Algerian oil field. *Applied Clay Science*, *49*(4), 383–393. doi:10.1016/j.clay.2010.06.008

King, G. E. (2010). *Thirty years of gas shale fracturing: What have we learned?* Paper presented at the SPE Annual Technical Meeting and Exhibition. Florence, Italy.

King, G. E. (2012). *Hydraulic fracturing 101: What every representative, environmentalist, regulator, reporter, investor, university researcher, neighbor and engineer should know about estimating frac risk and improving frac performance in unconventional gas and oil wells*. Paper presented at the SPE Hydraulic Fracturing Technology Conference. The Woodlands, TX.

King, G. E., & Lee, R. M. (1988). Adsorption and chlorination of mutual solvents used in acidizing. *SPE Production Engineering*, *3*(2), 205–209. doi:10.2118/14432-PA

Kingsborough, R. H., Lohec, W. E., Hempkins, W. B., & Nini, C. J. (1985). *Multivariate statistical analysis of stuck drill pipe situations*. Paper presented at the 60th Annual Technical Conference and Exhibition of the SPE. Las Vegas, NV.

Kinik, K. (2012). *Risk of well integrity failure due sustained casing pressure*. (Master Thesis). Louisiana State University, Baton Rouge, LA.

Kinik, K., & Wojtanowicz, A. K. (2011). *Identifying environmental risk of sustained casing pressure*. Paper presented at the HSSE Conference. Houston, TX.

Klementich, E. F., & Jellison, M. J. (1986, April). A service-life model for casing strings. *SPE Drilling Engineering*, 141–152.

Kletz, T. A. (1999). HAZOP and HAZAN, identifying and assessing process industry hazards. Warwickshire, UK: Institution of Chemical Engineers.

Kocijan, E. M., Melford, R. N., Hilbert, L. B., & Kalil, L. A. (1990). *Compressive load casing design*. Paper presented at the SPE/IADC Drilling Conference. Houston, TX.

Kontogiannis, T. (2003). A petri net-based approach for ergonomic task analysis and modelling with emphasis on adaptation to system changes. *Safety Science*, *41*(10), 803–835. doi:10.1016/S0925-7535(02)00035-8

Kontogiannis, T., Leopoulos, V., & Marmaras, N. (2000). A comparison of accident analysis techniques for safety-critical man-machine systems. *International Journal of Industrial Ergonomics*, *25*, 327–347. doi:10.1016/S0169-8141(99)00022-0

Kosmovski, K. T. (2006). Functional safety concept for hazardous systems and new challenges. *Journal of Loss Prevention in the Process Industries*, *19*, 298–305. doi:10.1016/j.jlp.2005.06.003

Krausman, E. (2010). *Analysis of Natech risk reduction in EU member states using a questionnaire survey*. Retrieved from http://publications.jrc.ec.europa.eu/repository/bitstream/111111111/16071/1/jrc_eu_natech.pdf

Kristafor, Z., Golub, M., & Matanovic, D. (1992). *What else about sizing accumulator volume of BOP control systems*. Paper presented at the Mediterranean Offshore Conference. Tripoli, Libia.

Krol, A. D. (1981). *Laboratory evaluation of stuck pipe spotting fluid effectiveness.* Paper presented at the 56th Annual Fall Technical Conference and Exhibition of Society of Petroleum Engineers of AIME. San Antonio, TX.

Kumamoto, H. (2007). *Satisfying safety goals by probabilistic risk assessment.* London, UK: Springer-Verlag.

Kutchko, B. G., Strzisar, B. R., Dzombak, D. A., Lowry, G. V., & Thaulow, N. (2007). Degradation of well cement by CO_2 under geologic sequestration conditions. *Environmental Science & Technology*, *41*, 4787–4792. doi:10.1021/es062828c PMID:17695930

Labenski, F., Reid, P., & Santos, H. (2003). *Drilling fluids approaches for control of wellbore instability in fractured formations.* Paper presented at SPE/IADC Middle East Drilling Technology Conference & Exhibition. Abu Dhabi, Saudi Arabia.

Lagreca, A. J., Damski, C., Peden, J. M., & Nakagawa, E. Y. (2008, December). The role of knowledge, tools and systems for drilling analysis. *SPE Drilling & Completion*, 330-337.

Lake, L. W. (Ed.). (2006). *Petroleum engineering handbook – Drilling engineering* (Vol. 2). Society of Petroleum Engineers.

Lake, L. W., & Warner, H. R. Jr., (Eds.). (2006). *Petroleum engineering handbook, emerging and peripheral technologies.* Richardson, TX: Society of Petroleum Engineering.

Lal, M. (1999). *Shale stability: Drilling fluid interaction and shale strength.* Paper presented at SPE Latin American and Caribbean Petroleum Engineering Conference. Caracas, Venezuela.

Landau, K., Rohmert, M., & Brauchler, R. (1998). Task analysis, part 1 – Guidelines for the practitioner. *International Journal of Industrial Ergonomics*, *22*(1-2), 3–11. doi:10.1016/S0169-8141(97)00064-4

Lane, G., Laleian, A., Marsh, G., Nadeau, P. H., Needoba, W., Nicholson, S., et al. (2011). *Final report on the investigation of the Macondo well blowout.* Deepwater Horizon Study Group. Retrieved from http://ccrm.berkeley.edu/pdfs_papers/bea_pdfs/dhsgfinalreport-march2011-tag.pdf

Lawson, J. D., & Brill, J. P. (1975). *Improving surface safety valve reliability – A progress report on API sponsored research.* Tulsa, OK: University of Tulsa.

Le Guen, Y., Le Gouevec, J., Chammas, R., Gerard, B., Poupard, O., Oxand, S. A., & Van Der Beken, A. (2008). *CO_2 storage – Managing the risk associated with well leakage over long timescales.* Paper presented at the 2008 SPE Asia Pacific Oil and Gas Conference and Exhibition. Perth, Australia.

Le Guen, Y., Meyer, V., Poupard, O., Houdu, E., Chammas, R., & Oxand, S. A. (2009). *A risk-based approach for well integrity management over long term in a CO_2 geological storage project.* Paper presented at the 2009 SPE Asia Pacific Oil and Gas Conference and Exhibition. Jakarta, Indonesia.

Leak Detection Services . (n.d.). Retrieved from http://www.tracerresearch.com

Lee, J. W., & Wattenbarger, R. A. (1996). *Gas reservoir engineering.* Richardson, TX: SPE.

Leggett, R. B., Griffith, C. A., & Wesson, H. R., Jr. (1991). *Snubbing unit applications in potentially high-rate gas wells: A case study of the Anschutz Ranch East Unit, Summit County, Utah.* Paper presented at the 66th Annual Technical Conference and Exhibition of the SPE. Dallas, TX.

Lewis, S. (2007). *Risk criteria – When is low enough good enough?* Risktec Solutions Limited. Retrieved from http://www.risktec.co.uk/media/43520/risk%20criteria%20-%20when%20is%20low%20enough%20good%20enough%20-%20saudi.pdf

LNG Carriers in Service or Under Construction . (n.d.). Retrieved from http://shipbuildinghistory.com/today/highvalueships/ lngactivefleet.htm

Long, B., & Garner, B. (1998). *Guide to storage tanks&equipment.* London: Professional Engineering Publishing.

Luketa, A. et al. (2008). *Breach and safety analysis of spills over water from large liquefied natural gas.* Albuquerque, NM: Sandia National Laboratories.

Manta-to-Richmond Simops Matrix. (n.d.). Retrieved from http://ebookbrowse.com/manta-to-richmond-simops-matrix-final-pdf-d459244888

Marhavilas, P. K., Koulouriotis, D., & Gemeni, V. (2011). Risk analysis and assessment methodologies in the work sites: On a review, classification and comparative study on the scientific literature of the period 2000-2009. *Journal of Loss Prevention in the Process Industries*, *24*, 477–523. doi:10.1016/j.jlp.2011.03.004

MARPOL. 73/78. (2009). Regulations for the prevention of pollution by oil. International Maritime Organization.

Martin, D.E. (1998). *Methods of prevention, detection and control of spillages in European oil pipelines*. Brussels: EU.

Martin, J. R., & Van Arnam, W. D. (2003). *The development of a corrosion-resistant coiled tubing product*. Paper presented at the Coiled Tubing Conference. Houston, TX.

Masi, S., Molaschi, C., Zausa, F., & Michelez, J. (2010). *Managing circulation losses in a harsh drilling environment: Conventional solution vs. CHCD through a risk assessment*. Paper presented at the IADC/SPE Drilling Conference and Exhibition. New Orleans, LA.

Massive Explosion Sparked by 2 SMOKERS on Top of Oil Tank Risk Engineering Position Paper. (n.d.). *Atmospheric storage tanks*. Retrieved from http://usa.marsh.com/NewsInsights/ThoughtLeadership/Articles/I

Matanovic, D., & Gaurina-Medjimurec, N. (1993). *Some knowledge about killing gas production deep wells in Croatia*. Paper presented at the 22nd Petroleum Itinerary Conference and Exhibition. Tihany, Hungary.

Matanovic, D., Gaurina-Medjimurec, N., Rajkovic, D., & Kristafor, Z. (2001). *Economy and technology balance in well life cycle management*. Paper presented at the SPE 3rd International Symposium on Oilfield Scale. Aberdeen, UK.

Mathieson, D., Rogers, J., Dajagopalan, S., & McManus, R. (2003). *Reliability assurance, managing the growth of intelligent completion technology*. Paper presented at the SPE Annual Technical Conference and Exhibition. Denver, CO.

Mazzucchelli, L., Pizzorni, D., Comini, A., & Scanarotti, N. (2002). *Rap project: An innovative approach to the risk assessment of pipeline*. Paper presented at the SPE International Conference on Health, Safety and Environment in Oil and Gas Exploration and Production. Kuala Lumpur, Malaysia.

McAndrews, K. L. (2011). *Consequences of Macondo: A summary of recently proposed and enacted changes to US offshore drilling safety and environmental regulation*. Paper presented at the SPE Americas E&P Health, Safety, Security and Environmental Conference. Houston, TX.

McLean, M. R., & Addis, M. A. (1990). *Wellbore stability analysis: A review of current methods of analysis and their field application*. Paper presented at IADC/SPE Drilling Conference. Houston, TX.

McLellan, P. (1994). *Assessing the risk of wellbore instability in horizontal and inclined wells*. Paper presented at SPE/CIM/CANMET International Conference on Recent Advances in Horizontal Well Applications. Calgary, Canada.

McLellan, P. J. (1996). Assessing the risk of wellbore instability in horizontal and inclined wells. *The Journal of Canadian Petroleum Technology*, *35*(5), 21–32.

McSpaden, A. R., & Glover, S. (2009). *Analysis of complex wellhead load events for conductor and surface casing strings*. Paper presented at the SPE/IADC Drilling Conference and Exhibition. Amsterdam, The Netherlands.

Medley, E. L. (1978). *Experience with surface controlled sub-surface safety valves*. Paper presented at European Petroleum Conference and Exhibition. London, UK.

Meschi, M. R., Shahbazi, K., & Pordel Shahri, M. (2010). *A new criteria to predict stuck pipe occurrence*. Paper presented at the SPE North Africa Technical Conference and Exhibition. Cairo, Egypt.

Metcalf, S., Delorey, J., & Allen, S. (2003). *Coiled tubing acid corrosion protection: Laboratory vs. field performance*. Paper presented at the Asia Pacific Oil and Gas Conference and Exhibition. Jakarta, Indonesia.

Meyer, B. R., Bazan, L. W., Jacot, R. H., & Lattibeaudiere, M. G. (2010). *Optimization of multiple transverse hydraulic fractures in horizontal wellbores*. Paper presented at the SPE Unconventional Gas Conference. Pittsburgh, PA.

Meyer, J. (2007). *Summary of carbon dioxide enhanced oil recovery (CO2EOR) injection well technology*. Washington, DC: American Petroleum Institute.

Michel, C. M. (1991). *Methods of detecting and locating tubing and packer leaks in the Western operating area of Prudhoe Bay field*. Paper presented at the 1991 Production Operations Symposium. Oklahoma City, OK.

Michel, J., Christopherson, S., & Whipple, F. (1994). *Mechanical protection guidelines*. Retrieved March 8, 2012, from http://response.restoration.noaa.gov/sites/default/files/mechanical-protection-guidelines.pdf

Millheim, K., Maidla, E., & Kravis, S. (1998). *An example of the drilling analysis process for extended reach wells*. Paper presented at the 1998 SPE Annual Technical Conference and Exhibition. New Orleans, LA.

Miri, R., Sampaio, J., Afshar, M., & Lourenco, A. (2007). *Development of artificial neural networks to predict differential pipe sticking in Iranian offshore oil fields*. Paper presented at the International Oil Conference and Exhibition in Mexico. Veracruz, Mexico.

MMS 30 CFR Part 250. (2010). *Annular casing pressure management for offshore wells, code of federal regulations, title 30, chapter ii, minerals management service*. Washington, DC: United States Dpt. of Interior.

Mohiuddin, M. A., Khan, K., Abdulraheem, A., Al-Majed, A., & Arifullah, V. (2005). *Field based criteria for the design of safe mud weight window*. Paper presented at SPE Technical Symposium of Saudi Arabia Section. Dhahran, Saudi Arabia.

Montgomery, J. K., Keller, S. R., Krahel, N., & Smith, M. V. (2007). *Improved method for use of chelation to free stuck pipe and enhance treatment of lost returns*. Paper presented at the 2007 SPE/IADC Drilling Conference. Amsterdam, The Netherlands.

Moos, D., Peska, P., Ward, C., & Brehm, A. (2004). *Quantitative risk assessment applied to pre-drill pore pressure, sealing potential, and mud window predictions from seismic data*. Paper presented at Gulf Rocks 2004, the 6th North America Rock Mechanics Symposium (NARMS). Houston, TX.

Moos, D., Pavel, P., Finkbeiner, T., & Zoback, M. (2003). Comprehensive wellbore stability analysis utilizing quantitative risk assessment. *Journal of Petroleum Science Engineering*, *38*, 97–109. doi:10.1016/S0920-4105(03)00024-X

Morris, A. J., & Pringle, R. E. (1988, August). Ultra-high-pressure tubing-retrievable annular safety valve. *SPE Production Engineering*, 361–365.

Muhlbauer, W. K. (2004). *Pipeline risk management manual* (3rd ed.). Oxford, UK: Elsevier Inc.

Muniz, E. S., da Fontoura, S. A. B., & Lomba, R. F. T. (2005). *Rock-drilling fluid interaction studies on the diffusion cell*. Paper presented at SPE Latin American and Caribbean Petroleum Engineering Conference. Rio de Janeiro, Brazil.

Muqeem, M. A., Weekse, A. E., & Al-Hajji, A. A. (2012). *Stuck-pipe best practices – A challenging approach to reducing stuck pipe cost*. Paper presented at the SPE Saudi Section Technical Symposium and Exhibition. Al-Khobar, Saudi Arabia.

Murillo, A., Neuman, J., & Samuel, R. (2009). *Pipe sticking prediction and avoidance using adaptive fuzzy logic and neural network modeling*. Paper presented at the 2009 SPE Production and Operations symposium. Oklahoma City, OK.

Myers, E. P. (1997). *Aboveground storage tanks*. New York: McGraw-Hill Companies, Inc.

NACE RP0502-2002. (2002). *Standard recommended practice: Pipeline external corrosion direct assessment methodology*. Author.

Nasr-El-Din, H. A., & Metcalf, A. S. (2004). *Workovers in sour environments: How do we avoid coiled tubing failures?* Paper presented at the 1st International Symposium on Oilfield Corrosion. Aberdeen, UK.

Natech Accident Database. (2010). Retrieved from http://enatech.jrc.ec.europa.eu

National Commission on the BB Deepwater Horizon Oil Spill and Offshore Drilling. (2011). *Deep water - The gulf oil disaster and the future of offshore drilling*. Author.

National Commission. (2011). *National commission on the BP deepwater horizon oil spill and offshore drilling*. Retrieved from http://www.oilspillcommission.gov/sites/default/files/documents/DEEPWATER_ReporttothePresident_FINAL.pdf

National Energy Technology Laboratory – NETL. (2009). *Monitoring, verification, and accounting of CO_2 stored in deep geologic formations*. National Energy Technology Laboratory.

Nelson, E. B., & Guillot, D. (2006). *Well cementing*. Sugar Land, TX: Schlumberger.

Nezhad, S. A. T., Mojarad, M. R. R., Paitakhti, S. J., Moghadas, J. S., & Farahmad, D. R. (2006). *Experimental study on applicability of water-alternating-CO_2 injection in the secondary and tertiary recovery*. Paper presented at First International Oil Conference and Exhibition. Cancun, Mexico.

Nguyen, V., Abousleiman, Y., & Hoang, S. (2007). *Analyses of wellbore instability in drilling through chemically active fractured rock formations: Nahr Umr shale*. Paper presented at the 15th SPE Middle East Oil & Gas Show and Conference, Bahrain International Exhibition Centre. Kingdom of Bahrain.

Nilsen, T., & Karlsen, H. K. (2008). *Risk based decision support for the planning of a challenging HPHT drilling operation*. Paper presented at the International Petroleum Technology Conference. Kuala Lumpur, Malaysia.

Nooijen, H., & Clinton, P. (2012). *Design of compressor suction scrubbers*. Paper presented at the TEKNA-Conference. Oslo, Norway.

NORSOK Standard D-010. (2004). *Well integrity in drilling and well operations*. Lysaker, Norway: Standards Norway.

Norwegian Petroleum Directorate (NPD). (2002). *Guidelines to regulations relating to management in the petroleum activities (the management regulation)*. Retrieved from http://www.npd.no/regelverk/r2002/frame_e.htm

O'Brien, D. E., & Chenevert, M. E. (1973, September). Stabilizing sensitive shales with inhibited, potassium-based drilling fluids. *Journal of Petroleum Technology*, 1089–1100.

OGP. (2012). *Recommendations for enhancements to well control training, examination and certification*. Association of Oil and Gas Producers.

Øien, K., & Nielsen, L. (2012). *Proactive resilience based indicators: The case of the deepwater horizon accident*. Paper presented at the SPE/APPEA International Conference on Health, Safety and Environment in Oil and Gas Exploration and Production. Perth, Australia.

Oilfield . (n.d.). Retrieved from http://www.glossary.oilfield.slb.com

Olita, A. et al. (2012). Oil spill hazard and risk assessment for the shorelines of a Mediterranean coastal archipelago. *Ocean and Coastal Management*, *57*, 44–52. doi:10.1016/j.ocecoaman.2011.11.006

Omrcen, B., Rakic, L., Rajkovic, J., Aslanovski, N., & Omrcen, D. (2011). The gulf oil disaster – Lessons learned for present and future. *Naftaplin*, *31*(63-64).

Onan, D. D. (1984). *Effects of supercritical carbon dioxide on well cements*. Paper presented at the 1984 Permian Basin Oil Gas Recovery Conference. Midland, TX.

OSPAR. (2007). *OSPAR guidelines for risk assessment and management of storage of CO_2 streams in geological formations*. OSPAR 07/24/1-E.

Ostebo, R., Tronstad, L., & Fikse, T. (1991) *Risk analysis of drilling and well operations*. Paper presented at the SPE/IADC Drilling Conference. Amsterdam, The Netherlands.

Osuji, C. E., Chenevert, M. E., & Sharma, M. M. (2008). *Effect of porosity and permeability on the membrane efficiency of shales*. Paper presented at SPE Annual Technical Conference and Exhibition. Denver, CO.

Ottesen, S., Zheng, R. H., & McCann, R. C. (1999). *Borehole stability assessment using quantitative risk analysis*. Paper presented at the 1999 SPE/IADC Drilling Conference. Amsterdam, The Netherlands.

Oudeman, P., terAvest, D., Grodal, E. O., Asheim, H. A., & Meissner, R. J. H. (1994). *Bull heading to kill live gas wells*. Paper presented at the European Petroleum Conference. London, UK.

Outmans, H. D. (1974, July 15). Spot fluid quickly to free differentially stuck pipe. *Oil & Gas Journal*, 65–68.

Padron, T., Luft, B., Kee, E., & Tipton, S. M. (2007). *Fatigue life of coiled tubing with external mechanical damage*. Paper presented at the Coiled Tubing and Well Intervention Conference and Exhibition. The Woodlands, TX.

Pal, K., Mangla, V. K., Joshi, N. P., Tewari, P. P., & Bose, U. N. (2000). *A new approach solves differential sticking problems in Geleki and Ahmedabad fields*. Paper presented at the 2000 SPE Annual Technical Conference and Exhibition. Dallas, TX.

Palmer, I. D., King, N. S., & Sparks, D. P. (1991). *The character of coal fracture treatments in oak grove field, black warrior basin*. Paper presented at the SPE Annual Technical Conference and Exhibition. Dallas, TX.

Parliament, E. U. (2011). *Proposal on safety of offshore oil and gas prospection, exploration and production activities, 2011/0309 (COD)*. Bruxelles: Author.

Pasic, B., Gaurina-Medjimurec, N., & Matanovic, D. (2007). Wellbore instability: Causes and consequences. *The Mining-Geological-Petroleum Engineering Bulletin*, *19*, 87–98.

Patton, D., & Abbott, W. A. (1979, June). Well completion and workover: Part 3 selecting the best well completion. *Petroleum Engineering International*, 64-74.

Payne, M. L., Zerbi, J. L., & Sims, D. C. (1995). *Tubular design optimization for ARCO China's Yasheng 13-1 development*. Paper presented at the International Meeting on Petroleum Engineering. Bejing, China.

Payne, M. L., & Swanson, J. D. (1990, December). Application of probabilistic reliability methods to tubular designs. *SPE Drilling Engineering*, 299–305.

Peden, J. M., & Leadbetter, A. (1986). *Rationality in completion design and equipment selection in the north sea*. Paper presented at the SPE European Petroleum Conference. London, UK.

Perrin, D., Caron, M., & Gaillot, G. (Eds.). (1999). Well completion and servicing- oil and gas development techniques. Paris, France: Ed.s Technip, Institut Français du Pétrole Publications.

Petroleum. (n.d.). Retrieved from http://www.Standard.no/en/sectors/Petroleum

Pinceratto, E., & Casey, J. (1998). *Environmental risk assessment- Case study of an offshore petroleum development*. Paper presented at the 1998 SPE International Conference on Health, Safety and Environment in Oil and Gas Exploration and Production. Caracas, Venezuela.

Pipeline Leak Detection. (n.d.). Retrieved from http://www.wavealert.com

Power, D., Ivan, C. D., & Brooks, S. W. (2003). *The top 10 lost circulation concerns in deepwater drilling*. Paper presented at the SPE Latin American and Caribbean Petroleum Engineering Conference. Port-of-Spain, West Indies.

Prasad, T., Castillo, H. C., & Elshahawi, H. (2012). *Effective mitigation of tool sticking risk in formation testing and fluid sampling operations*. Paper presented at the SPE Annual Technical Conference and Exhibition. San Antonio, TX.

Price, J. M. et al. (2003). Overview of the oil spill risk analysis (OSRA) model for environmental impact assessment. *Spill Science & Technology Bulletin*, *8*(5-6), 529–533. doi:10.1016/S1353-2561(03)00003-3

Price, J., & Smith, B. (2008). *Geological storage of carbon dioxide, staying safely underground (booklet). The International Energy Agency (IEA)*. Greenhouse Gas R&D.

Psarros, G., Skjong, R., & Vanem, E. (2011). Risk acceptance criterion for tanker oil spill risk reduction measures. *Marine Pollution Bulletin*, *62*, 116–127. doi:10.1016/j.marpolbul.2010.09.003 PMID:20888015

Quintesa. (2010). *Quintesa database*. Retrieved from http://www.quintessa.org/co2fepdb/

Rahmawati, S. D. (2012). *Integrated field modeling and optimization*. (Doctoral Thesis). Norwegian University of Science and Technology, Trondheim, Norway. Retrieved from www.iocenter.no/lib/exe/fetch.php?media=news:phds:finalthesis

Rana, S. (2008). *Facts and data on environmental risks- Oil and gas drilling operations*. Paper presented at the 2008 SPE Asia Pacific Oil & Gas Conference and Exhibition. Perth, Australia.

Rassenfoss, S. (2011, August). Growing offshore water production pushes search for subsea solutions. *JPT*.

Rausand, M., & Høyland, A. (2004). *System reliability theory: Models, statistical methods and applications* (2nd ed.). Hoboken, NJ: John Wiley & Sons.

Reason, J. (1997). *Managing the risks of organizational accidents*. Burlington, UK: Ashgate Publishing Company.

Reid, P. I., Meeten, G. H., Way, P. W., Clark, P., Chambers, B. D., Gilmour, A., & Sanders, M. W. (2000). Differential-sticking mechanisms and a simple wellsite test for monitoring and optimizing drilling mud properties. *SPE Drilling & Completion*, *15*(2), 97–104. doi:10.2118/64114-PA

Reniers, G. L. L., Dullaert, W., Ale, B. J. M., & Soudan, K. (2005). Developing an external domino prevention framework. Hazwim. *Journal of Loss Prevention in the Process Industries*, *18*, 127–138. doi:10.1016/j.jlp.2005.03.002

Renni, E., Krausmann, E., & Cozzani, V. (2010). Industrial accidents triggered by lightning. *Journal of Hazardous Materials*, *184*, 42–48. doi:10.1016/j.jhazmat.2010.07.118 PMID:20817399

Rispler, K., McNichol, J., Matiasz, K., & Rheinlander, M. (2001). *Using an artificial neural network to develop a wall-loss model for coiled tubing fracturing operations*. Paper presented at the Coiled Tubing Roundtable. Houston, TX.

Rojas, J. C., Clark, D. E., & Zhang, J. (2006). *Stressed shale drilling strategy-water activity design improves drilling performance*. Paper presented at SPE Annual Technical Conference and Exhibition. San Antonio, TX.

Roland, T. H. (2000). *New risk assessment approach shows significant reduction in oil blowout risk*. Paper presented at the SPE International Conference on Health, Safety, and the Environment in Oil and Gas Exploration and Production. Stavanger, Norway.

Romero, S. N., Monroy, R. R., Johnson, C., Cardenas, F., & Abraham, G. A. T. (2006, September). Preventing lost circulation by use of lightweight slurries with reticular systems: Depleted reservoirs in southern Mexico. *SPE Drilling & Completion*, 185-192.

Rosine, R. S., Blanco, I. L., & Bailey, M. (2008). *Comparison of computational fluid dynamics of erosion in coiled tubing to field and test data*. Paper presented at the Coiled Tubing and Well Intervention Conference and Exhibition. Woodlands, TX.

Ross, D. (2007). *Planning and communicating for conducting safe SIMOPS*. Paper presented at the Dynamic Positioning Conference. Houston, TX.

Sadlier, A., Wolfe, C., Reese, M., & Says, I. (2011). *Automated alarms for managing drilling pressure and maintaining wellbore stability – New concepts in while –Drilling decision making*. Paper presented at the SPE Annual Technical Conference and Exhibition. Denver, CO.

Saeed, S., Lovorn, R., & Knudsen, K. A. (2012). *Automated drilling systems for MPD – The reality*. Paper presented at the IADCE/SPE Drilling Conference and Exhibition. San Diego, CA.

Saintpère, S., Faget, J.B., Payer, N., Vergnot, D., Baranthol, C. (2005). *Simultaneous drilling-production-construction operations: Managing safety vs. deliverability on the Peciko gas field*. Indonesia: SPE 96152. Chevron. (2012). *Simultaneous operations*. Chevron.

Salamy, S. P., Aminian, K., Koperna, G. J., & Locke, C. D. (1991). *Pre- and post-stimulation well test data analysis from horizontal wells in the devonian shales*. Paper presented at the SPE Eastern Regional Meeting. Lexington, KY.

Salehi, S., & Nygaard, R. (2012). *Numerical modeling of induced fracture propagation: A novel approach for lost circulation materials (LCM) design in borehole strengthening applications of deep offshore drilling*. Paper presented at the SPE Annual Technical Conference and Exhibition. San Antonio, TX.

Sanders, M. W., Scorsone, J. T., & Friedheim, J. E. (2010). *High-fluid-loss, high-strength lost circulation treatments*. Paper presented at the SPE Deepwater Drilling and Completions Conference. Galveston, TX.

Sanders, W. W., Williamson, R. N., Ivan, C. D., & Powell, D. (2003). *Lost circulation assessment and planning program: Evolving strategy to control severe losses in deepwater projects*. Paper presented at the SPE/IADC Drilling Conference. Amsterdam, The Netherlands.

Santos, H. (2000). *Differentially stuck pipe: Early diagnostic and solution*. Paper presented at the IADC/SPE Drilling Conference. New Orleans, LA.

Santra, A., Reddy, B. R., Liang, F., & Fitzgerald, R. (2009). *Reaction of CO_2 with Portland cement at downhole conditions and the role of pozzolanic supplements*. Paper presented at the 2009 SPE International Symposium on Oilfield Chemistry. The Woodlands, TX.

Sarian, S., Varkey, J., Protasov, V., & Turner, J. (2013). *Polymer-locked, crush-free wireline composite cables reduce tool sticking and HSE risk in emerging deepwater reservoirs*. Paper presented at the North Africa Technical Conference & Exhibition. Cairo, Egypt.

Sas-Jaworsky, A., Blount, C., & Tipton, S. M. (2006). Coiled –tubing well intervention and drilling operations. In *Petroleum engineering handbook: Drilling engineering* (pp. 687–742). Richardson, TX: Society of Petroleum Engineering.

Sato, T., Hori, M., & Ichiara, T. (1985). *Finite element analysis of premium connections for tubing and casing*. Paper presented at the 8th Energy-Sources Technology Conference & Exhibition. Dallas, TX.

Sattler, J. P. (2013). *BOP performance-development and consequences in a post-Macondo world.* Paper presented at the SPE/IADC Drilling Conference and Exhibition. Amsterdam, The Netherlands.

Schlemmer, R., Friedheim, J. E., Growcock, F. B., Bloys, J. B., Headley, J. A., & Polnaszek, S.C. (2003, December). Chemical osmosis, shale, and drilling fluids. *SPE Drilling & Completion,* 318-331.

Schmidt Etkin, D. (2001). Analysis of oil spill trends in the united states and worldwide. In *Proceedings of International Oil Spill Conference* (vol. 2, pp. 1291-1300). Washington, DC: American Petroleum Institute.

Shafagi, A. (2008). Equipment failure rate updating – Bayesian estimation. *Journal of Hazardous Materials, 159*, 87–91. doi:10.1016/j.jhazmat.2008.01.042 PMID:18294769

Shoham, O. (2006). *Mechanistic modeling of gas-liquid two-phase flow in pipes*. Richardson, TX: Technology & Engineering, Society of Petroleum Engineers.

Shoraka, S. A. R., Shadizadeh, S. R., & Pordel Shahri, M. (2011). *Prediction of stuck pipe in Iranian south oil fields using multivariate statistical analysis.* Paper presented at the Nigeria Annual International Conference and Exhibition. Abuja, Nigeria.

Simangunsong, R. A., Villatoro, J. J., & Davis, A. K. (2006). *Wellbore stability assessment for highly inclined wells using limited rock mechanics data*. Paper presented at the 2006 SPE Annual Technical Conference and Exhibition. San Antonio, TX.

Simon, K., Gaurina-Medjimurec, N., & Pasic, B. (2005). Drilling fluids differential sticking tendency determination. *The Mining-Geological-Petroleum Engineering Bulletin, 17*, 31–35.

Singh, S. K., Subekti, H., Al-Asmkh, M., & Al-Samarraie, L. (2012). *An integrated approach to well integrity evaluation via reliability assessment of well integrity tools and methods: Results from Dukhan Field, Qatar.* Paper presented at SPE International Production and Operations Conference and Exhibition. Doha, Qatar.

Siruvuri, C., Nagarakanti, S., & Samuel, R. (2006). *Stuck pipe prediction and avoidance: A convolutional neural network npproach.* Paper presented at the IADC/SPE Drilling Conference. Miami, FL.

Sisak, W. J., & Crawford, D. J. (1994). *The fatigue life of coiled tubing*. Paper presented at the International Association of Drilling Contractors/Society of Petroleum Engineers Drilling Conference. Dallas, TX.

Sklet, S. (2002). *Methods for accident investigation.* Trondhaim, Norway: Norwegian University of Science and Technology.

Slideshare. (n.d.). Retrieved from http://www.slideshare.net/lifecombo/ngisimops

Smith, M. B., Rosenberg, R. J., & Bowen, J. F. (1982). *Frac width – Design vs. measurement*. Paper presented at the SPE Annual Conference and Exhibition. New Orleans, LA.

Somers, T., & Baugh, B. F. (1993). ROV capabilities for TLP support, including API 17D tooling system. In *Proceedings of Energy Technology and Exhibition Conference (ETCE).* ASME.

Spec, A. P. I. 17D. (1992). Specifications for subsea wellhead and Christmas tree equipment. Washington, DC: American Petroleum Institute.

Spec, A. P. I. 5B. (1996). Specification for threading, gauging and thread inspection of casing, tubing and line pipe threads (14th ed.). Washington, DC: American Petroleum Institute.

Spec, A. P. I. 5CT (metric). (1995). Specification for casing and tubing (metric units) (5th ed.). Washington, DC: American Petroleum Institute.

Spec, A. P. I. 5D. (1999). Specification for drill pipe (4th ed.). Washington, DC: American Petroleum Institute.

Spec, A. P. I. 6A. (1999). Specification for wellhead and Christmas tree equipment (7th ed.). Washington, DC: American Petroleum Institute.

Spec, A. P. I. 6AV1. (1996). Specification for verification test of wellhead surface safety valves and underwater safety valves for offshore service. Washington, DC: American Petroleum Institute.

Spec, A. P. I. 6D. (1994). Specification for pipeline valves (gate, ball, and check valves) (21st ed.). Washington, DC: American Petroleum Institute.

Spec, A. P. I. 7. (1997). Specification for rotary drill stem elements (39th ed.). Washington, DC: American Petroleum Institute.

Springett, F., & Franklin, D. (2007). *Advancements in subsea BOP stack equipment and controls*. Paper presented at the IADC Well Control Conference. Singapore.

Stahl, B., & Baur, M. P. (1983, November). Design methodology for offshore platform conductors. *JPT*.

Standard, N. O. R. S. O. K. (1998). *Drilling & well operations, D-010, Rev.2*. Author.

Statistical Review of World Energy. (2013). Retrieved from www.bp.com/en/global/corporate/about-bp/statistical-review-of-world-energy-2013.html

Std, A. P. I. 1160. (2001). Managing system integrity for hazardous liquid pipelines. Washington, DC: American Petroleum Institute.

Stewart, M., & Arnold, K. (1998a). Surface production operations, volume 1, design of oil-handling systems and facilities. New York: Elsevier Science.

Stewart, M., & Arnold, K. (1998b). Surface production operations, volume 2, design of gas-handling systems and facilities. New York: Elsevier Science.

Suddle, S. (2009). The weighted risk analysis. *Safety Science*, *47*(5), 668–679. doi:10.1016/j.ssci.2008.09.005

Suman, G. O., Jr. (1974, November). Sand control, part 1. *World Oil*.

Sutton, I. (2013). *Summarize the deepwater horizon/Macondo reports*. Paper presented at the Offshore Technology Conference. Houston, TX.

Suyan, K. M., Sharma, V., & Jain, V. K. (2009). *An innovative material for severe lost circulation control in depleted formations*. Paper presented at the SPE/IADC Middle East Drilling Technology Conference & Exhibition. Manama, Bahrain.

Tan, C. P., Wu, B., Mody, F. K., & Tare, U. A. (2002). *Development and laboratory verification of high membrane efficiency water-based drilling fluids with oil-based drilling fluid-like performance in shale stabilization*. Paper presented at SPE/ISRM Rock Mechanics Conference. Irving, TX.

Tare, U. A., Mody, F. K., & Tan, C. P. (2002). *New generation of membrane efficient water-based drilling fluids: Pragmatic and cost-effective solutions to borehole stability problems*. Paper presented at the Petroleum Society's Canadian International Petroleum Conference. Calgary, Canada.

Tarmoom, I., Bin Thabet, H., Samad, S., Chishti, K., Hussain, A., & Arafat, M. (2007). *A comprehensive approach to well-integrity management in Adma-Opco*. Paper presented at the Middle East Oil & Gas Show and Conference. Bahrein International Exhibition Centre, Kingdom of Bahrein.

Teodoriu, C., Reinecke, K. M., Fichter, C., & Wehling, P. (2010). *Investigations on casing-cement interaction with application to gas and CO_2 storage wells*. Paper presented at the SPE EUROPEC/AEGE Annual Conference and Exhibition. Barcelona, Spain.

Thiercelin, M. J., Baumgarte, C., & Guillot, D. (1998). *A soil mechanics approach to predict cement sheath behavior*. Paper presented at the SPE/ISMR Eurock '98. Trondheim, Norway.

Thyer, A. M., Hirst, I. L., & Jagger, S. F. (2002). Bund overtopping – The consequence of catastrophic tank failure. *Journal of Loss Prevention in the Process Industries*, *15*, 357–363. doi:10.1016/S0950-4230(02)00023-2

Thyer, A. M., Jagger, S. F., Atherton, W., & Ash, J. W. (2009). A review of catastrophic failures of bulk liquid storage tanks. *Loss Prevention Bulletin*, 205.

Torres, M. E., Frydman, M., Casalis, D., Ramirez, A., León, M. F., & Villalba, E. (2005). *3D analysis for wellbore stability: reducing drilling risk in Oriente basin, Ecuador.* Paper presented at the SPE Latin American and Caribbean Petroleum Engineering Conference. Rio de Jeneiro, Brazil.

Transmission and Distribution . (n.d.). Retrieved from www.ipt.ntnu.no/~jsg/undervisning/.../Chap8.TRANSMISSIONANDDISTRIBUTION.pdf

Tsuru, E., Maruyama, K., & Yazaki, Y. (1990, March). Influence of compound grease on the performance of premium connections. *SPE Drilling Engineering*, 71–78.

Tutuncu, A. N., Geilikman, M., Couzens, B., & van Duyvenboode, F. (2005). *Integrated wellbore quality and risk assessment study guides successful drilling in Amazon jungle.* Paper presented at SEG/Houston 2005 Annual Meeting. Houston, TX.

U.S. Department of Energy, Office of Fossil Energy. (2005). *Environmental benefits of advanced oil and gas.* Retrieved from www.fe.doe.gov

Underhill, W. B., Moore, L., & Meeten, G. H. (1998). *Model-based sticking risk assessment for wireline formation tools in the US Gulf Coast.* Paper presented at the SPE Annual Technical Conference and Exhibition. New Orleans, LA.

UNEP/IPCS. **(1999).** *Training module No. 3, Section B: Environmental Risk Assessment.* Retrieved October 18, 2012, from http://whqlibdoc.who.int/hq/1999/WHO_PCS_99.2_eng.pdf

United States Environmental Protection Agency. (2008). *Vulnerability evaluation framework for geologic sequestration of CO_2.* Retrieved from http://www.epa.gov/climatechange/emissions/downloads/VEF-Technical_Document_072408.pdf

Vachon, G., Vega, P., & Lee, J. (2006). *Closing the well-centric production optimization loop using intelligent completions.* Paper presented at the First International Oil Conference and Exhibition. Cancun, Mexico.

van der Voort, M. M., Klein, A. J. J., de Maaijer, M., van den Berg, A. C., van Deursen, J. R., & Versloot, N. H. A. (2007). A quantitative risk assessment tool for the external safety of industrial plants with a dust explosion hazard. *Journal of Loss Prevention in the Process Industries*, *20*(4-6), 375–386. doi:10.1016/j.jlp.2007.04.024

van Egmond, B. (2006). *Developing a method to screen and rank geological CO_2 storage site on the risk of leakage.* (Master Thesis). Utrecht University, Utrecht, The Netherlands.

van Oort, E., Hale, A. H., & Mody, F. K. (1995). *Manipulation of coupled osmotic flows for stabilisation of shales exposed to water-based drilling fluids.* Paper presented at SPE Annual Technical Conference & Exhibition. Dallas, TX.

van Oort, E., Hale, A. H., Mody, F. K., & Roy, S. (1996, September). Transport in shales and the design of improved water-based shale drilling fluids. *SPE Drilling & Completion*, 137-146.

Vanem, E., Endresen, Ø., & Skjong, R. (2008). Cost- effectiveness criteria for marine oil spill preventive measures. *Reliability Engineering & System Safety*, *93*, 1354–1368. doi:10.1016/j.ress.2007.07.008

Veeken, C. A. M., Davies, D. R., Kenter, C. J., & Koojima, A. P. (1991). *Sand production prediction review: Developing an integrated approach.* Paper presented at 66th Annual Technical Conference and Exhibition. Dallas, TX.

Vendrig, M., Spouge, J., Bird, A., Daycock, J., & Johnsen, O. (2003). *Risk analysis of the geological sequestration of carbon dioxide: No. R DTI/pub.* Department of Trade and Industry's Cleaner Coal Technology Transfer Programme.

Veneruso, A. F., Hiron, S., Bhavsar, R., & Bernard, L. (2000). *Reliability qualification testing for permanently installed wellbore equipment.* Paper presented at the SPE Annual Technical Conference and Exhibition. Dallas, TX.

Veneruso, A. F., Kohli, H., & Webster, M. J. (2003). *Towards truly permanent intelligent completions: Lifelong system survivability through a structured reliability assurance process*. Paper presented at the SPE Annual Technical Conference and Exhibition. Denver, CO.

Vinnem, J.E., Aven, T., Hundseid, H., Vassmyr, K.A., Vollen, R., & Øien, K. (2003). Risk assessments for offshore installations in the operational phase. In *Proceedings of ESREL 2003*. ESREL.

Wang, H., Sweatman, R., Engelman, B., Deeg, W., & Whitfill, D. (2005). *The key to successfully applying today's lost-circulation solutions*. Paper presented at the 2005 SPE Annual Technical Conference and Exhibition. Dallas, TX.

Warpinski, N. R. (2011). Hydraulic fracture height in gas shale reservoirs. In *Proceedings of SEG Annual Meeting*, (pp. 3705-3706). San Antonio, TX: SEG.

Warpinski, N. R. (1985). Measurement of width and pressure in a propagating hydraulic fracture. *SPE Journal, 25*(1).

Weiner, R. F., & Matthews, R. A. (2003). *Environmental engineering* (4th ed.). New York: Elsevier Science.

White, C. M., Straziser, B. R., Granite, E. J., Hoffman, J. S., & Pennline, H. W. (2003). Separation and capture of CO_2 from large stationary sources and sequestration in geological formations - Coalbeds and deep saline aquifers. *Journal of the Air & Waste Management Association, 53*, 645–715. doi:10.1080/10473289.2003.10466206 PMID:12828330

Whitfill, D. (2008). *Lost circulation material selection, particle size distribution and fracture modeling with fracture simulation software*. Paper presented at the IADC/SPE Asia Pacific Drilling Technology Conference and Exhibition. Jakarta, Indonesia.

Whitfill, D. L., & Hemphill, T. (2003). *All lost-circultion materials and systems are not created equal*. Paper presented at the SPE Annual Technical Conference and Exhibition. Denver, CO.

Whitfill, D. L., Wang, M., Jamison, D., & Angrove-Rogers, A. (2007). *Preventing lost circulation requires planning ahead*. Paper presented at the 2007 International Oil Conference and Exhibition. Veracruz, Mexico.

Whitfill, D., & Wang, H. (2005). *Making economic decisions to mitigate lost circulation*. Paper presented at the 2005 SPE Annual Technical Conference and Exhibition. Dallas, TX.

Wildenborg, T., Chadwick, A., Deflandre, J-P., Eiken, O., Mathiesone, A., Metcalfef, R., … Wollenwebera, J. (2013). Key messages from active CO_2 storage sites, GHGT-11. *Energy Procedia*, 1-9.

Williams, B., & Holbrook, J. (2001). *Adopting reliability practices from the space industry into intelligent wells*. Paper presented at the Offshore Technology Conference. Houston, TX.

Wilson, M. J., & Frederick, J. D. (Eds.). (1999). *Environmental engineering for exploration and production activities*. Richardson, TX: Society of Petroleum Engineers.

Wojtanowicz, A. K., Nishikawa, S., & Xu, R. (2000). *Diagnosis and remediation of sustained casing pressure in wells*. Retrieved from http://www.bsee.gov/

Woodhill . (n.d.). Retrieved from http://www.woodhill.co.uk/technical/os_cso.htm

Woodroof, R. A., Asadi, M., & Warren, M. N. (2003). *Monitoring fracturing fluid flowback and optimizing fluid cleanup using frac tracers*. Paper presented at the SPE European Formation Damage Conference. The Hague, The Netherlands.

Woods, I. T., Shirley, D., & Bottelberghs, P. H. (1991). *Environmental risk in the offshore industry*. Paper presented at the First International Conference on Health, Safety and Environment. The Hague, The Netherlands.

Woodward, J. L., & Pitblado, R. M. (2010). *LNG risk based safety: Modeling and consequence analysis*. Hoboken, NJ: John Wiley & Sons, Inc. doi:10.1002/9780470590232

Woodyard, A. (1980). *Risk analysis of well completion systems*. Paper presented at the 55th Annual Fall Technical Conference and Exhibition of the DPE of AIME. Dallas, TX.

Woodyard, A. (1981, April). Determining blowout risks for producing wells. *World Oil*, 87-91.

Xu, R. (2002). *Analysis of diagnostic testing of sustained casing pressure in wells*. (PhD Dissertation). Louisiana State University, Baton Rouge, LA.

Xu, R., & Wojtanowicz, A. K. (2001). *Diagnosis of sustained casing pressure from bleed-off/buildup testing patterns*. Paper presented at the SPE Production Operations Symposium. Oklahoma City, OK.

Yang, S.-H., & Yang, L. (2005). Automatic safety analysis of control systems. *Journal of Loss Prevention in the Process Industries*, *18*, 178–185. doi:10.1016/j.jlp.2005.04.003

Yanting, Z., & Liyun, X. (2011). Research on risk management of petroleum operations. *Energy Procedia*, *5*, 2330–2334. doi:10.1016/j.egypro.2011.03.400

Yarim, G., Ritchie, G. M., & May, R. B. (2008). *A guide to successful backreaming: Real-time case histories*. Paper presented at the 2008 SPE Annual Technical Conference and Exhibition. Denver, CO.

Yarim, G., Uchytil, R., May, R. B., Trejo, A., & Church, P. (2007). *Stuck pipe prevention – A proactive solution to an old problem*. Paper presented at the 2007 SPE Annual Technical Conference and Exhibition. Anaheim, CA.

Yokel, F. Y., & Simiu, E. (Eds.). (1985). *Application of risk analysis to offshore oil and gas operations – Proceedings of an international workshop*. Gaithersburg, MD: National Bureau of Standards.

Zhang, J., Chenevert, M. E., Al-Bazali, T., & Sharma, M. M. (2004). *A new gravimetric-swelling test for evaluating water and ion uptake in shales*. Paper presented at SPE Annual Technical Conference and Exhibition. Houston, TX.

Zhang, J., Clark, D. E., Al-Bazali, T. M., Chenevert, M. E., Rojas, J. C., & Ong, S. (2006b). *Ion movement and laboratory technique to control wellbore stability*. Paper presented at the AADE 2006 Fluids Conference. Houston, TX.

Zhang, J., Yu, M., Al-Bazali, T. M., Ong, S., Chenevert, M. E., Sharma, M. M., & Clark, D. E. (2006a). *Maintaining the stability of deviated and horizontal wells: Effects of mechanical, chemical, and thermal phenomena on well designs*. Paper presented at SPE International Oil & Gas Conference and Exhibition. Beijing, China.

Zhang, X. J. (1993). Statistical leak detection in gas and liquid pipelines. *Pipes & Pipelines International*, 26-29.

About the Contributors

Davorin Matanovic is a full time professor at the Faculty of Mining, Geology, and Petroleum Engineering, University of Zagreb in Croatia. He holds the Master Engineer (ME) degree in Petroleum Engineering, Master of Science (MS) in Chemical Engineering, and PhD in Petroleum Engineering. His current major study and specialization is well completion and workover, but he also teaches drilling technique, well stimulation, and technical risk analysis. As an ex-oilfield engineer working for INA-Naftaplin, Comp., he gained practical experience as a drilling engineer, earning the state engineer certificate. Apart from the oilfield and teaching experience, he is an editor in chief of The Mining, Geology, and Petroleum Engineering Bulletin, member of the Croatian Academy of Engineering, etc. He is the author of four books and numerous papers.

Nediljka Gaurina-Medjimurec is a professor at the Faculty of Mining, Geology, and Petroleum Engineering at University of Zagreb, specializing in drilling, drilling fluids, wellbore stability, cementing, and environmental protection in petroleum engineering. She holds a BS and PhD degree in petroleum engineering from the University of Zagreb, and MS degree in petroleum engineering from the University of Belgrade. She is a certified petroleum engineer. She has published over one hundred technical and scientific papers. She was head of Petroleum Engineering Department from 1996-1998, and Vice Dean for Finance and Administration of Faculty of Mining, Geology, and Petroleum Engineering from 2005-2009. She is a member of Society of Petroleum Engineers (SPE), Croatian Academy of Engineering (HATZ), and Croatian Academy of Science and Art (HAZU) – Scientific Council for Oil.

Katarina Simon is a Professor at the Petroleum Engineering Department of the Faculty of Mining, Geology, and Petroleum Engineering University of Zagreb. She has held the PhD title from the same University since 2000. Her area of interests are Oil and Gas Processing and Transportation as well as Directional Drilling. She was the Head of PE Department from 2005 to 2007. Currently she is a main researcher and a collaborator on a few scientific and professional projects in petroleum engineering.

* * *

Marin Cikes is associate professor at the University of Zagreb, Faculty of Mining, Geology, and Petroleum Engineering, Petroleum Engineering Department, Zagreb, Croatia. He lectures Petroleum Production Engineering, Well Test Analysis, and Well Stimulation at the graduate and postgraduate courses of the petroleum engineering study. He was also working in the national oil company INA, in the period from the 1973 until 2001, as a project engineer for production operations, especially for hydraulic fracturing. He holds PhD degree, MS, and BS degree in petroleum engineering from University of Zagreb.

Svijetlana Dubovski is a Chief Expert for Gathering and Transportation Systems in Croatia, INA Plc., Production and Field Development Sector. After graduating in 1989, she has worked in INA at different positions. From 2009 to 2012, she was managing Gathering and Processing Technology Business Unit. In 2008, she earned her PhD at Faculty of Mining, Geology and Petroleum Engineering University of Zagreb, Croatia. She is member of Croatian Society of Petroleum and Geology Engineers, Croatian Metrological Society, and the Scientific Council for Crude Oil of Croatian Academy of Science and Arts. In addition, she is member of Technical Committees 28 and 513 in State Office for Standardization and member of Team Winners Silver Medal for Pipeline Leak-detection System at 8th Croatian Exhibition of Innovations in 2000. During her career, she has published a number of professional papers.

Lidia Hrncevic has 13 years of working experience in research and development in field of environmental protection in petroleum industry. She graduated from Faculty of Mining, Geology, and Petroleum Engineering, University of Zagreb in 2000 (obtained Master's Degree in Petroleum Engineering), and at the same faculty in 2008, she obtained her Ph. D. degree. From 2000 to 2009, she worked at the same Faculty as a teaching assistant. In 2009, she was elected Assistant Professor. She is author and co-author of more than 40 scientific and professional papers and co-worker in 30 research projects and studies in the field of environmental protection in petroleum engineering and energetics.

Sonja Koscak Kolin is assistant at the University of Zagreb, Faculty of Mining, Geology, and Petroleum Engineering, Petroleum Engineering Department, Zagreb, Croatia. She lectures Petroleum Production Engineering, Well Test Analysis, and Well Stimulation at the graduate and postgraduate courses of the petroleum engineering study. She holds MS degree and BS degree in petroleum engineering from University of Zagreb.

Zdenko Kristafor is a Graduated Petroleum Engineer from the University of Zagreb, master of technical sciences in Petroleum Engineering (MSc) from University of Belgrade, and PhD in Petroleum Engineering from University of Zagreb. He has specialized in Drilling Engineering, Offshore Activities, Well Control, and Well Planning. He was also a member of scientific or technical committee at several international conferences; member of panel of judges in student paper contest (PhD division) at EUROPEC 2000 and Offshore Europe 2009, author of 60 scientific and professional papers. Professionally, he was a Dean of the Faculty of Mining, Geology and Petroleum Engineering, University of Zagreb, Head of Postgraduate Studies, Head of Petroleum Engineering Department, Head of the Administration Unit and Invigilator for International Well Control Forum (Crosco Well Control Training Centre). In addition, he is am member of the Society of Petroleum Engineers (SPE), Scientific Petroleum Council of Croatian Academy of Sciences and Arts (HAZU), Member of Croatian Society of Energy (HED), Croatian Association of Petroleum Engineers and Geoscientists (HUNIG).

Bojan Moslavac is a teaching assistant and PhD student at the Faculty of Mining, Geology, and Petroleum Engineering, University of Zagreb in Croatia. His current major study and specialization is well completion and workover, but he also teaches well stimulation and technical risk analysis as well. As an ex-oilfield engineer working for KCADeutag and Crosco-Integrated Drilling and Well Services, he gained practical experience in drilling and workover operations. Apart from the oilfield experience, he attended some worldwide conferences and exhibitions and is a co-author of few textbooks and scientific papers.

Borivoje Pasic is senior research assistant at Petroleum Engineering Department at the Faculty of Mining, Geology, and Petroleum Engineering (University of Zagreb), where he is engaged in teaching process. He holds a PhD degree in petroleum engineering. During past six years at the faculty, Borivoje Pašić has been actively engaged in class performance relating drilling, drilling fluids, and wellbore stability, and was involved in certain projects. As an SPE member, in 2010 he won ''Young Member Outstanding Service and Contribution to the Industry'' award. Borivoje Pašić is an author/co-author of 24 scientific and professional papers.

Andrew K. Wojtanowicz is a Texaco-Endowed Environmental Chair and Professor in the Craft and Hawkins Petroleum Engineering Department at the Louisiana State University and has co-authored over 200 publications and 8 books addressing environmental and productivity aspects of petroleum technology. A United Nations expert in drilling engineering and registered petroleum and environmental engineer in Louisiana, he was the 1992-93 Conoco Environmental Fellow, 2003-04 SPE Distinguished Lecturer, and served two terms as Editor-in Chief of ASME Transactions Journal of Energy Resources Technology from 2000 to 2011. Wojtanowicz received the 1996 Special Meritorious Award for Engineering Innovation, the SPE 2003 Health Environment and Safety Award, and the SPE 2008 Distinguished Achievement Award for Petroleum Engineering Faculty.

Index

A

ALARP (As Low as Reasonably Practicable) 1, 9, 22, 97
Axial Line System 246

B

Blowout Preventers 103, 117, 131, 138, 205, 328
Boiling Liquid Expanding Vapor Explosion (BLEVE) 301
Boiling Point 232, 282, 289, 301
Bullheading 200-201, 217

C

Casing Head 137, 153, 155, 163, 166, 169-172, 175-176, 201
Christmas Tree 134, 140-141, 143, 153-154, 235, 328, 330
Clay Minerals Hydration 26, 33, 46
Close Vicinity 114
CO2 Leakage 322, 325, 346, 348, 357
Coal-Bed Methane 198
Coiled Tubing 13, 199-200, 202-204, 206, 211, 217
Crude Oil 232, 237, 261, 265-268, 281, 283, 296, 298, 301, 303-310, 321, 362, 368-371, 378, 380
Cryogenic Liquids 301
Crystalline Swelling 33, 46

D

Decision Matrix 7, 104-106, 109, 112, 114
Differential Sticking 47, 49-51, 53, 59, 65-66, 72
Dogleg 54, 72
Due Diligence 246, 255
Dynamic Positioning (DP) 105, 114

E

Environmental Impact 219, 238, 240-241, 244, 247-248, 252, 254, 259-261, 263, 298, 302, 341, 358-359, 366-370, 372-373, 381, 383
Environmental Risk 16, 164, 257, 263, 358-360, 365, 378-379, 381-383, 387
Equivalent Circulating Density (ECD) 74, 94

F

Fiber Lost Circulation Material 94
Filter Cake 32, 47, 49-52, 72, 74, 77, 336
Fishing 48-49, 66-67, 72, 202, 209-210, 242, 373
Flake Lost Circulation Material 95
Flowback 187-188, 198

G

gas emissions 257, 260, 323, 378
gas migration 155, 168, 170, 173, 175-176

Gathering System 134, 141, 218-220, 235, 246
Geologic CO2 Storage 357
Gunk Plug (Or Diesel-Oil Plug) 95
Gunk Squeeze 95

H

Habitat Change 263
Hydraulic Fracturing 76, 181-183, 187, 190-191, 198, 202

I

Inert Gas 234, 286, 291, 293, 302
Injection Well Mechanical Integrity 357

K

Key Seat 53-54, 72

L

Liqufied Natural Gas 321
LNG Storage Tank 321
Lost Circulation (Lost Returns) 95
Lost Circulation Material 80, 94-95
Lost Circulation Zone 81, 95

M

Macondo 252 115-116, 119, 126, 131
Managed Pressure Drilling Systems (MPD) 123, 131
Mechanical Earth Model (MEM) 46
Mechanical Sticking 47, 53, 72
Mitigation 22, 88, 111, 210-211, 248, 322, 325, 346, 357, 359
Mud Weight Window 38, 46

O

Occupational Safety and Health 263
Oil Spill 97, 100, 116, 126, 253, 259, 306, 309, 360, 362-365, 369, 371-372, 374, 377-382, 387
Oil Toxicity 374, 387
Osmotic Swelling 33, 46

P

Packer 136-137, 140-141, 143, 145, 153, 158, 160, 200, 202, 210, 330-331, 337-340, 357
Packer Fluid 137, 153, 331
Pack Off 55-56, 60, 72
Permeability 31-32, 35, 50, 73, 75-76, 79, 85, 170, 183, 190, 198, 325, 333-338, 345
Permit-to-Work 104, 112, 114
Person-in-Charge 107-108, 112, 114
Petroleum Industry 1-2, 4, 12-15, 17, 23-24, 39, 47, 66, 74, 102, 163, 219, 358-360, 362, 365, 367-368, 370, 372-373, 378-379, 381-383, 387
petroleum wells 155-156
Pill 52, 72, 83, 85, 95
Pollutant 239, 359, 365-366, 380, 387
Portland Cement 333-336, 357
Primary Barrier 86, 103, 118, 123, 134, 138, 140, 200-202, 291, 302, 340
Proppant 198, 202, 204

R

Radial Line System 246
Receptor 357, 380
Remote Terminal Unit (RTU) 302
Resilience Based Early Warning Indicator (REWI) Method 125, 131
Risk Management 1-2, 6, 15, 22, 102, 119, 211, 254, 271-272, 346
Rock in-Situ Stress 46

S

Safety Valve 117, 134, 136, 140, 145, 154, 156, 200, 202, 210, 331, 340
SCADA 233, 254, 279, 297-298, 302

Secondary Barrier 103, 124, 134, 138, 200-201, 210, 288-291, 302, 340
Seepage Loss 95
Shale 24, 26, 31-36, 38-39, 46, 54-55, 79, 85, 181-183, 185-188, 191, 193, 198, 327, 336, 366
Shale Swelling 46
Shear Rams 122, 126, 131, 206, 208
Sidetrack 48, 57, 72
Simultaneous Operations 96-101, 103-112, 114
Snubbing Unit 199, 206-208, 211, 217
Spotting Fluid 51-52, 58, 72
Storage Site 324-325, 357
Storage Tank 303-306, 308-311, 315-316, 321, 369
Stuck 47-56, 58-67, 72, 74, 161, 199-200
Surface Facility 24, 246
sustained casing pressure 155, 157-159, 166-167, 340

T

Tank Farm 237, 311, 321
Thief Zone 73-74, 79, 81-82, 95
Tight Sands 198
Total Lost Returns 95
Tubing Hanger 140-141, 145, 154, 200, 210
Tubing Head 133, 154, 210

V

Vapours 304-305, 307, 310, 316, 321
Vugular Rock 95
Vulnerability 5, 7, 22, 323, 342-343, 357, 359, 379

W

Wait on Cement 116, 118, 126, 131
Wellbore Instability 23-27, 31-32, 34, 36-37, 39, 46-48, 54, 60, 66
well cementing 328, 357
Well Integrity 116, 119, 134, 140, 164, 166, 200-202, 209, 212, 217, 328, 336-337, 339-340, 346, 348
Wireline 37, 39, 48, 59-60, 141, 143, 199-200, 208-210, 217, 348
Workover 7, 23-24, 74, 96, 103, 133, 140, 170, 199-202, 206, 211, 217, 251, 347-348

Z

zonal isolation 74, 137, 202, 332-335, 339, 346

CPSIA information can be obtained
at www.ICGtesting.com
Printed in the USA
BVHW01*0814191018
530083BV00037B/202/P

9 781466 647770